Lecture Notes in Electrical Engineering

Volume 291

Board of Series Editors

Leopoldo Angrisani, Napoli, Italy
Marco Arteaga, Coyoacán, México
Samarjit Chakraborty, München, Germany
Jiming Chen, Hangzhou, P. R. China
Tan Kay Chen, Singapore, Singapore
Rüdiger Dillmann, Karlsruhe, Germany
Gianluigi Ferrari, Parma, Italy
Manuel Ferre, Madrid, Spain
Sandra Hirche, München, Germany
Faryar Jabbari, Irvine, USA
Janusz Kacprzyk, Warsaw, Poland
Alaa Khamis, New Cairo City, Egypt
Torsten Kroeger, Stanford, USA
Tan Cher Ming, Singapore, Singapore
Wolfgang Minker, Ulm, Germany
Pradeep Misra, Dayton, USA
Sebastian Möller, Berlin, Germany
Subhas Mukhopadyay, Palmerston, New Zealand
Cun-Zheng Ning, Tempe, USA
Toyoaki Nishida, Sakyo-ku, Japan
Federica Pascucci, Roma, Italy
Tariq Samad, Minneapolis, USA
Gan Woon Seng, Nanyang Avenue, Singapore
Germano Veiga, Porto, Portugal
Junjie James Zhang, Charlotte, USA

For further volumes:
http://www.springer.com/series/7818

About this Series

"Lecture Notes in Electrical Engineering (LNEE)" is a book series which reports the latest research and developments in Electrical Engineering, namely:

- Communication, Networks, and Information Theory
- Computer Engineering
- Signal, Image, Speech and Information Processing
- Circuits and Systems
- Bioengineering

LNEE publishes authored monographs and contributed volumes which present cutting edge research information as well as new perspectives on classical fields, while maintaining Springer's high standards of academic excellence. Also considered for publication are lecture materials, proceedings, and other related materials of exceptionally high quality and interest. The subject matter should be original and timely, reporting the latest research and developments in all areas of electrical engineering.

The audience for the books in LNEE consists of advanced level students, researchers, and industry professionals working at the forefront of their fields. Much like Springer's other Lecture Notes series, LNEE will be distributed through Springer's print and electronic publishing channels.

Harsa Amylia Mat Sakim
Mohd Tafir Mustaffa
Editors

The 8th International Conference on Robotic, Vision, Signal Processing & Power Applications

Innovation Excellence Towards Humanistic Technology

 Springer

Editors
Harsa Amylia Mat Sakim
Mohd Tafir Mustaffa
School of Electrical and Electronic
 Engineering
Universiti Sains Malaysia
Nibong Tebal
Malaysia

ISSN 1876-1100 ISSN 1876-1119 (electronic)
ISBN 978-981-4585-41-5 ISBN 978-981-4585-42-2 (eBook)
DOI 10.1007/978-981-4585-42-2
Springer Singapore Heidelberg New York Dordrecht London

Library of Congress Control Number: 2014932536

© Springer Science+Business Media Singapore 2014
This work is subject to copyright. All rights are reserved by the Publisher, whether the whole or part of the material is concerned, specifically the rights of translation, reprinting, reuse of illustrations, recitation, broadcasting, reproduction on microfilms or in any other physical way, and transmission or information storage and retrieval, electronic adaptation, computer software, or by similar or dissimilar methodology now known or hereafter developed. Exempted from this legal reservation are brief excerpts in connection with reviews or scholarly analysis or material supplied specifically for the purpose of being entered and executed on a computer system, for exclusive use by the purchaser of the work. Duplication of this publication or parts thereof is permitted only under the provisions of the Copyright Law of the Publisher's location, in its current version, and permission for use must always be obtained from Springer. Permissions for use may be obtained through RightsLink at the Copyright Clearance Center. Violations are liable to prosecution under the respective Copyright Law.
The use of general descriptive names, registered names, trademarks, service marks, etc. in this publication does not imply, even in the absence of a specific statement, that such names are exempt from the relevant protective laws and regulations and therefore free for general use.
While the advice and information in this book are believed to be true and accurate at the date of publication, neither the authors nor the editors nor the publisher can accept any legal responsibility for any errors or omissions that may be made. The publisher makes no warranty, express or implied, with respect to the material contained herein.

Printed on acid-free paper

Springer is part of Springer Science+Business Media (www.springer.com)

Preface

The *8th International Conference on Robotic, Vision, Signal Processing & Power Applications* (RoViSP2013) was held in Penang Malaysia from 10 to 12 November 2013. This conference hosted an electronic paper submission process for areas that include:

- Robotics, Control, Mechatronics, and Automation
- Vision, Image, and Signal Processing
- Artificial Intelligence and Computer Applications
- Electronic Design and Applications
- Telecommunication Systems and Applications
- Power System and Industrial Applications

RoViSP2013 is the latest conference held by the School of Electrical and Electronic Engineering, Universiti Sains Malaysia (USM) following its series of successful conferences. Almost 200 papers from around the world have been submitted; of these 60 papers were accepted for oral presentation. This proceeding gives a picture of the latest scientific and practical activities carried out in the field of Robotic, Vision, Signal Processing, and Power Applications.

The editors acknowledge the time and effort of all reviewers and technical committee members in ensuring high quality technical papers for RoViSP2013. The committee would also like to express our gratitude to Springer and the Institution of Engineers Malaysia (IEM) for their technical supports.

Harsa Amylia Mat Sakim
Mohd Tafir Mustaffa

Organization

Committees

General Chairman

Mohd Zaid Abdullah

Organizing Secretary

Zalina Abd. Aziz

International Advisory

Chee Peng Lim
Hale Kim
Ali Yeon Md Shakaff
Wuqiang Yang
John Billingsley
Han-Ping David Shieh
Kamal Zuhairi Zamli
Masuri Othman
Raghied Atta
Salina Abdul Samad
Mohsin M. Jamali
Spanial Otto
Tech Chaw Ling
Junzo Watada
Yoshikazu Miyanaga
Tarik R. Al-Khateeb

Mashkuri Yaacob
Supavadee Aramvith
Gabrielle Peko
Vangalur Alagar

Secretariat

Dzati Athiar Ramli
Nor Muzlifah Mahyuddin
Soo Siang Teoh
Bakhtiar Affendi Rosdi
Rosni Hassan
Sarina Razak
Sumariamah Mohd Radzi
Zaleha Jamaluddin
Suraya Sapian
Mohd Rahmat Arifin
Jamaluddin Mohamad
Roslina Hussin
Abdul Latiff Abdul Tawab
Ahmad Ahzam Latib

Technical and Publication

Harsa Amylia Mat Sakim
Mohd Tafir Mustaffa
Wan Amir Fuad Wajdi Othman
Tun Zainal Azni Zulkifli
Mohd Fadzli Mohd Salleh
Aiezaal Azman Abd Wahab
Bakhtiar Affendi Rosdi
Arjuna Marzuki
Rafidah Ahmad
Khairul Anuar Ab. Razak
Husna Mohd Yusof
Nor Azhar Zabidin
Amir Hamzah
Jamaludin Che Amat
Mohd Amin Mat Desa
Kamarulzaman Abu Bakar

Logistic and Accommodation

Zuraini Dahari
Mohamad Kamarol Mohd Jamil
Wan Mohd Yusof Rahiman Wan Abdul Aziz
Ruzaida Che Din
Fauziatun Dahari
Jamaluddin Mohamad

Financial and Sponsorship

Junita Mohamad Saleh
Zulfiqar Ali Abd Aziz
Bakhtiar Affendi Rosdi
Ruzaida Che Din
Khairul Anuar Ab. Razak

Facilities and Publicity

Norizah Mohamad
Ahmad Nazri Ali
Nurfishah Mat Saad
Mohammad Nazer Abdul Hadi
Ahmad Ahzam Latib
Abdul Latiff Abdul Tawab

Reviewers

Mohd Tafir Mustaffa
Nor Muzlifah Mahyuddin
Khairul Salleh Mohamed Sahari
Fawnizu Azmadi Hussin
Mohd Zaid Abdullah
Asrulnizam Abd Manaf
Mohamad Kamarol Mohd Jamil
Norizah Mohamad
Mohd Fadzil Ain
Umi Kalthum Ngah
Arjuna Marzuki
Kamal Zuhairi Zamli
Patrick Goh

Zuraini Dahari
Bakhtiar Affendi Rosdi
Wan Amir Fuad Wajdi Othman
R. Badlishah Ahmad
Mohd Fadzli Mohd Salleh
Aiezaal Azman Abd Wahab
Bee Ee Khoo
Muhammad Fahmi Miskon
Rosmiwati Mohd-Mokhtar
Mohd Razali Md Tomari
Harsa Amylia Mat Sakim
Norlaili Mohd Noh
Ahmad Nazri Ali
Dzati Athiar Ramli
Lea-Tien Tay
Wan Mohd Yusof Rahiman Wan Abdul Aziz
Farid Ghani
Siti Fatimah Abdul Razak
Junita Mohamad Saleh
Soo Siang Teoh
Zainah Md Zain
Khairunizam Wan
Mohd Alauddin Mohd Ali
Samsul Bahari Mohd Noor
Mohd Nazri Mohd Warip
Theam Foo Ng
Muhammad Khusairi Osman
Suhaidi Shafie
Soib Taib
Noor Izzri Abdul Wahab
Siti Anom Ahmad
Nur Syazreen Ahmad
Syed Sahal Nazli Alhady
Zaipatimah Ali
Mohd Rizal Arshad
Asral Bahari
Rizalafande Che Ismail
Hong Siang Chua
Mohammad Ghulam Rahman
Aini Hussain
Haidi Ibrahim
Shahid Iqbal
Addie Irawan
Mohamad Khairi Ishak
Zool Ismail

Haslina Jaafar
Noor Ain Kamsani
Faisal Nadeem Khan
De Xing Lioe
Aftanasar Md Shahar
W. Mimi Diyana W. Zaki
Yufridin Wahab
M. L. Dennis Wong
Mohamad Noh Ahmad
Phaklen Ehkan
Anwar Hasni Abu Hasan
Dahaman Ishak

Contents

Part I Robotics, Control, Mechatronics and Automation

1   **Adaptive Discrete Sliding Mode Control for a Non-minimum Phase Electro-Hydraulic Actuator System** 3
    Rozaimi Ghazali, Yahaya Md Sam,
    Mohd Fua'ad Rahmat and Zulfatman Has

2   **Musculoskeletal Robotics Modeling and Simulation** 15
    Ahmad Zaki bin Hj Shukor, Fariz bin Ali@Ibrahim,
    Muhammad Fahmi bin Miskon, Mohd Shakir bin Md Saat
    and Mohd Khairi bin Mohamed Nor

3   **Analysis of Finger Movement for Robotic Hand (MAPRoh-1) by Using Motion Capture and Flexible Bend Sensor** 23
    M. Hazwan Ali, Khairunizam Wan, Y. C. Seah,
    Nazrul H. Adnan and Juliana A. Abu Bakar

4   **Corrosion Detection Using LabVIEW for Robotic Inspection of Boiler Headers** 31
    Nadiah Amalina Zulkifli, Khairul Salleh Mohamed Sahari,
    Adzly Anuar and Mohd Azwan Aziz

5   **Kinematic Analysis of Walking with Scottish Rite Orthosis** 39
    Mahboubeh Keyvanara, Marzieh Mojaddarasil,
    Mohammad Jafar Sadigh and Mohammad Taghi Karimi

6   **Color and Thermal Image Fusion for Augmented Reality in Rescue Robotics** 47
    Ludek Zalud, Petra Kocmanova, Frantisek Burian and Tomas Jilek

7 Horizontal Distance Identification Algorithm for Sit to Stand
Joint Angle Determination for Various Chair Height
Using NAO Robot 57
Mohd Bazli Bahar, Muhammad Fahmi Miskon,
Norazhar Abu Bakar, Ahmad Zaki Shukor and Fariz Ali

8 Enhancing Wheelchair's Control Operation of a Severe
Impairment User 65
Mohd Razali Md Tomari, Yoshinori Kobayashi and Yoshinori Kuno

9 Noise Cancelation From ECG Signals Using Householder-RLS
Adaptive Filter....................................... 73
Shazia Javed and Noor Atinah Ahmad

10 Robotic Arm Control Based on Human Arm Motion.......... 81
Dickson Neoh Tze How, Chan Wai Keat, Adzly Anuar
and Khairul Salleh Mohamed Sahari

11 Elevator Riding of Mobile Robot Using Sensor Fusion......... 89
Jaehong Lee, Xuenan Cui, Hyoungrae Kim,
Seungjun Lee and Hakil Kim

12 Safe Global Path Planning of Mobile Robots Based
on Modified A* Algorithm 99
Jong-Hun Park, Jin-Hong No and Uk-Youl Huh

13 Review of Research in the Area of Agriculture Mobile Robots... 107
Sami Salama Hussen Hajjaj and Khairul Salleh Mohamed Sahari

14 Design and Development of Small and Simple Dual Robot
for Inspecting Pipe with Perpendicular Entry 119
Nur Shahida Roslin, Adzly Anuar, Khairul Salleh Mohamed Sahari
and Mohd Nazrin Afzan Abu Bakar

Part II Vision, Image and Signal Processing

15 Near Infrared Face Recognition: A Comparison
of Moment-Based Approaches............................ 129
Sajad Farokhi, Siti Mariyam Shamsuddin,
U. U. Sheikh and Jan Flusser

16	**Acceleration of Retinex Algorithm for Image Processing on Android Device Using Renderscript** .... Duc Phuoc Phat Dat Le, Duc Ngoc Tran, Fawnizu Azmadi Hussin and Mohd Zuki Yusoff	137
17	**Image Transmission Over Noisy Wireless Channels Using HQAM and Median Filter** .... Md. Abdul Kader, Farid Ghani and R. Badlishah Ahmad	145
18	**Frog Identification System Based on Local Means K-Nearest Neighbors with Fuzzy Distance Weighting**................ Haryati Jaafar, Dzati Athiar Ramli, Bakhtiar Affendi Rosdi and Shahriza Shahrudin	153
19	**A New Robust Reversible Watermarking Method in the Transform Domain**............................. Rasha Thabit and Bee Ee Khoo	161
20	**A Review for Image Segmentation Approaches Using Module-Based Framework** .... Guannan Jiang, Chin Yeow Wong, Stephen Ching-Feng Lin and Ngaiming Kwok	169
21	**Adaptive Image Super-Resolution with Neural Networks**....... Kah Keong Chua and Yong Haur Tay	181
22	**Fingerprint Recognition Based on Multi-Resolution Histogram of Gradient Descriptors**...................... Munalih Ahmad Syarif, Thian Song Ong and Connie Tee	189
23	**A Phase Linearization Method for IF Sampling Digitizer**....... Eric Chan	197
24	**Driving Circuitry of Complementary Metal Oxide Semiconductor (CMOS) Area Image Sensor for Optical Tomography Instrumentation System** .... Suhaila Mohd Najib, Mariani Idroas and Muhammad Nasir Ibrahim	205
25	**A Novel Technique for Mammogram Mass Segmentation Using Fractal Adaptive Thresholding** .... P. Shanmugavadivu and V. Sivakumar	213

26 Vehicle Classification Using Visual Background Extractor
 and Multi-class Support Vector Machines 221
 Lee Teng Ng, Shahrel Azmin Suandi and Soo Siang Teoh

27 A Robust Fuzzy-Based Modified Median Filter
 for Fixed-Value Impulse Noise 229
 P. Shanmugavadivu and P. S. Eliahim Jeevaraj

Part III Artificial Intelligence and Computer Applications

28 Vision-Based Human Gesture Recognition Using
 Kinect Sensor...................................... 239
 Huong Yong Ting, Kok Swee Sim, Fazly Salleh Abas
 and Rosli Besar

29 3D Facial Expression Classification Using 3D Facial
 Surface Normals.................................... 245
 Hamimah Ujir, Michael Spann and Irwandi Hipni Mohamad Hipiny

30 An Orchestrated Survey on *T*-Way Test Case Generation
 Strategies Based on Optimization Algorithms 255
 AbdulRahman A. Al-Sewari and Kamal Z. Zamli

31 Adaptive Rate Mechanism for WLAN IEEE 802.11
 Based on BPA-Artificial Neural Network.................. 265
 Jiwa Abdullah and A. M. I. Okaf

32 Traffic Sign Detection and Classification for Driver
 Assistant System.................................... 277
 Nursabillilah Mohd Ali, Nur Maisarah Mohd Sobran,
 M. M. Ghazaly, S. A. Shukor and A. F. Tuani Ibrahim

33 Comparison of Multilayer Perceptron and Radial Basis
 Function Neural Networks for EMG-Based Facial
 Gesture Recognition 285
 Mahyar Hamedi, Sh-Hussain Salleh, Mehdi Astaraki,
 Alias Mohd Noor and Arief Ruhullah A. Harris

34 The Effect of Bat Population in Bat-BP Algorithm 295
 Nazri Mohd. Nawi, Muhammad Zubair Rehman and Abdullah Khan

35	**Synthesizing *Asli* Malay Song: Transforming Spoken Voices into Singing Voices** .... Nurmaisara Za'ba, Nursuriati Jamil, Siti Salwa Salleh and Nurulhamimi Abdul Rahman	303
36	**Filled Pause Classification Using Energy-Boosted Mel-Frequency Cepstrum Coefficients**................................... Raseeda Hamzah, Nursuriati Jamil and Noraini Seman	311
37	**Developing a Rule Base for Recommending Insurance Products**..................................... Siti Fatimah Abdul Razak, Shing Chiang Tan and Way-Soong Lim	321
38	**Modeling and Path Planning Simulation of a Robotic Arc Welding System**...................................... Muhammad Hafidz Fazli Bin Md Fauadi, Fairul Azni Jafar, Lau Ong Yee, Mohd Nazmin Bin Maslan and Saifudin Hafiz Yahya	327
39	**A New Optimized Cuckoo Search Recurrent Neural Network (CSRNN) Algorithm** .... Nazri Mohd Nawi, Abdullah Khan and Muhammad Zubair Rehman	335
40	**Using Time Proportionate Intensity Images with Non-linear Classifiers for Hand Gesture Recognition** .... Omar Ahmad, Basilio Bona, Muhammad Latif Anjum and Ikramullah Khosa	343
41	**Applying a Multi-Agent Classifier System with a Novel Trust Measurement Method to Classifying Medical Data**........ Mohammed Falah Mohammed, Chee Peng Lim and Umi Kalthum bt Ngah	355

Part IV Electronic Design and Applications

42	**High-Speed Transmitter Designs for DDR3 SDRAM Memory Interfaces**...................................... Lim Zong Zheng, Mohd Tafir Mustaffa and Ch'ng Siew Sin	365
43	**Wideband Low Noise Amplifier Design for 2.3–2.4 GHz WiMAX Application**..................................... Nor Zaihar Yahaya, Shahrul Yazid, Anwar Osman and Helmi Huzairi	373

44 Development and Implementation of Synchronous DC–DC
    Buck Converter for Photovoltaic Power Generation   385
    Muhammad Hafeez Mohamed Hariri, Norizah Mohamad
    and Syafrudin Masri

45 An Investigation of Raindrop Size in Raindrop Energy
    Harvesting Application via Photography and Image
    Processing Approach   393
    Chin-Hong Wong, Joanne Neoh, Zuraini Dahari
    and Asrulnizam Abd Manaf

46 Latency Insertion Method with MNA Blocks
    Via Node Tearing   401
    Patrick Goh

47 A Fully-Integrated Dual-Band Concurrent CMOS LNA
    for 2.45/5.25 GHz Applications   409
    Hamidreza Ameri Eshghabadi, Mohd Tafir Mustaffa,
    Norlaili Mohd Noh, Asrulnizam Abd Manaf and Othman Sidek

48 Three Dimensional Through-the-Wall Imaging Using
    Ultrawideband (UWB) Sensors with Enhanced
    Delay-and-Sum Algorithm   419
    N. S. N. Anwar and M. Z. Abdullah

49 A Wideband LNA for Cognitive Radios   427
    Chee Han Cheong, Norlaili Mohd. Noh and Harakrishnan Ramiah

50 Direct Digital Synthesizer Based Clock Source
    for ADC Sampled System   435
    Desmond Tung and Rosmiwati Mohd-Mokhtar

Part V Telecommunication Systems and Applications

51 An Enhanced ITU-R 837-6 Rain Rate Prediction Model
    for Tropical Region   449
    Folasade Abiola Semire, Rosmiwati Mohd-Mokhtar, Widad Ismail,
    Norizah Mohamad and J. S. Mandeep

52	Study on the Effect of Dielectric Structure to the Cylindrical Dielectric Resonator Antenna (DRA)........... 457
	Mohamadariff Othman, Mohd Fadzil Ain, Ubaid Ullah, Seyi S. Olokede, Mohd Zaid Abdullah, Arjuna Marzuki, Wan Fahmin Faiz Wan Ali, Zainal Arifin Ahmad, Julie Juliewatty and Srimala Sreekantan
53	Feed Coupling Comparative Assessment of Selected Microstrip Patch Antenna.......................... 463
	Seyi Stephen Olokede, Mohd Fadzil Ain, Ubaid Ullah, Arjuna Marzuki, Julie J. Mohammed, Srimala Sreekantan, Sabar D. Hutagalung, Zainal A. Ahmad and Mohd Z. Abdullah
54	Design of a 5.8 GHz Bandstop Filter Using Split Ring Resonator Array 473
	Nor Muzlifah Mahyuddin and Nur Farah Syazwani Ab. Kadir
55	Optic Flow Based Occlusion Analysis for Cell Division Detection................................. 483
	Sha Yu and Derek Molloy

Part VI Power System and Industrial Applications

56	Design of Coreless PCB Transformer for DC/DC Converter Applications............................ 495
	Mohammadali Hashemi, Mohd Fadzil Ain and Majid Rafiee
57	Vector Control of Induction Motor Using Neural Network...... 501
	Azuwien Aida Bohari, Wahyu Mulyo Utomo, Zainal Alam Haron, Nooradzianie Muhd. Zin, Sy Yi Sim and Roslina Mat Ariff
58	Autonomous Dual Axis Solar Tracking System Using Optical Sensor and Sun Trajectory............................ 507
	Wong Yoong Wai Adrian, Vickneswari Durairajah and Suresh Gobee
59	Image Acquisition System for Boiler Header Inspection Robot... 521
	Dickson Neoh Tze How, Khairul Salleh Mohamed Sahari, Adzly Anuar, Mohd Zafri Baharuddin, Muhammad Fahmi Abdul Ghani and Mohd Azwan Aziz

60	**Some New Findings on Gauss–Seidel Technique for Load Flow Analysis**...........................	529
	Lea Tien Tay, Tze Hoe Foong and Janardan Nanda	

Author Index 537

Part I
Robotics, Control, Mechatronics and Automation

Chapter 1
Adaptive Discrete Sliding Mode Control for a Non-minimum Phase Electro-Hydraulic Actuator System

Rozaimi Ghazali, Yahaya Md Sam, Mohd Fua'ad Rahmat and Zulfatman Has

Abstract This paper presents an adaptive robust control technique based on discrete sliding mode control (DSMC) for an electro-hydraulic actuator (EHA) system. A new adaptive control strategy with the enhancement of DSMC with two-degree-of-freedom (2-DOF) structure is proposed. The control scheme that will render uncertain system and time-varying in EHA system's parameters can be obtained by the integration of recursive system identification technique. A comprehensive performance evaluation with quantitative measures and validation of the tracking performance is presented. In the experimental studies, Optimal Linear Quadratic Regulator (LQR) and Proportional-Integral-Derivative (PID) are implemented to be compared with the proposed robust controller. The results showed that robust system performance is achieved with DSMC for various system conditions while capable to reduce the control effort and gives better tracking performance as compared to the conventional LQR and PID controllers.

Keywords Discrete sliding mode control · 2-DOF controller · Perfect tracking control · Feed-forward controller · Electro-hydraulic actuator system

---

R. Ghazali (✉)
Advanced Mechatronics Research Group (AdMIRE), Department of Mechatronics and Robotics Engineering, Faculty of Electrical and Electronic Engineering, Universiti Tun Hussein Onn Malaysia, 86400 Batu Pahat, Malaysia
e-mail: rozaimi@uthm.edu.my
URL: http://www.springer.com/lncs

R. Ghazali · Y. M. Sam · M. F. Rahmat · Z. Has
Department of Control and Mechatronic Engineering, Faculty of Electrical Engineering, Universiti Teknologi Malaysia, 81310 Skudai, Malaysia

1.1 Introduction

There are many unique elements and advantages of EHA system over rival actuators such as pneumatic and electrical motor in the market these days. The main advantages of fluid power, which is led to its prominent feature, is the good ratio between forces delivered by the actuator over the weight and its size. However, it is difficult to establish or identify an accurate dynamic models where the EHA system inherently have many uncertainties. Non-linear flow and pressure characteristics, backlash in control valve, actuator friction, variations in the trapped fluid volume due to piston motion and fluid compressibility are major sources of non-linearity in the actuation system. These difficulties have motivated the researchers and academia to conduct further investigations on the actuator performance before the implementation of various potential applications in the industries. To solve those engineering issues, several research works focusing on control strategies of EHA system have been carried out.

It is necessary to design the appropriate controller that will be realized in real-time and gives satisfactory performance during the trajectory tracking control. Therefore, discrete-time control scheme should be considered in this study. Moreover, in order to reduce the control effort by the robust controller, adaptive mechanism must be integrated in the control strategy. This adaptive robust control scheme will assist the controller to adapt with any changes in the system's parameters while reducing the control effort in achieving the high tracking performance. Thus, the scheme will compromise the control energy and tracking accuracy.

1.2 Control Strategies of EHA System

A number of control strategies have been proposed to overcome the non-linearities problem in EHA system. Over the years, there has been widespread interest in SMC or known as variable structure control (VSC). EHA system positioning problem had been extended with a flexible load in [1]. In the attempt to assess the robustness of the SMC strategy, two cases had been carried out in the simulation and experimental studies by varying the load and spring stiffness. The results showed that the robust control is capable to cope with the changes occurred during the tracking control. Cho and Edge [2] dealt with unknown non-linear frictions and model error using discrete-time SMC to ensure the good position tracking was achieved and its robustness was guaranteed. Tracking error was found to be reduced significantly when the adaptive scheme was used as a control strategy.

In a rather different scope of work, [3] tackled the problem of position control in the presence of friction non-linearities using VSC approach. Due to variation in manufacturing tolerances and material properties, identification process has to be conducted to optimize the capability of the proposed controller although it will be time consuming. From the simulation and experimental results in [4], the VSC with integral action is not only a theoretical rule with good simulation results but also a practical controller with good experimental performances using CPLD chips. A continuous control action designed in SMC to make a boundary layer attractive was proposed in [5] using varying boundary layer. The varying boundary layer with SMC in that works has reduced the chattering in the control output while increasing the accuracy and robustness to various set points during the position control. In a similar scope of work, [6] proposed an adaptive SMC for EHA system since previous adaptive control methods always assumed that the original control volumes are certain and known, which can guarantee that all system unknown parameters occur linearly. [7] introduced the combination of H_∞ control and SMC with Block Control (BC) technique to force an EHA system which is driven by a servo valve to track a chaotic reference trajectory.

Recent works published in [8] introduced a variable structure filter (VSF) as a state estimator combined with SMC in discrete-time. The integration of VSF and SMC is referred as a sliding mode controller and filter (SMCF) capable to handle general uncertainties including mismatched uncertainties problems that usually occur as conservative SMC scheme. In [9], two second order sliding mode observers with finite time convergence are developed in the simulation study for a PUMA560 robot manipulators. In more recent works, [10] further analysed the effect of uncertain friction and considered the bounded region of the non-linear friction in pump-controlled EHA system. The sliding surface in the SMC scheme is determined using linear matrix inequality technique. Even though the tracking performance gives high precision tracking, there still exists a small error due to time delay during the transient response. In trajectory tracking control under the influence of uncertainties, non-linearities and disturbances, it is known that, the controller should be highly robust and adaptable to any changes in the system's parameters. In the literature, it shows that SMC has a great potential dealing with those problems existed in EHA system. However, most of the works are conducted in simulation environment and the algorithm is mostly developed in continuous-time. Therefore, SMC implementation in discrete-time is necessary to be studied. Recently, [11] also discussed in details concerning the state-of-the-art of computer-controlled in SMC research study.

1.3 Modelling of EHA System

In practical, the insight obtained is always used to reduce and simplify the model where possible. In this phase, linearisation plays an important role, as it provides much insight in the dynamics characteristics of the system. Then, experimental validation as a crucial procedure can be conducted by analysing and comparing between the developed linear model and the actual model [12].

A mathematical model of EHA system in Fig. 1.1 can be developed by neglecting the non-linearities such as internal or external leakage and dynamics of

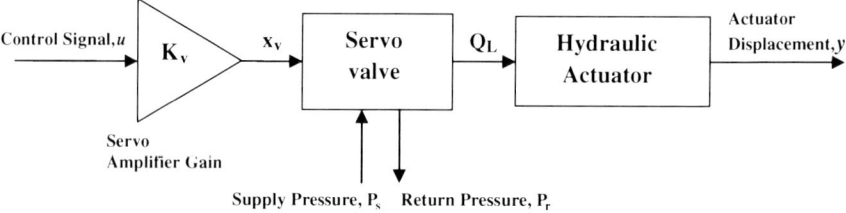

Fig. 1.1 Block diagram of EHA system

the valve as reported in [13]. The equation relates the input signal u (either voltage or current), servo valve gain K_v and the spool valve position x_v is given by

$$x_v = K_v u \quad (1.1)$$

Thus, the dynamics of the EHA system for the total oil flow Q_L are derived from a Taylor series linearisation by

$$Q_L = K_q x_v - K_c P_L \quad (1.2)$$

where K_q is the flow-gain coefficient and K_c is the flow-pressure coefficient.

Defining the load pressure, P_L as the pressure across the actuator piston, its derivative is given by the total oil load flow Q_L through the actuator divided by the fluid capacitance as given by

$$\dot{P}_L = \frac{4\beta_e}{V_t}(Q_L - C_{tp} P_L - A_p \dot{y}) \quad (1.3)$$

where β_e is the effective bulk modulus, V_t is the total compressed oil volume, C_{tp} is the total leakage coefficient, A_p is the surface area of the piston and y is the position of the piston. The force of the actuator F_a that generates from a total mass M_t attached to the end of the piston can be determined as

$$F_a = A_p P_L = M_t \ddot{y} \quad (1.4)$$

Substituting (1.2) and (1.3) into the derivative of (1.4) and taking a Laplace transform that yields

$$\frac{Y(s)}{U(s)} = \frac{K_a \omega_a^2}{s(s^2 + 2\zeta_a \omega_a s + \omega_a^2)} \quad (1.5)$$

where $K_a = \frac{K_q K_v}{A_p}$, $\omega_a = A_p \sqrt{\frac{4\beta_e}{V_t M_t}}$ and $\zeta_a = \frac{\sqrt{\frac{4\beta_e M_t}{V_t}}(K_c + C_{tp})}{2A_p}$.

Open loop transfer function of the EHA system in (1.5) relates between the control signal from the computer as a controller and position of the hydraulic actuator.

1 Adaptive Discrete Sliding Mode Control

The corresponding discrete-time model follows by transforming the continuous-time model in (1.5) with zero-order-hold is expressed as

$$G(z) = \frac{y(k)}{u(k)} = \frac{b_1 z^2 + b_2 z + b_3}{z^3 + a_1 z^2 + a_2 z + a_3} \quad (1.6)$$

The discrete-time transfer function in (1.6) can be represented in state-space control canonical form as

$$x(k+1) = \Phi x(k) + \Gamma u(k) \quad (1.7)$$

$$y(k) = \Psi x(k) \quad (1.8)$$

where $\Phi = \begin{bmatrix} -a_1 & -a_2 & -a_3 \\ 1 & 0 & 0 \\ 0 & 1 & 0 \end{bmatrix}$, $\Gamma = \begin{bmatrix} 1 & 0 & 0 \end{bmatrix}^T$ and $\Psi = \begin{bmatrix} b_1 & b_2 & b_3 \end{bmatrix}$.

The nominal state space model of EHA system in (1.7) also can be represented using third-order linear model incorporating uncertainties and disturbances as

$$x(k+1) = (\Phi \pm \Delta\Phi)x(k) + (\Gamma \pm \Delta\Gamma)u(k) + Pf(k) \quad (1.9)$$

where $f(k)$ consist of uncertain non-linearities such as external disturbances, non-linear friction and internal leakage. Φ and Γ are the nominal system parameters. The bounded uncertainties $\Delta\Phi$, $\Delta\Gamma$ and $Pf(k)$ are the uncertainties exist from the unmodelled dynamics. The third-order of the EHA system can be rearranged as

$$x(k+1) = \Phi x(k) + \Gamma u(k) + L(k) \quad (1.10)$$

where the lumped uncertainty is defined by

$$L(k) = \pm \Delta\Phi \pm \Delta\Gamma + Pf(k) \quad (1.11)$$

The state space model of EHA system in (1.10) will be used throughout this study and assumption of lumped uncertainty in (1.11) will be considered during the controller design.

1.4 2-DOF Adaptive Robust Tracking Control

The problem of controlling uncertain dynamical systems that are subjected to external disturbances is an important issue and has attracted the control engineering practitioners. One of the established approaches to solve these problems is by using the variable structure control (VSC) strategy called sliding mode control (SMC). Rapid further study on sliding mode in discrete-time confirmed that the discretisation of CSMC may cause instability and the implementation is quite complicated. Moreover, direct implementation of CSMC for discrete-time system will result several issues such as the finite amplitude oscillations persist in the system dynamics that caused an error magnitude of the order of the sampling period. Therefore, it is more appropriate to design the SMC based on a discrete-time and controller synthesis in discrete-time framework is necessary. Supposed that the uncertain and disturbed plant can be modelled as

$$x(k+1) = (\Phi + \Delta\Phi)x(k) + \Gamma u(k) + Ed(k) \quad (1.12)$$

where $\Delta\Phi$ is the uncertainties and $d(k)$ is the disturbances. Designing a control law with the sliding surface using reaching law method gives

$$\begin{aligned} u_{smc}(k) = &-(C_{smc}\Gamma)^{-1}[C_{smc}\Phi x(k) \\ &-(1-qT_s)C_{smc}x(k) + \eta T_s sign(S) \\ &+ C_{smc}\Delta\Phi x(k) + C_{smc}Ed(t)] \end{aligned} \quad (1.13)$$

This control law is not implementable because the $\Delta\Phi$ and $f(k)$ are generally unknown. However, the assumption can be made where the upper and lower bounds of $C_{smc}\Delta\Phi x(k) + C_{smc}Ed(t)$ are known as

$$-L_d T_s < C_{smc}\Delta\Phi x(k) + C_{smc}Ed(t) < L_d T_s \quad (1.14)$$

where L_d is the lumped uncertainties and disturbances boundary in the systems. Thus, the control law can be modified as

$$\begin{aligned} u_{smc}(k) = &-(C_{smc}\Gamma)^{-1}[C_{smc}\Phi - (1-qT_s)C_{smc}]x(k) \\ &-(C_{smc}\Gamma)^{-1}[\eta T_s sign(S) + L_d T_s sign(S)] \end{aligned} \quad (1.15)$$

The control law in (1.15) only guarantees that $S(k)[S(k+1) - S(k)] < 0$ which is not sufficient for a quasi-sliding mode. It is pointed out in [14] that the control law only guarantees for a discrete sliding mode if

$$L_d T_s < -\frac{\eta T_s}{2(1-qT_s)} \quad (1.16)$$

Therefore, the resulting control law in DSMC can be obtained as

$$u_{smc}(k) = -K_{smc}x(k) - K_{sw}sign(S) \quad (1.17)$$

where $K_{smc} = (C_{smc}\Gamma)^{-1}[C_{smc}\Phi - (1-qT_s)C_{smc}]$, $K_{sw} = (C_{smc}\Gamma)^{-1}\rho_{sw}T_s$ and $\rho_{sw} = \eta + L_d$.

The potential of recursive identification as been discussed briefly in [15] emphasized that many industrial control applications, such as adaptive control, adaptive prediction and adaptive filtering required the information of the system to be available recursively while the system is in operation. Therefore, system identification with on-line estimation technique is always performed as an adaptive mechanism which augmented with the robust controller. The potential of recursive identification such recursive least square (RLS) technique also has attracted researchers in various applications [16].

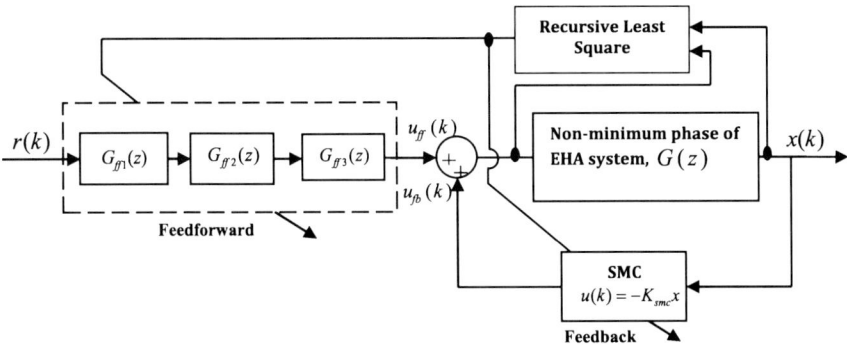

Fig. 1.2 Adaptive DSMC with feed-forward controller design

Even though there are a number of publications using DSMC with parameter estimation algorithm as an adaptive mechanism such as Kalman filter [17], RLS [18, 19], weight least square estimator [20], MRAC [21] and gradient-descent method [22], most of the works were conducted in simulation environments and not much work has been studied with the 2-DOF structure. Several paper studies that work on adaptive schemes with DSMC to study the effect of time delay has been initiated by [17]. In the literature, it is reported that most of the tracking control system is typically necessary in minimising gain and phase error from the output to the desired trajectory [23, 24]. One of the most successful feed-forward controller designs is the zero phase error tracking controller (ZPETC), which has wide application in advanced manufacturing equipment particularly for non-minimum phase discrete-time system.

Due to the dependent on parameters of the feed-forward and DSMC controller to the plant parameters as illustrated in Fig. 1.2, adaptive strategy is meaningful. Adaptive schemes can be applied for non-minimum phase models. Even though the controller to be designed is robust to the uncertainties and disturbances, the adaptive scheme may reduce the control effort of the robust controller. Generally, high control gain is needed to ensure the robustness of the controller and this usually will make the control effort range is large to be robust towards non-linear uncertainties and disturbances. By introducing adaptive scheme, the controller is able to be self tuning according to the nominal point or current operating range of the systems. Furthermore, by augmenting with the RLS with varying forgetting factor and covariance resetting technique, the robust controller with feed-forward scheme is recursively updated and improved.

1.5 Results and Discussion

An EHA system workbench with instrumented experimental system as in Fig. 1.3 has been developed in the laboratory for experimental evaluation and verification. The workbench is consists of hydraulic actuator, servo valve with amplifier,

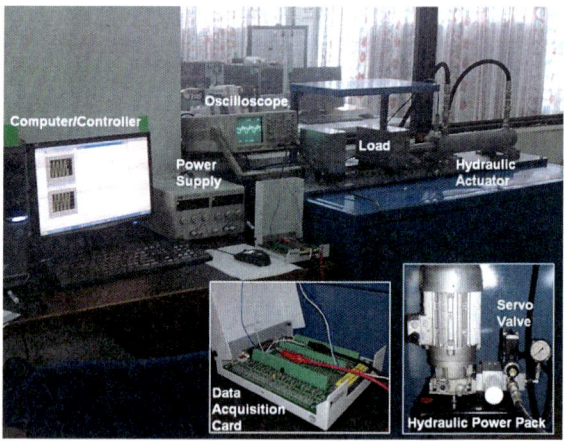

Fig. 1.3 Experimental workbench

measuring devices and a digital processor with interfacing system. For the tracking performance analysis, a more practical approaches which are mean positioning accuracy (MPA), absolute positioning accuracy (APA), weight positioning accuracy (WPA) and robustness index (RI) will be carried out as presented in [25] and [26].

The non-recursive estimation process based on the grey box approach is performed in [27] for the linear model of third order system. The discrete-time of EHA model can be represented as

$$G(z) = \frac{-0.03093z^2 + 0.3836z - 0.2738}{z^3 - 1.8870z^2 + 1.0560z - 0.1695} \qquad (1.18)$$

For the case of fast sampling time which caused non-minimum phase in the identified EHA system model, ZPETC with DSMC is employed experimentally. All the parameters of the controllers which have been determined in the simulation study are implemented directly where the parameters of DSMC are computed based on the theoretical and fulfil the necessary condition. A number of trajectories were designed to show the effectiveness of the proposed controller in tracking control of the EHA system. In this paper, the chaotic trajectory design will be presented for DSMC and adaptive DSMC control performances. For an appropriate comparison, the dominant poles location as given by LQR controller is used in the determination of the DSMC parameters. The closed loop characteristics can be achieved by setting the desired reference model with $\omega_n = 157.5113$, $\zeta = 0.707$ and $T_p = 0.0038$. From the closed loop poles, reaching law parameters are determined as $\rho = 300$, $q = 147.0254$ and the switching gain can be obtained as

$$C_{smc} = \begin{bmatrix} 1 & -1.0082 & 0.3284 \end{bmatrix} \qquad (1.19)$$

which fulfill the reachability condition.

1 Adaptive Discrete Sliding Mode Control

Table 1.1 MPA, APA, WPA and RI for chaotic trajectory

Control technique	PID	LQR	DSMC
MPA (NOM)	1.8071	1.2543	1.0192
MPA (VAR)	2.5167	1.8528	1.1240
APA (NOM)	10.8658	10.3420	9.7428
APA (VAR)	12.9075	12.0537	9.8043
WPA (NOM)	3.3516	3.2536	3.1816
WPA (VAR)	3.7817	3.5278	3.2167
RI	0.3927	0.4772	0.1028

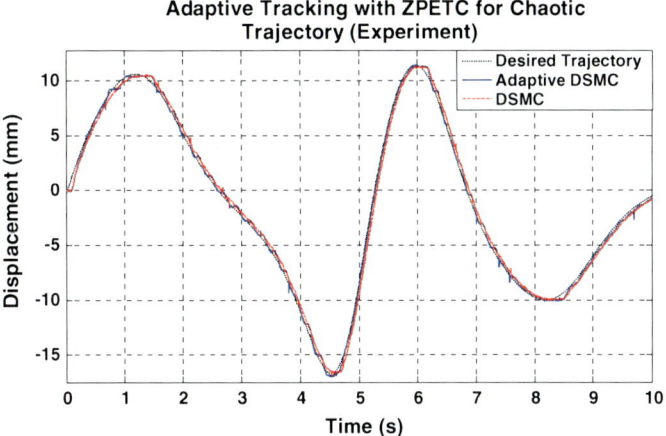

Fig. 1.4 Adaptive tracking performance with ZPETC for chaotic trajectory

In the experimental exercises, the quantitative measures for the robustness test under supply pressure variations are evaluated. It showed that the LQR and PID have demonstrated significant results in the APA evaluation as compared to the DSMC. In the evaluation of RI value, the DSMC showed better in maintaining the performance even though the WPA of these controllers look identical. PID controller performance as proven in the simulation studies is the worst comparing to others. As in theoretical aspect, the robustness of DSMC still can be increased by tuning the switching gain to a suitable value. Nevertheless, a high controller performance has been produced by the DSMC in the tracking of chaotic trajectory which subjected to uncertainties and disturbances. The value of MPA, APA, WPA and RI are tabulated in Table 1.1.

The effectiveness of the adaptive DSMC strategy was also investigated due to the pressure supply variations which caused changes in the plant parameters. In the adaptive control strategy, the experimental results as in Fig. 1.4 showed that the improvement in tracking performance is insignificant in achieving it precisely. However, the control efforts and tracking error of the adaptive scheme is slightly improved as compared to DSMC. It can be seen that the actuator was compelled to

Fig. 1.5 Quantitative measure with ZPETC for chaotic trajectory under adaptive scheme

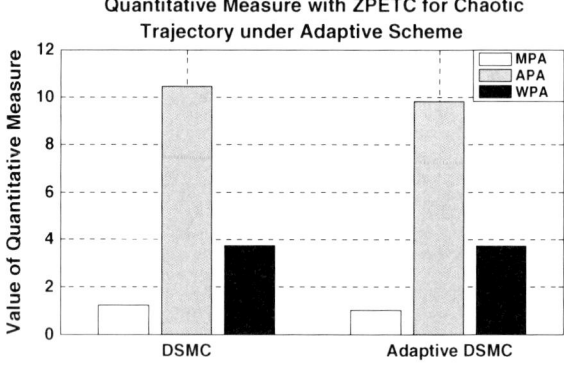

Table 1.2 MPA, APA and WPA under adaptive scheme

Control technique	DSMC	Adaptive DSMC
MPA	1.2311	1.0227
APA	10.4709	9.8234
WPA	3.7492	3.7418

be produced by the adaptive DSMC scheme slightly higher than its counterpart. Thus, more power is needed by the scheme to accomplish the task. The adaptive DSMC scheme is more effective in reducing the control efforts. In general, the adaptive DSMC method gives better performance as compared to the DSMC. Notice that the control signals as it is fluctuating during the first 10 s. Figure 1.5 summarised the performance of each parameter estimator in terms of the MPA, APA and WPA of the tracking performance while the values are tabulated in Table 1.2.

1.6 Conclusion

A new adaptive control technique based on the DSMC with feed-forward approach that render non-minimum phase problem in the EHA system is successfully developed with only input and output measurements. With the 2-DOF structure, the conventional method of designing controller through a heuristic method can be avoided and hence making the design process easier. By using the reaching law approach, it is shown that the adaptive DSMC assured that the uncertainties and disturbances are ultimately bounded stable during the sliding mode. As a conclusion, it is shown that the proposed adaptive DSMC improvement the performance of the EHA system with the ability to keep the actuator stable and robust even on the different operating condition. It also been shown through these works that the proposed new controller is capable of improving the performance of the position tracking control while reducing the phase lag problem.

Acknowledgments The authors would like to thank Universiti Teknologi Malaysia (UTM) for the Research University Grant Vot 05H02, Universiti Tun Hussein Onn Malaysia (UTHM) and Ministry of Science, Technology and Innovation (MOSTI) Malaysia for their support.

References

1. Liu Y, Handroos H (1999) Sliding mode control for a class of hydraulic position servo. Mechatronics 9(1):111–123
2. Cho SH, Edge KA (2000) Adaptive sliding mode tracking control of hydraulic servosystems with unknown non-linear friction and modelling error. Proc Inst Mech Eng, Part I: J Syst Control Eng 214(4):247–257
3. Bonchis A, Corke PI, Rye DC, Ha QP (2001) Variable structure methods in hydraulic servo systems control. Automatic 37:589–595
4. Chuang C, Shiu L (2004) CPLD based DIVSC of hydraulic position control systems. Comput Electr Eng 30(7):527–541
5. Chen H, Renn J, Su J (2005) Sliding mode control with varying boundary layers for an electro-hydraulic position servo system. Int J Adv Manuf Technol 26(1–2):117–123
6. Guan C, Pan S (2008) Adaptive sliding mode control of electro-hydraulic system with nonlinear unknown parameters. Control Eng Pract 16(11):1275–1284
7. Loukianov AG, Rivera J, Orlov YV, Teraoka EYM (2009) Robust trajectory tracking for an electrohydraulic actuator. IEEE Trans Industr Elect 56(9):3523–3531
8. Wang S, Burton R, Habibi S (2011) Sliding mode controller and filter applied to an electrohydraulic actuator system. ASME J Dyn Syst, Measur, Control 133(2):1–7
9. Van M, Kang H, Suh Y (2013) Second order sliding mode-based output feedback tracking control for uncertain robot manipulators. Int J Adv Rob Syst 10(16):1–9
10. Lin Y, Shi Y, Burton R (2013) Modeling and robust discrete-time sliding-mode control design for a fluid power electrohydraulic actuator (EHA) system. IEEE ASME Trans Mechatron 18(1):1–10
11. Yu X, Wang B, Li X (2012) Computer-controlled variable structure systems: the state-of-the-art. IEEE Trans Industr Inf 8(2):197–205
12. Jelali M, Kroll A (2003) Hydraulic servo-systems: modelling, identification and control. Springer, London
13. Knohl T, Unbehauen H (2000) Adaptive position control of electrohydraulic servo systems using ANN. Mechatronics 10(1–2):127–143
14. Bartoszewicz A (1996) Remarks on 'Discrete-time variable structure control systems'. IEEE Trans Industr Electron 43:235–238
15. Ghazali R, Sam YM, Rahmat MF, Zulfatman Z (2011) Recursive parameter estimation for discrete-time model of an electro-hydraulic servo system with varying forgetting factor. Int J Phys Sci 6(30):6829–6842
16. Shifeng O, Xianyun W, Ying G (2013) Adaptive improved RLS algorithm for blind source separation. Przeglad Elektrotechniczny 3b/2013:81–83
17. Sha DD, Bajic VB (2000) Robust discrete adaptive input-output-based sliding mode controller. Int J Syst Sci 31(12):1601–1614
18. Park YM, Kim W (1996) Discrete-time adaptive sliding mode power system stabilizer with only input/output measurements. Electr Power Energy Syst 18(8):509–517
19. Wu B, Li S, Wang X (2009) Discrete-time adaptive sliding mode control of autonomous underwater vehicle in the dive plane, intelligent robotics and applications, Lecture notes in computer science, pp 157–164
20. Yoshimura T (2008) Adaptive sliding mode control for a class of non-linear discrete-time systems with mismatched time-varying uncertainty. Int J Model Ident Control 4(3):250–259

21. Semba T, Furuta K (1996) Discrete-time adaptive control using a sliding mode. Math Probl Eng 2:131–142
22. Akpolat ZH, Gokbulut M (2002) Discrete time adaptive reaching law speed control of electrical drives. Electr Eng 85:53–58
23. Tomizuka M (1987) Zero phase error tracking algorithm for digital controller. ASME J Dyn Syst, Measur, Control 109(1):65–68
24. Wang J, Van Brussel H, Swevers J (2003) Robust perfect tracking control with discrete sliding mode controller. ASME J Dyn Syst, Measur, Control 125(1):27–32
25. Bonchis A, Corke PI, Rye DC (2002) Experimental evaluation of position control methods for hydraulic systems. IEEE Trans Control Syst Technol 10(6):876–882
26. Choux M, Hovland G (2010) Adaptive backstepping control of nonlinear hydraulic-mechanical system including valve dynamics. Model, Ident Control 31(1):35–44
27. Ghazali R, Sam YM, Rahmat MF, Zulfatman AW, Hashim IM (2012) Simulation and experimental studies on perfect tracking optimal control of an electrohydraulic actuator systems. J Control Sci Eng 2012:1–8

Chapter 2
Musculoskeletal Robotics Modeling and Simulation

Ahmad Zaki bin Hj Shukor, Fariz bin Ali@Ibrahim, Muhammad Fahmi bin Miskon, Mohd Shakir bin Md Saat and Mohd Khairi bin Mohamed Nor

Abstract This paper presents the kinematics and dynamics of a musculoskeletal model inspired by humans/animals where the bones (links) are actuated by muscles (linear actuators). Starting from a single-link musculoskeletal structure, the kinematics were addressed to link the joint and muscle variables. From the kinematic constraints, dynamics of the structure were also presented. Later the 3D model of a two-link structure was constructed in ROCOS software to simulate the effect of gravity and environment (floor). In the simulations, the constraints were correctly defined as snapshots show the bones and muscle are intact before and after contact with environment.

Keywords Biarticular · Monoarticular · Musculoskeletal · Closed-chain · Redundant

2.1 Introduction

Conventional robotic designs utilize joint actuation using rotational actuators, geared or other transmission mechanism to achieve angular motion that moves arm-like structures [1, 2]. These designs have shown vast applications in industrial

---

A. Z. bin Hj Shukor (✉) · F. bin Ali@Ibrahim · M. F. bin Miskon
Fakulti Kejuruteraan Elektrik, Universiti Teknikal Malaysia Melaka, Durian Tunggal, Malaysia
e-mail: zaki@utem.edu.my

M. S. bin Md Saat
Fakulti Kejuruteraan Elektronik, Universiti Teknikal Malaysia Melaka, Durian Tunggal, Malaysia

M. K. bin Mohamed Nor
Fakulti Kejuruteraan Mekanikal, Universiti Teknikal Malaysia Melaka, Durian Tunggal, Malaysia

robotic arms used in manufacturing such as in spray painting and assembly for cars, as well as mobile/legged robotics in various fields, such as space exploration [3] and concrete pipe inspection [4]. In recent times, robotic structures have also been inspired by the musculoskeletal arrangements of humans/animals. In short, the muscles actuate the motion of the bones/skeleton by extending or contracting, causing joint movement of the arms/legs. Another interesting property of musculoskeletal robotics is the redundant actuation of the muscles which results in improved shape and magnitude of the force ellipsoid of the end effector, for example musculoskeletal biarticular actuation as compared to monoarticular actuation [5].

In terms of muscle actuation, researchers venture into the use of different actuators for implementation. For the four-legged robot HyQ [6], pneumatic cylinders were used in their leg design for flexion motion. Biwi, the wire driven robot arm [7] uses six motors to pull/extend six cables to represent the six muscles from shoulder to elbow. Infinity norm approach was used to distribute forces from the end effector to the six muscles (motors). Another design uses rotational motors with planetary gears to represent redundant actuation of the muscles on the joints [8] and utilize biarticular muscle torque.

2.2 Kinematics of a Musculoskeletal Structure

The musculoskeletal structure in this paper refers to the representation of a muscle with a linear extension/contraction. Consider a single link planar musculoskeletal robotic structure as shown in Fig. 2.1.

The single link (bone) structure is composed of a muscle, rotational joint, and bones. The motion desired is on the bones connected to θ_1, but derived from the extension/contraction of muscle. The angle between the base and muscle, θ_2 also changes with θ_1. However focus is on θ_1 because it is the position where most joint actuators are placed in conventional robotics. The kinematics linking rotational angle θ_1 and muscle length l_1 can be derived from trigonometry (2.1).

$$\begin{aligned} l_1 &= \sqrt{(a_1 + a_2)^2 + (a_3 \sin \theta_1)^2} \\ &= \sqrt{(a_1 + a_3 \cos \theta_1)^2 + (a_3 \sin \theta_1)^2} \\ &= \sqrt{a_1^2 + 2a_1 a_3 \cos \theta_1 + a_3^2} \end{aligned} \quad (2.1)$$

The kinematics is essential to determine the angle from the muscle length and vice versa. This means that a linear actuator equipped with a linear position encoder representing the muscle does not require a joint position encoder to measure the joint angle. A further differentiation of the length with respect to joint angle would result in the Jacobian linking the muscle and joint velocities.

Fig. 2.1 Musculoskeletal single link structure

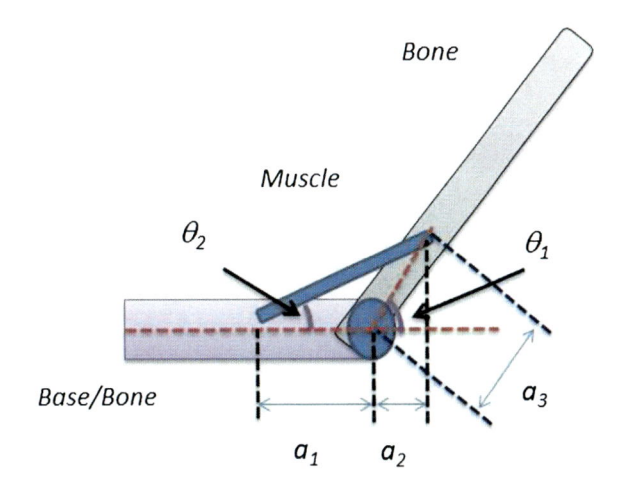

2.3 Dynamics of a Musculoskeletal Structure

A generalized equations of motion of an open-link robotic mechanism can be defined as in (2.2) where τ is the generalized force/torque vector, M is the mass matrix, C is the Centrifugal/Coriolis term and G is the gravity term which could be obtained by any dynamic modeling technique (i.e. Lagrange–Euler, Newton–Euler).

$$\tau = M(\theta)\ddot{\theta} + C(\theta,\dot{\theta})\ddot{\theta} + G(\theta) \quad (2.2)$$

The structure in Fig. 2.1 is not an open-link robot, because there exist a closed-kinematic loop/chain between the muscle and the bone. Thus for a closed-link structure, its dynamics are governed by (2.3) and (2.4):

$$\tau + J_c^T \lambda = M(\theta)\ddot{\theta} + C(\theta,\dot{\theta})\ddot{\theta} + G(\theta) \quad (2.3)$$

$$J_c \ddot{\theta} + \dot{J}_c \dot{\theta} = 0 \quad (2.4)$$

J_c is the Jacobian of the constraint equation (closed kinematics) while λ is the Lagrange multiplier. The term $J_c^T \lambda$ represents the constraint (internal) forces of the muscle loop. As an example, the constraint equations of Fig. 2.1 are depicted as K_1 and K_2 as shown.

$$a_1 + a_3 \cos\theta_1 - l_1 \cos\theta_2 = K_1 = 0 \quad (2.5)$$

$$a_3 \sin\theta_1 - l_1 \sin\theta_2 = K_2 = 0 \quad (2.6)$$

By differentiating the constraint equations with respect to the motion variables of the structure (θ_1, θ_2 and l_1), the following constraint Jacobian (J_c) can be obtained.

$$\begin{bmatrix} \dfrac{\partial K_1}{\partial \theta_1} & \dfrac{\partial K_1}{\partial \theta_2} & \dfrac{\partial K_1}{\partial l_1} \\ \dfrac{\partial K_2}{\partial \theta_1} & \dfrac{\partial K_2}{\partial \theta_2} & \dfrac{\partial K_2}{\partial l_1} \end{bmatrix} = \begin{bmatrix} -a_3 \sin \theta_1 & l_1 \sin(\theta_2) & -\cos \theta_2 \\ a_3 \cos \theta_1 & -l_1 \cos(\theta_2) & -\sin \theta_2 \end{bmatrix} \quad (2.7)$$

The dynamics of the structure stated in (2.3) and (2.4) can also be constructed in matrix form as shown in (2.8).

$$\begin{bmatrix} M & -J_c^T \\ -J_c & 0 \end{bmatrix} \begin{bmatrix} \ddot{\theta} \\ \lambda \end{bmatrix} = \begin{bmatrix} \tau - C\dot{\theta} \\ \dot{J}_c \dot{\theta} \end{bmatrix} \quad (2.8)$$

From the equations of motion of closed-kinematics musculoskeletal structure shown in (2.8), the movement of the structure can be simulated/visualized. Note that the Lagrange multiplier λ also changes with time, along with the generalized accelerations. For simulation of closed-kinematic structure with environmental effects (i.e. floor), the term $J^T F_e$ can be added to (2.3). The term represents the reaction forces of the structure when in contact with the floor.

2.4 Simulation

From the single-link structure shown, two other structures were considered for simulation. This is shown in Fig. 2.2. The simulation software used is the Robot Control Software (ROCOS) developed by Yasutaka Fujimoto at Yokohama National University [9]. Note that the increase in the number of linear actuators incurs redundancy in the biarticular two-link structure. Kinematic equations derived for the biarticular muscle using trigonometry (l_3) shows that the shoulder (θ_1) and elbow joint angles (θ_2) are both present in the muscle equation (2.9).

$$l_3 = \sqrt{a_{31}^2 + a_{32}^2 + a_1^2 + 2a_{31}a_1 \cos \theta_1 + 2a_{32}a_1 \cos \theta_2 + 2al_{31}al_{32} \cos(\theta_1 + \theta_2)} \quad (2.9)$$

Some of the initial conditions that need to be pre-defined are the link parameters such as mass and length. Other important parameters include the connection length and angles between the muscles and the links (bones). These muscle lengths and joint angles at initial condition must be correctly defined using the muscle-joint kinematics.

Both structures were simulated with the effect of gravity and without any control to demonstrate it falling with gravity (Fig. 2.3). In these conditions, the structure (links and muscle) falls to the floor while the base is static. This shows that the closed-link pre-defined initial conditions (lengths and angles) were set correctly.

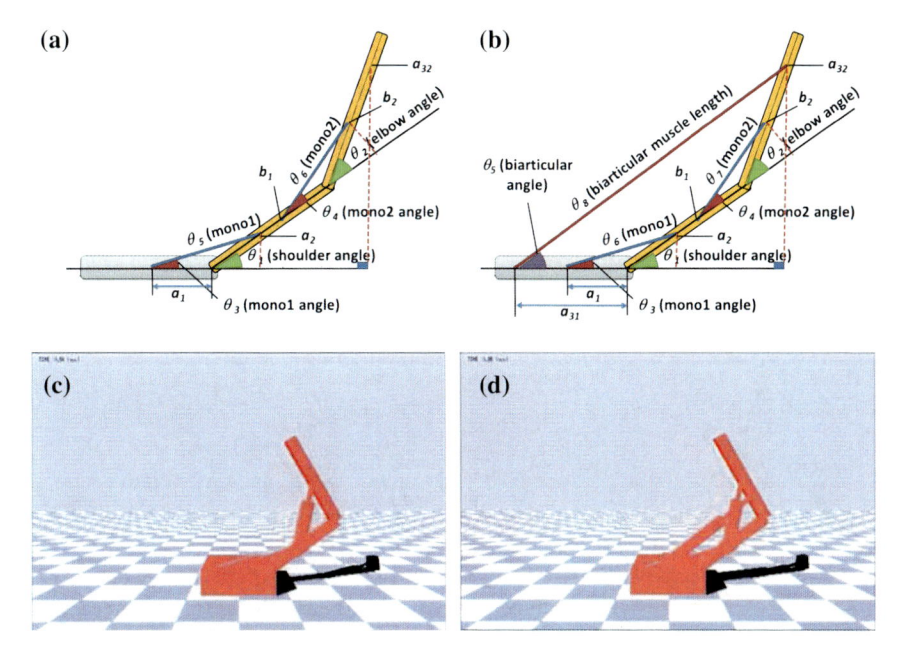

Fig. 2.2 **a** Monoarticular two-link structure. **b** Biarticular two-link structure. **c** Rocos model of (**a**) and **d** Rocos model of (**b**)

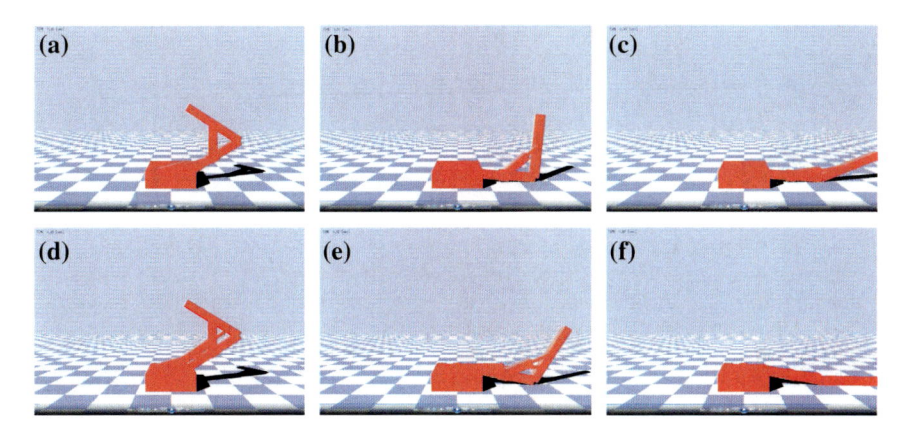

Fig. 2.3 Monoarticular (**a**–**c**) and biarticular (**d**–**f**) structure falling with gravity

2.5 Discussion

At first, the musculoskeletal structure proposed was explained in terms of its kinematic properties. The muscle kinematics relate muscle length and joint angles for motion. Then, the closed-kinematic structure incur a different dynamics representation due to the muscles forming a closed loop within the structure inducing constraint equations and Jacobian. In the simulations, two structures were modeled, one structure with monoarticular (single-joint actuation) muscles and one structure with biarticular (two-joint actuation) muscle. The simulation tests show the success in modeling in ROCOS software because the structures fall with gravity to the floor. The intention of the simulations is to verify the correct definitions of constraints in both monoarticular and biarticular structure and not to compare performance results between the two. Thus the results in this paper are only the snapshots of the models with gravity and environmental effects.

2.6 Conclusion

A new musculoskeletal robotic structure incurs new kinematic and dynamic properties. To model the structure, kinematic relations induces constraints that must be included in the dynamics. A correct definition of kinematic properties would result in successful motion in dynamics simulation software. The success in modeling in software would induce further work which includes application of muscle position and force control on the musculoskeletal structure to explore the advantages of redundant actuation.

Acknowledgments Authors wish to acknowledge Universiti Teknikal Malaysia Melaka (UTeM) for the research grant (PJP/2013/FKE(5D)/S01171) provided for this project. Also to Prof. Dr. Yasutaka Fujimoto from Yokohama National University for the robot simulation software.

References

1. Calanca A, Capisani LM, Ferrara A, Magnani L (2011) MIMO closed loop identification of an industrial robot. IEEE Trans Control Syst Technol 19(5):1214–1224
2. Bonilla I, Mendoza M, Gonzalez-Galvan EJ, Chavez-Olivares C, Loredo-Flores A, Reyes F (2012) Path-tracking maneuvers with industrial robot manipulators using uncalibrated vision and impedance control. IEEE Trans Syst Man Cybern Part C Appl Rev 42(6):1716–1729
3. Bartsch S, Birnschein T, Cordes F, Kuehn D, Kampmann P, Hilljegerdes J, Planthaber S, Roemmermann M, Kirchner F (2010) SpaceClimber: Development of a six-legged climbing robot for space exploration. In: IEEE Proceedings on 2010, 41st international symposium on robotics (ISR) and 2010, 6th German conference on robotics (ROBOTIK), Bremen, pp 1–8
4. Saenz J, Elkmann N, Stuerze T, Kutzner S, Althoff H (2010) Robotic systems for cleaning and inspection of large concrete pipes. In: IEEE Proceedings on 2010, 1st international conference on applied robotics for the power industry (CARPI), Montreal, pp 1–7 (2010)
5. Shukor AZ, Fujimoto Y (2011) Workspace control of biarticular manipulator. In: IEEE Proceedings on 2011, international conference on mechatronics (ICM), pp 415–420
6. Focchi M, Boaventura T, Semini C, Frigerio M, Buchli J, Caldwell DG (2012) Torque-control based compliant actuation of a quadruped robot. In: IEEE Proceedings of 12th international workshop on advanced motion control, IEEE, pp 1–6

7. Salvucci V, Kimura Y, Sehoon Oh, Hori Y (2013) Force maximization of biarticularly actuated manipulators using infinity norm. IEEE/ASME Trans Mechatron 18(3):1080–1089
8. Sehoon Oh, Kimura Y, Hori Y (2010) Force control based on biarticular muscle system and its application to novel robot arm driven by planetary gear system. In: Proceedings on 2010, IEEE/RSJ international conference on intelligent robots and systems (IROS), IEEE, pp 4360–4365
9. Fujimoto Y, Kawamura A (1998) Simulation of an autonomous biped walking robot including environmental force interaction. IEEE Robot Autom Mag 5(2):33–42

Chapter 3
Analysis of Finger Movement for Robotic Hand (MAPRoh-1) by Using Motion Capture and Flexible Bend Sensor

M. Hazwan Ali, Khairunizam Wan, Y. C. Seah, Nazrul H. Adnan and Juliana A. Abu Bakar

Abstract Since the beginning of twentieth century, human–computer interaction (HCI) and humanoid robot has been the trend of advance countries to show their achievement in the technology. Thus, it is rituals for others develop countries to tail their footstep. By means of self-construct robotic hand based on human hand behaviors, an experiment was conducted to investigate the characteristic of robotic finger movements. The purpose of this paper is to analyze the correlation between angle produced by motion capture system (MOCAP) and voltage produced by the flexible bend sensor attached to the robotic hand. At the end of the project, the relationship regarding both angle and voltage will be clarified by using regression method and the preliminary result indicates that voltage and angle variation is possibly linear to each other based on correlation of coefficient outcome.

Keywords MOCAP · Flexible bend sensor · Robotic hand

M. H. Ali (✉) · K. Wan · Y. C. Seah
Advanced Intelligent Computing and Sustainability Research Group, School of Mechatronic, Universiti Malaysia Perlis Kampus Pauh Putra, 02600 Arau, Perlis, Malaysia
e-mail: hazwan_hafiz89@yahoo.com

K. Wan
e-mail: khairunizam@unimap.edu.my

N. H. Adnan
Bahagian Sumber Manusia, Tingkat 17 & 18, Ibu Pejabat MARA Jalan Raja Laut, 50609 Kuala Lumpur, Malaysia
e-mail: nazrulhamizi.adnan@gmail.com

J. A. Abu Bakar
Department of Multimedia, School of Multimedia Tech and Communication, College of Arts and Sciences, Universiti Utara Malaysia, 06010 Sintok, Kedah, Malaysia
e-mail: liana@uum.edu.my

3.1 Introduction

According to Richard Harper et al., HCI is a term used to refer to an understanding and designing of differences relationship between people and computer [1]. This relationship could occur at various features such as command line, menus, natural language, direct manipulation, and form fill [2]. Through direct manipulation, gesture of human body that contains meaningful information [3] will be interpreted through pointing device/graphical display [2]. Robotic hand is an example of relationship between human and computer where robotic hand allows a direct mapping from a human movement to robot. Robotic hand is an autonomous robot which control by auto, semi or manual command to move according to operator input. This characteristic of robotic hand produces a varied possibility in industry and academic purpose.

This research paper organized as subsequent; Sect. 3.2 encompasses literature review of the related research and approach toward robotic hand data acquisition. Section 3.3 presents the methodologies of applied procedures. Section 3.4 is divided into 2 sections, first section states about experiment setup where second section demonstrations the results of experiments. Finally Sect. 3.5 expresses the conclusions over current research.

3.2 Literature Review

The development of robotic hand in this research is based on construct dataglove known as *GloveMAP*. *GloveMAP* on the contrary is input device construct using strain gauge sensor to measure finger flexion [4] used to direct mapping from a human movement to robotic hand. *GloveMAP* has accomplished numerous successions such as virtual interaction [5] and PCA-based finger movement and grasping classification development [6]. *GloveMAP* is constructed based on human hand characteristics and presents voltage rate as output data. Angle contradictory to *GloveMAP* required MOCAP for angle analysis.

Robotic hand branded as MAPRoh-1 is another enhancement to the project understanding HCI. MAPRoh-1 is designed to reproduce human hand motions on finger movements. Robotic hand isn't new in HCI whereby the technology already develop by collaboration of University of Southern California and University of Novi-Sad patented as The Belgrade/USC hand used to mimic human hand functionality [7] where the control architecture are based on the principle of non-numerical or reflex control to drive the hand [8, 9]. University of Utah also has developed robot hand registered as The Utah/MIT hand equipped with end effector for function as a general purpose research tool [10].

MOCAP by Qualisys Track Manager Software (QTM) [11] otherwise present an alternative approach in angle analysis. Notable usage of MOCAP is in the study of skeletal parameter by Kirk et al. [12]. Jonathan Maycock et al. [13] also manipulates MOCAP and dataglove on robust tracking of human hand postures for the robot teaching. While on International Joint Conference 2006, Oh et al. [14] display a promising research in low cost MOCAP for PC-based immersive Virtual Environment (PIVE) system.

3.3 Methodology

3.3.1 Robotic Hand (MAPRoh-1)

MAPRoh-1 is a UniMAP Robot Hand Version 1, which is developed at Advanced Intelligent Computing and Sustainability Research Laboratory. The robotic hand is built based on understanding of human finger bending joints where Metacarpophalangeal (MP), Proximal Interphalangeal (PIP) and Distal Interphalangeal (DIP). Since PIP and DIP joint mostly affecting each other, PIP and DIP joint is built mechanically coupled and driven by one actuator. DC motor is selected as actuator to drive finger movements connected with steel string. While DC motor with steel string capable to initiate robotic finger bending, additional force is needed in order to return robotic finger into straight position. Therefore, a spring is attached to each finger joint where it will maintain the straightness of robotic finger as shown in Fig. 3.1. Material for robotic hand design is also vital as it will effect on the increasing of product weight hence affecting on motor torque and power usage requirement. Aluminum is chosen for this project as it is light weight, stainless, cheaper and easier to form the shape of robotic hand.

3.3.2 Flexible Bend Sensor

Flexible bend sensor is a variation form of variable resistance where resistance of the sensor intensified with the increasing distance between each of carbon resistivity element inside thin strip flex sensor. Voltage outputs of flexible bend sensor can be calculated by referring to (3.1), where V_{out} is output voltage, R_{sensor} is resistance of flexible bend sensor, R is resistance of voltage divider, V_{in} is input voltage from power supply, R_F is resistance of feedback resistor, R is the resistance of inverting resistor and gain is oobtained by referring to (3.2).

$$V_{out} = \left[\frac{R_{sensor}}{R_{sensor} + R}\right] \times V_{in} \times Gain \quad (3.1)$$

$$Gain = 1 + \frac{R_F}{R} \quad (3.2)$$

Fig. 3.1 Framework of MAPRoh-1

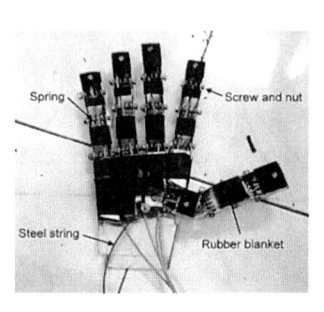

3.3.3 Regression Analysis

Regression Analysis is widely used in mathematical model to predict dependent variable y on dependent variable x. Regression Analysis coefficients calculation allowing the quantitative analysis of significant data. Regression analysis suitable in forecast purpose of interpolates and extrapolates as well as indicating association between variable. First degree of polynomial regression, linear polynomial is selected to display relationship between angles and voltage variable. Linear polynomial equation can be obtained by referring to (3.3) where y represents as robotic hand angle and x represents as robotic hand voltage for index finger.

$$y = mx + c \qquad (3.3)$$

3.4 Experiments

3.4.1 Experiment Setup

An experiment was conducted in MOCAP environment with 4 passive markers placed into robotic hand index finger segment while the finger is straightened. Upper portion of MAPRoh-1 was attached with flexible bend sensor to provide voltage analysis for the robotic finger as shown in Fig. 3.2. In the experiments, both MOCAP and Flexible bend sensor data acquisition start at the same time (Fig. 3.3).

3.4.2 Experiment Results

Figure 3.4 shows the angle and voltage signal in-phase while the index finger of the robotic hand moves started from the full bend position to the straighten position. From the graph, although voltage range variation smaller contrast to

Fig. 3.2 Robotic hand in MOCAP environment

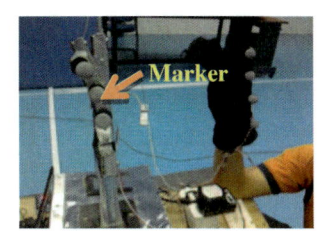

Fig. 3.3 Finger's angle

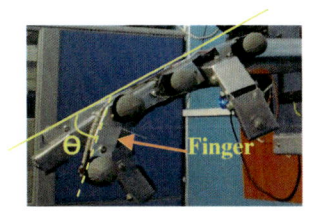

angle range variation, the angle and voltage signal achieve synchronization with each other where the angle and voltage raise time increment interval and the fall time decrement interval correspond to each other. To further investigate the rise time and fall time characteristics, a scatter plot of angle versus voltage was constructed. Slopes graft was analyzed using linear regression methods to find the correlation of coefficient for interpolate motion. Correlation coefficient (R^2) indicating linearity relationship between the correlation coordinate points of x-axis to coordinate points of y-axis. R^2 ranges from 0.85 to 1 suggests solid correlation or relationship between both variables. Correlation of coefficient concerning variable angle to voltage for Fig. 3.5a was 0.9901 for the rise time signal where correlation of coefficient concerning variable angle to voltage for Fig. 3.5b was 0.9034 acclaimed that the angle was direct correlation with the voltage. Due to R^2 value for the rise time signal closes to 1 when comparing to the fall time signal, it was selected to represent the robotic hand index finger behavior for a universal formula to calculate the relation between them. For universal formula construction, in order to determine data was in normal range or abnormal range, the null-hypothesis (H0) was set as normal range, where hypothesis (H1) was set as abnormal range. By using mathematic calculation, the P value was 0.0126* which was more than the significance level (α) 0.01. From the calculation, the null-hypothesis (H0) was acceptable and rejects the hypothesis (H1). Data that provided were in range of normal behavior. Thus, creating the universal formula for robotic hand index finger by referring to (3.4) whereas y represents angle and x represents voltage.

$$y = -71x + 254 \qquad (3.4)$$

Fig. 3.4 Angle and voltage graft for MAPRoh-1

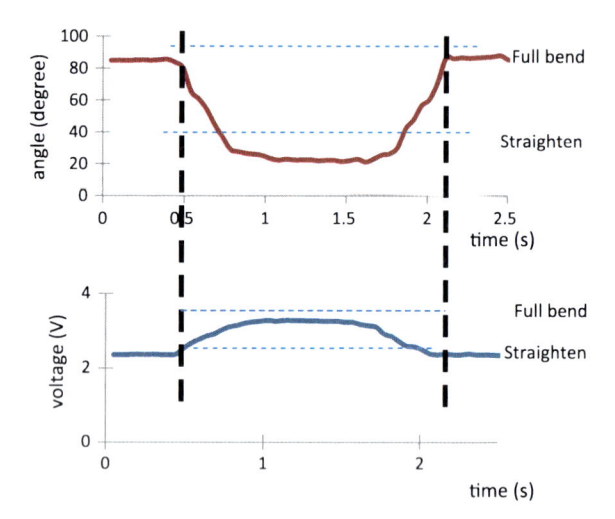

Fig. 3.5 Scatter plot of angle versus voltage **a** rise time **b** fall time

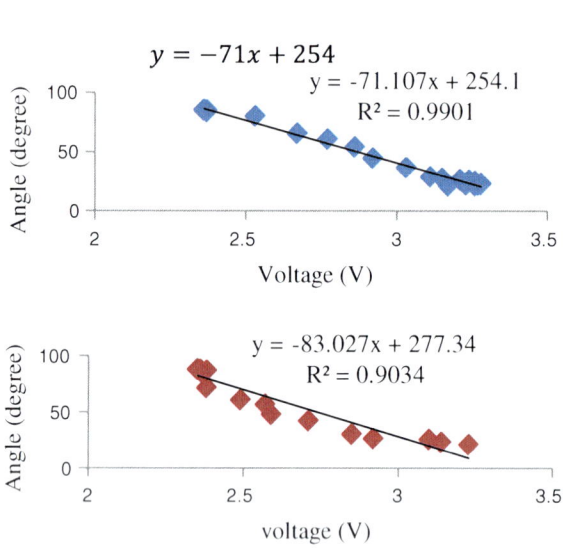

$$y = -71x + 254$$

y = -71.107x + 254.1
R² = 0.9901

y = -83.027x + 277.34
R² = 0.9034

3.5 Conclusions

As a conclusion, the research works are based on motion acquisition analysis for robotic finger MAPRoh-1. The MOCAP system is used to measure bending angle where flexible bend sensor provides voltage of robotic hand index finger. The analysis is executed to correlate the output angle and voltage using linear regression method. Preliminary experiment verifies that angle variable is directed with voltage variable thus construction of solitary equation that can be used to represent robotic hand index finger motion is accomplished. Furthermore, slopes study of angle in contrast to voltage signal show that bending and straighten movement have similarity characteristic.

References

1. Harper R et al (2008) Being human: human–computer interaction in the year 2020. Microsoft Research Ltd, p 43
2. Bechhofer S (2010) Human computer interaction. Lecture note, University of Manchester, pp 13–16
3. Billinghurst M (2011) Gesture based interaction. Lecture note, Chapter 14, p 1
4. Adnan NH, Wan K (2012) Accurate measurement of the force sensor for intermediate and proximal phalanges of index finger. Int J Comput Appl 45(15):59–65
5. Adnan NH et al (2012) Measurement of the flexible bending force of the index and middle fingers for virtual interaction. International symposium on robotics and intelligent sensors 2012 (IRIS 2012). Procedia Eng 41:388–394
6. Adnan NH et al (2013) PCA-based finger movement and grasping classification using data glove "glove map", 3rd edn. Int J Innovative Technol Explor Eng (IJITEE) 2:66–71
7. Bekey GA, Tomovic GR, Zeljkovic I (1990) Control architecture for the belgrade/USC hand. In: Venkataraman ST, Iberall T (eds) Dextrous robot hands. Springer, New York, pp 136–149
8. Tomovic R (1983) Control of assistive systems by external reflex arcs. In: Proceedings IFAC conference on bioengineering control systems, pp 555–562
9. Bekey GA, Tomovic R (1986) Robot control by reflex actions. In: Proceedings 1986 IEEE international conference on robotics and automation, San Francisco, pp 240–247
10. Jacobsen SC, Iversen EK, Knutti DF, Johnson RT, Biggers KB (1986) Design of the Utah/M.I.T. Dextrous hand. In: Proceeding IEEE international conference on robotics and automation, pp 1520–1532
11. Qualisys Software Information. http://www.qualisys.com/products/ Software/qtm/
12. Kirk AG, O'Brien JF, Forsyth DA (2005) Skeletal parameter estimation from optical motion capture data. In: IEEE Computer Society conference on computer vision and pattern recognition (CVPR 2005), vol 2
13. Maycock J, Steffen J, Ritter H (2011) Robust dataglove mapping for recording human hand postures. In: Proceedings of 4th international conference (ICIRA 2011), Part II, Germany, pp 34–45
14. Oh YI, Jo K-H, Lee J (2006) Low cost motion capture system for PC-based immersive virtual environment (PIVE) system. In: International joint conference 2006 (SICE-ICASE), Korea, pp 3527–3530

Chapter 4
Corrosion Detection Using LabVIEW for Robotic Inspection of Boiler Headers

Nadiah Amalina Zulkifli, Khairul Salleh Mohamed Sahari, Adzly Anuar and Mohd Azwan Aziz

Abstract Boiler header is the backbone of piping system inside thermal power plant which is used to tie multiple steam mains to one boiler. It is important to check the header for any signs of defects. This paper presents an image based approach to detect cracks and corrosions inside a boiler header using LabVIEW software. After an image of the boiler header inner wall is captured, thresholding technique is applied to manage background variation of the acquired image. Then, the boundaries of the corroded area are identified by using edge detection algorithm. The last step is to apply particle analysis for parameters measurement. Experiments are carried out on a 360° view of a cross-section of a boiler header for inspecting the surface defects of the boiler headers. The result from the experiment shows a reasonable success rate of correctly identifying corrosion inside the header.

Keywords Boiler headers · Visual inspection · Image processing · Corrosion · LabVIEW

4.1 Introduction

This paper documents the work done in corrosion detection using LabVIEW for robotic inspection of boiler headers. The aim of image processing is to detect signs of potential defects such as corrosion inside the boiler header which could lead to

N. A. Zulkifli (✉) · K. S. M. Sahari · A. Anuar
Centre for Advanced Mechatronics and Robotic, Universiti Tenaga Nasional, Jalan IKRAM-UNITEN, 43000 Kajang, Selangor, Malaysia
e-mail: nadiahamalinaz@gmail.com

M. A. Aziz
TNB Research Sdn Bhd, Lorong Ayer Itam, Kawasan Institusi Penyelidikan, 43000 Kajang, Selangor, Malaysia

failure of the boiler header's operation. Boiler header is the backbone of piping system inside thermal power plant in which work to tie multiple steam mains to one boiler, or multiple boilers to one or more steam mains. Recently, many problems occur in boiler headers which are caused by aging, corrosion, cracks, and mechanical damages from the third parties [1]. Therefore, boiler header inspection is necessary to check for signs of defects especially corrosion, before it fails. A robotic device has been developed for inspection of boiler header to remove labor intensive and to act in unreachable environment. In order to detect corrosion, the robot is equipped with camera to acquire the image of internal surface of the boiler headers. Then, image processing algorithm to detect the corrosion is developed using LabVIEW platform.

4.2 Background

Corrosion is the breaking down or destruction of material, especially a metal through oxidation or chemical reactions [2]. Corrosion is one of the leading causes of failures in boiler header and can cause huge damage to property and environment. This is due to the exposure of the internal wall of boiler header with surroundings which contains water and other contaminants such as oxygen, carbon dioxide or chloride. Image on the left side in Fig. 4.1 shows the surface of boiler header in normal condition when no operation is performed. The smooth surface indicates that the boiler header is clear from being oxidize, compared to the image on the right side in which the surface of boiler header is already exposed to the chemical reactions with environment.

4.3 Image Processing

Image processing algorithm is done by using Laboratory Virtual Instrument Engineering Workbench (LabVIEW), which is a graphical programming language that used icons instead of text based programming approach to create applications [3]. Image Acquisition (IMAQ) Vision is one of the development modules in LabVIEW which is used for image processing and contains more than 400 functions for displaying and building imaging systems [4]. LabVIEW is chosen in this experiment because it is suitable to be used for acquiring, processing and displaying data [5]. Furthermore, code use in LabVIEW is much easier to debug than the other language programming software such as MATLAB. The block diagram used in the LabVIEW as shown in Fig. 4.2, makes the programming easy to understand due to the simple construction of the graphical user interface.

IMAQ Vision is divided into three categories which are Vision Utilities, Image Processing, and Machine Vision [6]. We give more attention to image processing in this report for better inspection. Basically, image processing refers to the

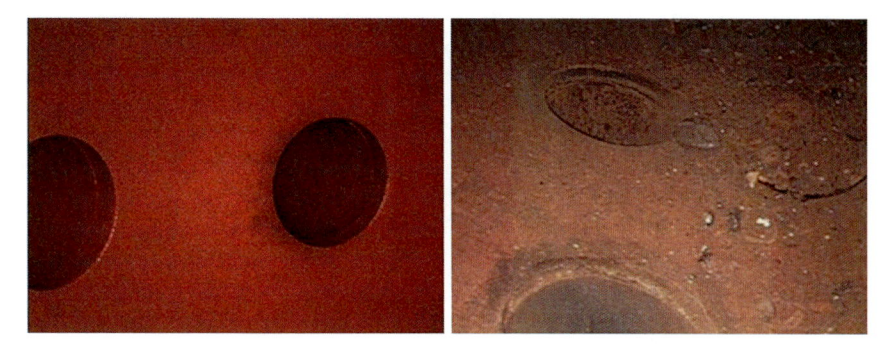

4 Corrosion Detection Using LabVIEW for Robotic Inspection of Boiler Headers 33

Fig. 4.1 Inner surface of boiler header [*source* Tenaga Nasional Berhad Reseacrh (TNBR) 2011]

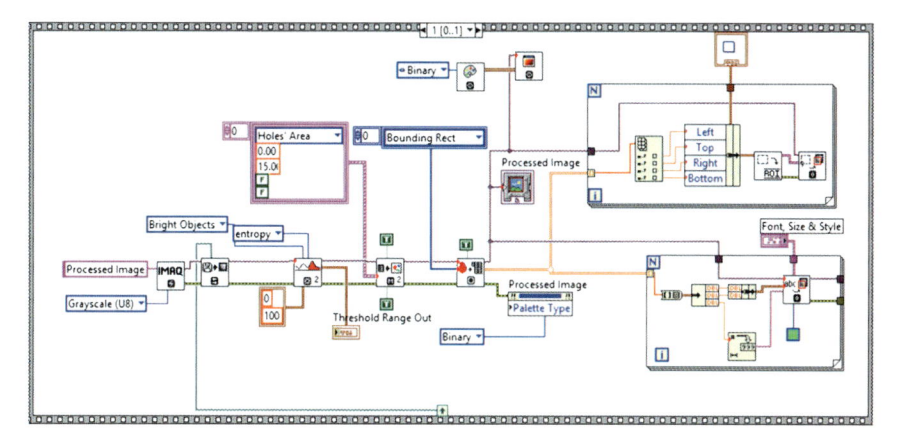

Fig. 4.2 Block diagram of the algorithm using LabVIEW

manipulation of a digitized image in order to enhance its quality [7]. Image processing allows us to analyze, filter and process image according to the information that we need. Following is the important image processing operations use in this report.

4.3.1 Image Acquisition

Image acquisition involves retrieving an image from a source whether real-time image or non real-time image [8]. In this project, the image is acquired by loading the images from a disk captured by the robot that equipped with a camera.

4.3.2 Thresholding

Thresholding is basically the first step in any image segmentation approach, where the objects can be separated from the background by adjusting the size and tolerance of the filter limits. The intensity values that above a threshold value correspond to foreground and all the remaining pixels correspond to the background [9]. A thresholding image is defined as:

$$g(x,y) = \begin{cases} 1 & \text{if } f(x,y) > T \\ 0 & \text{if } f(x,y) \leq T \end{cases} \quad (4.1)$$

Firstly, an image in grayscale or color image is converted to a binary image. The initial estimation is set for threshold (T) value to see the effects on the image processing. Then, the value is adjusted until the required result is obtained. After several experiments have been done, the threshold limit value is set to be 100 in order to get a better outcome. The corrosion area is relatively bright compared with the rest of the image. Hence, a method called "Look For (Bright Objects)" is used to detect the image represented by pixels with values greater than the value computed by the threshold limit value. However, the thresholding can be expected to be successful in controlled environments where the surrounding lighting is important for a better image processing. The lighting either from the camera or from the surrounding plays an important role in thresholding method.

4.3.3 Edge Detection

Edge detection is used in this project to extract the contours in gray-level values. Important features like corners, lines and curves can be extracted by using edge detection in order to get the boundaries of the corroded area.

4.3.4 Particle Analysis

Particle analysis is performed to find statistical information such as area, pixel value, location and presence of particles [10]. In this process, IMAQ Particle Analysis returns the number of particles detected in a binary image and commonly measurement parameters including the particle area and bounding rectangle can be reported.

Fig. 4.3 Actual image inside boiler header for experiment 1

Fig. 4.4 Actual image inside boiler header for experiment 2

4.4 Experimental Results

The binary image of the processing image is analyzed to find the possible defects. The defects inspection is achieved using thresholding method where the bright object expected as corrosion is separated from the background. Besides, the unwanted particles are likely dusts in the boiler header are discarded because of their small size. For each component, the pixel values on the bright objects are calculated based on its image RGB values. There are several challenges in analyzing these boiler header images. The main obstacle in the image processing is the algorithm hard to inspect the defects due to lighting factor comes from the camera

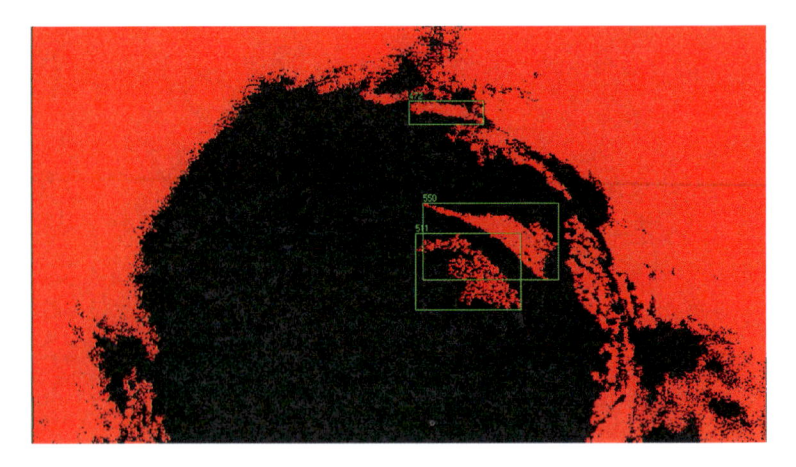

Fig. 4.5 Identification of corrosion in boiler header for experiment 1

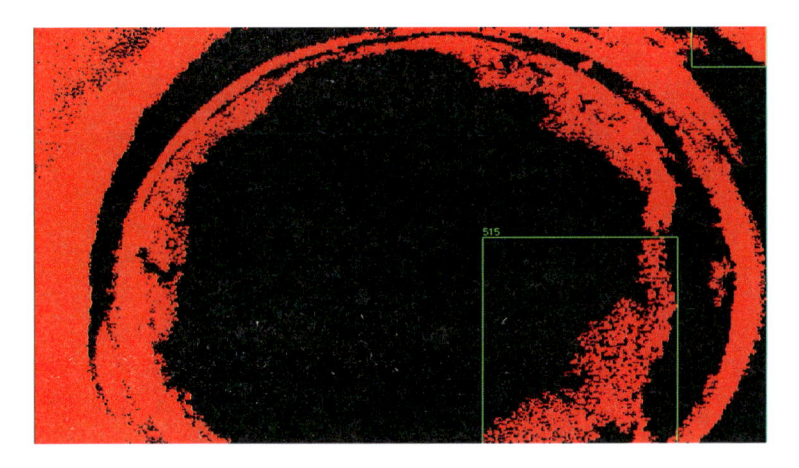

Fig. 4.6 Identification of corrosion in boiler header for experiment 2

and environment. Figures 4.3 and 4.4 show the actual images of boiler header which predicted to have corrosions.

The image processing has been executed to detect the defects relevant to the corrosion. The results are shown as in Figs. 4.5 and 4.6. The corroded areas are successfully detected and highlighted by bounding rectangles with their pixel values. The segmentation results are said to be successful when a comparison has been done manually with the original images. A threshold value can be chosen by several of experiments for a number of samples images and then tested more widely over a wider class of images. In this experiment, the threshold value is limited to 100, which also allows small size defects to be discarded, are likely dust on the boiler header surface.

The results of the experimental testing showed that out of twenty images given by TNBR, twelve images which have the possibility of corrosion are successfully detected. It is found that the other images are having a little bit of difficulty in image processing due to high brightness. Further improvement for this image processing will be enhancing the light intensity to obtain a better result.

4.5 Conclusion

Boiler headers are one of the important medium in power plant to carry steams which may be threaten by corrosion that leads to formation of crack if the corroded part is not being given attention. Thus, an image processing algorithm is developed using LabVIEW to inspect the corrosion inside the boiler header. The image loaded is being processed with several operations to extract the object from the background. Thresholding approach is then used to identify the corrosion, but it is difficult to identify the corrosion due to the severity of lighting in which effects the quality of image. The corroded areas are extracted and further analyzed to compute the pixel value of the defects. The bounding rectangle is used to mark the possible corrosion areas that occur in the boiler header. The experiment result shows reasonable success. Further research is needed to improve the image processing technique for precise detection of corrosion identification.

Acknowledgments Authors thank Tenaga Nasional Berhad Research (TNBR) for providing funding for this research project.

References

1. Choi HR, Roh SG (2007) In-pipe robot with active steering capability for moving inside pipelines. Bioinsp Robot Walk Climb Robot 23:375
2. The free dictionary by Farlex. http://www.thefreedictionary.com/corrosion
3. LabVIEW fundamentals (2007) National Instruments Corporation
4. Panayi GC, Rajashekar U, Bovik AC (2000) Image processing for everyone. In: First signal processing education workshop
5. Tadej T, Darko L, Frančišek T, Jörg E (2012) Comparison of LabVIEW and MATLAB for scientific research. Annals Faculty Eng Hunedoara: Int J Eng, ISNSN 1584–2673, 389–394
6. Boyik AC (2009) The essential guide to image processing. Academic, New York
7. Silviu F (2011) Practical applications and solutions using LabVIEW software. Intech Open Access Publisher, Croatia
8. Adamo F, Attivissimo F, Di Nisio A, Savino M (2009) An online defects inspection system for satin glass based on machine vision. In: Instrumentation and measurement technology conference I2MTC'09 (IEEE)
9. Fisher R, Perkins S, Walker A, Wolfart E (2003) Image Processing Learning Resources HIPR2. Adaptive thresholding. http://homepages.inf.ed.ac.uk/rbf/HIPR2/hipr_top.htm
10. IMAQ Vision for LabVIEW User Manual (2004) National Instruments Corporation, Austin

Chapter 5
Kinematic Analysis of Walking with Scottish Rite Orthosis

Mahboubeh Keyvanara, Marzieh Mojaddarasil,
Mohammad Jafar Sadigh and Mohammad Taghi Karimi

Abstract Leg Calve Perthes Disease (LCPD) is a condition in which the blood supply of the femoral head is closed and as a result the bone temporarily dies. Depending on the severity of the disease this leads to irritability of the hip bone and finally the head of the femur is deformed. Various types of methods have been used to treat LCPD which are divided into operative and non operative methods; where in the non-operative method different orthosis are prescribed (Karimi et al. in Evaluation of the gait of subjects with avascular necrosis of hip joint: a study of long term orthotic use. Submitted to: Physiotherapy Research International, [1]). These orthosis are designed base on the idea of reducing the stress applied on the bone. To facilitate study of effect of different orthosis and their design parameters on the force applied to the bone; we tried to develop a mathematical model of such patients during walking. To this end a kinematic analysis is given here. The model is then verified by numerical simulation and comparison of the results with those obtained from experiments done by a Gait Analyzing System that uses QTM software to capture motions and Visual 3D software to derive the tests' data.

Keywords Perthes disease · Forward kinematics · Inverse kinematics · Biped motion · Three dimensional model

---

M. Keyvanara · M. Mojaddarasil · M. J. Sadigh (✉)
Department of Mechanical Engineering, Isfahan University of Technology, Isfahan, Iran
e-mail: jafars@cc.iut.ac.ir

M. Keyvanara
e-mail: m.keyvanara@me.iut.ac.ir

M. Mojaddarasil
e-mail: m.mojaddarasil@me.iut.ac.ir

M. T. Karimi
Musculoskeletal Research Centre, Isfahan University of Medical Sciences, Isfahan, Iran
e-mail: Mohammad.karimi.bioengineering@gmail.com

5.1 Introduction

LCPD is an illness where the blood supply on the hip joint doesn't meet the needs of the bone and therefore it is said that the bone turns black and temporarily dies. Although original description of Perthes disease returns to nearly 100 years ago, the treatment remains controversial and there is a lack of agreement regarding the best treatment protocol for the patients. This disease happens in children with ages between 5 and 12 years old and its incidence varies between 0.45 for black and 10.8 per 100,000 for white children [2, 3]. Problems such as decreasing hip joint range of motion, alternation in growth of femoral bone and pain during walking are associated with Perthes disease [4, 5]. Furthermore there are a few studies regarding the gait performance of the subjects with Perthes disease [6, 7]. One of the main concerns in the treatment of this illness is reducing the forces tolerated by the patients in their hip joints. The treatment prescribed in the non- operative field, is using different orthosis; the most famous one called Scottish Rite (see Fig. 5.1). It is claimed that this orthosis reduces the forces on the hip joint by giving an extra abduction to each leg. In other words when the legs of a patient are abducted enough, the femoral head of the hip bone is placed so correctly inside the accetabulum, that the force tolerated on the hip joint is reduced. However recent results [1, 8] have shown that the result of this treatment may not be as good as expected. In order to have a better understanding of how this orthosis affects the hip joint, and to be able to study effect of parameters of orthosis on applied force to the hip joint, we may use a mathematical model to study that. Such model must provide equations of motion governing motion of a biped walking under the constraint of the orthosis. Then we need to develop both forward kinematics needed for calculations of velocity and acceleration, and inverse kinematics, needed for path planning. To this end we first tried to propose a model which with minimum degrees-of-freedom gives us the closest results to the real walking of a man who uses the Scottish rite orthosis.

The main objective of this paper is to present and validate model and the kinematic equations of motion for the system. The result of this effort is then used in a dynamic model to evaluate applied forces during walking. This model gives us the chance to try different parameters on a person's walking without the need of having anybody try those particular circumstances.

There are five parts in this paper. After this introduction we first concentrate on presenting the ideal model by working on the essential degrees-of-freedom, which is discussed in Sect. 5.2. In the third section, we go through the inverse kinematic problem. And Sect. 5.4 is the model's verification which is done by comparing the results of the forward kinematics with those derived from the measurements of walking man. The last part of the paper is the conclusions.

Fig. 5.1 Scottish rite orthosis

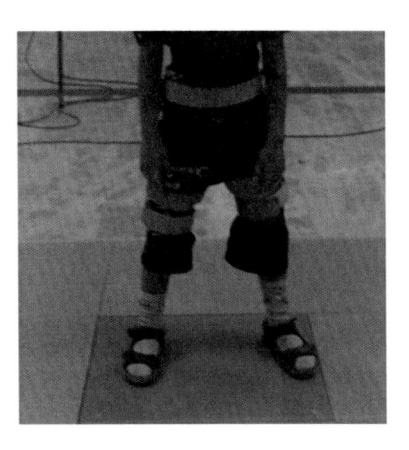

5.2 Direct Kinematics

The design of a humanoid model is naturally inspired by the functional motions of the human body, where the complex nature of the human structure cannot be exactly reproduced in modeling. Therefore, in modeling the lower limb and the pelvis, the number of the mobility in the human body should be limited to the essentials.

The anthropomorphic model presented in this article is a seven linked one with an overall of 18 degrees-of-freedom (DOF). A foot, a shank and a thigh on each leg which make the lower limb, plus a T-shaped link considered as the pelvis, all together making up the total seven links. Six DOFs is considered for the pelvis movements which consists of three degrees for its linear and three other for its rotational displacements; all six defined with respect to the inertial frame. The hip articulation has three revolute degrees of mobility, one DOF is defined for the knee joint, and for each ankle we have considered two DOF.

In Figs. 5.2, 5.3 and 5.4 the model and its DOFs have been pictured in the three main planes of motion.

Having defined the generalized coordinates of the system we should find position, velocity and acceleration of the mass centre of each link, needed for direct dynamics, and also position of each joint, needed for model verification. To this end, we use basic homogeneous transformations [9]. For instance we may find position of joint 1 using the following formula:

$$\mathbf{H}_1^0 = \text{trans}_{x,q_1} * \text{trans}_{y,q_2} * \text{trans}_{z,q_3} * \text{rot}_{z,q_4} * \text{rot}_{x,q_5} * \text{rot}_{y,q_6} \quad (5.1)$$

In which "trans" and "rot" stand for translational and rotational transformation matrix as defined in [9].

Fig. 5.2 The model in the transversal plane

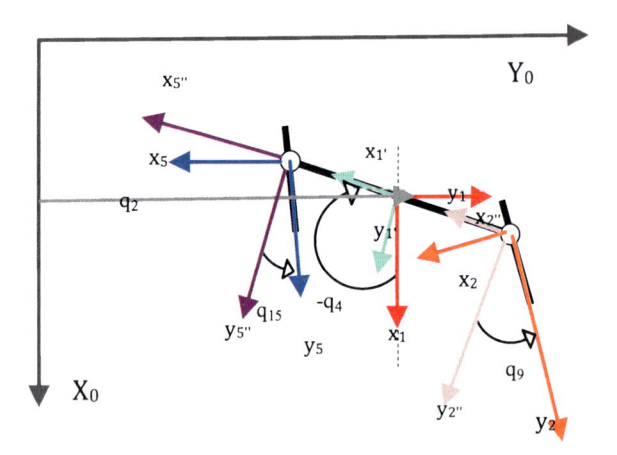

Fig. 5.3 The model in the frontal plane

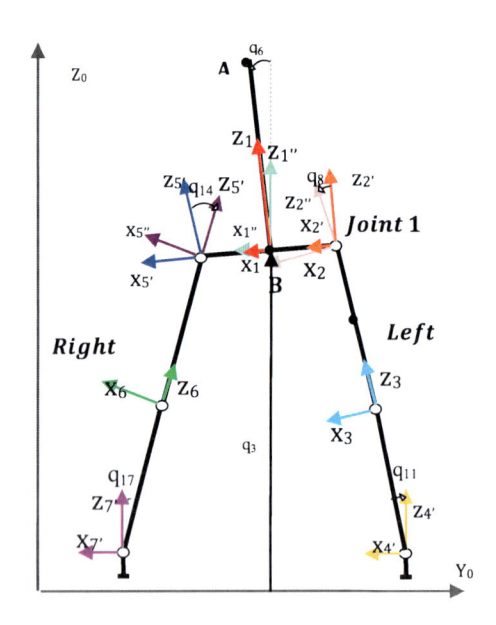

5.3 Inverse Kinematics

Inverse kinematic solution is an essential part of path planning which itself is needed in the procedure of evaluating forces applied to the hip joints due to change of parameters of the orthosis. Such solution should provide algebraic equations which give generalized coordinates of functions of position of markers. In this particular problem the values of q_1, q_2, and q_3 can be easily calculated as the mean value of corresponding values for left and right hip.

Fig. 5.4 The model in the sagittal plane

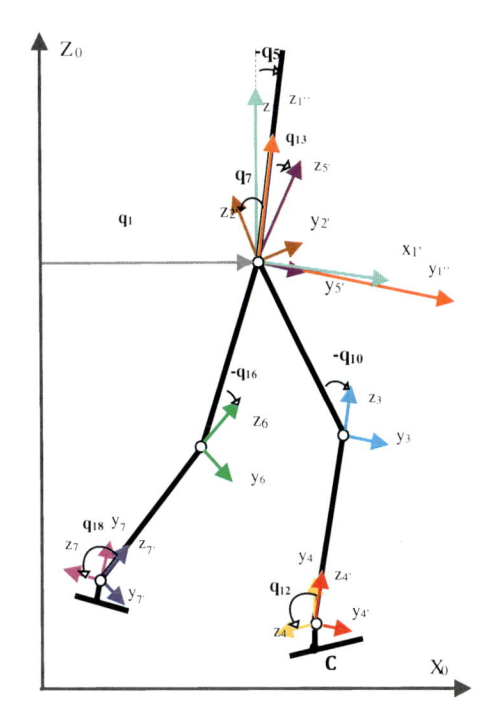

Considering the effect of the constraint applied by the brace, the gate of motion for patients who use the brace is three dimensional and has considerable portion of out-of-sagittal plane motion. Investigation of the experimental data shows that q_6 is small enough to ignore coupling effect of that with the other angles describing the upper body motion ($-3° < q_6 < 8°$).

An assumption which helps to solve inverse kinematics is that two links of shank and thigh remain in a plane. Considering the above mentioned assumptions; we may calculate joint angles which result in the values presented in Table 5.1 (point A, B and C can be seen in Figs. 5.3 and 5.4) q_{13}, q_{14}, ... and q_{18} which are related to right leg can similarly be calculated.

5.4 Model Verification

In order to verify the developed model, we arranged an experiment with a gait analysis system, where the kinematic variables of the movements of a walking patient of a height 1.29 m who is using the Scottish Rite brace and compare this data with those obtained by the model. The gait analysis system is one composed of a motion capture system, Qualysis, and two software, Qualysis Track Manager (QTM) and Visual 3D, which make it possible to record the motion and produce a three-dimensional analysis of the movements of a polyarticulated system such as

Table 5.1 Inverse kinematic result

DoF	Equivalent angle
q_1	$(X_{\text{Left hip}} + X_{\text{Right hip}})/2$
q_2	$(Y_{\text{Left hip}} + Y_{\text{Right hip}})/2$
q_3	$(Z_{\text{Left hip}} + Z_{\text{Right hip}})/2$
q_4	$270° + \tan^{-1}\left(\frac{X_{\text{Left hip}} - X_{\text{Right hip}}}{Y_{\text{Left hip}} - Y_{\text{Right hip}}}\right)$
q_5	$\tan^{-1}\left(\frac{X_A - X_B}{Z_A - Z_B}\right)$
q_6	$\tan^{-1}\left(\frac{Z_{\text{Left hip}} - Z_{\text{Right hip}}}{Y_{\text{Left hip}} - Y_{\text{Right hip}}}\right)$
q_7	$90° - q_5 - \gamma_{\text{Left Leg}}, \ \gamma = \tan^{-1}\left(\cos\theta \frac{x_{\text{knee}} - x_{\text{hip}}}{y_{\text{knee}} - y_{\text{hip}}}\right)$
q_8	$90° - \theta_{\text{Left Leg}}, \ \theta = \tan^{-1}\frac{Z_{\text{hip}} - Z_{\text{ankle}}}{Y_{\text{ankle}} - Y_{\text{hip}}}$
q_9	$\tan^{-1}\left(\frac{X_{\text{Left hip}} - X_{\text{Right hip}}}{Y_{\text{Left hip}} - Y_{\text{Right hip}}}\right) + \tan^{-1}\left(\frac{X_{\text{Metatarsus}} - X_{\text{Left hip}}}{Y_{\text{Metatarsus}} - Y_{\text{Left hip}}}\right)$
q_{10}	$-\alpha_{\text{Left Leg}}, \ \alpha = \cos^{-1}\left(\frac{r^2 - L_{\text{shank}}^2 - L_{\text{thigh}}^2}{2 * L_{\text{shank}} * L_{\text{thigh}}}\right), \ r^2 = (x_{\text{hip}} - x_{\text{ankle}})^2 + \left(\frac{y_{\text{hip}} - y_{\text{ankle}}}{\cos\theta}\right)^2$
q_{11}	$\theta_{\text{Left Leg}} - 90°$
q_{12}	$\beta + \tan^{-1}\left(\frac{X_{\text{metatarsus}} - X_C}{Z_{\text{metatarsus}} - Z_C}\right),$ $\beta = \tan^{-1}\left(\frac{x_{\text{hip}} - x_{\text{ankle}}}{y_{\text{hip}} - y_{\text{ankle}}} \cos\theta\right) - \tan^{-1}\left(\frac{L_{\text{thigh}} * \sin\alpha}{L_{\text{thigh}} + L_{\text{thigh}} * \cos\alpha}\right)$

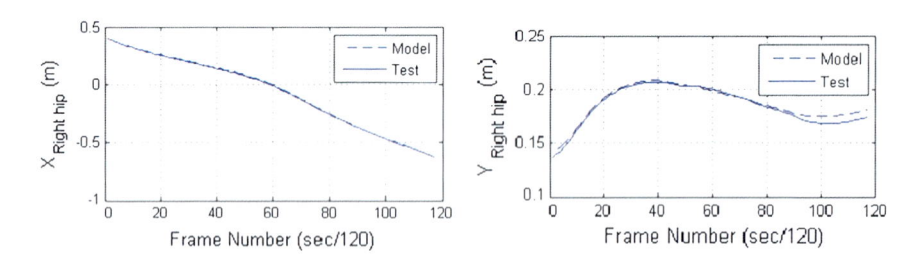

Fig. 5.5 X and Y displacements of right hip

the human body. The subject, whose data is being received from the system, has to be covered in certain points on the body with markers of special kind that can be recognized by the cameras of the system.

For the model verification to be done in correct conditions, working on a certain test we first collect the data of position of markers. Then the inverse kinematics is solved by QTM to obtain time history of joints angles. These joint angles are then

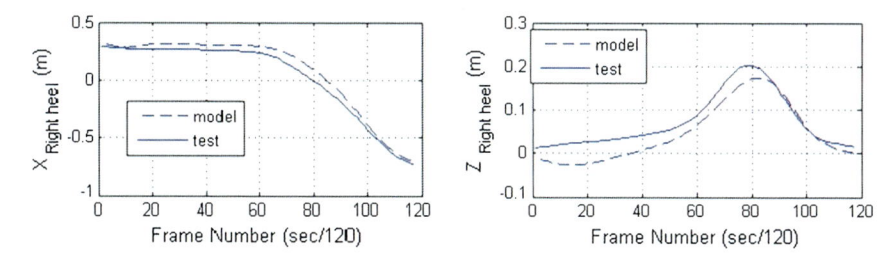

Fig. 5.6 X and Z displacements of right heel

used in forward kinematics model to calculate the position of markers. Agreement of results verifies model in two ways: first, the selected DOF, for modeling is sufficient to our purpose, and the second it proves correctness of mathematical modeling. The following figures show the comparison of these two set of data for the hip and the heel of the right leg (Figs. 5.5 and 5.6).

As it is noticed from the above figures, either the two diagrams are totally fitted or at some points they have little differences, which is a good support for model's validity.

5.5 Conclusion

Leg Calve Perthes Disease is an illness that happens in children of aged between 5 and 12. A common treatment of this illness is to use the so called Scottish Rite brace to reduce the force applied to the hip joint.

To establish a mathematical model for walking of a patient with this orthosis, we first proposed a seven-link model with 18 DOFs capable of resembling all motions of such patient during walking. The mathematical model of forward and inverse kinematic of the system are then developed the resulting model is verified by test result for the same subject.

References

1. Karimi M, Sadigh J, Mojaddarasil M, Keyvanara M, Sadjadifar F, McGarry T (2012) Evaluation of the gait of subjects with avascular necrosis of hip joint: a study of long term orthotic use. Physiother Res Int (Submitted)
2. Pillai A, Atiya S, Costigan PS (2005) The incidence of Perthes' disease in Southwest Scotland. J Bone Joint Surg 87(11):1531–1535
3. Rowe SM, Jung ST, Lee KB, Bae BH, Cheon SY, Kang KD (2005) The incidence of Perthes' disease in Korea: a focus on differences among races. J Bone Joint Surg 87(12):1666–1668

4. Eijer H, Berg RP, Berg RP, Haverkamp D, Pecasse GA (2006) Hip deformity in symptomatic adult Perthes' disease. Acta Orthop Belg 72(6):683–692
5. Evans DL (1958) Legg-Calve-Perthes' disease: a study of late results. J Bone Joint Surg 40:168–181
6. Westhoff B, Martiny F, Reith A, Willers R, Krauspe R (2012) Computerized gait analysis in Legg-Calve-Perthes disease–analysis of the sagittal plane. Gait Posture 35(4):541–546
7. Yoo WJ, Choi IH, Cho TJ, Chung CY, Park MS, Lee DY (2008) Out-tocing and in-toeing in patients with Perthes disease: role of the femoral hump. J Pediatr Orthop 28(7):717–722
8. Karimi M, Sadigh J, Fatoye F (2012) Evaluation of gait performance of a participant Perthes disease while walking with and without a Scottish-Rite orthosis. Prosthet Orthot Int 37:233
9. Spong MW, Hutchinson S, Vidyasagar M (2006) Robot modeling and control, 1st edn. Wiley, New York

Chapter 6
Color and Thermal Image Fusion for Augmented Reality in Rescue Robotics

Ludek Zalud, Petra Kocmanova, Frantisek Burian and Tomas Jilek

Abstract At the beginning of this article, the authors address the main problems of todays remotely-operated reconnaissance robots. The reconnaissance robots Orpheus-AC, Orpheus-AC2 and Orpheus-Explorer, made in the Department of Control and Instrumentation (DCI), are then shortly described. Since all the described robotic systems use visual telepresence as the main control technique, visual information from the robots surroundings is essential for the operator. For this reason, the authors make a fusion of data from a Charge-Coupled Device (CCD) color camera, and a thermovision camera to provide the operator with data in all visibility conditions, such as complete darkness, fog, smoke, etc.

Keywords Robot · User interface · Telepresence · Augmented reality

6.1 Introduction

The reconnaissance of dangerous areas is one of the most challenging tasks for todays robotics. According to many indications, e.g. from the Robocup Rescue League community where the DCI team is involved [1, 2], it seems that nowadays

L. Zalud (✉) · P. Kocmanova · F. Burian · T. Jilek
CEITEC Central European Institute of Technology, Technicka 10, 61600 Brno, Czech Republic
e-mail: ludek.zalud@ceitec.vutbr.cz

P. Kocmanova
e-mail: petra.kocmanova@ceitec.vutbr.cz

F. Burian
e-mail: frantisek.burian@ceitec.vutbr.cz

T. Jilek
e-mail: tomas.jilek@ceitec.vutbr.cz

the development of practical and usable reconnaissance robots [3] is aimed at the following tasks:

- A larger number of robots controlled by one operator.
- Easy and intuitive human-to-robot interface.
- For many kinds of reconnaissance missions it would be highly beneficial if the user interface would somehow emphasize alive people.

The authors propose a possible solution of the abovementioned problems through an advanced user interface program called CASSANDRA and show its application on Orpheus reconnaissance robots. Three different robots: Orpheus-AC, Orpheus-AC2 and Orpheus-Explorer, completely built in our laboratory, are roughly described in the Sect. 6.2, providing an example of remotely controlled robotic systems possibly controlled by CASSANDRA, while the described data-fusion is nowadays present on Orpheus-Explorer.

6.2 Orpheus Robots

Orpheus robots have been developed at Department od Control and Instrumentation (DCI) since 2003. The first version was called simply Orpheus, and our team was quite successful in Robocup Rescue 2003 world competition in Padova, Italy—we won the competition (see [2]). In 2003–2006 we improved/rebuilt the robot to the version Orpheus-X2 (see [4]). In 2006 we were asked to make a military version of the robot. The prototype was finished in 2007 and named Orpheus-AC (Army and Chemical) [5]. In 2009 we started development of second generation, based on Orpheus-A2 platform. We decided to make two basic modifications—Orpheus-AC2 for chemical and nuclear contamination measurements and Orpheus-Explorer for more general reconnaissance missions and victim search (Fig. 6.1).

The hereinafter described fusion is done on Orpheus-Explorer robot, which is equipped with novel sensory head containing wide-FOV (Field of View) overview camera AXIS 1114 with 2.3 mm lens, wide FOV thermal camera NEC C100C with 8 mm lens, and high-resolution reconnaissance camera AXIS Q1755 with "night vision" IR mode, see Fig. 6.2.

6.3 Visible Spectrum and Thermovision Data Fusion

The goal of the research described in this subsection is to improve the user interface to be a system that: makes robot control possible virtually in all visibility conditions [6, 7, 8, 9], such as fog, smoke or darkness; displays the most appropriate data or fused data in a convenient and intuitive way, visually emphasizes living victims, permits the use of the same data for digital map building and self-localization. This is done through data-fusion from CCD color camera and thermal imager [10, 11].

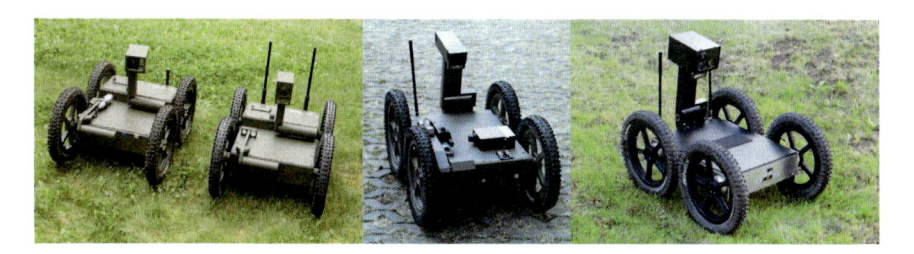

Fig. 6.1 Orpheus robots (*from left*): Orpheus-AC prototype, Orpheus-AC, Orpheus-AC2, Orpheus-Explorer

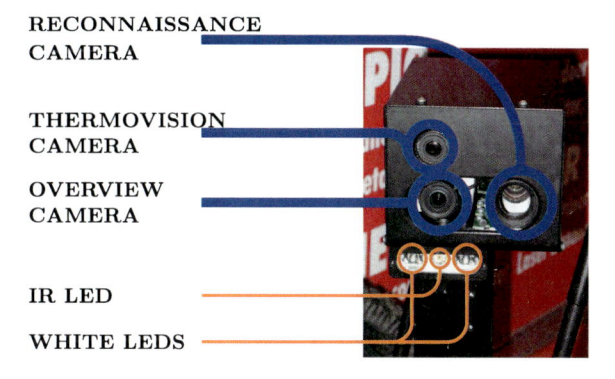

Fig. 6.2 Orpheus-Explorer sensory head

RECONNAISSANCE CAMERA

THERMOVISION CAMERA

OVERVIEW CAMERA

IR LED

WHITE LEDS

The basic problem of this data fusion is that the goal is to mix the data from the visible spectrum sensed by a color CCD camera and LWIR (long wave infrared) spectrum measured by the thermovision imager to the visible spectrum shown on a display [12, 13]. Since the spectrum to be displayed is evidently wider than the one we have at our disposal, it seems this task may not be done without some kind of compression of the color data.

We will use HSV color model to describe the colors. If we express a CCD camera image in the HSV mode [14] and make histograms of it, we clearly see that the distribution of S and V is not balanced and there are few (if any) pixels with $S = 1$ and $V = 1$. This is a result of the restrictions of CCD and their data processing.

From this we can conclude that if we add pixels with colors that have an arbitrary H parameter and $S = 1$ and $V = 1$, they will thus clearly be perceptible by the operator. In other words, it could be said we are adding a rainbow of colors to the ordinary image, which while perhaps uncommon (or even present) in an image, may well be used to emphasize parts of the image.

Regarding the previous text, the whole procedure of data mixing described as follows:

- The image from a thermovision camera is digitalized, if it is not already digital.
- The temperatures that are not near human body temperatures are filtered off.

- The pixels that are to be displayed are recalculated—H corresponds to frequency, S = 1 and V = 1.
- This thermovision image is rendered (e.g. alpha-blended) over the CCD camera image.

So we can write the color transformation:

$$H' = \frac{H}{60} = 6 \times (1 - R_{TH}) \qquad (6.1)$$

$$H_A = H' - \lfloor H' \rfloor \qquad (6.2a)$$

$$H_B = 1 - H_A \qquad (6.2b)$$

$$(R, G, B) = \begin{cases} (1, H_A, 0) & \text{if } \lfloor H' \rfloor = 0 \\ (H_B, 1, 0) & \text{if } \lfloor H' \rfloor = 1 \\ (0, 1, H_A) & \text{if } \lfloor H' \rfloor = 2 \\ (0, H_B, 1) & \text{if } \lfloor H' \rfloor = 3 \\ (H_A, 0, 1) & \text{if } \lfloor H' \rfloor = 4 \\ (1, 0, H_B) & \text{if } \lfloor H' \rfloor = 5 \\ (0, 0, 0) & \text{otherwise} \end{cases} \qquad (6.3)$$

Now, we have to remove the pixels that do not correspond to the temperatures, we are interested in 6.4.

$$(R, G, B) \cdot (0.30, 0.59, 0.11) < K, \qquad (6.4)$$

where K ia a constant that corresponds to a temperature. In our case we do not have so called "calibrated" thermal camera, but we have made our own calibration using Peltier, and K corresponds approximately to human body temperature, while the parameter is user selectable, even during mission.

The CCD camera and thermovision camera images spatial correspondence scaling is yet to be mentioned and still to be done. This results from the fact that the field-of-view of the cameras is different, they have different positions, they are not aligned to be perfectly parallel, and each of the sensors has a different resolution.

6.3.1 CCD and Thermocamera Alignment

In the Fig. 6.3, the problem of CCD camera and thermovision camera misalignment is defined, with all the important features.

All the important parameters of the two-camera-system are summarised in the Table 6.1. For these calculations and experiments we use AXIS M1114 CCD network camera with Computar 2.3 mm DC lens, and NEC C100C thermal camera OEM module with 8 mm lens inserted in Orpheus-Explorer.

Fig. 6.3 Cameras position, resolution and chip size—parameter definition

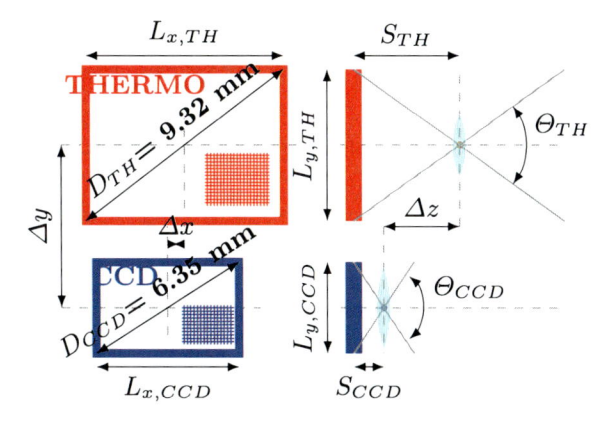

Table 6.1 Parameters of both sensor chips

	Resolution				Chip size		FOV	
	D [mm]	S [mm]	R_x [px]	R_y [px]	L_x [mm]	L_y [mm]	Φ [°]	Θ [°]
Thermo	9.32	8.0	320	240	7.46	5.60	50.0	37.5
CCD	6.35	2.3	1280	800	5.38	3.36	98.9	72.3

In the Table 6.1 the chip diameter D, focal length S, and chip resolution R are known from specification of cameras. The chip sizes L_x and L_y were calculated from camera specifications by simple formula 6.5.

$$L_x = \frac{R_x \cdot D}{\sqrt{R_x^2 + R_y^2}}, \quad L_y = \frac{R_y \cdot D}{\sqrt{R_x^2 + R_y^2}} \quad (6.5)$$

Next columns in the table, the field of view (Θ for the vertical axis, and Φ for the horizontal axis) of the cameras can be calculated using Eq. 6.6. The L_x and L_y parameters in these equations contain appropriate dimensions of image sensor chip, solved previously.

$$\tan\left(\frac{\Phi}{2}\right) = \frac{L_x}{2 \cdot S}, \quad \tan\left(\frac{\Theta}{2}\right) = \frac{L_y}{2 \cdot S} \quad (6.6)$$

Suppose that the camera system with optics is linear. In this case, we can draw Fig. 6.4. The observed point P placed at position (P_x, P_y, P_z) is projected on the sensor chip with distance P'_y from the center axis of projection.

We can now solve the problem of projection by using simple formula, revealed in the Eq. 6.7. If we continue with the transformation, we can write the result of Eq. 6.7 in the pixel coordinate system (as H_y) in Eq. 6.8

Fig. 6.4 To calibration of two cameras

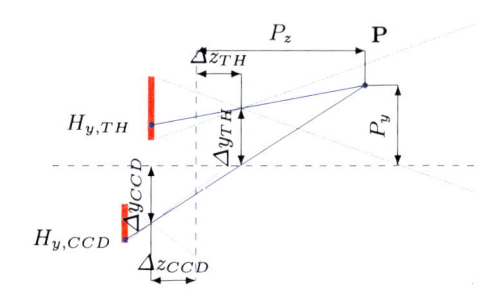

$$P'_x = S \cdot \frac{P_x}{P_z}, \quad P'_y = S \cdot \frac{P_y}{P_z} \tag{6.7}$$

$$H_x = R_x \left(\frac{1}{2} - \frac{P'_x}{L_x} \right), \quad H_y = R_y \left(\frac{1}{2} - \frac{P'_y}{L_y} \right) \tag{6.8}$$

After combining Eqs. 6.7 and 6.8, we can write final projection transformation between original axis system (with center in the focal spot of the camera optics) and camera pixel system (with top-left center zero) in vector Eq. 6.9.

$$\mathbf{H} = 0.5 \cdot \mathbf{R} - \left(\frac{R_x \cdot S \cdot P_x}{L_x \cdot P_z}, \frac{R_y \cdot S \cdot P_y}{L_y \cdot P_z} \right) \tag{6.9}$$

$$\mathbf{P}_{TH} = \mathbf{P} - \Delta_{TH}, \quad \mathbf{P}_{CCD} = \mathbf{P} - \Delta_{CCD} \tag{6.10}$$

In the next step, we have two cameras, and we can parametrise their alignment in fixture as Δ, by distance of focus point to the center of the fixture in all axes parallell with camera's axes. The $\Delta = (\Delta_x, \Delta_y, \Delta_z)$ can be assumed in Eq. 6.9 to apply this shifts. The pixel transforms will be the same as in Eq. 6.11.

The final equation of transformation of point P to pixel index H is shown in Eq. 6.11.

$$\mathbf{H} = 0.5 \cdot \mathbf{R} - \left(\frac{R_x \cdot S \cdot (P_x - \Delta_x)}{L_x \cdot (P_z \Delta_z)}, \frac{R_y \cdot S \cdot (P_y - \Delta_y)}{L_y \cdot (P_z - \Delta_z)} \right) \tag{6.11}$$

Let's assume that we can control the calibration process, and we can trim the offset of focus points of cameras in z axis to zero ($\Delta_z \to 0$). The Eq. 6.11 will be slightly optimized to resulting Eq. 6.12b. The $\mathbf{A}$ vector is constant, and contains only parameters of the sensor. The value of $\mathbf{A}$ is shown in Eq. 6.16.

$$\mathbf{H} = 0.5 \cdot \mathbf{R} - \left(\frac{R_x \cdot S \cdot (P_x - \Delta_x)}{L_x \cdot P_z}, \frac{R_y \cdot S \cdot (P_y - \Delta_y)}{L_y \cdot P_z} \right) \tag{6.12a}$$

$$\mathbf{H} = 0.5 \cdot \mathbf{R} - \frac{1}{P_z}(A_x P_x, A_y P_y) + \frac{1}{P_z}(A_x \Delta_x, A_y \Delta_y) \tag{6.12b}$$

Fig. 6.5 Aligning thermo picture inside CCD picture

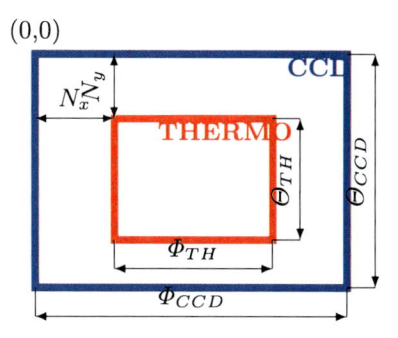

$$\mathbf{A} = \left(\frac{R_x \cdot S}{L_x}, \frac{R_y \cdot S}{L_y} \right) \tag{6.13}$$

Now we will assume that we are like to display both informations in one display. This can be seen on Fig. 6.5. Let the screen resolution of the monitor is $R_{x,SCR}$, $R_{y,SCR}$, and assume that every pixel is rectangular. We can map the camera signal with bigger field of view onto screen, like as stretched. The image of camera view will be placed at point (0, 0) and if the assumption, that $R_{x,SCR}$, $R_{y,SCR}$ is made, it will be stretched with coefficient K_{CCD} shown on Eq. 6.14a.

$$K_{CCD} = \frac{R_{x,SCR}}{R_{x,CCD}} \tag{6.14a}$$

$$K_{TH} = \frac{R_{x,SCR}}{R_{x,CCD}} \cdot \frac{\Phi_{TH}}{\Phi_{CCD}} \tag{6.14b}$$

After that, the thermometer signal will be mixed with CCD data in the same way, but the stretching coefficient will be slightly more complicated, and is solved in Eq. 6.14b.

The offset of thermocamera signal in the screen space can be explained mathematically from 6.15a and 6.15b.

$$N_x = \frac{1}{P_z} \cdot \frac{R_{x,SCR}}{R_{x,CCD}} \cdot \frac{1}{\Phi_{CCD}} \\ \cdot \left(A_{x,CCD} \cdot \Delta_{x,CCD} \cdot \Phi_{CCD} - A_{x,TH} \cdot \Delta_{x,TH} \cdot \Phi_{TH} \right) \tag{6.15a}$$

$$N_y = \frac{1}{P_z} \cdot \frac{R_{x,SCR}}{R_{x,CCD}} \cdot \frac{1}{\Phi_{CCD}} \\ \cdot \left(A_{y,CCD} \cdot \Delta_{y,CCD} \cdot \Phi_{CCD} - A_{y,TH} \cdot \Delta_{y,TH} \cdot \Phi_{TH} \right) \tag{6.15b}$$

For clarity, the Eqs. 6.15a and 6.15b can be simplified to form 6.16, where $\mathbf{Q}$ is constant.

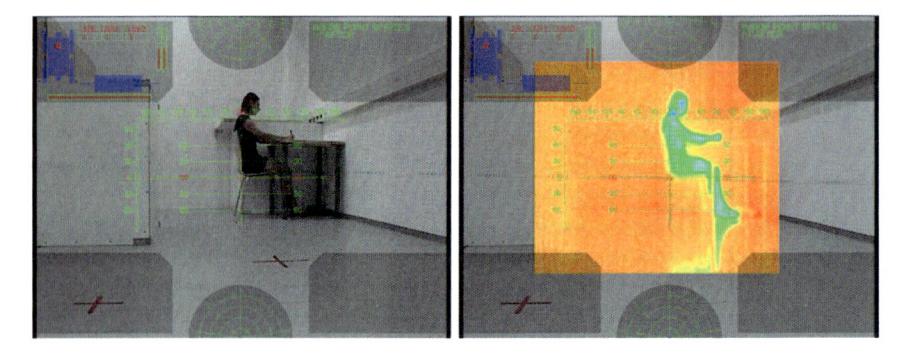

Fig. 6.6 CCD image (*left*), corresponding thermo-image placed over it (*right*)

Fig. 6.7 Visible spectrum and thermovision data fusion

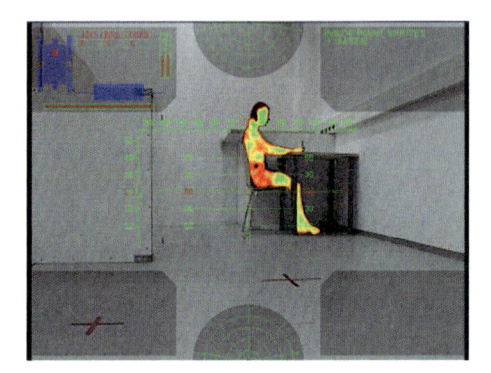

$$\mathbf{N} = \frac{1}{P_z}(Q_x, Q_y) = \frac{1}{P_z}\mathbf{Q} \qquad (6.16)$$

The example of the described technique can be seen on Figs. 6.6 and 6.7. These figures are screenshots from Cassandra. The whole thermovision layer (i.e. RGB-to-HSV conversion, temperature filtration) is done by pixel-shaders in .NET, so the calculations are done on graphical card [15, 16].

To conclude this subsection, we can say there is almost a philosophical question how to display the frequencies of electromagnetic spectra that humans cannot see. The problem is much more complex in our case, where the image made by these imperceptible colors is mixed with the one we know (visible spectra). The technique described heretofore takes advantage of the imperfection of CCD imagers and commonly used displays together with the ability of our brain to adjust to something that does not represent reality perfectly, and to more clearly distinguish something uncommon.

6.4 Conclusions

The CCD camera and thermovision camera data-fusion has been commonly used on Orpheus-Explorer, and was demonstrated e.g. on IDET 2011 world exhibition. Although the user experience is very good, our team already works on more sophisticated method that includes CCD camera stereovision pair, thermal imager pair and TOF camera.

Acknowledgments This work was supported by the project CEITEC—Central European Institute of Technology (CZ.1.05/1.1.00/02.0068) from European Regional Development Fund.

References

1. Jacoff A, Weiss B, Messina E (2003) Evolution of a performance metric for urban search and rescue robots. In: Performance metrics for intelligent systems workshop, Aug 2003. Gaithersburg, MD
2. Zalud L (2004) Rescue robot league—1st place award winner. In: RoboCup 2003: robot soccer world cup VII. Springer, Germany. ISBN 3-540-22443-2
3. Wise E (1999) Applied robotics. Prompt Publications, USA. ISBN 0-7906-1184-8
4. Zalud L (2001) Universal autonomous and telepresence mobile robot navigation. In: 32nd international symposium on robotics. ISR 2001, pp 1010–1015, Seoul, Korea
5. Zalud L (2005) ORPHEUS reconniaissance teleoperated robotic system, In: 16th IFAC world congress, pp 1–6, Prague, Czech Republic
6. Martin CM, Moravec HP (1996) Robot evidence grids. The Robotics Institute Carnegie Melon University, Pittsburgh, 15213
7. Mullet K, Sano D (1995) Designing visual interfaces communication oriented techniques. Sun Microsystems Inc, USA. ISBN 0-13-303389-9
8. Oyama E, Tsunemoto N, Tachi S, Inoue S (1993) Experimental study on remote manipulation using virtual reality. Presence 2(2):112–124
9. Sheridan TB (1992) Telerobotics, automation, and human supervisory control. MIT Press, Cambridge
10. Ayache N (1991) Artificial vision for mobile robots stereo vision and multisensory perception (translation). The MIT Press, Cambridge. ISBN 0-262-01124-7
11. Gonzalez G, Woods RE (2002) Digital image processing, 2nd edn. Prentice Hall, Englewood Cliffs. ISBN 0-201-18075-8
12. Everett HR (1995) Sensors for mobile robots, theory and applications. AK Peters Ltd, USA. ISBN 1-56881-048-2
13. FlirSystems (2007) Retrieved Mar 12 2007. from http://www.flirthermography.com
14. Wyszecki G, Stiles WS (2000) Color science concepts and methods, quantitative data and formulae. Wiley-Interscience, New York. ISBN 0-471-02106-7
15. LaMothe A (2003) Tricks of the 3D game programming gurus advanced 3D graphics and rasterization. SAMS Publishing, USA. ISBN 0-672-31835-0
16. Luna DF (2003) Introduction to 3D game programming with DirectX 9.0. Wordware Publishing Inc, USA. ISBN 1-55622-913-5

Chapter 7
Horizontal Distance Identification Algorithm for Sit to Stand Joint Angle Determination for Various Chair Height Using NAO Robot

Mohd Bazli Bahar, Muhammad Fahmi Miskon,
Norazhar Abu Bakar, Ahmad Zaki Shukor and Fariz Ali

Abstract This paper presents the development of an autonomous Sit To Stand (STS) motion using NAO robot. NAO robot hip limitation will emulate the limitation faced by people having STS problem. To perform the motion, three main steps have been developed (1) *horizontal distance identification*, (2) *joint angle determination*, and (3) *stability control*. This step was developed based on Alexander STS technique. Results show that NAO robot is able to achieve halfway stand up from chair height between 9.6 and 12.7 cm automatically. The robot's best performance is at 12.7 cm height with swinging time of 0.52 s in experiment and 0.2914 in simulation. The developed system will contribute to the development of exoskeleton, rehabilitation and evolution of humanoid robot. This system also enhances NAO's ability for medical study on STS motion or as tools in searching for the best chair design.

Keywords Aautonomous · STS · NAO · Alexander technique · Hip limitation

---

M. B. Bahar (✉) · M. F. Miskon · N. A. Bakar · A. Z. Shukor · F. Ali
Faculty of Electrical Engineering, Universiti Teknikal Malaysia Melaka, Melaka, Malaysia
e-mail: bazlibahar96@yahoo.com

M. F. Miskon
e-mail: fahmimiskon@utem.edu.my

N. A. Bakar
e-mail: norazhar@utem.edu.my

A. Z. Shukor
e-mail: zaki@utem.edu.my

F. Ali
e-mail: fariz@utem.edu.my

7.1 Introduction

The study of sit to stand motion (STS) gives high impact to the robotics field particularly in rehabilitation [1], exoskeleton [2] as well as humanoid robots. In humanoid robotics, the STS study has not been given much emphasis until recently [3]. Only four groups studying STS using humanoid were identified; Mistry et al. [3], Qi et al. [4], and Pchelkin et al. [5]. Sugisaka [6], in his study on STS control system used different humanoid robot equipped with artificial muscles.

However, no study has been conducted with regards to STS while considering the hip bend limitation and a system that can identify and estimate the joint position when facing a multi chair height. For this reason, this project is proposed to design and develop an algorithm and methods using NAO humanoid robot that mimics 'biological system' in limbs actuation and body posture. Limitation of the robot will represent the limitation to bend waist faced by people with STS problems.

The algorithm will contribute in development of exoskeleton, rehabilitation and evolution of humanoid robot. This system also enhances NAO's ability for medical study on STS motion or in searching for the best chair design. The applications of the study includes experimental tool for medical STS studies on rehabilitation, exoskeleton development, and diagnostic study that will help 26.5 % of the total Malaysian population [7] that suffers from obesity. The study will also benefit in the advancement of humanoid robot motion to increase the capability for long term energy usage and human interaction application for example in security field and domestic robotics.

7.2 System Overview

The motion is divided into three phase which are (1) initial position identification, (2) centre of mass (CoM) transfer technique, and (3) standing motion. In initial position identification, the normal position of each joint has to be set. The normal position of each joint is 90° to each other as shown in Fig. 7.1. The robot hip limitation was taken into account at CoM transfer technique phase. This is because the CoM must be inside the stability region before the Stabilization Strategy phase takes place.

This paper presents a new method by theoretically identifying the initial position of x_i, i.e. the distance of head-arm-torso (HAT) CoM to a predefined stability region. This paper also presents a method to shift HAT CoM autonomously to predefined stability region. The stability region was defined by Jining and colleagues as the area between the tiptoe and heel [8]. However, in this paper the region was set to 0.03 m back from the ankle joint (x_{min}) to the tiptoe. x_{min} becomes the reference point for x_{new} where x_{new} was the length between CoM and x_{min}. HAT

7 Horizontal Distance Identification Algorithm

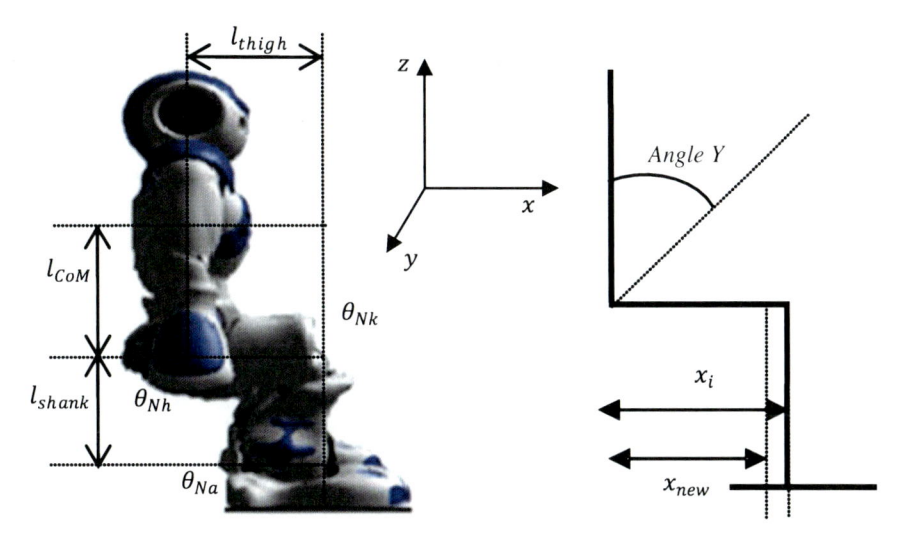

Fig. 7.1 The normal position of NAO robot at sitting position

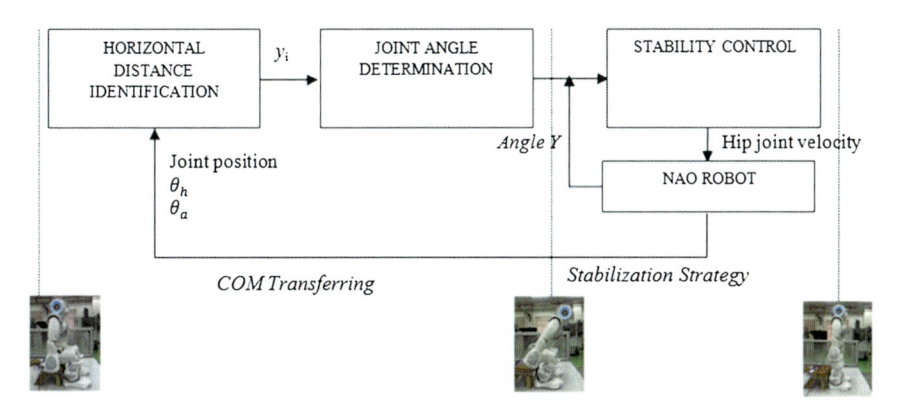

Fig. 7.2 Overall system overview for autonomous sit to stand motion

CoM was set at the centre of the robot body as mentioned by Prinz and his colleagues [9].

Step 1, Horizontal distance identification and Step 2, Joint angle determination as shown in Fig. 7.2, are the focus of this paper. Feedback from the robot at sitting position is the first input for the system. The objective of this study is to develop an autonomous initial position monitoring algorithm and CoM transfer algorithm based on Alexander technique (AT) for the robot to perform STS human-like motion. This technique was used by Wang et al. [10].

7.3 Initial Position Identification

From the transformation matrix of ankle joint to the CoM, T_{ankle}^{CoM} the translation at x axis is used to identify the distance of CoM and ST. The motor angle was referred to the robot setting, as shown in Fig. 7.1 where at this position the hip angle is $-75.57°$, knee angle is $90°$, and ankle is $-7°$. At normal position, the distance between the CoM and ankle joint, x is 10 cm. From this normal position, it becomes as reference to identify the initial position of the robot. Using translation of x-axis, x_i was defined as:

$$x_i = \pm[\alpha_h] + [\alpha_k] \pm [\alpha_a] \tag{7.1}$$

where

$$\begin{aligned}
\alpha_h &= \sin(|diff(\theta_{Rh}, \theta_{Nh})|) \times l_{COM} \\
\alpha_k &= \cos(|diff(\theta_{Rk}, \theta_{Nk})|) \times l_{thigh} \\
\alpha_a &= \sin(|diff(\theta_{Ra}, \theta_{Na})|)_{shank} \\
\theta_{hnew} &= \theta_{Nh} + [\theta_{Nk} - \theta_{Rk}]
\end{aligned} \tag{7.2}$$

In (7.1), $\theta_{Rh,k,a}$ is a reading of joint angle at the initial position and $\theta_{Nh,k,a}$ is the normal joint value that has been defined before. The different between these two angles is represented by *diff*. l_{COM}, l_{thigh}, and l_{shank} represent length of robot CoM from hip joint, thigh length and shank length. For hip and ankle joint, if the joint reading is smaller than the normal angle of hip and ankle, negative sign will be added to the value. In normal, human sit with body straight upward before starting to stand. To do so, the method was modified to make sure the distance, x_i was calculated at moment where the NAO robot sit with the body perpendicular to the ground. From the normal position, each knee joint angle that is added or reduced will give change at hip joint according to (7.2).

7.4 Joint Angle Determination

Value x_i was used to identify angle change at each joint. The joint angle was identified to make sure that the HAT CoM is in the predefine stability region (SR). The method stressed on hip and ankle joint to shift the HAT CoM to the SR as mentioned in AT. The technique first brings the body to the front. The maximum rotation range for the hip joint to go is $-89°$. Using (7.3), the needed joint angle change was calculated. Result from (7.3) was observed to make sure the robot does not exceed the hip joint limitation. If the value is smaller than $-89°$, the new hip joint value was set at $-89°$. However, if θ_{hneed} was larger than that $\theta_{hnew1} = \theta_{hneed}$. Ankle joint was another joint that will react if the HAT CoM still do not reached the

7 Horizontal Distance Identification Algorithm

SR edge. The limitation at hip joint leads to the need of ankle joint change. At this point, remaining distance between HAT CoM and SR edge was calculated as in (7.4). The remaining distance, x_{remain} would determine whether the ankle joint change was needed or not. If $x_{remain} = 0$, the system proceed to another step. However, if x_{remain} has a positive value (negative was never happen) a new ankle joint was calculated using (7.5). After both hip and ankle joint has its value, the system will move the robot to the desired position starting with hip followed by ankle joint.

$$\theta_{h_{need}} = \theta_{Rh} - \left[90 - abs(\cos^{-1}(x_{new}/15))\right] \quad (7.3)$$

$$x_{remain} = x_{new} - abs[(15 \times \cos(90 - (\theta_{Rh} - \theta_{h_{new1}})))] \quad (7.4)$$

$$\theta_{a_{new1}} = \theta_{Ra} + \left[-\left(\sin^{-1}(x_{remain}/10.3)\right) - abs(\theta_{Na} - \theta_{Ra})\right] \quad (7.5)$$

7.5 Result and Discussion

The algorithm was tested in the simulation environment using (NAOSim) and NAO robot version 3.2. Knee joint was set at 11 different angles to validate the developed algorithm. The experiment was repeated 5 times for each knee joint angle. Change at knee joint was made to represent the chair height more precise. The increasing of knee angles in degree will represent the decreasing of the chair height. Reading from Force Sensitive Resistor (FSR) at each foot was recorded to observe the centre of pressure (CoP) position. In simulation, normal position is at 0.807 and −0.03 m for the robot where it is at rest and stable. If the robot falls down, the reading should be zero. Another feedback used to monitor the system is the angle y. To test whether the technique is able to transfer CoM into the SR, experiment was done only until the robot at half way to stand or at the end of the CoM transferring phase (target) as in Fig. 7.2. This was done to avoid the influence of the movement in the stabilization strategy phase that might cause fall.

The method is considered working when the CoP is in the range of −0.03 to 0.01 cm and angle y reading ended as straight line with ±5 % from the planned angle y trajectory. Table 7.1 shows the experimental results. With a new ankle joint, $\theta_{a_{new1}}$ and hip joint, $\theta_{h_{new1}}$ calculated by the system, the robot was able to achieve the target position without falling within chair height at 12.7–10.65 cm for simulation and 12.7 cm until 11 cm during experiment. Ankle joint change time, t_a was increased after that to ensure NAO achieved the target position when chair height was decreased. If t_a was too fast this change nearly has straight line trajectory. However when time, t_a is increased, the decreasing of the body momentum in the motion happen because of slow velocity produce at the end of the motion. The falls are categorized into two which are sitback and fall front. The fall front means that motion momentum was too high. In experiment, the robot

Table 7.1 The first result which at the top with height of chair of 12.7 cm and the bottom result with 9.25 cm for experiment with data of angle y reading and CoP reading

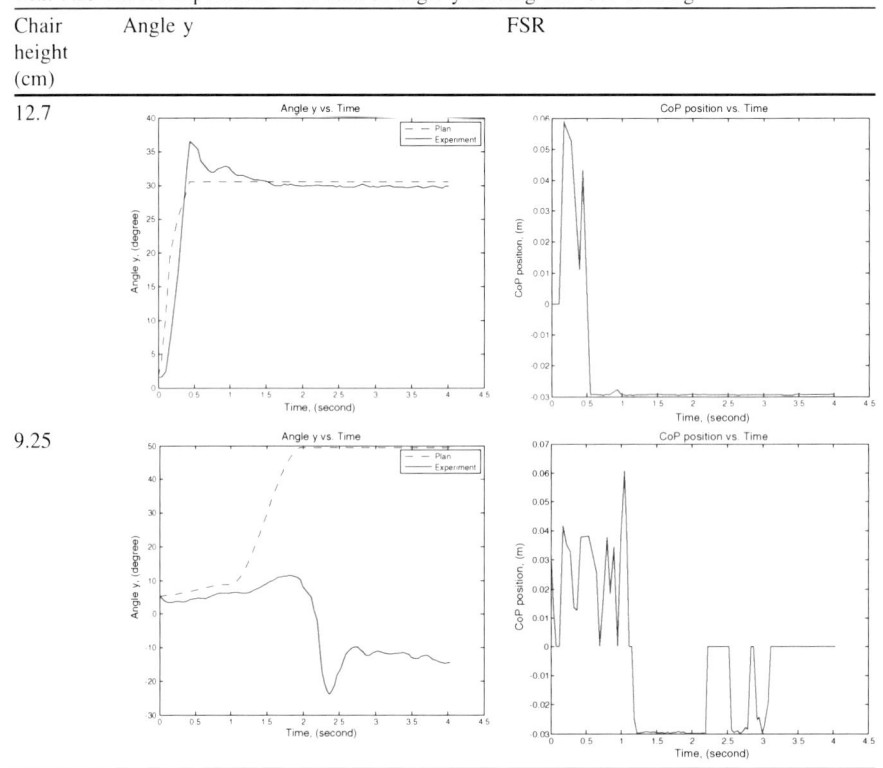

Chair height (cm)	Angle y	FSR
12.7		
9.25		

experienced sitback once at chair height of 9.25 cm. This was due to the weight that was not concentrated to the front before the ankle joint starts to move. In Table 7.1, when chair height at 9.25 cm the CoP reading was concentrated to the back of the feet after 1 s operation and then fall down based from the direction of falling shown by angle y reading.

7.6 Conclusion

From the results, the developed system was able to automatically calculate the horizontal distance between the CoM and ankle joint before determining the new joint angle for hip and ankle joint with respect to the Alexander technique approach in STS motion. It was proved by the CoP reading at -0.03 m at final position and angle y reading nearly the same with the plan trajectory. Increased of ankle joint change time, t_a was done only to lower the momentum created by the motion. The method should able to perform the same result with system that has the same configuration where all joint of the lower body is align to each other at standing position. The difference in certain variable such as link length, hip joint limitation, and mass should not affect the performance of the system. The author recommends a new automatic method to verify the suitable trajectories before t_a is set in the joint angle determination to reduce momentum creates by the motion. Another problem faced in this paper is the body limitation of the robot which shows that the robot can only sit on chairs with 9.6 cm height and above.

References

1. Chuy O et al (2006) Approach in assisting a sit-to-stand movement using robotic walking support system. In: IEEE/RSJ international conference on intelligent robots and systems, pp 4343–4348
2. Strausser KA, Kazerooni H (2011) The development and testing of a human machine interface for a mobile medical exoskeleton. In: IEEE/RSJ international conference on intelligent robots and systems (IROS), pp 4911–4916
3. Mistry M et al (2010) Sit-to-stand task on a humanoid robot from human demonstration. In: 10th IEEE-RAS international conference on humanoid robots (Humanoids), pp 218–223
4. Qi K et al (2009) Analysis of the state transition for a humanoid robot SJTU-HR1 from sitting to standing. In: International conference on mechatronics and automation. ICMA, pp 1922–1927
5. Pchelkin S et al (2010) Natural sit-down and chair-rise motions for a humanoid robot. In: 49th IEEE conference on decision and control (CDC), pp 1136–1141
6. Sugisaka M (2007) A control method for soft robots based on artificial musles. In: ICM 2007 4th IEEE international conference on mechatronics, pp 1–3
7. Ismail MN, Nawawi H, Yusoff K, Lim TO, James WP (2002) Obesity in Malaysia. US National Library of Medicine, National Institutes of Health
8. Liu J, Kamiya Y, Seki H, Hikizu M (2010) Weightlifting motion generation for a stance robot with repeatedly direct kinematics. Intell Control Autom, 20–27
9. Prinz R et al (2007) Development of a fuzzy-based sit-to-stand controller. In: Canadian conference on electrical and computer engineering, CCECE, pp 1631–1634
10. Wang F-C et al (2007) Optimization of the sit-to-stand motion. In: IEEE/ICME international conference on complex medical engineering, CME, pp 1248–1253

Chapter 8
Enhancing Wheelchair's Control Operation of a Severe Impairment User

Mohd Razali Md Tomari, Yoshinori Kobayashi and Yoshinori Kuno

Abstract Users with severe motor ability are unable to control their wheelchair using standard joystick and hence an alternative control input is preferred. However, using such an input, undoubtedly the navigation burden for the user is significantly increased. In this paper a method on how to reduce such a burden with the help of smart navigation platform is proposed. Initially, user information is inferred using an IMU sensor and a bite-like switch. Then information from the environment is obtained using combination of laser and Kinect sensors. Eventually, both information from the environment and the user is analyzed to decide the final control operation that according to the user intention, safe and comfortable to the people in the surrounding. Experimental results demonstrate the feasibility of the proposed approach.

Keywords Wheelchair · Severe impairment user · Control operation

---

M. R. M. Tomari (✉)
Advanced Mechatronics Research Group (ADMIRE), Department of Mechatronics and Robotics Engineering, Faculty of Electrical and Electronic Engineering, Universiti Tun Hussein Onn Malaysia, 86400 Parit Raja, Batu Pahat, Johor, Malaysia
e-mail: mdrazali@uthm.edu.my

Y. Kobayashi · Y. Kuno
Graduate School of Science and Engineering, Saitama University, 255 Shimo-Okubo, Sakura-Ku, Saitama 338-8570, Japan
e-mail: yosinori@cv.ics.saitama-u.ac.jp

Y. Kuno
e-mail: kuno@cv.ics.saitama-u.ac.jp

Y. Kobayashi
Japan Science Technology Agency, PRESTO, 4-1-8 Honcho Kawaguchi, Saitama 332-0012, Japan

8.1 Introduction

In recent years, numerous methods have been introduced for developing smart platform of wheelchairs system to accommodate the needs of people with severe disabilities. These users' limitations to control the wheelchair may be due to several reasons such as cerebral palsy, cognitive impairment, or being fatigue prone [1]. There has been a great deal of research devoted in this area and some recent results can be found in [2, 3], which highlights the setups of various individual systems and their strategies used for assisting the user. The development trend can be broadly classed into three main areas [4]: (1) Improvements to assistive technology mechanics, (2) Improvements to user-machine physical interface, (3) Improvements to shared control between the user and the machine. Users with severe motor impairment (e.g. spinal cord injury) generally lack muscle control, and in the worst case they are unable to command the movement of arms and legs. For such a patient, input devices based on cues or actions generated from the head (e.g. facial, brain, gaze, tongue, and bite) can be possible media at all levels of injuries [5]. The medium should provide the users with ability to control the direction of wheelchair, and to initiate/terminate such tasks.

Even though the alternative medium can enable the user to steer the wheelchair; yet some clinical studies found that their patient still find it hard or impossible to operate it [6], especially for avoiding obstacle [7]. These clinical findings provide insights for the importance of devising a computer-controlled platform to assist the users during navigating. Under this framework, the user input along with the environmental information will be seamlessly analyzed for providing necessary assistive tasks. The amount of given assistance usually varies depending on how severe the users' impairments and the assistance can be categorized into three main levels: shared-control, semi-autonomous control, and autonomous control [8]. The level of given assistance should be decided by considering the maximum of users' abilities to control the wheelchair and the computer only complement the loophole [9].

In this paper, the main objective is to improve the wheelchair controllability for easing the severe impairment user to operate the wheelchair. When the user relying on alternative input medium, they have limited control command and also need more effort to issue such a command. For that reason, the computer will take over part of the responsibilities for navigating especially to avoid any possible threat and also guide the wheelchair into the desired direction of travel.

8.2 System Overview

The proposed system has been implemented on a standard electrical wheelchair (TT-Joy, Matsunaga Corporation) mounted with a forward-looking RGBD camera (Kinect, Microsoft), laser range finder and two IMU sensors as shown in Fig. 8.1.

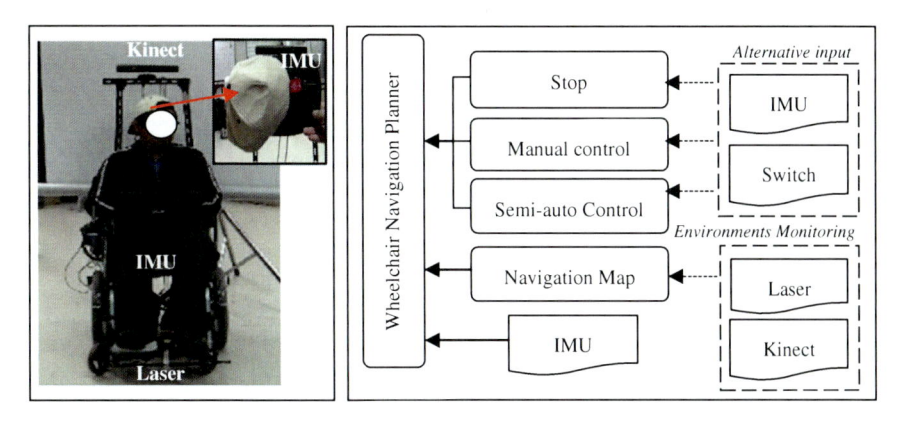

Fig. 8.1 Sensors arrangement on the wheelchair (*left*) and how each sensors contribute to enhance the wheelchair control system (*right*)

The alternative input consists of a switch and a IMU sensor for receiving the user's commands. The former is responsible for initiating/terminating several control modes (i.e., stop, manual and semi-auto) while the latter provides the direction to move in.

For the environments monitoring module, the combination of a laser range sensor and a Kinect is used to sense the environment for potential obstacles and deliver a safety map to the computer. The laser sensor is located in the front part of the wheelchair at a height of 20 cm above the ground, covering an angle of 270°. The Kinect on the other hand is installed 1.3 m above the ground and can perceive 3D information within 57° × 43° of the angle view.

8.3 Alternative Input

The alternative interface is designed to control the wheelchair in two different control modes: manual mode and semi-auto mode. Because of that reason, a switch is needed in order to distinguish such operations. The switch must be simple and easily reachable by the user. It can be realized by various mediums, such as motion of facial parts (e.g., eye blinking or shaking), voices or a button switch. Physiological features will impose many burdens on the user when she/he needs to issue a command frequently, especially when she/he navigates in a limited space. Therefore, a simple bite-like switch button (i.e., not an actual bite switch but imitating the nature of the switch operation) is employed. The switch will be used for selecting the operating mode while the gaze orientation will be used to instruct the direction of movement. With such a setup, not only can easing the user to select the control operation, but also the gaze actions which are intended to steer the wheelchair or just look around can be easily distinguished by the central controller.

When the switch receives a cycle of momentary pattern (low → high → low), the "stop" or "semi-auto" command is executed depending on the current state (i.e., when the current state is stop, the system turns into semi-auto and when the current state is semi-auto, the system turns into stop). When it continuously receives a high signal, the system enters the "manual" mode until a low signal is issued. Every time the user exits from the manual mode, the system stops and waits for the next command.

For obtaining the gaze data, an IMU sensor that mounted at the back of user's head is used. Under the manual mode, the direction that the user looks at will be proportional to the direction that the wheelchair moves. Means that when the user looks to the left/right, the wheelchair will turn left/right and when the user looks to the front, the wheelchair move straight. When in the "semi-auto" mode, the gaze direction is only considered during initializing the goal direction and avoiding collision. Except in both situations, the gaze direction does not cause any effect on the wheelchair's motion. This can give freedom to the user to enjoy the surroundings while navigating. For initializing the goal, the user just needs to look at the area where she/he would like to go and presses the switch. The direction where the user looks is then regarded as the goal direction. During obstacle avoidance, the gaze direction is also used to determine the paths to be avoided. The area that the user looks at during obstacle avoidance is automatically assigned as the direction to avoid if it is collision free.

8.4 Environment Monitoring

Smart wheelchair either driven by the user or the computer should capable to detect and avoid the threat during navigating. In indoor environment various potential hazardous threats exist including obstacle, drop-off and hanging objects. To perceive such information a combination of laser and Kinect camera that can complement each sensor's limitation is proposed. The laser can detect most of the obstacles in a wide view; while the Kinect will refine the laser reading along with detect areas that cannot be sensed directly by the laser. Apart from that, data from the Kinect also will be used to detect and tracking humans' head orientation for assigning appropriate personal space model [10]. Data from both sensors will output a navigation map which can be defined as the best estimate of the 3D environment in a 2D representation. Each cell in the navigation map encodes the possible of the thread location and useful to guide the smart wheelchair in various navigating mode (i.e. manual and semi-auto). The effectiveness of the system has been tested with real indoor environments and shown the feasibility of the navigation map to portray the traversabilty region [11]. Figure 8.2 shows sample of the obtained navigation map in indoor navigation scenario.

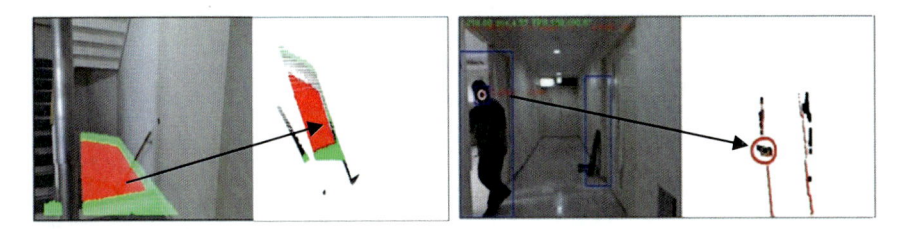

Fig. 8.2 Two samples of navigation map. (*left*) *Red region* denotes negative obstacle region while green region represent ground plane. (*right*) Human detection and tracking with the personal space assigned (*red circle*)

8.5 Wheelchair Navigation Planner

To model the wheelchair navigation planner, an extension of Vector Field Histogram (VFH) method namely VFH+ [12] is adopted. VFH uses the two-dimensional Cartesian histogram known as the active window that is built around the surrounding of the robot with square shape. Each active cell in the histogram grid is treated as an obstacle vector namely primary Polar Obstacle Density (*POD*). *POD* is represented by the direction and magnitude from the cell to the sensor centre point. A sample of the generated *POD* is given in Fig. 8.3, where the RGB image, the navigation map and the *POD* distribution image are shown from left to right. In the *POD* image, the yellow and green bars represent the low and high *POD* regions, respectively, while the red bars show high risk obstacle locations.

When under the manual mode the POD will assist the user by terminating any user command that may hurt him/her. For example, if the user commands the wheelchair to a direction that subject to collision, the computer overrides its command and causes the wheelchair to brake. It will remain stopped and only resumes if the user issues a safe command. On the other hand, when under semi-auto mode, the *POD* will assist the user by reactively avoiding any obstacle in the scene. When it reaches the safe zone, the wheelchair then fixates its orientation towards the goal direction. Note that, the goal direction can always be set and reset by the user at any time by using the switch as explained previously. More detail explanation about the planner operation can be found in [11].

8.6 Result and Discussion

For evaluating the feasibility of the wheelchair control system, an operation experiments is performed. The experiments were conducted in a typical laboratory and the performance was evaluated based on the ability of the wheelchair to navigate successfully in the confine space environment. The navigation task is considered success if the wheelchair able to avoid obstacle and moves according to the user direction.

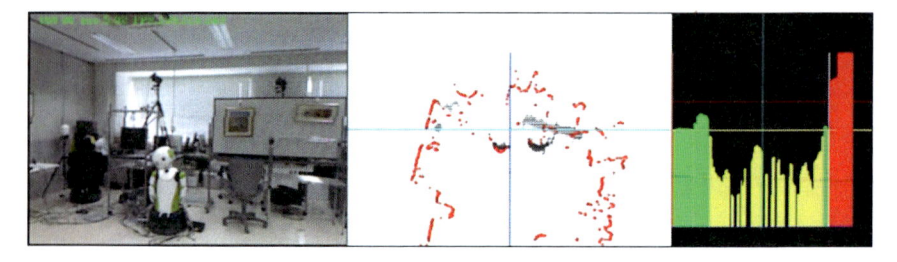

Fig. 8.3 System monitoring view: (*left*) RGB Image. (*middle*) navigation map. (*right*) POD image

The first experiment was performed to demonstrate the wheelchair's ability in manual control mode to move according to the user command and to terminate any user command that may bring harm to him/her. Figure 8.4 shows the results. It shows the wheelchair movement from the initial positions (Fig. 8.4a) until it reaches the obstacle as shown in Fig. 8.4b. At that condition, the wheelchair stop moving since the user still commanding the wheelchair to move straight, which is unsafe to the user. When the user issued a control command to the left (Fig. 8.4c), since it is safe the wheelchair executes the command and move accordingly. The path generated in the experiments is schematically illustrated in Fig. 8.4d, where the dash lines denote the motor command, thin line represent the head orientation and thick line indicate the wheelchair orientation. In the figure, the driving signal that is sent to the wheelchair normally and closely follows the user's head direction (e.g., when the user looks to the left, the wheelchair also turns left). However as seen in frames #40 to #100, the system does not execute the user's command (i.e., motor command = 0) since it is not safe and the wheelchair remains in stationary form. Later on, the user commands to the left (frames #100 to #110), since the steering command shows no possibility to collision, the system properly responds and executes it. Such a verification process is continuously carried out as long as the user is under the manual mode.

The next experiment was performed to confirm the system's ability when in semi-auto mode for moving to the direction according to the user intention automatically and for avoiding any obstacles during navigation. Experimental results are shown in Fig. 8.5. The figure shows that the wheelchair can navigate to the goal direction assigned by the user successfully (Fig. 8.5a, b). When facing with obstacle (Fig. 8.5c), the wheelchair reactively avoid it until reaching the safe zone, and then re-orientate its heading towards the goal destination (Fig. 8.5d). Figure 8.5e illustrates the wheelchair motion. In this figure, when the wheelchair faces with an obstacle seen in frame #40, the planner reactively steers to the left to maintain safety. Once reached an appropriate space, in frames #50 to #70, the wheelchair corrects the heading direction to face the goal location. This result highlights the benefit of the semi-auto controller for providing help to the user by automatically passing through the obstacles safely while seeking the goal.

8 Enhancing Wheelchair's Control Operation

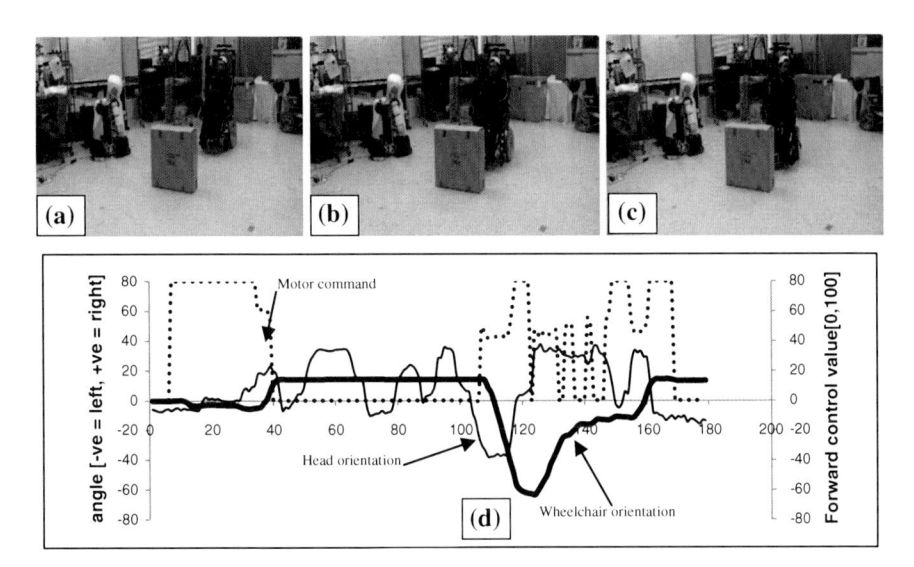

Fig. 8.4 Snapshot of wheelchair movement under manual control mode **a** Initial Position. **b** Encounter with obstacle. **c** Avoiding obstacle. **d** Recorded wheelchair's motion

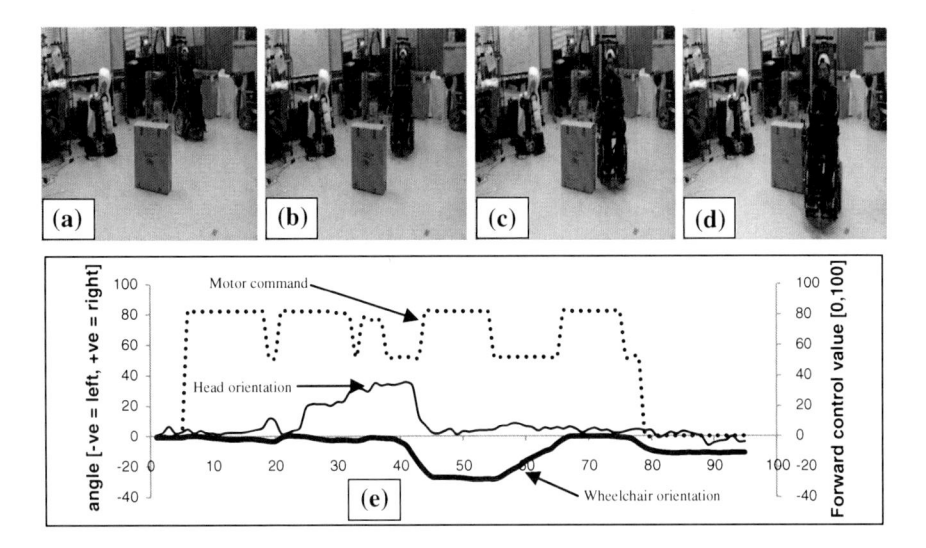

Fig. 8.5 Snapshot of wheelchair movement under semi-auto control mode **a** Initial Position. **b** Encounter with obstacle. **c** Avoiding obstacle. **d** Re-orientate wheelchair's heading. **e** Recorded wheelchair's motion

8.7 Conclusion

In this paper, a framework for enhancing the wheelchair controllability to cater the need of users with severe motor disabilities is proposed. With the use of multiple alternative inputs, the user can easily steer the wheelchair in manual and semi-auto control mode. By incorporating the safety map, apparently collision can be avoided in both modes, i.e., manual and semi-auto, and hence may reduce user burden of continuously monitoring the surrounding while maneuvering.

References

1. Simpson RC, LoPresti EF, Cooper RA (2008) How many people would benefit from a smart wheelchair? J Rehabil Res Dev 45(1):53–72
2. Simpson RC (2005) Smart wheelchairs: a literature review. J Rehabil Res Dev 42(4):423–436
3. Grasse R, Morere Y, Pruski A (2010) Assisted navigation for person with reduced mobility: path recognition through particle filtering (condensation algorithm). J Intell Robot Syst 60:19–57
4. Cowan RE, Fregly BJ, Boninger ML, Chan L, Rodgers MM, Reikensmeyer DJ (2012) Recent trends in assistive technology for mobility. J NeuroEng Rehabil 9:20
5. Bates RA (2002) Computer input device selection methodology for users with high-level spinal cord injuries. In: Proceeding of the 1st Cambridge workshop on universal access and assistive technology
6. Fehr L, Langbein W, Skaar S (2000) Adequacy of power wheelchair control interfaces for persons with severe disabilities: a clinical survey. J Rehabil Res Dev 37(3):353–360
7. Wan LM, Tam E (2010) Power wheelchair assessment and training for people with motor impairment. In: Proceedings of 12th international conference on mobility and transport for elderly and disabled person
8. Perrin X (2009) Semi-autonomous navigation of an assistive robot using low throughput interfaces. PhD Thesis, ETH Zurich
9. Nisbet PD (2002) Who's intelligent? wheelchair, driver or both? In: Proceedings of IEEE international conference on control and application, pp 760–765
10. Tomari R, Kobayashi Y, Kuno Y (2012) Empirical framework for autonomous wheelchair systems in human-shared environments. In: Proceedings of IEEE international conference on mechatronic and automation (ICMA), pp 493–498
11. Tomari R, Kobayashi Y, Kuno Y (2013) Enhancing wheelchair maneuverability for severe impairment user. Int J Adv Robot Syst 10:1–13
12. Ulrich I, Borenstein J (1998) VFH+: reliable obstacle avoidance for fast mobile robots. In: Proceedings of IEEE international conference on robotics and automation, pp 1572–1577

Chapter 9
Noise Cancelation From ECG Signals Using Householder-RLS Adaptive Filter

Shazia Javed and Noor Atinah Ahmad

Abstract In this paper an adaptive recursive least squares filter is used for removal of artifacts from clinical ECG signals. The householder RLS (HRLS) algorithm is an efficient algorithm which recursively updates an arbitrary square-root of the input data correlation matrix and naturally provides the LS weight vector. A data dependent householder matrix is applied for such an update. In this paper an adaptive noise canceler (ANC) is designed for ECG denoising using HRLS algorithm. The promising characteristic of proposed ANC is its flexibility in choosing the reference signals, because it has a trade off between the correlation properties of the noise and the reference signals. Simulation results show the efficiency of RLS based algorithms in ECG denoising.

Keywords Electrocardiography · Adaptive filter · Noise cancelation

9.1 Introduction

Adaptive filters are self-designing systems that rely on a recursive algorithm to become able to perform adequately in an environment where knowledge of the relevant data signals is not available. They have ability to detect time varying potentials and to track the dynamic variations of biomedical data signals. Many adaptive filtering techniques have been presented to denoise ECG signals and to track their dynamic variations. These techniques are of two types: iterative and

S. Javed (✉) · N. A. Ahmad
School of Mathematical Sciences, Universiti Sains Malaysia, 11800 George Town, Penang, Malaysia
e-mail: shaziafateh@hotmail.com

N. A. Ahmad
e-mail: atinah@cs.usm.my

direct. Iterative methods include LMS based algorithm, such as the one in [1] having deterministic functions as reference signals, and sign based LMS algorithm in [2] for removal of noise from ECG signals. But these LMS based algorithms are highly sensitive to the correlation properties of the reference signals, and their performance becomes poor under highly colored noise [3, 4]. Direct methods are not very common in ECG denoising, these include rapidly convergent RLS based algorithms and their performance is less dependent on the correlation properties of input signals as compared with iterative methods. Conventional RLS algorithm is famous for its convergence speed, but the problem is its computational complexity and numerical instability. A better implementation is Householder reflections based HRLS algorithm [5]. The performance of the algorithm depends on the orthogonalization capabilities of the householder transformation used to process the input for the next update. Fast convergence rate of HRLS algorithm make it numerically robust [6], and its performance estimation in [7] has motivated us towards its application in ECG denoising.

In this paper we examine the performance of the HRLS algorithm in removal of baseline wander noise from clinical ECG signals. Baseline wander (BW) is caused by variable contact between the electrodes and skin, and interrupts clear recording of respiration. For this purpose an adaptive noise canceler (ANC) is designed having a novel feature for trade off between the correlation properties of noise and the reference signals. Simulations include comparison of HRLS and conventional RLS algorithms, which results in performance of HRLS algorithm in removing WB noise from real ECG signals.

9.2 HRLS Algorithm Based Adaptive Noise Canceler

An adaptive noise canceler (ANC) is designed in this section for removal of noise from simulated ECG signals. The block diagram of proposed ANC is shown in Fig. 9.1. Here signal $x(n)$ consists of the desired clean signal $ecg(n)$ and contaminated BW noise $g(n)$, such that

$$x(n) = ecg(n) + g(n) \qquad (9.1)$$

The noise source produces a noise signal which is recorded simultaneously with noise $g(n)$ and the reference signal $u(n)$, obtained by passing noise $g(n)$ through an unknown filter with frequency response: $H(z) = \frac{\sqrt{1-\alpha^2}}{1-\alpha z^{-1}}$, where $|\alpha| < 1$, α is a correlation parameter and controls the spectral properties of reference signals. $\alpha = 0$ corresponds to the case when noise and reference signals are same. The eigenvalue spread of reference signals increases with an increase in the value of α, which intern increases the difference between the correlation of the noise and reference signals. This novel property of proposed ANC presents a trade-off between the correlation properties of noise signal and the reference signal, which helps in the performance study of different adaptive filtering algorithms in application of ECG

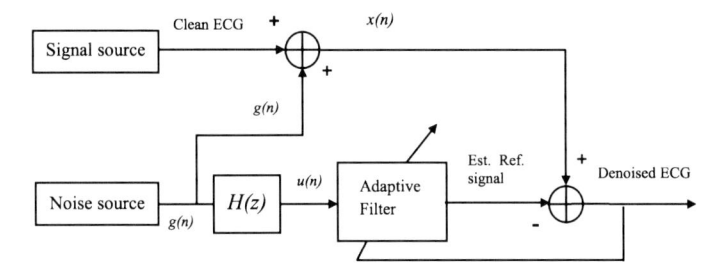

Fig. 9.1 Proposed adaptive noise cancelation (ANC) model

denoising. The reference signal, so obtained, is used to calculate the filter coefficients via an adaptive filter of length N, such that

$$\hat{g}(n) = \sum_{i=0}^{N-1} w_n(i) u(n-i) = \mathbf{a}_n^T \mathbf{w}_n \quad (9.2)$$

where $\mathbf{w}_n = [w_n(0), w_n(2), \ldots, w_n(N-1)]^T$ is the coefficient vector of the filter, $\mathbf{a}_n = [u(n), u(n-1), \ldots, u(n-N+1)]^T$ referenced input vector and $\hat{g}(n)$ the estimated reference signal at time n. The filtered ECG is obtained by subtracting $g(n)$ from $\hat{g}(n)$. The denoised biomedical signal $\widehat{ecg}(n)$ is then, obtained by subtracting (9.2) from (9.1), i.e.,

$$\widehat{ecg}(n) = x(n) - \hat{g}(n) \quad (9.3)$$

It is worthwhile to say that the adaptive filter extracts the ECG signal by minimizing the cost function, defined by:

$$C(n) = E[\widehat{ecg}^2(n)] = E[ecg^2(n)] + E[\eta^2(n)] \quad (9.4)$$

for $\eta(n) = g(n) - \hat{g}(n)$. The above relation makes sense because signal and noise are uncorrelated. The minimization of $C(n)$ is in fact concerned with the minimization of second term on right hand side of Eq. (9.4), which is the mean square error (MSE) of the noise signal.

The adaptive algorithm used for the removal of noise from ECG signals is the Householder's reflections based HRLS algorithm in [5]. HRLS algorithm is famous for its robustness and better control on the eigenvalue spread of input correlation matrix. This algorithm is briefly described below.

9.2.1 The HRLS Algorithm

Consider an adaptive least squares problem of the form:

$$\min_{\mathbf{w}_n \in \mathbf{R}^N} J_n(\mathbf{w}_n) = \sum_{i=1}^{n} \lambda^{n-i} (\mathbf{w}_{i-1}^T \mathbf{a}_i - g(i))^2 \qquad (9.5)$$

where $g(i) \in \mathbf{R}$ is the output signal, and $\hat{g}(i) = \mathbf{w}_{i-1}^T \mathbf{a}_i$ is the prediction of $g(i)$ for $1 \le i \le n$, while n is the current time value.

For a transversal finite impulse response (FIR) adaptive filter, vectors $\mathbf{a}_i \in \mathbf{R}^N$ are formed by the reference signal $u(i)$, such that

$$\mathbf{a}_i = [u(i) \ u(i-1) \ \ldots \ u(i-N+1)]^T$$

and vector $\mathbf{w}_n \in \mathbf{R}^N$ is an estimate of the filter tap vector, which is updated by minimizing the sum of squared error cost function $J_n(\mathbf{w}_n)$. The constant $\lambda \in [0, 1]$ is known as the forgetting factor.

Define the $n \times N$ data matrix $\mathbf{A}_n$ by

$$\begin{pmatrix} u(1) & 0 & \ldots & 0 \\ u(2) & u(1) & \ldots & 0 \\ \vdots & \vdots & \ddots & \vdots \\ u(n-1) & u(n-2) & \ldots & u(n-N) \\ u(n) & u(n-1) & \ldots & u(n-N+1) \end{pmatrix}$$

and the diagonal matrix Λ_n by

$$\Lambda_n = diag[\sqrt{\lambda^{n-1}}, \sqrt{\lambda^{n-2}}, \ldots, \sqrt{\lambda}, 1]$$

The above definitions allow us to write the minimization problem in (9.5) as

$$J_n(\mathbf{w}_n) = \|\Lambda_n \mathbf{d}_n - \Lambda_n \mathbf{A}_n \mathbf{w}_n\|_2^2 \qquad (9.6)$$

where

$$\mathbf{d}_n = [g(1) \ g(2) \ldots g(n)]^T + v$$

v is the white Gaussian noise added in the system output. If R_n denotes the correlation matrix of input data $\mathbf{A}_n$, then following relation holds for the square-root factor S_n of R_n,

$$R_n = S_n^T S_n$$

Let us define $N \times 1$ vector $\mathbf{x}_n$ by considering new data vector $\mathbf{a}_n$ and the square-root factor of the previous instant as:

$$\mathbf{x}_n = \frac{S_{n-1}^T \mathbf{a}_n}{\sqrt{\lambda}}$$

At this stage we consider the $(N+1) \times (N+1)$ orthogonal matrix P_n, presented in [5], to formulate the equation for updating S_{n-1}^{-T} to S_n^{-T}, i.e.,

$$P_n \begin{pmatrix} \mathbf{x}_n & \lambda^{-\frac{1}{2}}S_{n-1}^{-T} \\ 1 & 0^T \end{pmatrix} = \begin{pmatrix} 0 & S_n^{-T} \\ \delta_n & \mathbf{u}_n^T \end{pmatrix} \qquad (9.7)$$

where $\delta_n = \sqrt{1 + \|\mathbf{x}_n\|^2} = \frac{\sqrt{\lambda + \mathbf{a}_n^T R_{n-1}^{-1} \mathbf{a}_n}}{\sqrt{\lambda}}$.

The orthogonal matrix P_n, used to annihilate the vector $\mathbf{x}_n$ in Eq. (9.7), is a Householder matrix [5]. Here,

$$\mathbf{u}_n = \frac{R_{n-1}^{-1} \mathbf{a}_n}{\lambda \delta_n}$$

is a scaled version of the Kalman gain vector [8], which can be used to update the filter tap-weight vector.

Householder RLS(HRLS) algorithm can be deduced by computing the a priori error $\eta(n)$, and then updating the filter tap-weight $\mathbf{w}_n$. The important equations of the algorithm are:

$$\eta(n) = (g(n) + v(n)) - \mathbf{w}_{n-1}^T \mathbf{a}_n \qquad (9.8)$$

$$\mathbf{w}_n = \mathbf{w}_{n-1} - \frac{\eta(n)}{\delta_n} \mathbf{u}_n$$

9.3 Simulation Results

In order to analyze the performance of this adaptive filter as a noise canceler, an artificial sine wave $g(n) = A \sin(2\pi f_{bw} n)$ is added to the real ECG signals, obtained from MIT-BIH arrhythmia database, to simulate the baseline wander (BW) effects, where f_{bw} is the high frequency component of respiratory sinus arrhythmia (RSA), and A is amplitude of wandering from baseline. MIT-BIH arrhythmia database is a benchmark for ECG recordings, and consists of 48 half hour samples of two channel ambulatory ECG recordings. These recordings are digitized at 360 samples per second per channel with 11-bit resolution over a 10 mV range.

In our simulation, we Choose data of signal 106 for 5 seconds duration with time $n = 0 : 0.002778 : 5$. Signal 106 represents ECG recording of a 24 years old female, having heartbeat in the range from 49 to 87. Figure 9.2 shows 5 s waveform of real ECG signal 106 and noisy one. Setting $\alpha = 0.5$, and filter length is $N = 4$ in adaptive noise canceler of previous section, learning curves of MSE are recorded using conventional RLS and HRLS algorithms for $f_{bw} = 0.25$. During

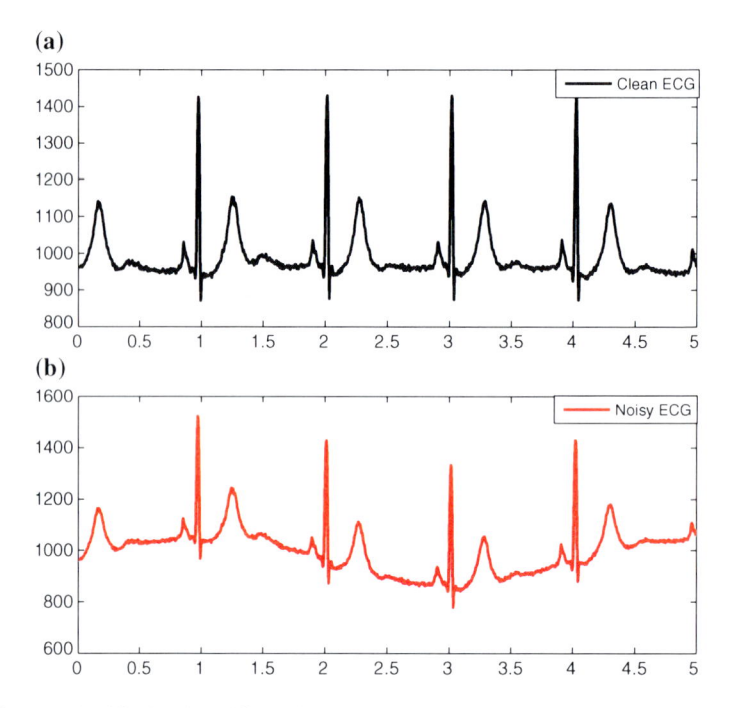

Fig. 9.2 a Real ECG signal, and **b** ECG signal contaminated by BW noise

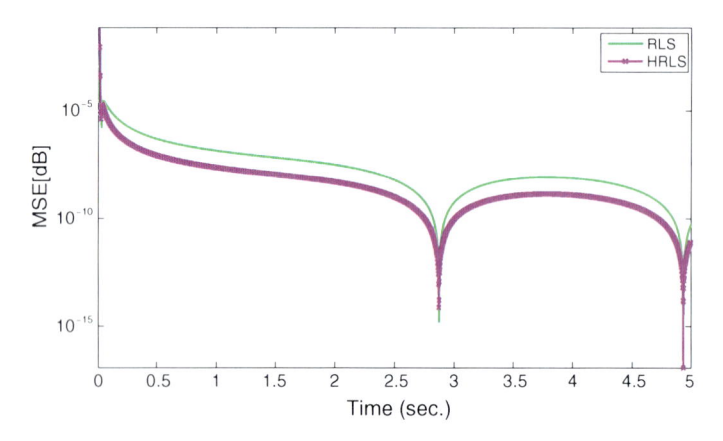

Fig. 9.3 Learning curves of MSE of RLS and HRLS algorithms for $\alpha = 0.5$

these computations, $\lambda = 1$, while $\delta_n = 0.1$ are kept fixed. Their efficiency is clear from the learning curves of MSE in Fig. 9.3, which show preference of HRLS algorithm in this comparison.

9.4 Conclusion

In this paper an adaptive noise canceler is designed to check the efficiency of HRLS adaptive filter in removal of baseline wander noise from real ECG signals. The proposed noise canceler is able to tackle with correlated noise signals and has shown capability to remove noise from ECG signal by minimizing the mean squares error.

Acknowledgments The authors would like to acknowledge the financial support of Universiti Sains Malaysia for registration and attendance of the conference by short term grant.

References

1. Thakor NV, Zhu YS (1991) Applications of adaptive filtering to ECG analysis: noise cancellation and arrhythmia detection. IEEE Trans Biomed Eng 38(8):785–794
2. Rahman MZU, Shaik RA et al (2011) Efficient sign based normalized adaptive filtering techniques for cancelation of artifacts in ECG signals: application to wireless biotelemetry. Signal Process 91(2):225–239
3. Ahmad NA (2005) Comparative study of iterative search method for adaptive filtering problems. In: International conference on applied mathematics
4. Farhang-Boroujeny B (1998) Adaptive filters: theory and applications. Wiley, New York
5. Apolinario JA Jr (2009) QRD-RLS adaptive filtering. Springer, Berlin
6. Douglas SC (2000) Numerically robust $O(N)^2$ RLS algorithms using least squares prewhitening. IEEE
7. Javed S, Ahmad NA (2012) An estimation of the performance of HRLS algorithm. World Academy of Science, Engineering and Technology
8. Haykin S (1991) Adaptive filter theory, 2nd edn. Prentice Hall, New Jersey

Chapter 10
Robotic Arm Control Based on Human Arm Motion

Dickson Neoh Tze How, Chan Wai Keat, Adzly Anuar and Khairul Salleh Mohamed Sahari

Abstract This paper discusses the development of a robotic arm whereby its motion is controlled based on the movements of the human arm. There are three major components in this project. They are a wearable control device, microcontroller and robotic arm. The wearable control device acts as a sensor to translate the physical movement of the human arm into electrical signals. The translated signals in the form of positional data are sent to the microcontroller unit for processing. The microcontroller unit processes the signals and outputs it to the robotic arm as physical replicated movements. The wearable control device is capable of sensing six degrees of freedom by the human arm joints movements. This includes rotation of the shoulder, arm, bending of the shoulder, elbow, wrist, and gripping motion. The fabricated robotic arm is able to perform six different motions replicating and mimicking the motions of the human arm.

Keywords Robotic arm · Human wearable device

10.1 Introduction

Robotic arm is one of the many applications in the industry that revolutionized the automation industry. Because of greater efficiency, pinpoint accuracy and super speed, the robotic arm is often used in factories and production plants to boost

---

D. N. Tze How (✉) · C. W. Keat · A. Anuar · K. S. Mohamed Sahari
Centre of Advanced Mechatronics and Robotics, Universiti Tenaga Nasional,
Jalan IKRAM-UNITEN 43000 Kajang, Selangor, Malaysia
e-mail: dicksonN@uniten.edu.my; dickson.neoh@gmail.com

A. Anuar
e-mail: adzly@uniten.edu.my

K. S. Mohamed Sahari
e-mail: khairuls@uniten.edu.my

production outputs. Robotic arms also are becoming a trend in replacing humans especially to perform repetitive task [1]. According to Luo and Su [2] robotics can be divided into industrial and service robotic. International Federation of Robotics (IFR) defines a service robot as a robot which operates semi- or fully autonomously to perform services useful to the well-being of humans and equipment, excluding manufacturing operations [2]. In some cases, the control of the robotic arms requires human judgement and at the same time, requires the omni-functionality of robots. Hence, there is a need of human operated robot to perform task that is beyond human capability. In the field of robotic arm control based on human motion, it appears that many individuals have been developing better designs to replace the fixed motion of the robotic arm with the interface of human arm motion [3]. In other words, it means to produce an interactive robotic arm which is not programmed to perform certain task repeatedly but to perform task based on the input signal from the human arm motions. Although there are many different types of designs developed throughout the years, theoretically most of the existing designs are using the same principles where the sensors senses the motions of the human arm and transforms the mechanical moment into electrical signals. The signals are then transmitted to the microcontroller for signal processing purpose. The processed signal will be sent to the servo controller and the servo driving motor to control the movement of the robotic arm. Based on the similar working principals, more advanced and sophisticated hardware and software are being integrated to produce a much more robust wearable device and robotic arm.

Figure 10.1 shows the overall working principle of the project. The three major components are illustrated above that is; wearable device, microcontroller, and the robotic arm. The signal moves in a one-way direction in an open loop control system without any feedbacks to the user or the controller.

10.2 Mechanical Design of Arm and Wearable Device

The goal of this project is to produce a fully functional prototype of a robotic arm to be controlled based on human arm motion. The requirements and specifications of the robotic arm are as follows:

- To achieve six Degree of Freedom (DOF) on the robotic arm included the clamping motion of the gripper.
- To be able to detect the motions of a human arm and send it to the microcontroller.
- To achieve minimal visible time delay between the sensing of the motions of a human arm and the robotic arm.
- To achieve stability of the robotic arm while in motion or static conditions.

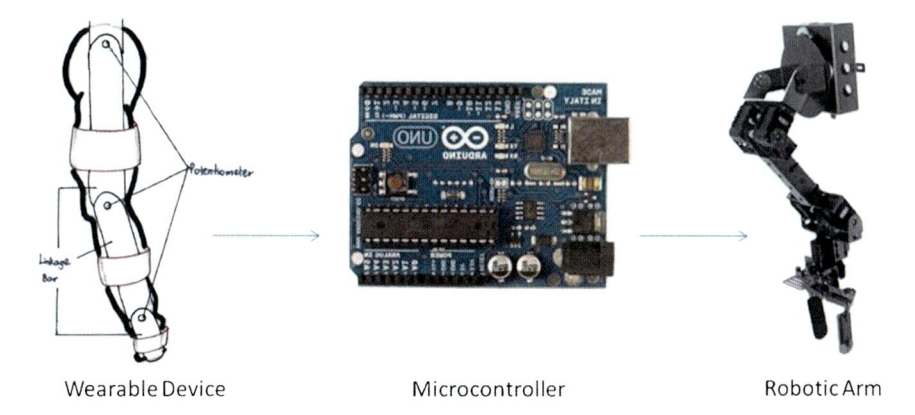

Fig. 10.1 General experimental setup [4, 5]

Figure 10.2a shows the design of the wearable device proposed. The joints are mounted with potentiometer which acts as a sensor to detect joint movements of the human arm. Figure 10.2b shows the mechanical design of the robotic arm. The designs are based on the structure of the human arm and it is one-to-one scale with the human arm. The drivers used in this design are the RC servos which are placed at each joint to move the different section accordingly. The design of the robotic arm enables movement of six degrees of freedom which included the rotational motion of the shoulder in both horizontal and vertical direction; bending motion of the elbow and wrist; rotation of the arm; and the gripping motion. The material used for the robotic arm is polycarbonate due to its flexibility, lightweight, appearance and relatively low cost.

10.3 Electronics Hardware and Control Algorithm

For the electronics hardware and design of the project, the Arduino UNO open source microcontroller is chosen. The main advantage of using this microcontroller is the open source nature and universality of the hardware and compiler. The hardware is simple to be used by students with no backgrounds in electronics and programming. Other advantages include cost efficiency as Arduino microcontrollers are relatively cheaper than its counterparts. Since the robotic arm involves the use of RC servos as actuators, a servo controller board is needed. The board that is utilized is the 8 Channel Servo Controller (SC08A) board by Cytron Technologies Sdn. Bhd [6]. The controller is capable of controlling over 8 units of RC servos independently. The control algorithm of the system is very straight forward. There are six inputs from the six potentiometers on the wearable device. Since all the inputs to the Arduino are in analog voltage form, the Analog to Digital Converter (ADC) modules on the Arduino UNO are used to sample the

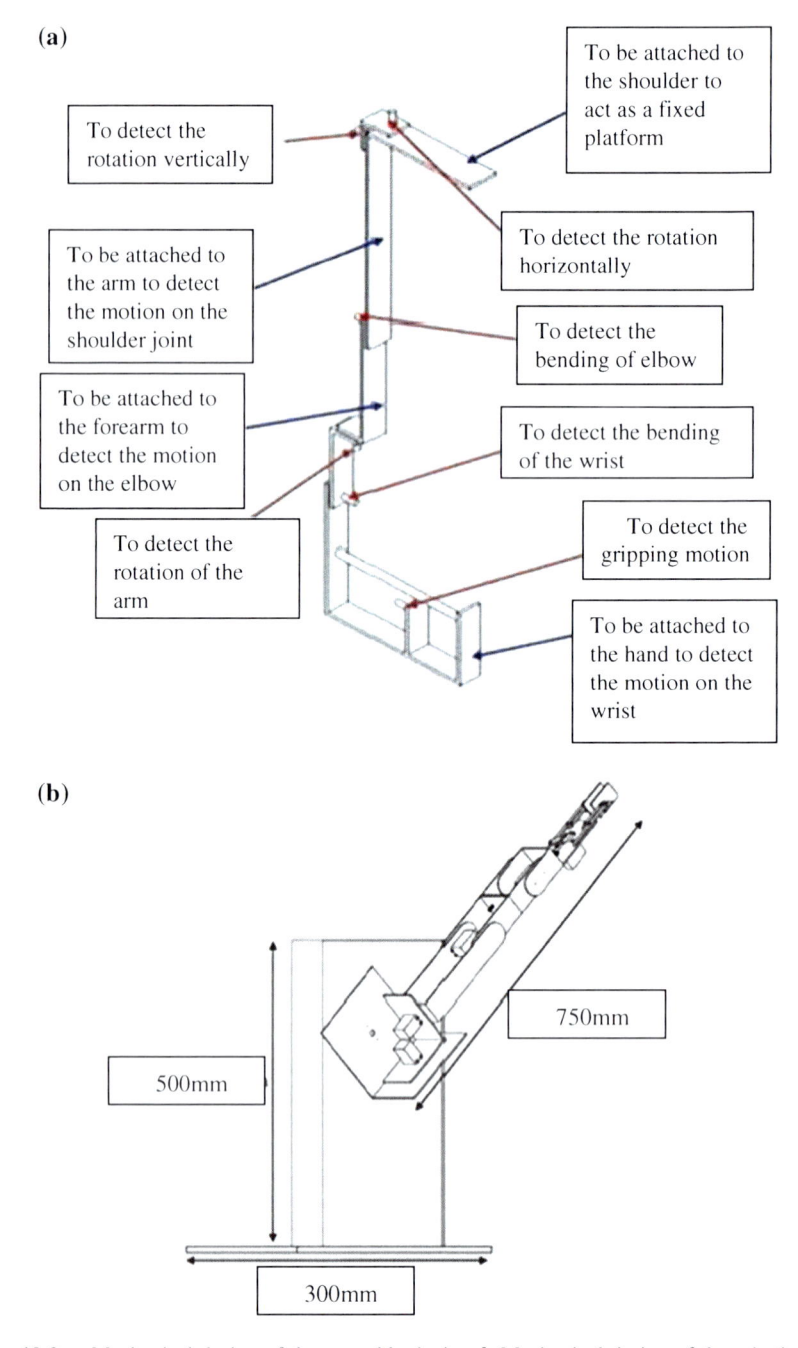

Fig. 10.2 a Mechanical design of the wearable device. b Mechanical design of the robotic arm

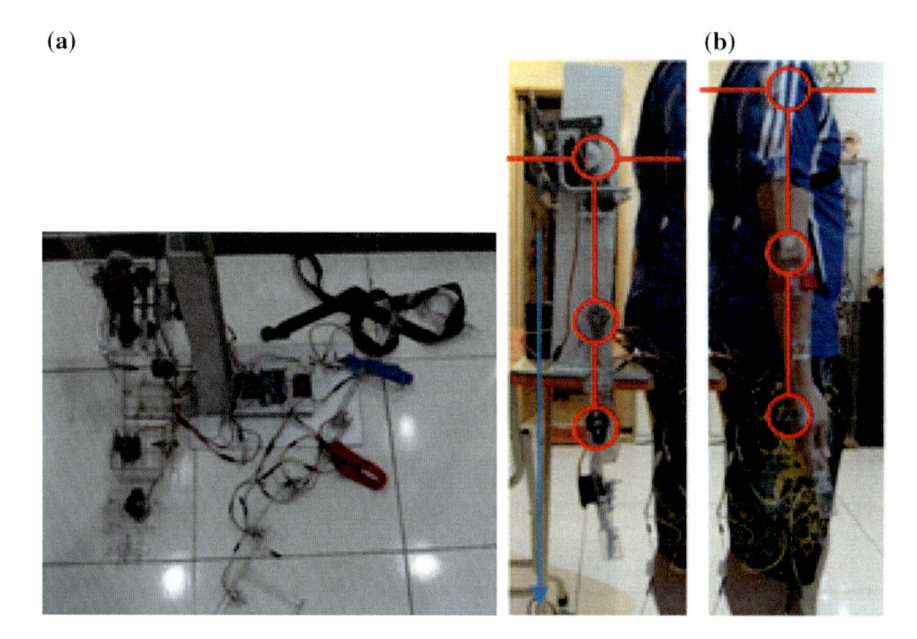

Fig. 10.3 Wearable device and robotic arm

analog voltage into digital form. The digital form signal is then mapped to fit the angle of movement of the servo motor which is on the robotic arm. A pre-calibration is also done on the wearable device and the RC servos on the robotic arm to ensure that the centre position of the potentiometer matches the centre position of the RC servo.

10.4 Results and Analysis

The results of the project are documented by comparing the movements of the wearable device and the actual movement produced on the robotic arm. Figure 10.3a shows the picture of the wearable device and the robotic arm.

The initial position of the wearable device and robotic arm is as Fig. 10.3b whereby both the wearable device and robotic arm are in straight position pointing to the ground.

Table 10.1 shows the results from the test performed on the fabricated system. The position of joints on the wearable device is compared to the position of the joints on the robotic arm by making simple movements on the wearable device. The error of the system is variable depending on the joint position. The error is highest at the shoulder joint portion and least at the wrist joint portion. The cause of the error is investigated and several possible causes are identified. First, the mechanical design of the robotic arm lacks support. This puts more burdens on the

Table 10.1 Comparison of the actual movement on the wearable device and the output movement on the robotic arm

Wearable device	Robotic arm	Percentage error (%)	Mean error (%)
Wrist joint (degrees)			
−135	−130	3.70	−6.17
−90	−90	0.00	
−45	−55	−22.22	
Elbow joint (degrees)			
−90	−90	0.00	−8.47
−70	−80	−14.29	
−45	−50	−11.11	
Arm rotation (degrees)			
90	50	44.44	37.04
45	30	33.33	
−45	−30	33.33	
Shoulder joint (degrees)			
−90	−90	0.00	−25.66
−70	−85	−21.43	
−45	−70	−55.56	

Fig. 10.4 Comparison of angular position pertaining to the wrist joint

Fig. 10.5 Comparison of angular position pertaining to the elbow joint

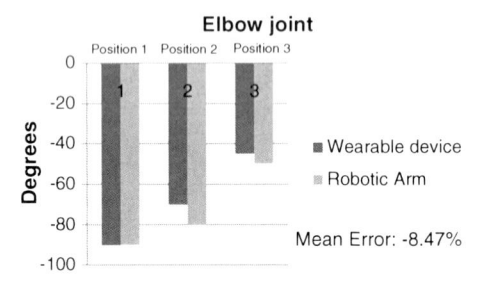

servo motors to actuate the joints. Due to the limitation in the servo torque, some of the position cannot be achieved accurately thus causing high error percentage between the wearable device and the robotic arm. Second, the analogue servo motor used is not sufficient in terms of holding torque and accuracy. The analogue

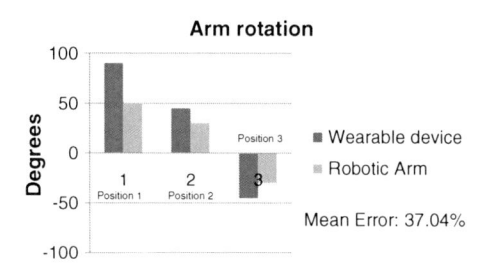

Fig. 10.6 Comparison of angular position pertaining to the arm rotation

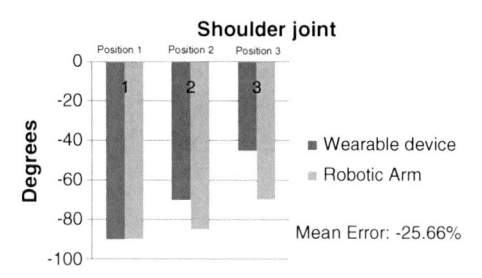

Arm rotation chart: Wearable device vs Robotic Arm across Position 1, 2, 3. Mean Error: 37.04%

Fig. 10.7 Comparison of angular position pertaining to the shoulder joint

Shoulder joint chart: Wearable device vs Robotic Arm across Position 1, 2, 3. Mean Error: -25.66%

servo motors must be replaced with digital servo motor which gives far greater torque value and better accuracy. This improves the accuracy and consistency of the robotic arm in achieving the desired position. However, the concept that is presented in this paper is proven to be working. By utilizing the cost efficient Arduino platform the cost of purchasing the electronic hardware and software is greatly reduced. Acrylic material proved sufficient for the current scale of the robotic arm. Improvements can be done on the mechanical design of the arm. Figures 10.4, 10.5, 10.6 and 10.7 graphically illustrates the angular positional data in graphical form from each joint of the arm. Data are taken in three positions from each joints and from there, the mean of error each respective joint is calculated.

10.5 Conclusion

This paper had discussed about the development of a tele-operated robotic arm. The tele-operation concept in this paper is carried out by reproduction of human movement on a robotic device by going through a microcontroller as a signal processor. By replicating movements on human wearable device, a more precise movement is expected as outcome of the robotic arm. This can be an advantage in situations that require accuracy and strengths that is beyond human capability but at the same time requires human gestures or movements as reference or input. The robotic arm can be seen as an amplifier to the human movements in terms of strength and accuracy. Future development of the project includes incorporating haptic feedback system to the wearable device in order for the user to gain feedback from the robotic arm. The haptic feedback will be a solution to further enhance the accuracy of the robotic arm movement.

Acknowledgments The authors would like to thank the Centre of Advanced Mechatronics and Robotics Application (CAMaRo), Universiti Tenaga National (UNITEN) for funding the project.

References

1. Kamaril Yusoff MA, Samin RE, Ibrahim BSK (2012) Wireless mobile robotic arm. In: International symposium on robotic and intelligent sensors, pp 1073–1078
2. Luo RC, Su KL (2003) A multi agent multi sensor based real-time sensory control system for intelligent security robot. In: IEEE international conference on robotics and automation, vol 2, pp 2394–2399
3. Schirmbeck EU, Haßelbeck C, Mayer H, Nágy I, Knoll A, Freyberger FKB, Popp M, Wildhirt SM, Lange R, Bauernschmitt R (2005) Evaluation of haptic in robotic heart surgery. In: Proceedings of the 19th international congress and exhibition on CARS 2005: computer assisted radiology and surgery, vol 128, pp 730–734
4. Crustcrawler Robotics Official Page. http://www.crustcrawler.com/products/AX-18F%20Smart%20Robotic%20Arm/images/smartarm%20016.jpg
5. The Arduino Official Webpage. http://arduino.cc/en/Main/arduinoBoardUno
6. Cytron Technologies Official Webpage. http://www.cytron.com.my/viewProduct.php?pcode=SC08A&name=8%20Channel%20Servo%20Controller

Chapter 11
Elevator Riding of Mobile Robot Using Sensor Fusion

Jaehong Lee, Xuenan Cui, Hyoungrae Kim, Seungjun Lee and Hakil Kim

Abstract Elevator riding is an essential skill for a mobile service robot to carry out various tasks. This paper proposes a framework for robot navigation based on sensor fusion of a laser range finder (LRF) and a vision camera to detect an elevator door and recognize its state. The state of the door is determined by calibrating and combining LRF-camera data. The indicator including the current floor number of the elevator and a direction arrow is recognized by a neural network classifier. The robot moves inside the elevator after verifying the state of the door being opened and the moving direction of the elevator. The robot confirms the target floor by an artificial landmark and localizes itself by the marker detection. The proposed method is implemented on a robot platform, and elevator riding is achieved as the experiment results.

Keywords Elevator riding · Sensor fusion · Artificial landmark · Localization · NN

J. Lee (✉) · X. Cui · S. Lee · H. Kim
School of Electronic Engineering, Inha University, Incheon, Korea
e-mail: jaehong@vision.inha.ac.kr; jurio3924@gmail.com

X. Cui
e-mail: xncui@vision.inha.ac.kr

S. Lee
e-mail: sjlee@vision.inha.ac.kr

H. Kim
e-mail: hikim@inha.ac.kr

H. Kim
School of Robot Engineering, Inha University, Incheon, Korea
e-mail: hrkim@vision.inha.ac.kr

11.1 Introduction

In company with development of the robot industry, intelligent environment recognition is needed. For a mobile robot service in public places such as hospitals and silver towns, elevator riding is a significant technique. If a robot is not able to move on other floors, its service is limited. In order to offer an intelligent service, the total process should be similar to what humans are doing as follows:

Step 1: The elevator door is recognized and detected whether it is opened or not.
Step 2: After checking the direction, people get in the elevator.
Step 3: Turning towards the door, they wait until the elevator arrives at the target floor.
Step 4: People get off after confirming the door is opened on the target floor.

To ride an elevator, its door recognition is the first step. So far, many previous studies about detecting doors using a laser range finder (LRF) or a camera sensor [1, 2] have been introduced. In the paper of Ma et al. [3], the elevator door recognition method using an LRF and a single camera is proposed.

In this paper, a framework for robot navigation including recognition of elevator door states, moving to the target floor and artificial marker-based localization is proposed. This paper is structured as follows: Sect. 11.2 presents the process of elevator door recognition and indicator recognition using a neural network classifier (NN). Localization using artificial markers is described in Sect. 11.3. Experiments using the robot platform and conclusion are described in Sects. 11.4 and 11.5, respectively.

11.2 Elevator Door Recognition

11.2.1 Elevator Door Detection and Verification

Figure 11.1a shows LRF scanning data in front of elevator door. Some line features are extracted by the points in front of the door. The line features are used to detect the elevator door state. Nguyen et al. [4] compared line extraction algorithms from the scan data, where Split-and-Merge algorithm showed relatively fast and accurate performance, and Iterative-End-Point-fit algorithm [5] is faster than the Split-and-Merge algorithm. To guarantee a real-time process in robot navigating conditions, the two algorithms are combined in this study. The detailed process is described in the previous study [3].

The elevator door state recognition with the verification is the second step to get into an elevator. Before the process, the states have to be defined. The interval space between both sides of elevator door is divided into 5 blocks. The counted number of blocks b_sum whether each block is filled with scanned data or not determines the outside door state S_t^o at time t in Fig. 11.2. If b_sum is 1 or 0, the space of door is considered empty and S_t^o becomes 1 which denotes OPENED. If all the blocks filled,

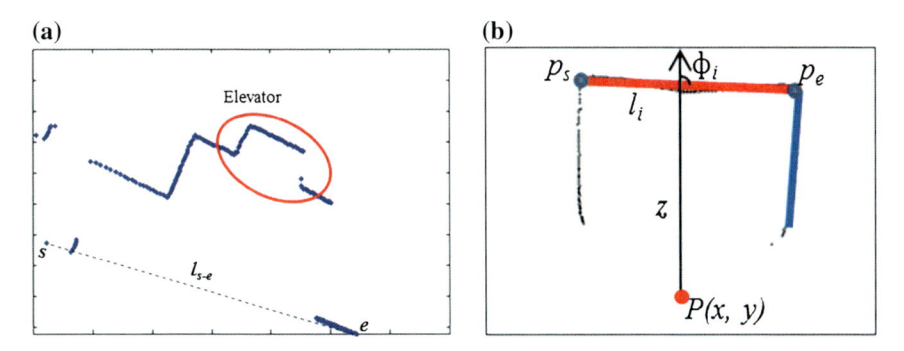

Fig. 11.1 Door detection of outside (**a**) and inside (**b**) of the elevator

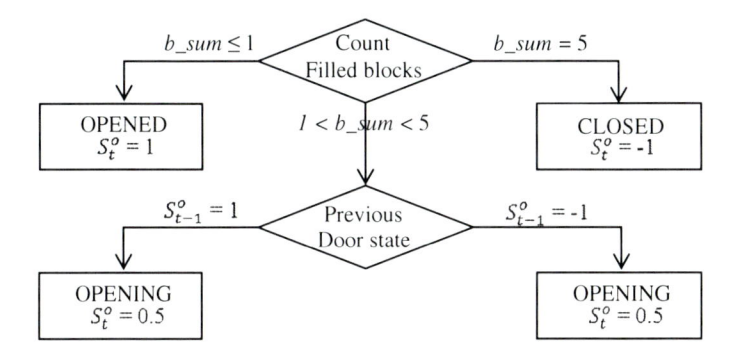

Fig. 11.2 Decision tree for an elevator door state

$S_t^o = -1$ denoting CLOSED. The previous state S_{t-1}^o at the time $t - 1$ is referred in other cases. $S_t^o = 0.5$ corresponds to OPENING and $S_t^o = -0.5$ to CLOSING. Figure 11.2 shows the process of deciding the current door state S_t^o.

11.2.2 Door Detection Inside of the Elevator

To arrive at the correct destination, the robot has to get off the elevator. Inside door recognition in the elevator is not easy to detect the door and its state. The confined space causes data distortion, and it is hard to detect the floor number indicator. The distance between the robot and the door is too short to acquire reliable LRF data. An average filter is applied to the LRF data before the state is estimated. But when the robot gets into an elevator, the door state is unpredictable because the LRF sensor and the front camera view is on the opposite side of the door. Thus the door state inside of the elevator S_t^i at time t is defined by two states, INSIDE_-DOOR_CLOSED ($S_t^i = -1$) and INSIDE_DOOR_OPENED ($S_t^i = 1$).

A line segment $l_i = \{p_s, p_e, \Phi_i\}$ consists of start/end indexed point p_s, p_e and the included angle Φ_i of straight forward vector z and l_i as shown in Fig. 11.1b. S_t^i is defined based on the predefined angle range Φ_{min} and the distance d_{max} as shown in

Fig. 11.3 Perspective transformation of indicator and floor number classification using NN

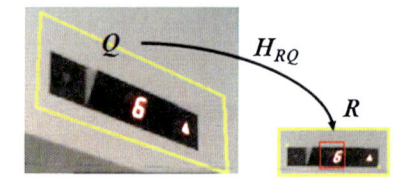

(11.1). INSIDE_DOOR_CLOSED is the state of scanning behind of the door or turning in the elevator, so l_i satisfies the condition of $S_t^i = -1$. Inversely, INSIDE_DOOR_OPENED means that l_i is not detected because the door had already opened. After the door is detected as being opened, the robot gets off the elevator by the localization process described in Sect. 11.3.

$$S_t^i = \begin{cases} -1, & when\ (|P-p_s|<d_{max})\ and\ (|P-p_e|<d_{max})\ and\ (\Phi_i<\Phi_{min}) \\ 1, & otherwise \end{cases}$$

(11.1)

11.2.3 Indicator Segmentation and Perspective Transform

After the elevator door recognition and verification, the current position of the elevator is estimated by indicator recognition. Region of Interest (ROI) is set on the indicator. After pre-processing, a segmented number is classified by a neural network classifier. For the simple process, the quadrangle indicator from various view angles has to be converted to a rectangle using a perspective transform.

The elevator indicator is segmented between left and right vertical edges of the elevator door. Assuming the indicator is above the elevator door, a quadrangle is searched. The quadrangle Q needs to be converted to a rectangle R, therefore, it simplifies the classifying step. Figure 11.3 shows the transformation H_{RQ} that maps the points from Q to R. The four correspondences of points determine the transform matrix.

In the rectangle which includes indicator, two vertex points are chosen in the gray scaled image. The histogram of R is scanned vertically and horizontally. Left-top and right-bottom that the value sharply descents determine the elevator indicator. The current floor number of elevator is displayed on the center of indicator.

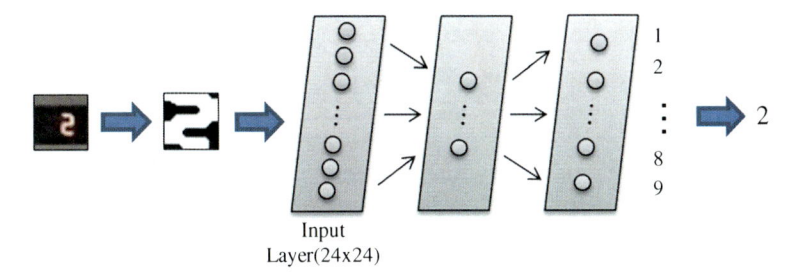

Fig. 11.4 Floor number classification using NN

11.2.4 Floor Number Classifying by NN and Direction Recognition

A neural network (NN) classifier is widely investigated in many research areas. The basic application of NN is character recognition and classification. To Apply NN in character recognition, two steps are necessary. First, resizing to 24 × 24 binary image that is fitted by the number. Let the normalized binary image $x_i = \{x_1, x_2, \ldots x_d\}^T$, ($d = 24 \times 24$) which is the input of the NN classifier. Preparing the weights u, v which are the result of training is the second step. In the training step, error back-propagation algorithm for Multi-Layer Perception (MLP) learning is used in Fig. 11.4. Mean squared error (MSE) is calculated for comparison of the threshold. The result just comes from the product of weights and input, therefore it enables classifying to real-time process.

The floor number that the elevator located is insufficient data for the robot to choose getting in or not. When the robot wants to go upstairs, the direction of elevator moving is to be upward. Direction recognition is somewhat a simple problem. Like the method of extracting the square in 3.1, squares on the left/right side of the indicator are extracted. After binarization with an adaptive threshold, the numbers of white pixels on both sides c_l, c_r are counted. The direction of elevator d_{elv} is determined by (11.2), where '1' means GOING_UP, '−1' means GOING_DOWN and '0' means STOP, and τ denotes the threshold value.

$$d_{elv} = \begin{cases} 1, & c_l, -c_r > \tau \\ -1, & c_r, -c_l > \tau \\ 0, & \text{otherwise} \end{cases} \qquad (11.2)$$

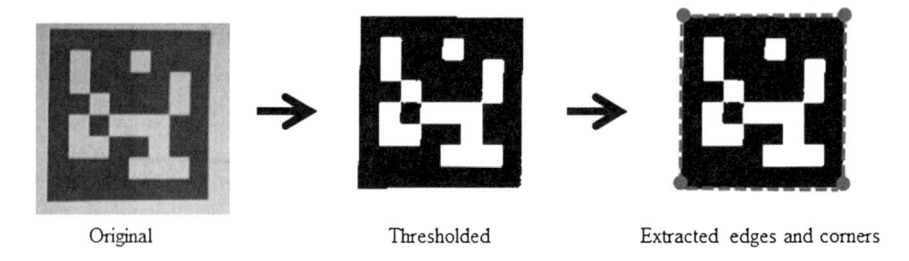

Original → Thresholded → Extracted edges and corners

Fig. 11.5 Process of marker detection

11.3 Localization Based on Fiducial Marker

11.3.1 The Shape of Fiducial Marker and Detection

In this paper, artificial markers and a single camera are used to indoor localization. ARTag markers are used since it achieved better performance than other artificial markers according to [6]. ARTag is correctly detected even large occlusions and challenging lighting. The marker detected, relative position of camera which considered as robot position is calculated by rotation and translation matrix. Therefore, world coordinates of the robot is determined by the marker position which installed in advance.

The marker used in this paper is a binary patterned square in Fig. 11.5. The patterned markers satisfies the criteria [7] for practical fiducial marker. Since each marker has its own ID, it is useful to access relevant information that contains location data. Also, 3D position and heading of the robot can be estimated by comparing detected marker.

In Fig. 11.5, pre-processing is the first step to detect markers. Input image from camera is binarized to 1 or 0 by thresholding method. Next, search contour before extract corner and edges. If the edge and corner points define a quadrangle, the marker is recognized and identified by inner pattern. The rotation and translation vector are estimated when the marker detected. The four corner points corresponding vertices are used to estimate position and heading in 3D space.

11.3.2 Localization

Once the marker is detected which is patched on the wall or ceiling, marker M to camera C rotation and translation are estimated. However, what we want know is the position of camera in the world coordinates. Therefore some transformation steps are proposed to localize the camera.

The final output is $^{W}H_{C}$, the transformation from the camera coordinates to the world coordinates. When the marker is detected, we get the inverse transform

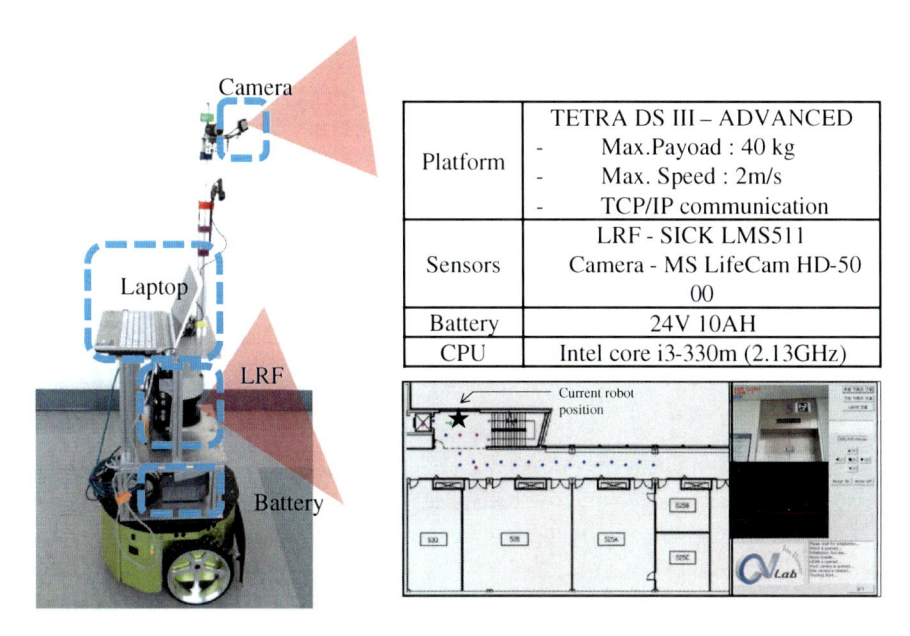

Fig. 11.6 Robot platform configuration and specification of components

matrix $^CH_M^{-1}$ from ^CH_M that means marker to camera conversion. Let the robot position 'M' in the marker coordinates be represented by P^M, and the camera position in camera coordinates 'C' be represented by P^C, and the position in world coordinates 'W' is represented by P^W. Then P^M is described in (11.3) below.

$$P^M = {}^CH_M^{-1} P^C \tag{11.3}$$

The marker to world coordinates transform matrix ^WH_M is derived from the database that contains each markers ID and rotation translation matrix. Finally, P^W is calculated by the result of (11.4).

$$P^w = {}^WH_M P^M = {}^WH_M {}^CH_M^{-1} P^C \tag{11.4}$$

11.4 Experiments

11.4.1 Environment

Experiments are performed by our robot platform, TETRA DSIII which is equipped with an MS LifeCam HD-5000, SICK LMS511 Laser Range Finder and battery for LRF as in Fig. 11.6. The camera is oriented upward since it is used for elevator recognition and calibrated with LRF in advance. Color images captured

Table 11.1 The average results of the experiment

		Get in		Get off
		Floor number (%)	Elevator direction (%)	Marker recognition (%)
FN	Incorrect	3.37	2.24	0.00
	Miss	0.00	0.00	4.54
TP	Correct	96.63	97.76	95.46

Table 11.2 The average process time

Process	Door	Door + Indicator	Door + Indicator + Localization
Average time (fps)	25.00	22.27	15.87

320×240 resolution. In the color images, markers which are patched on ceiling are detected in the range between 1.5 and 3.0 m from the robot.

Elevator door, indicator recognition, and localization by artificial landmark methods for elevator riding are proposed in this paper. Elevator door recognition is evaluated in the paper [3], and we assume that the robot operates in an elevator-visible area. Generally True Positive (TP), False Negative (FN), False Positive (FP), True Negative (TN) are used for evaluation. However, as assumed, only TP and FN are measured. FN is divided two terms 'incorrect' and 'miss'. 'Incorrect' corresponds classifying error and 'miss' means undetected. Performance is evaluated using the CVLab Robot Program in Fig. 11.6. The Robot is controlled by a button or autonomously drives. All the operations of the robot in Fig. 11.6 are monitored and positions are mapped on the window.

11.4.2 Experimental Results

Tables 11.1 and 11.2 show the result of the experiments. All processes are evaluated by averaging percentages. There are two main processes: 'get in' and 'get out'. To get in the elevator, the robot has to confirm the current floor of elevator and its direction. And to get off the elevator, the target floor number has to be known by marker detection.

Floor number classification had 96.63 % of accuracy and the incorrect rate is 3.37 %. Miss rate had 0 % since we assumed that floor number segmentation is not missed if the elevator door was detected. Elevator direction showed 97.76 % of accuracy. However, there were some incorrect results by occlusion of light. Door recognition in the elevator showed 100 % of accuracy because it just processed LRF data guaranteed by its confidence in a limited area. In the 'get off' step, marker recognition showed 95.46 % of accuracy. When the marker detected correctly, the robot got off the elevator successfully. The miss rate was 4.54 % and there was no inter-marker confusion. It is already analyzed in [7].

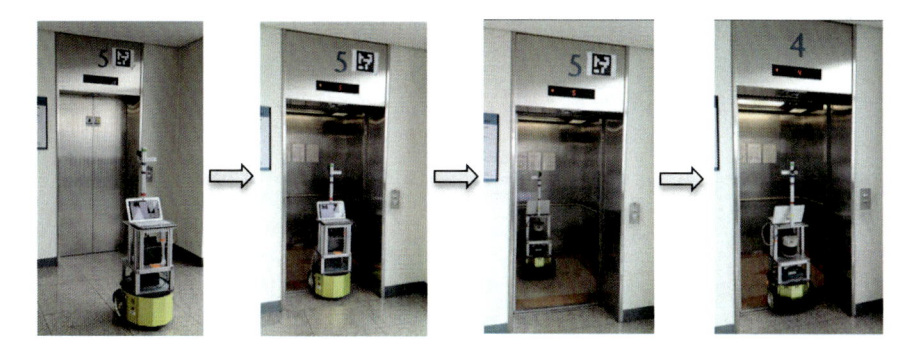

Fig. 11.7 Process of elevator riding: *First*, door state recognition. *Second*, floor number and direction classification. *Third*, get in the elevator and turn toward door. Finally get off the elevator after target floor confirming

In the paper of Ma et al. [3], the door recognition hit rate was 96.7–97.1 % and average processing time was 25 fps. In this method, the average processing time was 22.27 fps when elevator indicator recognition is added. When the localization method is added, the overall processing time is 15.87 fps (Fig. 11.7).

11.5 Conclusion

In this paper, a framework for robot navigation which includes recognition of elevator door states, moving to different target floor and artificial marker-based localization is proposed. Sensor fusion using a calibrated LRF and vision camera are used to detect the elevator door and estimation of its state. A single camera is used to verify the door detection and to recognize the indicator of elevator. The extracted floor number is classified by an NN classifier. The localization is performed using artificial landmarks. The robot is able to ride an elevator and move toward a destination on a different floor. Floor number classification showed 91.53 % of accuracy and correct direction estimation, 92.65 %. Marker detection rate was 95.46 %. In the future work, we will use natural landmarks instead of artificial landmarks and improve the performance in the dynamic environment. In addition, obstacle avoidance using the fused sensors would be enabled.

Acknowledgments This work was partially supported by "Development of mobile assisted robot and emotional interaction robot for the elderly (10038574)" under the Industrial Source Technology Development Programs of the Ministry of Knowledge Economy (MKE) of Korea and partially supported by the MOTIE (Ministry of Trade, Industry & Energy), Korea, under the Robotics-Specialized Education Consortium for Graduates support program supervised by the NIPA (National IT Industry Promotion Agency) (H1502-13-1001).

References

1. Murillo AC, Košecká J, Guerrero JJ, Sagüés C (2008) Visual door detection integrating appearance and shape cues. Robot Auton Syst 56(6):512–521
2. Kang JG, An SY, Choi WS, Oh SY (2010) Recognition and path planning strategy for autonomous navigation in the elevator environment. Int J Control Autom Syst 8(4):808 821
3. Ma S-W, Cui X, Lee H-H, Kimg H-R, Lee J-H, Kim H (2012) Robust elevator door recognition using LRF and camera. J Inst Control, Robot Syst 18(6):601–607
4. Nguyen V, Gächter S, Martinelli A, Tomatis N, Siegwart R (2007) A comparison of line extraction algorithms using 2D range data for indoor mobile robotics. Auton Robots 23(2):97–111
5. Borges GA, Aldon MJ (2004) Line extraction in 2D range images for mobile robotics. J Intell Rob Syst Theory Appl 40(3):267–297
6. Fiala M (2005) Comparing ARTag and ARToolkit plus fiducial marker systems. In: IEEE international workshop on haptic audio visual environments and their applications, Ottawa, Ontario, Canada, 1–2 Oct 2005
7. Fiala M (2010) Designing highly reliable fiducial markers. IEEE Trans Pattern Anal Mach Intell 32(7):1317–1324

Chapter 12
Safe Global Path Planning of Mobile Robots Based on Modified A* Algorithm

Jong-Hun Park, Jin-Hong No and Uk-Youl Huh

Abstract This study evaluated Safe Global Path Planning (SGPP) for mobile robot navigation. This model considers the potential risk as a safety index, whereas most studies of global path planning algorithms focused on the time and distance cost. Robots can encounter difficulties in reaching a target point due to the limited field of view behind obstacles. This paper presents a modified A* algorithm to reduce the collision risk by considering such occluded environments. According to the simulation results, the presented algorithm showed better performance in terms of safety and navigation time than the conventional A* algorithm.

Keywords Safe global path planning · Modified A* algorithm · Safety cost · Risk area

12.1 Introduction

Intelligent navigation is needed for autonomous mobile robots to move safely to their destination via an optimized path without collision. Path planning in intelligent navigation is divided mainly into global and local path planning. Global path-planning searches for the optimized path toward a target off-line and local path-planning aims to avoid obstacles in real time [1]. In path planning, three methods are used to draw a navigation map: cell decomposition, roadmap, and

---

J.-H. Park (✉) · J.-H. No · U.-Y. Huh
Robot Engineering, INHA-University, Incheon, Korea
e-mail: patch_hanl@naver.com

J.-H. No
e-mail: finiel@naver.com

U.-Y. Huh
e-mail: uyhuh@inha.ac.kr

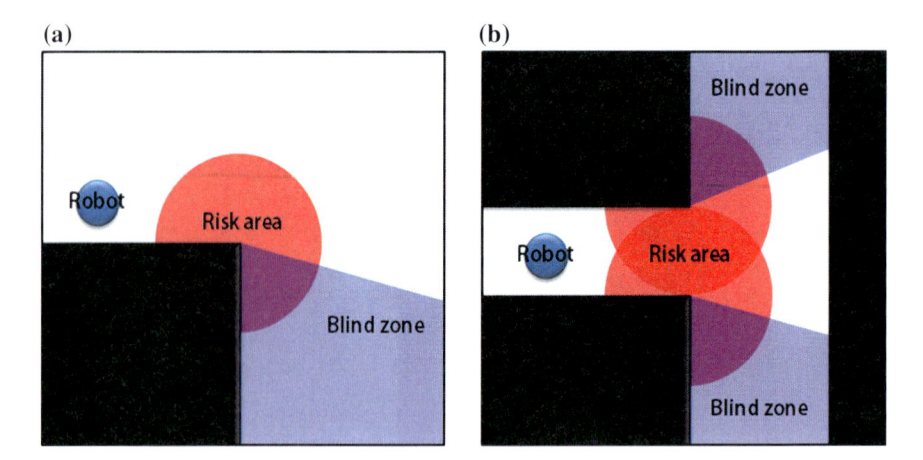

Fig. 12.1 Definition of risk areas

potential field methods [2, 3]. The potential field methods have the advantages of simplicity and smoothness over the others. The potential field method has some drawbacks: a robot in occluded spaces can sink into local minima and spend too much search time and memory. As a result, the robot is unable to reach the target place. A range of search algorithms have been developed to solve these problems. All these algorithms originated from Dijkstra, and A* algorithms deal mainly with distance- and time-related costs to acquire the two minimal costs [4]. Those algorithms are based on the situation that obstacles are fixed, whereas the situation including mobile obstacles will become more critical and unexpected, which is more common in the real world. For this reason, safety that refers to collision-less navigation of a robot while exploring a final target should also be considered under an occluded environment [5, 6]. In global path planning, the employment of this risk evaluation function suggested in this paper provides the advantages for safe navigation over previous risk evaluation functions in local path planning because a robot can reduce the potential collision risk by avoiding highly risk areas or moving slowly in advance. This study is considered in global path planning of the modified A* algorithm, which is called safe global path planning (SGPP). In particular, the modified A* algorithm includes a risk evaluation function considering the shape of obstacles, which is numerated by the angle, enabling the robot to reach a desired place safely without collision.

12.2 The Definition of Risk Area

This study focused on how to handle the potential risk from occluded areas, which needs to be considered for autonomous navigation [7]. Figure 12.1 shows two types of risk areas, such as corner and intersection. Figure 12.1a shows the risk area with a blind zone near the corner. Figure 12.1b shows the blind zones as a risk area with an increased collision potential risk at the junction.

Collision can occur when a mobile robot passes by a wall, particularly in places including convex edges, such as buildings and rooms. Although the robot has a sensor, it is difficult for the robot to recognize static or moving obstacles beyond a topographical feature. In the case of the intersection, the potential of collision will increase more than that in the case of a single corner, because of the larger number of corners. In defined risk areas, the robot is highly likely to fail to react promptly even if it has a sensor. Therefore, a safe path contemplating potential risk areas is established so that the robot can detour into other pathways around or move slowly while passing through risk areas.

12.3 Safe Global Path Planning Base on the Modified A* Algorithm

Safe global path planning based on the modified A* algorithm, which considers safety, was established using the following process. First, a map was constructed by introducing a risk evaluation function to evaluate areas with a limited field of view and developed safety-focused path planning with the modified A* algorithm.

12.3.1 A Risk Evaluation Function

In reality, the existence of static obstacles does not need to be considered, but other unexpected moving obstacles hidden behind a static obstacle need to be taken into account. To avoid this potential collision risk, the shape of the obstacle should also be considered with the distance in the risk evaluation function. The shape of the obstacle can be simplified as a convex edge with a range of angle values. With increasing angle of the convex edge, there is a decrease in the dead zone area behind the corner, where the sensor of a robot is unable to detect. The risk evaluation function is constructed in two steps. In the first step, the extraction process of topographical points, such as corners and intersection, from the mapped information was performed initially using an image processing technique [8], so that the risk force value with a risk evaluation function was calculated at those points. The next step involves the mapping process in global path planning for the design of safe path planning incorporating the risk areas produced by the risk evaluation function. The risk evaluation function can be defined as

$$R(n_c) = \alpha \frac{e^{-\theta}}{r^2} \qquad (12.1)$$

Fig. 12.2 Risk force at the corner

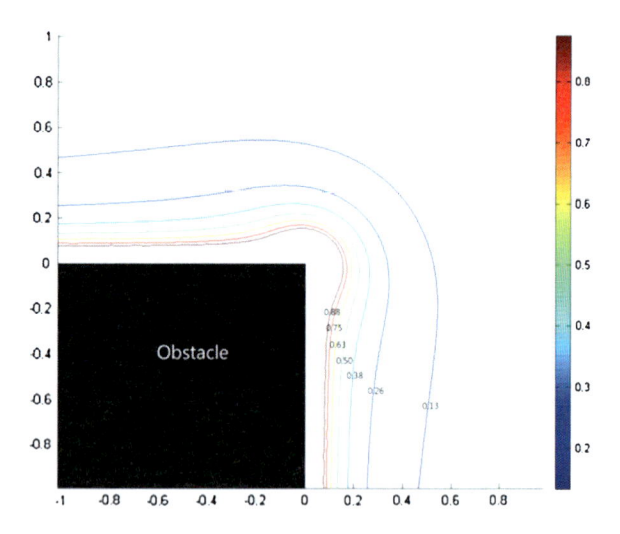

where $R(n_c)$ is a risk evaluation function, r is the distance between adjacent nodes (certain spots in a cell space), θ is the angle of the corner, and α is the constant value of the risk area. The range of the angle is $0 < \theta < \pi$. The magnitude of the risk force has a value within $0 < R(n_c) < 1$. Equation 12.1 suggests that the risk force increases with decreasing angle.

Figure 12.2 shows the relationship between the angle of a corner and a wall. The risk of collision increases if the robot approaches an area with corners with an angle <180°. As the angle of the wall decreases, the risk of collision will increase due to lack of the sensor information of the robot. If the angle reaches one, it can be treated as a static obstacle. The highest potential risk exists when the angle of the corner is 0°.

12.3.2 The Modified A* Algorithm Using the Risk Force

Equation (12.2) expresses the safety costs function of modified A* algorithm, which is based on the potential risk.

$$S(n) = C(n_s, n_c) + R(n_c) + G(n_c, n_g) \qquad (12.2)$$

The algorithm consists of three terms: $C(n_s, n_c)$ is a distance cost function of a starting node to current nodes; $R(n_c)$ is a risk function that evaluates the current risk level and its surrounding nodes; $G(n_c, n_g)$ is the estimated costs for reaching from the current node to target point. The lowest safety cost is selected as the best node.

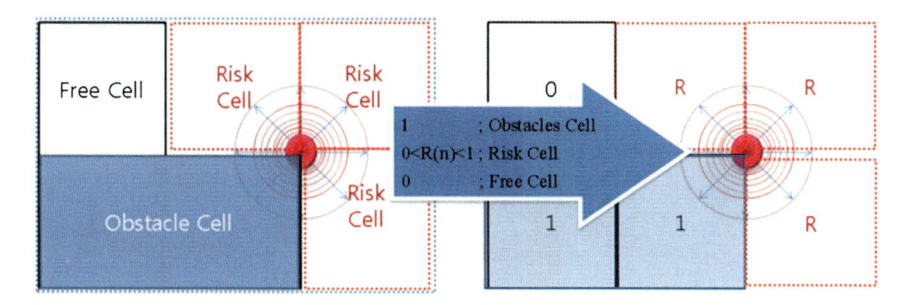

Fig. 12.3 A grid map of including the risk areas for SGPP

Fig. 12.4 Conventional A* path. It is minimal distance path planning without considering the corners and risk

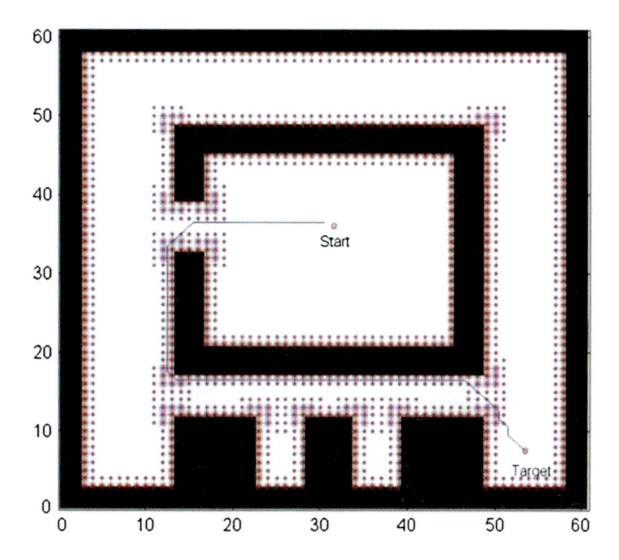

12.4 Simulation Results

A simulation was performed using Matlab. The grid map should include the risk areas to generate a global path, as shown in Fig. 12.3. To generalize the path planning problems in configuration space [1], the kinematic posture of the robot was represented by x, y in every possible position. Free-Cell obstacle Cell 0, 1 and risk cell has a between within $0 < R(n_c) < 1$.

The size of map is 60(Cells) × 60(Cells) and each cell is a square, which is 1 m corresponding to the size of the robot. The evaluation of the suggested algorithm carried out for navigation time and safety. Figures 12.4 and 12.5 compare the two algorithms in a more complex environment. The starting and ending points of the robot are (31, 36) and (53, 7), respectively.

Fig. 12.5 Modified A* path. The distance is not minimal, but it is the SGPP considering the corners and risk

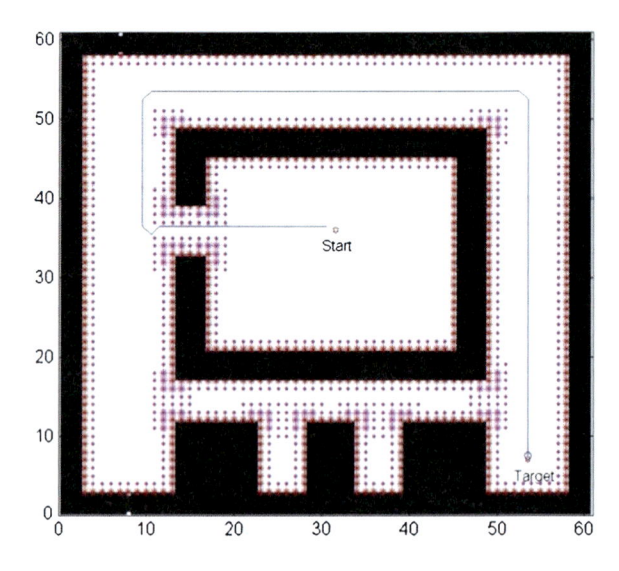

Table 12.1 Comparison of the two algorithms on GSPP

Performance	Algorithm	
	Conventional A*	Modified A*
Safety cost	14.5	1.1
Navigation time (s)	133	130
Length of path (m)	82	127
Execution time (s)	0.70	1.3

The modified A* algorithm enables the robot to turn soundly and find the optimal path. In contrast, the original A* algorithm ignores the risk areas during navigation. The cost of safety is the sum of the risk of the path. The navigation time is defined as

$$Navigation\ time = d \Big/ (1-R)^2 v_{max} \qquad (12.3)$$

where d is the length of the path and R is risk level. v_{max} is the maximum velocity of the mobile robot. The navigation times were determined depending on R and d. The max velocity was assumed to be 1 m per second.

The path costs of the conventional and modified A* algorithm were 82 and 127 m, respectively (Table 12.1). The safety cost and navigation time were 14.5/0.7 and 1.1/1.3 s, respectively. Table 12.1 shows excellent results in distance and execution time. On the other hand, the proposed algorithm showed excellent safety cost and navigation time. The safety cost of the conventional A* algorithm was 14 higher than that of the modified A* algorithms.

12.5 Conclusion

This study suggested a SGPP, which considers the potential risk area. Corners and intersections that the robot cannot see due to the limited field of view, are defined as risk area. The safe path planning was implemented by solving the problems of potential collisions on this global map using the modified A* algorithm. The conventional A* algorithm is excellent in distance and the time costs function but does not consider potential collisions. SGPP showed better performance in terms of the safety cost and navigation time than the conventional A* algorithm.

Acknowledgments This work was supported by the National Research Foundation of Korea (NRF) grant funded by the Korean Government (NRF-2011-0016212) and partially supported by the MOTIE (Ministry of Trade, Industry & Energy), Korea, under the Robotics-Specialized Education Consortium for Graduates support program supervised by the NIPA (National IT Industry Promotion Agency) (H1502-13-1001).

References

1. Siegwart R, Nourbakhsh IR (2004) Introduction to autonomous mobile robots. MIT Press, Cambridge
2. Giesbrecht J (2004) Global path planning for unmanned ground vehicles. Technical memorandum, DRDC Suffield TM 2004-272
3. Latombe J-C (1991) Potential field methods. In: Robot motion planning. Kluwer, Boston, pp 296–355
4. Hart PE (1968) A formal basis for the heuristic determination of minimum cost paths in graphs_A star, vol ssc-4, no. 2
5. Sadou M, Polotski V, Cohen P (2004) Occlusions in obstacle detection for safe navigation. In: IEEE intelligent vehicle symposium, pp 716–721
6. Anthony SM (2004) Occlusions in obstacle detection for safe navigation. In: IEEE intelligent vehicles symposium 2004, pp 716–721, 14–17
7. Choi M (2008) Safe and high speed navigation of a patrol robot in occluded dynamic obstacles. In: Proceedings of the 17th IFAC world congress, vol 17, part 1
8. Freeman H, Davis LS (1977) A corner-finding algorithm for chain-coded curves. IEEE Trans Comput 26(3):297–303

Chapter 13
Review of Research in the Area of Agriculture Mobile Robots

Sami Salama Hussen Hajjaj and Khairul Salleh Mohamed Sahari

Abstract Rise in demand for food worldwide has led the agriculture industry to shift towards *Corporate Agriculture*; major conglomerates operate huge lands with *Precision Farming*; maximizing outputs and utilization of resources while reduce waste and costs. This efficiency required the introduction of Automation and Robotics in Agriculture, which led to great technological challenges. This in turn sparked interest in research in the area of *Agriculture Mobile Robots* (*AMRs*). This paper reviews research in this area for the last 5 years; it highlights examples of robots *already* in action in fields around the world, identifies trends and important sub-topics, and finally outlines the direction of where research in Mobile Agriculture Robots is heading.

Keywords Agriculture mobile robots · Mobile robot navigation in agriculture · Image processing for agriculture · Tractor-trailer stability · Agribots

13.1 The Need for Agriculture Mobile Robots

Food security is a global concern, governments worldwide are facing unprecedented rise in demand for food, human population is growing rapidly, but land and agriculture resources remain the same, and in some cases it's even shrinking.

This led to the rise of *Corporate Agriculture*; major conglomerates operate large lands with *precision agriculture* philosophy; which focuses on maximizing output and productivity, while fully utilizing the available land resources [1].

Just as with any industry, production efficiency requires automation and elimination of human factor issues, which lead to great interest in incorporating

S. S. H. Hajjaj (✉) · K. S. M. Sahari
Centre for Advanced Mechatronics and Robotics, Universiti Tenaga Nasional, Jalan IKRAM-UNITEN, 43000 Kajang, Malaysia
e-mail: ssalama@uniten.edu.my

robotics into the agriculture industry. This has led to many technological and engineering challenges, which in turn led to increase in interest in research in the area of *Agriculture Mobile Robots* and *Precision Autonomous Farming*.

This research area has produced many successful stories; over the last five years, robots have been introduced—successfully—into the agriculture industry worldwide; robots tap rubber and work the land in India, pick citrus fruits in the United States, harvest tomatoes and pick strawberries in China [2–6].

Others include solar powered robots picking dates from palm trees in Saudi Arabia, robots transplanting rice and transporting other plants in Japan, and robot inspecting irrigation water in Thailand [7–11]. Mobile robots also harvested white asparagus and other crops in Greece [12].

13.1.1 An Overview of the Research on Agriculture Mobile Robots (AMRs)

Upon looking up the term *agriculture mobile robots* on IEEE-indexed research journals of the last five years, one can see that 42 % of this research focused on agriculture-specific navigation and control, while 33 % focused on agriculture-specific sensor technology and image processing, 21 % on robot stability and handling rough terrain, and finally only 5 % focused on hardware/software background systems needed to incorporate robotics in agriculture applications. This is shown in Fig. 13.1.

This paper reviews research completed in each of these areas, identifies any subgroups (if they existed) based on the topic of interest, and outlines important findings and breakthroughs in each area, all in the attempt to identify the direction research in Agriculture Mobile Robots is heading.

13.2 Agriculture-Specific Navigation and Control

Navigation is the challenge of preprogramming the robot with an algorithm to allow it to identify its surroundings, avoid any obstacles, and reach its target. This problem becomes even more challenging when the robot is outdoors, in a very dynamic environment, as in the case of agriculture robots.

13.2.1 Stand-Alone Frameworks

Mingjun et al. [13] tested their algorithm; the *Conditional Random Fields based Near-to-Far Perception* framework (CRFNFP), on agriculture robots and found

Fig. 13.1 Breakdown of research in agriculture mobile robots (*AMRs*)

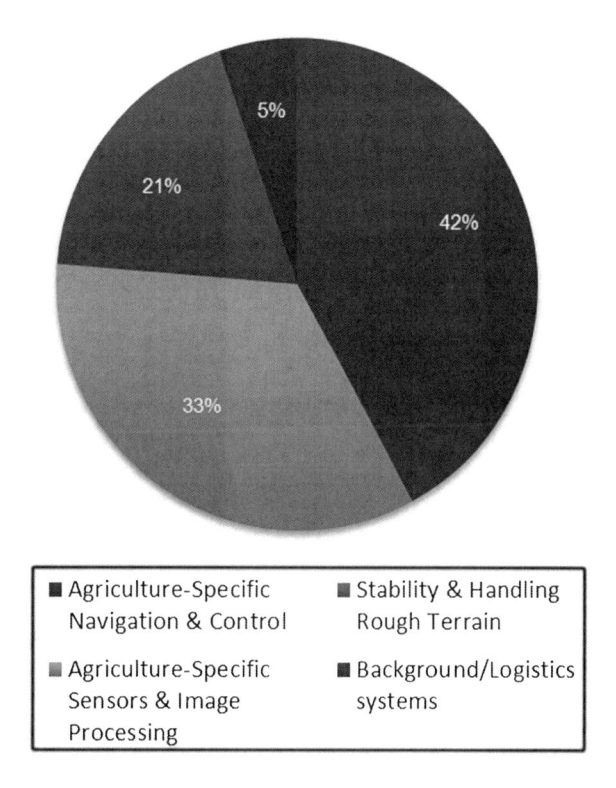

- Agriculture-Specific Navigation & Control
- Agriculture-Specific Sensors & Image Processing
- Stability & Handling Rough Terrain
- Background/Logistics systems

that it increased perception range better than the traditional local-map-based navigation.

Patino et al. addressed the same challenge by applying *Adaptive Critic Designs* (ACDs). Their simulation results showed that ACD-based intelligent controller did indeed learn to guide the AUV through a set of points autonomously [14].

Cheng, J. et al. applied the *Rapid-exploring Random Trees* (Dual RTT) algorithm to control a tractor-trailer mobile robot in an unknown environment. Results of simulation and prototypes showed that the tractor-trailer was able to quickly plan a feasible path under complex and unknown environments [15].

Long et al. developed a complete hardware/software agriculture robot system that combined *AHRS* and *DGPS-RTK* navigation algorithms, four ultrasonic sensors, and a C8051F040 microcontroller. The system showed high repeatability and accuracy [16].

Hansen et al. applied the *derivative free filters* method, which fuses odometry and gyro data with the surrounding fruit trees, which were modeled with lines created using a 2D laser scanner. Kalmtool toolbox (of Matlab®) was used for easy switching between different filters without changing the base structure of the system [17].

Piyathilaka and Munasinghe developed a vision-based outdoor localization system entirely of off-the-shelf components to guide a mobile robot in small agricultural field, and were able to control a two wheel tractor in the same field [18].

13.2.2 Artificial Intelligence Related

Cheng, F. et al. applied the traditional *artificial potential field* method due to its simplicity; simulation was used to model an apple growing and picking environment. Their results showed that productivity of apple picking robot was improved [19].

Yan-hong, D. et al. solved the problem of optimizing the drive servo PID controller of wheeled robots by first identifying the drive system by BP neural network, then optimizing the controller by the genetic algorithm based on the afore results [20].

Pazderski and Kozlowski applied the *Transverse Function* to control a multi-body vehicle which consists of a unicycle-like tractor with three trailers. Their algorithm was based on input transformation of an open-loop error dynamics [21].

Cariou et al. separated the navigation algorithm into three parts, steering, speed control, and reversing, and combined them to control the vehicle. Their robots were able to reverse automatically to connect the next reference track [22].

13.2.3 Fuzzy Controllers Related

de Sousa et al. combined simple fuzzy and non-fuzzy behaviors. Their strategy was to coordinate behaviors that operate in different stages on the robotic architecture to perform navigation actions. The outcome reflected on the feasibility of the approach [23].

Borrero et al. developed a fuzzy controller using a multivariable plant which incorporated simplified linear model of the lateral dynamics of a vehicle whose input were the linear combination of the rear and front steering angles [24].

Prema et al. developed fuzzy based micro-controller for their agriculture robot that was able to plug the land, plant seeds, and sense the soil moisture. Their robot was controlled remotely over the internet using LABVIEW [25].

13.2.4 GPS Localization Related

Jun, C. used inner information collected by inner sensors to control straight-line motion of an agriculture mobile robot pulling a tractor for a short distance. Results showed precision of position to be higher, allowing the robot to control itself momentarily even when GPS and visual signal where unstable or missing [26].

Jun Zhou, W. achieved maneuverability over rough terrains by individually actuating each wheel. He also combined machine vision, differential GPS, and an inertial measurement unit for input gathering. However, he also reported that precise control is difficult due to the over-constrained nature of the actuation [27].

Hamid et al. utilized a combination of GPS for navigation and Sonar for localization to develop a low cost agriculture mobile robot that could navigate through the desired waypoint and at the same time apply the obstacle avoidance rules [28].

13.3 Agriculture-Specific Sensor Technology and Image Processing

Agriculture robots need to know that the crops they working on are ripe enough for collection. They also need to identify obstacles and vegetation in the field. For all that to work, they need a robust sensor and image processing system.

13.3.1 Image Based Algorithms for Agriculture Robots

Between 2009 and 2012, Lulio et al. worked on several projects related to computer vision for agriculture robots. They developed a platform for image processing and segmentation through an omnidirectional vision system to agricultural mobile robots.

They first combined JSEG-based computational methods with BP-ANN for image classification, characterization, and recognition. Then they combined it with a customized BP multilayer perception (BP-MLP) algorithm in Matlab/Octave environments, as well as structured heuristics methods in Simulink environment [29–32].

Palipana et al. developed a semi-autonomous robot able to localize itself, based on wireless sensor network combined with the use of Zigbee protocol. Also, an extended Kalman filter and a fuzzy inference system were applied to filter the initial training data. Results showed the robot to be accurate within the range of 2–10 m [33].

Zhao C. J. and Jiang G. Q. developed a simple and a robust vision based mobile robot algorithm. Firstly, images captured were processed to obtain quasi navigation baseline, from which the navigation line was extracted to control the motion of the robot. They reported that the system was simple to develop and implement [34, 35].

13.3.2 Effects of Variance in Illumination

To overcome the influence of shadows on path recognition, Bo et al. developed a navigation method based on SOM neural networks for image processing. It worked as results showed the robot was able to navigate in the shadows accurately [36].

Rajendra et al. studied the impact of illumination intensity on area detection for a strawberries picking robot. They concluded that because of its shape and distance from the light source, illumination variation was almost negligible [37].

Morshidi et al. used BP neural Networks algorithms to segment color images from RGB (red, green, blue) color space and projected them into HSV (hue, saturation, value), to provide data points insensitive to illumination changes. Results showed the effectiveness of the algorithm as images were segmented reliably with less blobs [38].

Xiang et al. tackled the same problem, and they tested three algorithms of image segmentation; R-G segmentation, normalized R-G segmentation, and band ratio segmentation. Results showed that R-G segmentation was not adequate to counter the effects of illumination variance, while the other two were effective [39].

13.3.3 Other Agriculture-Specific Image Related Research

Joycy and Prabavathy developed an algorithm for image segmentation that helps agriculture robot identify the three items; green plants, soil, and sky, the main items found in the field. Their algorithm depended on thresholding approaches combined with adjusting the supervised fuzzy clustering [40].

Suzukiy et al. focused on the issue of human face detection and tracking, which is very helpful for agriculture robots as it would make them more human friendly. Their algorithm detected faces by focusing on skin color and eyes features [41].

13.4 Robot Stability and Handling Rough Terrain

In order to make agriculture mobile robots reliable in the field, the Robots must be able to handle rough terrain. In most cases, robots would have to work on surfaces that are uneven, covered with vegetation, contain animals and insects. The wheels of the robots could slip, or even get tangled in the soil. Finally, robots have to be sturdy enough to handle the unforgiving weather conditions.

13.4.1 Effect of Slip on Navigation in a Rough Terrain

Matveev et al. [42] tackled the problem path tracking when the robot is subjected to wheel slips by dividing the guidance into two specific laws, one for pure sliding controller, and other for a combined slip and motion controller.

Wang Y, et al. studied the same problem, but instead decided to incorporate nonlinear friction, center-of-gravity shifts, and load changes, which are prevalent under slip motion. To do so they proposed a digital acceleration control algorithm to compensate for nonlinear friction [43].

13.4.2 Stability of Tractor Trailer Motion

In agriculture applications, mobile robots would have to toe trailers, and control the stability of the trailers, by controlling the motion of the robot, is very important. Morales et al. [44] analyzed the effect of towing a single-axle trailer on static tip-over stability for field mobile robots on slopes.

They proposed an algorithm, the *Altered Supporting Polygons* (*ASP*), for both tractor and trailer. These ASPs were based on a force-torque static equilibrium analysis that reshaped the corresponding ground contact supporting polygons. That allowed them to tip-over stability for each unit by projecting its center of gravity onto its ASP, even when both bodies have different inclinations [44].

Zhe Leng and Minor proposed a two-tier controller that directly controls the curvature of the trailer's trajectory. It allowed the control input to be more directly related to path specification and handles path curvature discontinuity better. Results demonstrated good performance on modest side-slope [45].

This problem is further complicated if the motion is in reverse. Due to the instability of the reverse motion of the tractor-trailer mobile robot, Cheng J. constructed and utilized a fuzzy controller via line-of-sight. Simulation showed that the backward curve path tracking can be carried out successfully [46].

13.4.3 Tree Climbing Robots

Not all crops are within man's reach, so Guan, Y. et al. tackled this problem by developing biped climbing robot, or *Climbot*. Inspired by Inchworms, their robot consisted of five 1-DoF joint modules connected in series and two special grippers mounted at the ends. Results showed that not only Climbot was able to climb a variety of media, but also to grasp and manipulate objects [47].

Lun and Xu developed another tree climbing robot; *Treebot*. Their climbing algorithm is based on the use of minimal sensing resources. Once again inspired by inchworms, the algorithm reconstructed the shape of a tree by the use of tactile sensors. Then, with an efficient non-holonomic motion planning strategy, Treebot identified the optimal climbing path before climbing it [48].

13.5 Support/Logistics Systems Needed for Agriculture Robots

Incorporating robotics in the agriculture industry is a challenging undertaking; resources would need to be dedicated for a successful implementation.

Koshi et al. proposed the *Greenhouse Partner Robot System* which is agricultural support system in a greenhouse by cooperation between humans and robots, focusing mainly on harvesting and pest control. Their system consisted of Greenhouse Partner Robots, which was a four wheeled cart that ran autonomously and a human supervisor [49].

In this system, the supervisor controlled the Greenhouse Partner Robots without interference and the Greenhouse Partner Robots operated in accordance with supervisor's commands. The experimental results using two Greenhouse Partner Robots illustrated the validity of the system [49].

Ali and Oudheusden proposed an integer linear programming formulation to improve the utilization of agricultural vehicles during the crop harvesting process. Crops were harvested by combined harvesters. The harvested product is transferred to a tractor trailer every time the combine harvester's storage capacity is reached [50].

The proposed planning method specified optimal routes and interactions for the agricultural vehicles in the field. The planning model was based on the minimum-cost network flow problem and it minimized non-productivity [50].

13.6 Discussion

One can notice that when it comes to research in agriculture mobile robots, the main focus of research was on agriculture-specific navigation and control. Secondly, research focused on agriculture-specific image processing. These are summarized in Table 13.1.

The total number of literature reported is 50. From the table, one can also see that there were other sub-topics that are worth noting; the problem of tractor-trailer stability consisted of 10 % of the overall research. Similarly, the problem of variance in illumination, or the effect of shades on the quality of image processing of plants and crops also consisted of 10 % of overall research.

Finally, it is also worth noting that nearly 15 % of the research was on stand-alone frameworks, such as the *Conditional Random Fields based Near-to-Far Perception* framework (CRFNFP), and similar frameworks.

13.7 Conclusions

Table 13.1 can also give us an insight into where research in agriculture mobile robots is heading; agriculture specific navigation and image processing is primary focus and interest of researchers. Getting the robot to know its way around the field, and identifying its surroundings and the ripeness of the crops seems to be the main focus.

It is also worth noting that some other issues do interest researchers as well; the problem of tractor-trailer stability, the problem of variance of illumination (shades), seems to be of interest.

Table 13.1 Summary of research in the area of agriculure mobile robots (AMRs)

Category and sub-categories	Percentage of overall research (%)
Agriculture-specific navigation and control	42
Stand-alone frameworks	15
Artificial neural network related	10
Fuzzy controllers related	8
GPS localization related	8
Agriculture-specific sensor technology and image processing	33
Image based algorithms for agriculture robots	18
Effects of variance in illumination	10
Others	5
Robot stability and handling rough terrain	21
Effect of slip on navigation in a rough terrain	5
Stability of tractor-trailer motion	10
Tree climbing robots	5
Support/logistics systems needed for agriculture robots	5

The introduction of automation and robotics in the agriculture is expected to continue to grow. This means that research for efficient and cost effective agriculture mobile robots is expected to continue to grow and expand as well.

References

1. Eaton R et al (2008) Precision guidance of agricultural tractors for autonomous farming. In: 2nd annual IEEE systems conference, pp 1–8
2. Simon S (2010) Autonomous navigation in rubber plantations. In: 2nd international conference on machine learning and computing, pp 309–312
3. Gollakota A, Srinivas M (2011) Agribot—a multipurpose agricultural robot. In: IEEE India conference (INDICON), pp 1–4
4. Aloisio C et al (2012) Next generation image guided citrus fruit picker. In: IEEE international conference on technologies for practical robot applications (TePRA), pp 37–41
5. Jun W et al (2012) Design and co-simulation for tomato harvesting robots. In: 31st Chinese control conference (CCC), pp 5105–5108
6. Qingchun F et al (2012) Study on strawberry robotic harvesting system. In: IEEE international conference on computer science and automation engineering (CSAE), vol 1. pp 320–324
7. Shukla A, Jibhakate S (2011) Design and implementation of real time pollution free autonomous vehicle for harvesting on VI platform. In: International conference on electronics computer technology (ICECT), vol 1. pp 335–339
8. Aljanoobi A et al (2010) A setup of mobile robotic unit for fruit harvesting. In: 19th international workshop on robotics in Alpe-Adria-Danube region (RAAD), pp 105–108
9. Tamaki K et al (2009) A rice transplanting robot contributing to credible food safety system. In: IEEE workshop on advanced robotics and its social impacts, pp 78–79
10. Sakai S et al (2007) Robust control systems of a heavy material handling agricultural robot: a case study for initial cost problem. IEEE Trans Control Syst Technol 15(6):1038–1048

11. Ruangwiset A, Higashino S (2012) Development of an UAV for water surface survey using video images. In: IEEE/SICE international symposium on system ntegration (SII), pp 144–147
12. Chatzimichali AP et al (2009) Design of an advanced prototype robot for white asparagus harvesting. In: IEEE international conference on advanced intelligent mechatronics, pp 887–892
13. Mingjun W et al (2012) Study on long-range navigation behavior of agricultural robots. In: International conference on computing, measurement, control and sensor network, pp 409–412
14. Patino HD et al (2009) Adaptive critic designs-based autonomous unmanned vehicles navigation: application to robotic farm vehicles. In: IEEE symposium on adaptive dynamic programming and reinforcement learning, pp 233–237
15. Cheng J et al (2012) Motion planning algorithm for tractor-trailer mobile robot in unknown environment. In: 8th international conference on natural computation (ICNC), pp 1050–1055
16. Long Y et al (2011) A system for fruit tree canopy characters measuring based on CAN-bus. In: International conference on intelligent computation technology and automation, vol 2. pp 15–21
17. Hansen S et al (2011) Orchard navigation using derivative free Kalman filtering. In: American control conference (ACC), pp 4679–4684
18. Piyathilaka L, Munasinghe R (2011) Vision-only outdoor localization of two-wheel tractor for autonomous operation in agricultural fields. In: 6th IEEE international conference on industrial and information systems (ICIIS), pp 358–363
19. Cheng F (2012) Apple picking robot obstacle avoidance based on the improved artificial potential field method. In: 5th IEEE international conference on advanced computational intelligence, pp 18–20
20. Yan-hong D (2011) PID controller optimization of mobile robot servo system. In: IEEE 2nd international conference on computing, control and industrial engineering (CCIE), vol 1. pp 235–237
21. Pazderski D, Kozlowski K (2012) Control of a unicycle-like robot with three on-axle trailers using transverse function approach. In: IEEE/RSJ international conference on intelligent robots and systems (IROS), pp 395–401
22. Cariou C et al (2009) Motion planner and lateral-longitudinal controllers for autonomous maneuvers of a farm vehicle in headland. In: IEEE/RSJ international conference on intelligent robots and systems, pp 5782–5787
23. de Sousa RV et al (2011) A methodology for coordinating primitive fuzzy behaviors to guide mobile agricultural robots. In: 9th international conference on control and automation, pp 280–285
24. Borrero G et al (2012) Fuzzy control strategy for the adjustment of front steering angle of a 4WSD agricultural mobile robot. In: 7th Colombian computing congress (CCC), pp 1–6
25. Prema K et al (2012) Online control of remote operated agricultural robot using fuzzy controller and virtual instrumentation. In: International conference on advances in engineering, science and management (ICAESM), pp 196–201
26. Jun C (2011) On-tracking control of agricultural mobile robot based on inner information. In: International conference on computer distributed control and intelligent environmental monitoring (CDCIEM), pp 103–106
27. Jun Zhou W (2008) Coordinating control for an agricultural vehicle with individual wheel speeds and steering angles. Control Syst IEEE 28(5):21–24
28. Hamid MHA et al (2009) Navigation of mobile robot using global positioning system (GPS) and obstacle avoidance system with commanded loop daisy chaining application method. In: 5th international colloquium on signal processing & its applications, pp 176–181
29. Lulio LC et al (2009) JSEG-based image segmentation in computer vision for agricultural mobile robot navigation. In: IEEE international symposium on computational intelligence in robotics and automation, pp 240–245

30. Lulio LC et al (2010) ANN statistical image recognition method for computer vision in agricultural mobile robot navigation. In: International conference on mechatronics and automation, pp 279–283
31. Lulio LC et al (2010) Pattern recognition structured heuristics methods for image processing in mobile robot navigation. In: IEEE/RSJ international conference on intelligent robots and systems (IROS), pp 4970–4975
32. Lulio LC et al (2012) Cognitive-merged statistical pattern recognition method for image processing in mobile robot navigation. In: Robotics symposium and latin American robotics symposium (SBR-LARS), pp 279–283
33. Palipana S et al (2012) Localization of a mobile robot using ZigBee based optimization techniques. In: 6th IEEE international conference on information and automation for sustainability, pp 215–220
34. Zhao C, Jiang G (2010) Baseline detection and matching to vision-based navigation of agricultural robot. In: International conference on wavelet analysis and pattern recognition, pp 44–48
35. Jiang G, Zhao C (2010) A vision system based crop rows for agricultural mobile robot. In: International conference on computer application and system modeling, vol 11. pp 142–145
36. Zhao B et al (2010) Path recognition method of agricultural wheeled-mobile robot in shadow environment. In: International conference on digital ecosystems and technologies, vol 1. pp 284–287
37. Rajendra P et al (2011) Shading compensation methods for robots to harvest strawberries in tabletop culture. In: IEEE/SICE international symposium on system integration, pp 172–177
38. Morshidi MA et al (2008) Color segmentation using multi layer neural network and the HSV color space. In: International conference on computer and communication engineering, pp 1335–1339
39. Xiang R et al (2011) Research on image segmentation methods of tomato in natural conditions. In: 4th international congress on image and signal processing (CISP), vol 3. pp 1268–1272
40. Joycy JS, Prabavathy K (2012) Survey on automatic segmentation of relevant textures in agricultural images. In: International conference on advances in engineering, science and management (ICAESM), pp 26–32
41. Suzukiy S et al (2009) A human tracking mobile-robot with face detection. In: 35th IEEE annual conferecne on industrial electronics, pp 4217–4222
42. Matveev AS et al (2010) Mixed nonlinear-sliding mode control of an unmanned farm tractor in the presence of sliding. In: 11th international conference on control automation robotics & vision, pp 927–932
43. Wang Y et al (2011) Car-like mobile robot oriented digital acceleration control method. In: International conference on mechatronics and automation, pp 1491–1497
44. Morales J et al (2013) Static tip-over stability analysis for a robotic vehicle with a single-axle trailer on slopes based on altered supporting polygons. IEEE/ASME Trans Mechatron 18(2):697–705
45. Zhe L, Minor M (2011) A tractor-trailer backing control for path following with side-slope compensation. In: IEEE international conference on robotics and automation (ICRA), pp 2386–2391
46. Cheng J (2011) Curve path tracking control for tractor-trailer mobile robot. In: 8th international conference on fuzzy systems and knowledge discovery (FSKD), vol 1. pp 502–506
47. Guan Y et al (2011) Climbot: a modular bio-inspired biped climbing robot. In: IEEE international conference on intelligent robots and systems, pp 1473–1478
48. Lun T, Xu Y (2011) Climbing strategy for a flexible tree climbing robot—treebot. IEEE Trans Robot 27(6):1107–1117
49. Koshi K et al (2010) Greenhouse partner system. In: 41st international symposium on robotics, pp 1–8
50. Ali O, Van Oudheusden D (2009) Logistics planning for agriculture vehicles. In: IEEE international conference on industrial engineering and engineering management, pp 311–314

Chapter 14
Design and Development of Small and Simple Dual Robot for Inspecting Pipe with Perpendicular Entry

Nur Shahida Roslin, Adzly Anuar, Khairul Salleh Mohamed Sahari and Mohd Nazrin Afzan Abu Bakar

Abstract This paper presents the development of a small and simple dual robot system for pipe inspection activity. The objective is to propose small and simple dual robot in navigating perpendicular entry with sudden change in pipe diameter. Sudden change in diameter is referred to the drastic difference in size of the entry (80 mm in diameter) and the pipe (250 mm in diameter and above). The robot consists of two rectangular-shaped modules linked with a connector. The size of the body is 38 mm (width) × 55 mm (length) × 44 mm (height). Each module is driven by two micro metal gear motors. It uses caterpillar base for locomotion to provide larger surface contact to the pipe wall. The 2 degree-of-freedom (DoF) connector allows the robot to bend in pitch and yaw axes without any drive force required. Preliminary test result shows its ability to navigate through the expected geometry.

Keywords Dual robot · Pipe inspection robot · 2 DoF connector · Crawler

14.1 Introduction

Exploration of in-pipe robots is growing with numerous researches has been carried out from the past two decades, due to high demand especially in power plant industry, oil and gas industry and sewage industry [1]. These in-pipe robots

---

N. S. Roslin (✉) · A. Anuar · K. S. M. Sahari · M. N. A. A. Bakar
Centre for Advanced Mechatronics and Robotics, Universiti Tenaga Nasional, Jalan IKRAM-UNITEN 43000 Kajang, Selangor, Malaysia
e-mail: edaroslin@hotmail.com

A. Anuar
e-mail: adzly@uniten.edu.my

K. S. M. Sahari
e-mail: khairuls@uniten.edu.my

are developed for many reasons such as maintenance work, cleaning and inspection. The main function of the robot will influence its type. There are seven common types of in-pipe robot such as wheeled, caterpillar, screw, legged, pipe inspection gauge (PIG), wall press and inchworm type and three hybrid types such as wheeled wall press, caterpillar wall press and wheeled wall pressing screw type [2]. Most of the researchers are enthusiastic in designing wall press type robot for overcoming perpendicular pipe. An autonomous mircorobotic modular robot is one type of wheeled wall press robot which is developed for inspection and exploration of pipes range between 50 and 70 mm [3]. A miniature pipe inspection robot has been invented for pipe diameter range between 64 and 125 mm [4]. A caterpillar based wall press robot is developed for 80–100 mm pipelines [5]. Another pipe inspection robot is developed for larger pipe diameter which is between 125 and 180 mm [6]. There is also robot with wheeled wall pressing type with helical motion that is developed for 125 mm pipe with curves [7]. All of these robots are tested in one dimension pipe per testing. None of them have been tested for perpendicular entry with sudden diameter change. The objective of this research is to design and develop a robot that can maneuver perpendicular entry pipe with sudden diameter change.

14.2 Design Requirement

In this section, the design requirement for the robot based on the nature of the boiler headers is discussed. Figure 14.1 shows the perpendicular entry pipe with sudden diameter change. The entry of the pipe is 80 mm whereas the pipe diameters are 200, 250, and 300 mm. The increase of pipe diameter compared to entry diameter will increase the complexity of the design requirement.

14.3 Mechanical Design

14.3.1 Structure of Small and Simple Dual Robot

This small and simple device consists of two module track based robot connected with a linkage system. The rectangular shape is made of aluminum alloy. This material provides anti-corrosion behavior to the robot and also protects the motors in the body.

Modular Design. Two modules of robot are needed as this feature can avoid motion singularity issue especially when overcoming perpendicular entry. Motion singularity usually happened when one or more wheels of the robot lose their contact with the wall surface. Figure 14.2 shows the front and back robot module. Table 14.1 shows the robot specifications.

14 Design and Development of Small and Simple Dual Robot

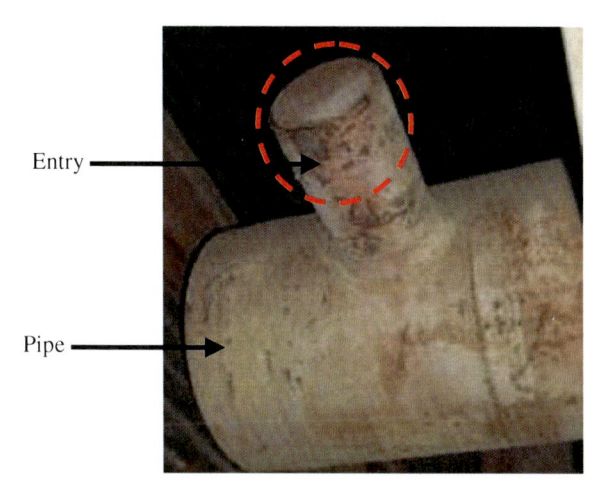

Fig. 14.1 Perpendicular entry pipe with sudden diameter change (*Source* TNB Research Sdn. Bhd)

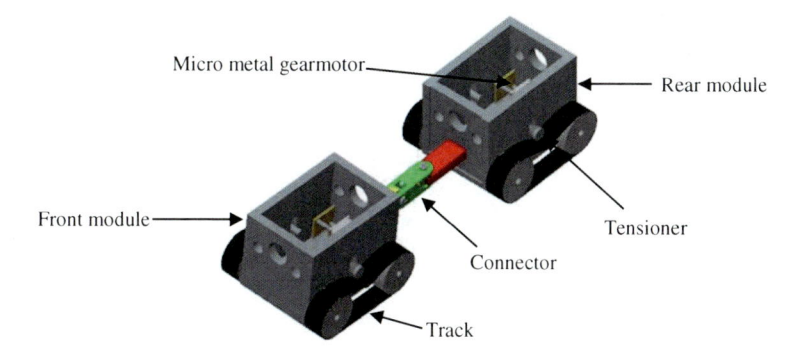

Fig. 14.2 Modular robot design

Table 14.1 Robot specifications

Parameter	Unit	Specification
Size per module	mm	38 (W) × 55 (L) × 44 (H)
Total length, L	mm	165
Wheel diameter, D	mm	21
Total mass, M	g	390
Maximum speed, V	ms^{-1}	0.186
Maximum slope angle, θ	deg	40

Traction Design. Tracked-wheel is the best solution in providing more contact area especially in a condition with uneven surface. Pulley is used as the wheel to reduce the tendency of the track from slip as it has tooth that matched with the

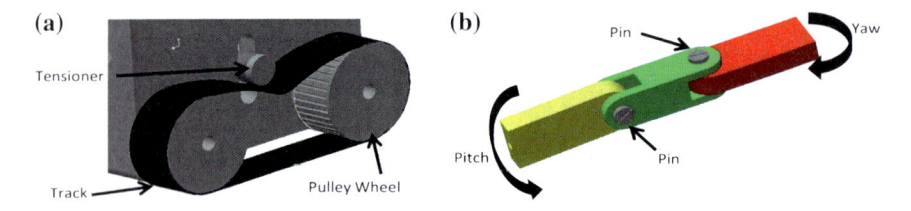

Fig. 14.3 a Traction design. b Linkage connector

track. A tensioner has been introduced in this design to provide reliability of the track from slipping. Figure 14.3a shows the traction design that helps the robot to overcome obstacles.

Connector Module. Another unique feature of this robot is its connector. The connector comprises of three links that give 2 DoF. This mechanism provides flexibility to the robot to bend in pitch and yaw axes passively as the robot negotiates the sudden diameter change. Figure 14.3b shows the robot connector.

14.3.2 Moving Principle of Small and Simple Dual Robot

Each of robot modules is driven by two micro metal gear motors. These motors are installed at two different axes, front and rear to reduce space consumption and provide more space in the robot body. However, this dual-axis motor can still provide differential drive characteristic which enables it to turn. As mentioned in Sect. 14.3.1, rear module will provide additional force by pushing the front module when entering perpendicular pipe. The front module will pull the rear module to enter perpendicular pipe. This pull and push behavior is essential in preventing motion singularity problem.

14.4 Analysis

14.4.1 Motor Calculation

Motor calculation analysis is done to select an appropriate motor. Using data from Table 14.1, the torque required is calculated based on the equation below:

$$T_{req} = \frac{(a + g \sin \theta) \times M \times r}{N} \times \text{eff}_m \times \text{eff}_g \times SF \qquad (14.1)$$

where a is the robot acceleration preliminary assumption of 0.015 ms^{-2}, N is the number of motor used, and eff$_m$ and eff$_g$ are motor and gearhead efficiency respectively with assumption value of 50 %. Using safety factor of two, the

Fig. 14.4 Slope inclination testing

calculation gives 53 mNm of torque requirement. Thus, micro metal gearmotor with a gear ratio of 75:1 is chosen to drive the robot. This motor has fulfilled the design requirement which is small, lightweight and has a maximum output torque of about 63 mNm. Estimated maximum velocity of the robot can be calculated based on the motor specification.

$$V = \frac{N_G \times \pi \times D}{60} \qquad (14.2)$$

Based on the parameters, the maximum velocity is 0.186 ms^{-1}.

14.5 Experimental Testing and Result

14.5.1 Slope Inclination Testing

This testing is carried out to measure the maximum slope angle that the dual robot can travel. The robot has to travel down the slope and turn left or right when it reached down. Figure 14.4 shows the sequence of the robot when travels down the slope ramp. This dual robot is placed at 57 mm height (Fig. 14.4a). This dual robot is able to travel on the slope up to 40° tilt angle (Fig. 14.4b). Front module turns left (Fig. 14.4c). The rear module is also able to turn left or right with the guidance from the front module (Fig. 14.4d).

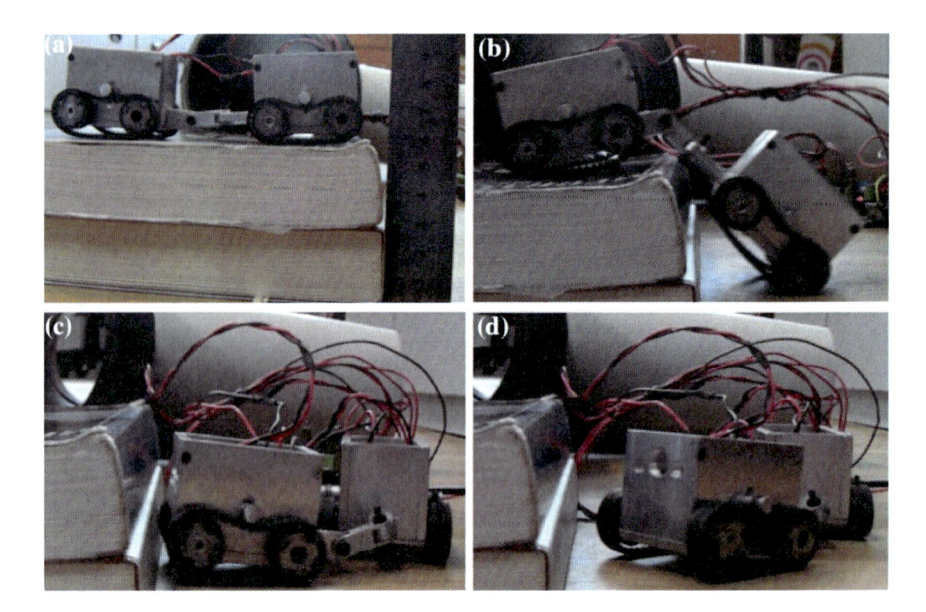

Fig. 14.5 Steep ramp testing

14.5.2 Steep Ramp Testing

This testing is carried out to investigate the ability of the robot to run down a steep ramp and then turn left or right. Steep ramp testing is the substitution to the sudden change diameter testing. Figure 14.5 shows the sequence of the robot when travels down the steep ramp. The front module travels down the ramp (Fig. 14.5a). It is observed that the connector plays its role to guide the rear module with pitch and yaw rotations (Fig. 14.5b and c). Both modules able to turn right after reached down the ramp (Fig. 14.5d).

14.6 Conclusion

The small and simple design dual robot presented in this paper has fulfilled its environment requirement to navigate through perpendicular entry with sudden diameter change. Individual drive modules help the robot to avoid motion singularity problem with push and pull behavior. This mechanism enables the robot to tackle sudden change diameter. To this end, experimental results show that this design is capable for that motion. From the analysis and experimental testing, it is proven that this micro metal gear motor used is already sufficient to drive the robot. This design has opened the opportunity to explore more on in-pipe robot structure other than wall press-based robot.

Acknowledgments The authors wish to thank Ministry of Higher Education Malaysia for supporting and funding the research through its ERGS research grant (ERGS/1/2011/TK/UNITEN/02/11).

References

1. Baharuddin MZ, Saad JM, Anuar A, Ismail IN, Hassan Basri NM, Roslin NS, Kaja Mohideen SS, Abdul Jalal MF, Mohamed Sahari KS (2012) Robot for boiler header inspection "LS-01". Procedia Eng 41:1483–1489
2. Roslin NS, Anuar A, Abdul Jalal MF, Mohamed Sahari KS (2012) A review: hybrid locomotion of in-pipe inspection robot. Procedia Eng 41:1456–1462
3. Tătar O, Cirebea C, Alutei A, Mândru D (2010) The design of adaptable indoor pipeline inspection robots. In: IEEE international conference on automation quality and testing robotics, vol 1. pp 1–4
4. Dertien E, Stramigioli S, Pulles K (2011) Development of an inspection robot for small diameter gas distribution mains. In: IEEE international conference on robotics and automation, pp 5044–5049
5. Kwon YS, Yi BJ (2012) Design and motion planning of a two-module collaborative indoor pipeline inspection robot. IEEE Trans Robot 28(3):681–696
6. Lim HO, Oki T (2009) Development of pipe inspection robot. In: ICROS-SICE international joint conference, pp 5717–5721
7. Kakogawa A, Ma S (2010) Mobility of an in-pipe robot with screw drive mechanism inside curved pipes. In: IEEE international conference of robotics and bimimetics, pp 1530–1535

Part II
Vision, Image and Signal Processing

Chapter 15
Near Infrared Face Recognition: A Comparison of Moment-Based Approaches

Sajad Farokhi, Siti Mariyam Shamsuddin, U. U. Sheikh and Jan Flusser

Abstract Moment based methods have evolved into a powerful tool for face recognition applications. In this paper, a comparative study on moments based feature extraction methods in terms of their capability to recognize facial images with different challenges is done to evaluate the performance of different type of moments. The moments include Geometric moments (GM's), Zernike moments (ZM's), Pseudo-Zernike moments (PZM's) and Wavelet moments (WM's). Experiments conducted on CASIA NIR database showed that Zernike moments outperformed other moment-based methods for facial images with different challenges such as facial expressions, head pose and noise.

Keywords Moments · Near infrared · Comparative study · Face recognition

---

S. Farokhi (✉) · S. M. Shamsuddin
Soft Computing Research Group, Faculty of Computing, Universiti Teknologi Malaysia, 81310 Johor Bahru, Johor, Malaysia
e-mail: fsajad2@live.utm.my

S. M. Shamsuddin
e-mail: mariyam@live.utm.my

U. U. Sheikh
Faculty of Electrical Engineering, Universiti Teknologi Malaysia, 81310 Johor Bahru, Johor, Malaysia
e-mail: usman@fke.utm.my

J. Flusser
Institute of Information Theory and Automation of the Academy of Sciences of the Czech Republic, 182 08 Prague, Czech Republic
e-mail: flusser@utia.cas.sz

15.1 Introduction

Face recognition beyond visible spectrum has received much more attention in computer vision society to improve the poor performance of the face recognition systems caused by illumination variations. The use of near infrared imagery introduces a new dimension for applications of visible lights for face recognition. Different methods have been introduced in this domain to propose accurate face recognition systems. Some of them can be found in [1, 2].

Moments are scalar quantities used to characterize a function and to capture its significant features. From the mathematical point of view, moments are "projections" of a function onto a polynomial basis [3]. They are powerful feature extractors which have received a lot of attention as pattern features in many applications such as face recognition, hand writing etc. Geometric moments, Zernike moments, Pseudo Zernike moments and Wavelet moments are the most popular moments which have been used in the field of face recognition with good results [4–8].

In this paper we compare the performance of the aforementioned moments in the near infrared domain and evaluate their performance in the presence of different challenges.

The remainder of the paper is organized as follows. In Sects. 15.2 and 15.3, brief reviews of four moments and regularized linear discriminant analysis are discussed. Experimental results and performance analysis are given in Sect. 15.4. The final conclusion is drawn in Sect. 15.5.

15.2 Different Types of Moments

15.2.1 Geometric Moments

Geometric moments (GMs) are widely used in image processing and pattern recognition. The $(p+q)$th geometric moment of an intensity function $f(x, y)$ in the continuous domain is given by:

$$m_{pq} = \int_{-\infty}^{\infty} \int_{-\infty}^{\infty} x^p y^q f(x, y) \, dx \, dy \qquad (15.1)$$

15.2.2 Zernike Moments and Pseudo-Zernike Moments

Zernike moment (ZMs) of order p and repetition q is defined by the following equation:

$$Z_{pq} = \frac{p+1}{\pi} \int_{\theta=0}^{2\pi} \int_{r=0}^{1} V_{pq}^{*}(r,\theta) f(r,\theta) r dr \, d\theta, \quad |r| \leq 1 \tag{15.2}$$

$$p = 0, 1, 2, \ldots, \quad q = -p, -p+2, \ldots, p$$

i.e. the difference $p - |q|$ is always even. The symbol $*$ is the sign of complex conjugate and V_{pq} denotes Zernike polynomial of order p and repetition q which is defined as follows:

$$V_{pq}(r,\theta) = R_{pq}(r) e^{\hat{j}q\theta}, \quad \hat{j} = \sqrt{-1} \tag{15.3}$$

The real-valued radial polynomial R_{pq} is expressed as follows:

$$R_{pq}(r) = \sum_{k=0}^{\frac{p-|q|}{2}} (-1)^k \frac{(p-k)!}{k! \left(\frac{p+|q|}{2} - k\right)! \left(\frac{p-|q|}{2} - k\right)!} r^{p-2k} \tag{15.4}$$

Pseudo-Zernike moments (PZMs) are derived from the ZMs by releasing from the condition $p - |q|$ is even [3]. In other words the Eq. (15.2) is valid for Pseudo-Zernike moments (PZMs) but with $q = -p, -p+1, \ldots, p$. The radial polynomial R_{pq} of PZMs is expressed as follows:

$$R_{pq}(r) = \sum_{k=0}^{p-|q|} (-1)^k \frac{(2p+1-k)!}{k!(p+|q|+1-k)!(p-|q|-k)!} r^{p-k} \tag{15.5}$$

15.2.3 Wavelet Moments

Wavelet moments (WMs) are another type of moments which have been used widely in pattern recognition [9]. The wavelet moments for continuous images are defined as follows:

$$W_{mnq} = \int_{\theta=0}^{2\pi} \int_{r=0}^{1} f(r,\theta) \psi_{mn}(r) r dr \, d\theta, \quad |r| \leq 1 \tag{15.6}$$

where the wavelet function defined along a radial axis in any orientation is denoted as:

$$\psi_{mn}(r) = 2^{m/2} \psi(2^m r - 0.5n) \tag{15.7}$$

The mother wavelet $\psi(r)$ of the cubic B-spline in Gaussian approximation form is given by:

$$\psi(r) = \frac{4a^{n+1}}{\sqrt{2\pi(n+1)}} \sigma_w \cos(2\pi f_0(2r-1)) \exp\left(-\frac{(2r-1)^2}{2\sigma_w^2(n+1)}\right) \quad (15.8)$$

15.3 Regularized Linear Discriminant Analysis

Linear Discriminant Analysis (LDA) is a well-known classification method which tries to find the optimal projection in which the within-class distance of data is minimized and the between-class distance is maximized simultaneously. Accordingly to achieve maximum discrimination, three scatter matrices, i.e., the within-class (S_w), between-class (S_b) and total scatter matrices (S_t) are included in LDA. S_t is a multiple of the sample covariance matrix and is required to be nonsingular which may not hold when the number of features is larger than the number of samples. This is called small sample size (SSS) or undersampled problem which is common in many applications such as face recognition, text mining etc. In the past few decades numerous solutions have been introduced to cope with singularity of S_t and to stabilize the sample covariance matrix estimation. One of the effectively proposed solution is to apply the idea of regularization to improve the classification performance of LDA and guarantee the singularity of S_t which is known as the Regularized Linear Discriminant Analysis (RLDA) [10]. This technique is implemented by adding a constant value $\lambda \succ 0$ (regularization parameter) to the diagonal elements of S_t as follows:

$$\widetilde{S}_t = S_t + \lambda I \quad (15.9)$$

It is easy to show that $\widetilde{S}_t$ is nonsingular. In a nutshell the objective function (optimal transformation) of RLDA is given as follows:

$$\gamma_{opt} = \arg\max_{\gamma} \frac{tr(\gamma^T S_b \gamma)}{tr(\gamma^T \widetilde{S}_t \gamma)} \quad (15.10)$$

where $tr()$ denotes matrix trace and γ is the transformation matrix.

15.4 Experimental Results and Performance Analysis

In this section, the performances of the different types of moments are tested and then comparison study is done to evaluate the performance of the moments. The face images of CASIA NIR database are used in our experiments which are composed of 3,940 NIR face images of 197 people [2]. Since some of the subjects do not include any challenges, for our experiment 100 persons with 10 images per

Table 15.1 Performance comparison of various methods on CASIA database (Mean ± Std-Dev in percentage)

Type of moments	Number of images in the training set		
	2	3	4
GM	78.52 ± 4.69	85.57 ± 3.52	89.24 ± 2.43
ZM	89.41 ± 3.24	92.21 ± 2.71	94.05 ± 1.89
PZM	87.21 ± 3.56	90.41 ± 3.64	93.63 ± 2.32
WM	77.21 ± 4.11	82.21 ± 4.25	88.26 ± 2.05

person which include normal images, images with facial expressions and head pose are selected. Hence the total number of images involved in our experiments is 1,000. The tolerance for head pose is utmost 5°. For image normalization, first face images are aligned by placing the two eyes at fixed position and then they are cropped manually to remove hair and background. Finally they are resized to 64 × 64 pixels with 256 gray levels to decrease the computational complexity. For feature extraction, in the case of Geometric moments, Zernike moments and Pseudo-Zernike moments, we calculate the moments up to order 10. For wavelet moments we choose $m = 0, 1, \ldots, 5$, $n = 0, 1, 2, \ldots, 2^{m+1}$ and $q = 0, 1, \ldots, 5$. As soon as the data are extracted, they are sent to RLDA to decrease the dimension of data and generate salient features. The regularization parameter Eq. (15.9) is set to 0.01 due to the best performance of the system by this value. To classify the testing image, nearest neighbor classifier is used in the proposed algorithm due to its simplicity as well as speed. Among three distance measures of Cosine distance, Euclidean distance and City-block distance, the Euclidean distance is selected as a criterion for nearest neighbor classifier due to the best result of the system by this criterion.

The following sets of experiments are carried out:

- Testing the performance of the system with different challenges such as facial expression and slight head pose.
- Testing the robustness against additive zero mean Gaussian white noise.

15.4.1 Testing the Performance of the System with Different Variations

In this experiment, the performance of the algorithm for face images with different challenges including facial expressions and head pose is evaluated. A random subset with l (= 2, 3, 4) images per person is taken with labels to form the training set and the rest is considered as the testing set. Since we have 1,000 images belonging to 100 persons in our database, the number of images in training set in the case of l (= 2, 3, 4) is 200, 300, and 400 respectively and the number of images in testing set is 800, 700 and 600 accordingly. For each l the average recognition rate along with standard deviations for over 25 splits is calculated and the results are tabulated in Table 15.1.

Table 15.2 Effect of noise on the performance of different methods (Mean ± Std-Dev in percentage)

Algorithm	Signal to noise ratio (SNR) level			
	No noise	21	18	16
GM	85.57 ± 3.52	52.55 ± 2.79	35.25 ± 3.30	26.45 ± 3.94
ZM	92.21 ± 2.71	82.16 ± 2.64	71.75 ± 2.53	61.26 ± 3.38
PZM	90.41 ± 3.64	78.31 ± 2.67	64.93 ± 3.25	56.72 ± 3.75
WM	82.21 ± 4.25	61.47 ± 2.73	46.85 ± 3.43	36.05 ± 3.48

From the experimental results obtained from the CASIA NIR database some observations can be drawn:

- From this data we can see that GM resulted in the lowest value of recognition rate which proves the low discrimination power of GMs in comparison with other moments. The underlying reason is that GMs are nonorthogonal and they are not able to classify facial images accurately.
- Contrary to the results presented in [5], WMs achieve lower recognition accuracy in comparison to ZMs and PZMs. The main reason is that ZMs and PZMs extract global features whereas WMs extract both local and global features which are included in one single factor. Hence the global information provided by the wavelet moments is lost when they are integrated to one single feature vector.

15.4.2 Testing the Robustness Against Additive Zero Mean Gaussian White Noise

In this set of experiments, the robustness of the different moments to additive zero mean Gaussian white noise is tested. Three images per subject are randomly selected for training and the remaining seven images are used for testing. While the training set includes images without noise, the testing set included noisy images with different signal to noise ratio (SNR) level. To simulate a real scenario first noise is added to raw cropped images, and then they are resized to 64 × 64 pixels. Each test is performed 20 times to obtain the average and the results are tabulated in Table 15.2.

From Table 15.2, the following observations can be made:

- The high sensitivity of GMs to noise can be seen in Table 15.2. On the other hand ZMs and PZMs are less sensitive to noise due to the orthogonality of the basis polynomial.
- WMs are more sensitive to noise in comparison with ZMs, PZMs. A possible explanation for this could be the existence of local features in WMs which are highly sensitive to noise.
- In contrast to earlier findings [4], however, ZMs perform better than PZMs in the presence of noise.

15.5 Conclusion

In this paper, we have evaluated four moments based methods: Geometric moments (GMs), Zernike moments (ZMs), Pseudo-Zernike moments (PZMs) and Wavelet moments (WMs). Comparative study of different moments based methods was performed on CASIA NIR database to verify their effectiveness and feasibility. The most significant findings to emerge from this study is that regarding to the discrimination power and noise immunity, ZMs perform the best among other moments and the superiority of PZMs to ZMs in the presence of noise in not true in common. The most important limitation of this study is that we compared only moment-based methods, however, there exist many other approaches which may in some situations perform even better than the best moment-based methods.

Acknowledgments The authors would like to thank Universiti Teknologi Malaysia (UTM) for the support in Research and Development, and Soft Computing Research Group (SCRG) for the inspiration in making this study a success, the Institute of Automation, Chinese Academy of Sciences (CASIA) for providing CASIA NIR database to carry out this experiment and Institute of Information Theory and Automation (UTIA) for providing MATLAB codes. This work is supported by the Ministry of Higher Education (MOHE) under Long Term Research Grant Scheme (LRGS/TD/2011/UTM/ICT/03- 4L805) and the Research Grant No (Q.J130000.2623.08J89). It is also partially supported by the Czech Science Foundation under the grant No. P103/11/1552.

References

1. Zhang B, Zhang L, Zhang D, Shen L (2010) Directional binary code with application to PolyU near-infrared face database. Pattern Recog Lett 31:2337–2344
2. Li SZ, Chu R, Liao S, Zhang L (2007) Illumination invariant face recognition using near-infrared images. IEEE Trans Pattern Anal Mach Intell 29:627–639
3. Flusser J, Suk T, Zitova B (2009) Moments and moment invariants in pattern recognition. Wiley, Chichester
4. Haddadnia J, Ahmadi M, Faez K (2002) An efficient method for recognition of human faces using higher orders pseudo Zernike moment invariant. In: Fifth IEEE international conference on automatic face and gesture recognition. IEEE Press, Washington DC, pp 330–335
5. Zhi R, Ruan Q (2008) A comparative study on region-based moments for facial expression recognition. In: Congress on image and signal processing. IEEE Press, Sanya, pp 600–604
6. Lajevardi SM, Hussain ZM (2010) Higher order orthogonal moments for invariant facial expression recognition. Digit Signal Process 20:1771–1779
7. Farokhi S, Shamsuddin SM, Flusser J, Sheikh UU (2012) Assessment of time-lapse in visible and thermal face recognition. Int J Comput Commun Eng 6:181–186
8. Farokhi S, Shamsuddin SM, Flusser J, Sheikh UU, Khansari M, Kourosh J-K (2013) Rotation and noise invariant near-infrared face recognition by means of Zernike moments and spectral regression discriminant analysis. J Electron Imaging 22:013030
9. Broumandnia A, Shanbehzadeh J (2007) Fast Zernike wavelet moments for Farsi character recognition. Image Vision Comput 25:717–726
10. Ye J, Wang T (2006) Regularized discriminant analysis for high dimensional, low sample size data. In: Proceedings of the 12th ACM SIGKDD international conference on knowledge discovery and data mining. ACM Press, Philadelphia, pp 454–463

Chapter 16
Acceleration of Retinex Algorithm for Image Processing on Android Device Using Renderscript

Duc Phuoc Phat Dat Le, Duc Ngoc Tran, Fawnizu Azmadi Hussin and Mohd Zuki Yusoff

Abstract The popularity and availability of Android devices have motivated researchers to implement their image processing systems on Android mobile platform. Retinex is considered as an effective method to restore the image's original appearance and used as a pre-processing step in many computer vision applications. It would give a lot of benefits to implement Retinex on such portable system and optimize it for real-time performance. This is a challenge because of limited computational power and memory of the portable system. This paper presents an implementation of rsRetinex, an optimized Retinex algorithm by using Renderscript technique. The experimental results show that rsRetinex could gain up to five times speedup when applied to different image resolution.

Keywords Retinex · Image enhancement · Image restoration · Android device Renderscript

16.1 Introduction

Retinex is a very popular and effective method to remove environmental light interferences which is used as a pre-processing step such as contrast enhancement in many image processing systems. It was introduced by Land and McCann [1, 2] and has a lot of updates, improvements by different researchers. The original Land and McCann work [3] described four steps for each iteration of Retinex calculation: ratio, product, reset and average [4]. Based on how the comparison pixels are chosen, the Retinex algorithm can be separated into three classes: path-based

D. Phuoc Phat Dat Le (✉) · D. N. Tran · F. A. Hussin · M. Z. Yusoff
Centre for Intelligent Signal and Imaging Research Laboratory, Electrical and Electronics Engineering Department, Universiti Teknologi Petronas, 31750 Bandar Seri Iskandar, Tronoh, Perak, Malaysia
e-mail: datduc007@ieee.org

Retinex algorithms [2, 3, 5], recursive Retinex algorithm [6], and center/surround Retinex algorithm [7]. Retinex processing of color image can be applied separately in R, G, and B spectral bands and combined together to get the final output image. The consumption cost of Retinex algorithm is very high for large image, which is especially taxing on mobile device. Therefore, this paper proposes a Renderscript acceleration of McCann99 Retinex called rsRetinex to achieve the real-time improvement on Android device. The experiment result shows that rsRetinex can achieve better real-time performance compared with Java code while achieving at least 4–5 times performance gain.

The organization of the rest of this paper is as follows: Sect. 16.2 presents a brief introduction of Retinex algorithm and Renderscript programming model. Section 16.3 presents detail of the rsRetinex implementation. The experimental results are reported and discussed in Sect. 16.4. Finally the conclusion is given in Sect. 16.5.

16.2 Background

16.2.1 Retinex Algorithm

The idea of the Retinex was conceived by Land [2] as a model of lightness and color perception of human vision. The basic form of multi-scale Retinex (MSR) [8–10] is given by

$$R_i(x_1, x_2) = \sum_{k=1}^{n} W_k (\log I_i(x_1, x_2) - \log[F_k(x_1, x_2) \otimes I_i(x_1, x_2)]) \quad i = 1, \ldots, N$$

(16.1)

where R_i is the output of the MSR process at the coordinates (x_1, x_2), the sub index i represents the ith spectral band, N is the number of spectral band, $N = 1$ for grayscale images and $N = 3$ for a typical color image. So in normal case, i = 3 I_i is the input image in the spectral band ith, F_k represents the Kth surround function, W_k are the weights associated with F_k, k is the number of surround functions, or scales, and $\otimes$ represents the convolution operator. The basic principles of Retinex are: color is obtain from three lightness computed separately for each of color channels; the ratios of intensities from neighboring locations are assumed to be illumination invariant; lightness in a given channel is computed over large regions based on combining evidence from local ratios; the location with the highest lightness in each channel is assumed to have 100 % reflectance within that channel's band. On the other hand, the Retinex computation of lightness at a given pixel is the comparison of the pixel's value to that of other pixels. One version of Retinex is McCann99 Retinex [11], which has been given standardized definitions in terms of Matlab code. This paper proposes rsRetinex, an optimized implementation of McCann99 in Android OS.

16.2.2 Renderscript Programming Model

Renderscript is a new programming framework and API for Android. Renderscript code is called during the runtime in native level and communicates with the Android Virtual Machine (VM) Dalvik, so the way a Renderscript application is set up is different from a pure VM application. An application that uses Renderscript is still a traditional Java application and runs in the VM, but the developers write the script and indicate which parts of the program requires to be run with Renderscript. The developers do not need to target the multiple architectures because it is platform independent which means that any device that supports Renderscript will run the code properly regardless of the architecture. Renderscript gives the applications the ability to run operations with automatic parallelization across all available processor cores.

16.3 rsRetinex Implementation

McCann99 has three main functions as shown in the following pseudo code segment. They are ComputeLayer, ImageDownResolution and CompareWithNeighbor.

```
program Retinex_McCann99(){
  Global OPE RRE Maximum
  nLayers = ComputeLayer (nrows, ncols);
  Maximum = max(InputImage(:)) ;
  OP = Maximum*ones([nrows, ncols]);
  For each layer, do
    RR = ImageDownResolution(InputImage,2^(nLayers - layer));
    RRE = Padded RR with zero
    OPE = Padded OP with zero
  For iter = 1:nIterations
    CompareWithNeighbor();
  End
  NP = OPE(2:(end-1), 2:(end-1));
}
```

In rsRetinex implementation, two main functions have been chosen to enhance the performance of the McCann99 Retinex algorithm in Renderscript: ImageDownResolution and CompareWithNeighbor because of their expensive pixel-wise computation. ImageDownResolution function averages image data to make a multi-resolution pyramid from the input image. Each pixel in the input image will be divided into many blocks; the number of pixels depends on the block size. After that the entire pixel in one block will be averaged to make a new pixel in the new layers. This task is pixel-wise, and the number of calculation steps depend on the block size, the smaller the block size is the more calculation steps to be performed.

Fig. 16.1 The ratio-product-reset-average operation

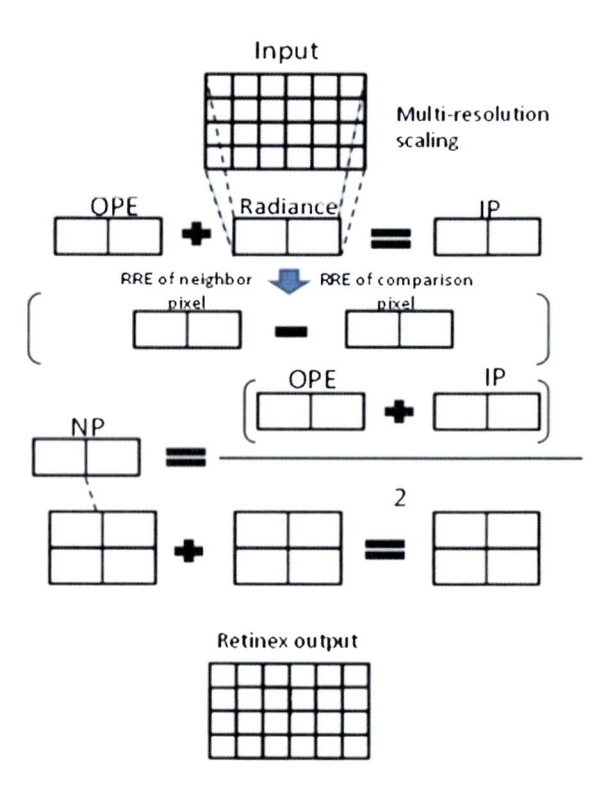

By applying Renderscript technique to this function, the number of calculation steps will keep the same regardless the block size, because each pixel is calculated simultaneously. At each level, the new product will be calculated by visiting each of its eight neighboring pixels in clockwise order, which is implemented by the CompareWithNeighbor function. This operation calculates the New Product (NP) for the input pixel, for example (x', y'). Figure 16.1 illustrates the ratio-product-reset-average operation. This operation starts with pixel (x, y) using the Old Product (OP). The entire OP is initialized with the maximum value for that wave band. It starts with subtracting the neighbor's log luminance called ratio step and then adding the result to the OP (the product step). If the radiance ratio, called Intermediate Product (IP), between pixel (x, y) and pixel (x', y') is greater than the maximum, the IP will be reset to the maximum (called reset step) and then averaged with the previous NP (called average step).

For each pixel in the current layer, the ratio-product-reset-average operation is performed on eight neighboring pixels in clockwise order. Because the Renderscript layer cannot access the memory directly, it needs to be allocated a memory location as a container to contain four data plane: OP, Radiance (RR), IP and NP by using allocation variables. In this implementation, only two allocation variables are used, inAlloc and outAlloc in order to reduce memory utilization. There are two main components in Renderscript code, init() and root(). The init() function is

16 Acceleration of Retinex Algorithm for Image Processing   141

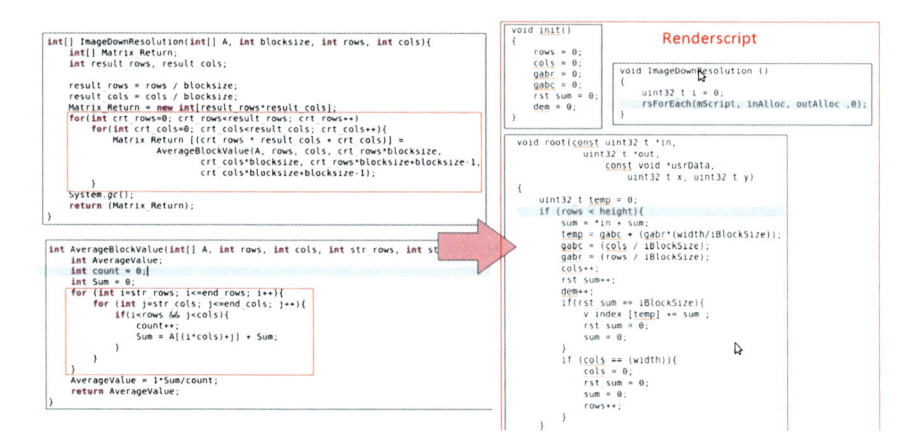

Fig. 16.2 Code fragment of ImageDownResolution function

to initialize the local variable for the script. The same calculations to be performed on each pixel of the input image are described in root() function. When the rsForEach() function is invoked and passed the inAlloc as the input, Renderscript will automatically perform the calculation to all the pixels in the input and parallelize all the tasks across all available processor core, while the Java version only can perform sequentially by using a loop. Figure 16.2 shows the difference of ImageDownResolution implementations between the Java version with many loops and the loop-free Renderscript.

16.4 Experimental Result

To evaluate the performance of the proposed method, an experiment is performed on emulator, which settings are: ARM Cortex A9 1 GHz, RAM 1 GB, to compare the executing time between two implementations: java-only and Renderscript.

Figure 16.3 shows that with the 968 × 648 bitmap image, ImageDownResolution and CompareWithNeighbor function are invoked four times during execution. As shown in Fig. 16.3, the rsImageDownResolution gained around 1.6–6 times speedup for each run. The total execution time of rsImageDownResolution and jvImageDownResolution are 6,163 ms and 24,636 ms, which shows that rsImageDownResolution function gained almost four times. rsCompareWithNeighbor function gained around 4–25 times for each run. When applying rsRetinex to different image size, the system also achieved the significant performance improvement as shown in Table 16.1. With the 968 × 648 and 1,024 × 768 resolution, system speedups are about three and five times in real-time performance.

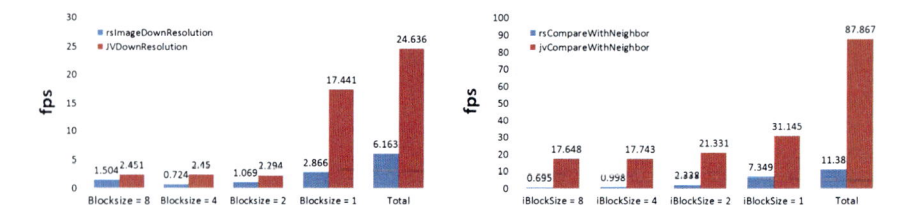

Fig. 16.3 Execution time of two parts in the Retinex algorithm applying to 968 × 648 bitmap image

Table 16.1 Experiment result with respect to image resolution

Image resolution	Java code execution time	rsRetinex execution time	Gain performance
284 × 177	18	4	4.5
290 × 439	43	10	4.3
340 × 148	17	3.8	4.4
968 × 648	150	32	4.6
1,024 × 768	446	121	3.68

16.5 Conclusion and Discussion

Building an image processing application on Android device is very useful because of mobility, portability and low cost. However the portable devices typically provide limited computational power, and memory capacity. In this paper, we have presented an optimization method using Renderscript. Two parts of McCann99 Retinex have been chosen for Renderscript implementation. The experimental results show that Renderscript can accelerate the McCann99 Retinex algorithm and rsRetinex gained at least 3–5 times speedup compared with the java-only implementation.

References

1. Ebner M (2007) Color constancy. Wiley, England, pp 143–153
2. Land E (1964) The Retinex. Amer Scient 52(2):247–264
3. Land EH, McCann JJ (1971) Lightness and Retinex theory. J Opt Soc Am 61:1–11
4. Land EH (1977) The Retinex theory of color vision. Sci Am 237:108–129
5. Land EH (1986) Recent advances in Retinex theory. Vision Res 26(1):7–21
6. Frankle J, McCann J (1983) Method and apparatus for lightness imaging. US Patent number 4384336
7. Land EH (1986) An alternative technique for the computation of the designator in the Retinex theory of color vision. Proc Nat Acad Sci 83:3078–3080

8. Wang Y-K, Huang W-B (2011) Acceleration of an improved Retinex algorithm. Computer vision and pattern recognition workshops (CVPRW). In: Conference on 2011 IEEE Computer Society, 20–25 June 2011, pp 72–77
9. Rahman Z, Jobson D, Woodell GA (1996) Multiscale retinex for color image enhancement. In: Proceedings of IEEE international conference on image processing, pp 1003–1006
10. Jobson DJ, Rahman Z, Woodell GA (1997) A multi-scale Retinex for bridging the gap between color images and the human observation of scenes. IEEE Trans Image Process Spec Issue Color Process 6(7):965–976
11. Funt B, Ciurea F, McCann J (2004) Retinex in Matlab. J Electron Imaging 13:48–57

Chapter 17
Image Transmission Over Noisy Wireless Channels Using HQAM and Median Filter

Md. Abdul Kader, Farid Ghani and R. Badlishah Ahmad

Abstract This paper considers the use of unequal error protection and median filtering for transmission of images over poor wireless channels usually encountered over cellular mobile networks. Hierarchical Quadrature Amplitude Modulation (HQAM) that provides Unequal Error Protection (UEP) to the transmitted image data is used at the transmitter. In HQAM, non-uniform signal constellation is used to provide different degrees of protection to the significant and non-significant bits in the image data at lower channel Signal to Noise Ratio (SNR). Median filter is employed at the receiver to remove the impulsive noise present in the received image. Simulation results show that the use of HQAM and median filtering provides a gain of PSNR over the more conventional Quadrature Amplitude Modulation (QAM).

Keywords Unequal error protection · HQAM · Impulse noise · PSNR · Wireless channel

17.1 Introduction

The rapid growth of wireless communications has a demand for robust multimedia transmission with better quality, coverage, and more power and bandwidth efficiency. The restriction of the wireless communication channels like

---

Md. A. Kader (✉) · F. Ghani · R. B. Ahmad
School of Computer and Communication Engineering, Universiti Malaysia Perlis (UniMAP), 02000 Kuala Perlis, Perlis, Malaysia
e-mail: kdr2k4@yahoo.com

F. Ghani
e-mail: faridghani@unimap.edu.my

R. B. Ahmad
e-mail: badli@unimap.edu.my

limited bandwidth increases the demand for more reliable image communication system that does not consume more bandwidth for achieving better image quality. Furthermore, real-time applications are important because it is wide spread. Implementation of reliable image communication with the real-time requirement needs low bit-rate, low power, low delay and low complexity maintaining good image quality. The solution of the problem of reliable real time transmission of high-quality images through wireless communication channels can be achieved by using Unequal Error Protection (UEP) techniques. Thus providing reliable image transmission with acceptable data rates and in the same time maintains the good image quality. UEP consists of two main parts; the first part is called data partitioning and the second part is applying UEP. Concerning with data partitioning, it is used to provide different levels of significance (importance) for the image source, so the data is classified into important and less important [1].

Concerning UEP, the degree of error-protection level will be assigned according to the significance of the image data; as the importance of data increases, the level of protection increases. UEP is performed in the channel coding stage. In particular, the more important data will be highly protected using high channel coding level (or rate) while the less important data will be channel coded by lower protection level.

This paper proposes the use of an asymmetric modulation method known as Hierarchical Quadrature Amplitude Modulation (HQAM) for the transmission of images over erroneous wireless channels. This is a simple and efficient approach in which non-uniform signal-constellation is used to give different degrees of protection to the transmitted bits. The advantage of this method is that different degrees of protection are achieved without an increase in bandwidth in contrast to channel coding that increases the data rate by adding redundancy to the transmitted signal [2–4]. Median filtering is used to increase the PSNR by removing the impulse noise from the received image. PSNR analysis is presented for different values of the modulation parameter and performance comparison is carried out through computer simulation using 16-HQAM with gray image as test image.

The paper is organized as follows: In Sect. 17.2, general model of the image transmission system is briefly described. Section 17.3 considers the effects of noise and distortion in image transmission. In Sect. 17.4 overview of HQAM is given. Based on computer simulation results, the performance of HQAM is considered in Sect. 17.5.

17.2 Model of Image Transmission System

The essentials of the image transmission system considered here are shown in Fig. 17.1. The source encoder encodes the source image using appropriate image compression technique. For the protection of coded image in Fig. 17.1 channel

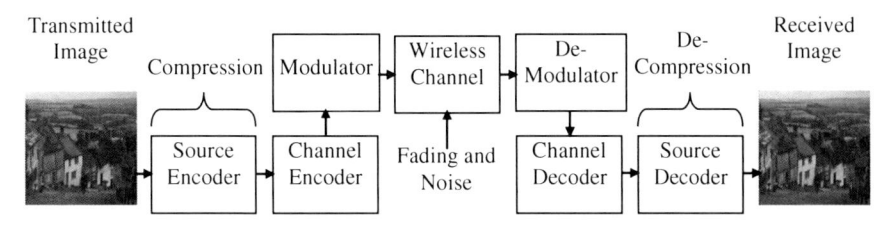

Fig. 17.1 Model of image transmission system

encoder adds redundancy to the coded image by using appropriate channel coding technique. Modulator modulates the coded image and transmits through the wireless channel. QAM is invariably used as the modulation technique [5]. The channel introduces noise and distortion to the transmitted image. The demodulator receives the image data with error and demodulates it. After channel decoding, the coded image is decompressed.

There are two major constraints in transmission of images over wireless channels. First, there are fluctuations in the channel bandwidth for this reason the image data must be compressed. Second, there is a high probability of channel error for this reason the image data must be protected from errors in order to maintain image quality.

17.3 Effects of Noise and Distortion in Image Transmission

The most common type of noise that is encountered in the image transmission system is the Salt and pepper noise (special case of impulse noise) [6, 7], where a certain percentage of individual pixels in digital image is randomly digitized into two extreme intensities (maximum and minimum). Faulty memory locations or transmission through erroneous channels can result in the received image being corrupted with this type of noise [8]. The effect of salt and pepper noise on the received image is shown in Fig. 17.2a.

Distortion in images occurs when errors cause local variations in image scale and coordinate location of the image pixels. The distortion is more severe when the errors occur in critical (significant/high priority) bits of the received signal. For errors in non-critical bits (sign bits/low priority bits and refinement bits/low priority bits) the distortion is not that severe. This can be seen from Fig. 17.2b, c and d. Figure 17.2b corresponds to the case when the errors are in significant bits while the Fig. 17.2c and d result when the error is in sign or refinement bits. Thus in image transmission it is necessary to give more protection to significant bits as compared to the insignificant bits rather than giving equal protection to all the bits.

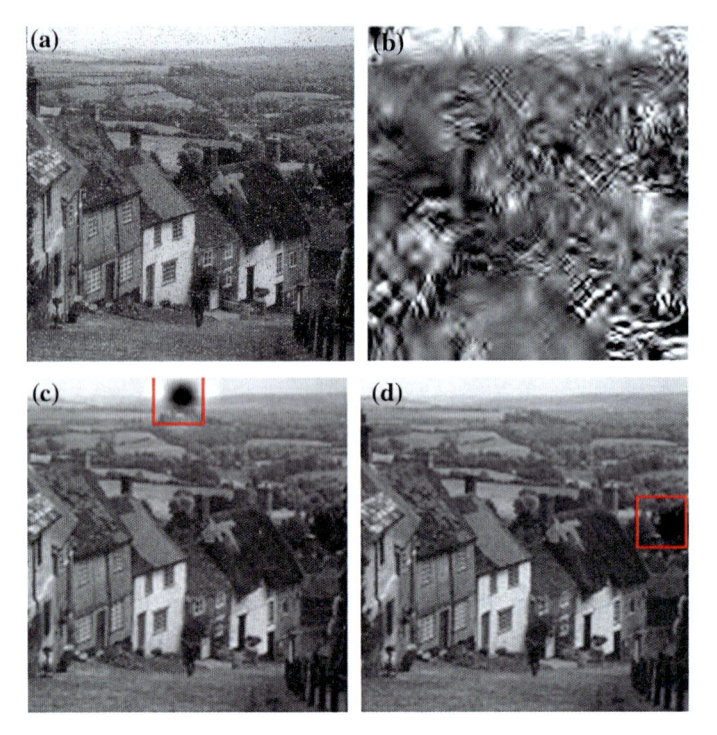

Fig. 17.2 **a** Effects of Salt and pepper noise, **b** Effects of error in critical bits, **c** Effects of error in sign bit (non-critical bit) and **d** Effects of error in refinement bits

17.4 Hierarchical Quadrature Amplitude Modulation

Hierarchical Quadrature Amplitude Modulation (HQAM) is the more spectrally efficient and dc-free modulation scheme [9]. It provides the different degree of protection to the transmitted data bits, in which the high priority (HP) data bits are mapped to the most significant bits (MSB), and the low priority (LP) data bits are mapped to the least significant bits (LSB) of the modulation constellation points. Using HQAM will, therefore, a result in improved image quality specially at low channel SNR conditions, since the highly sensitive HP data bits are mapped to the MSBs with low bit error rate (BER) in HQAM. For the sake of simplicity only 16-HQAM is considered in this paper.

In Hierarchical QAM, it is possible to give higher protection to the most important data (significant bits) by changing the value of the modulation parameter α. α is the ratio of the distance b between quadrants to the distance c between the points within a quadrant [10]. In the constellation diagram referring to Fig. 17.3a the modulation parameter $\alpha = b/c$. For a given transmitted signal power the sum of b and c should remain constant. The value of the modulation parameter should

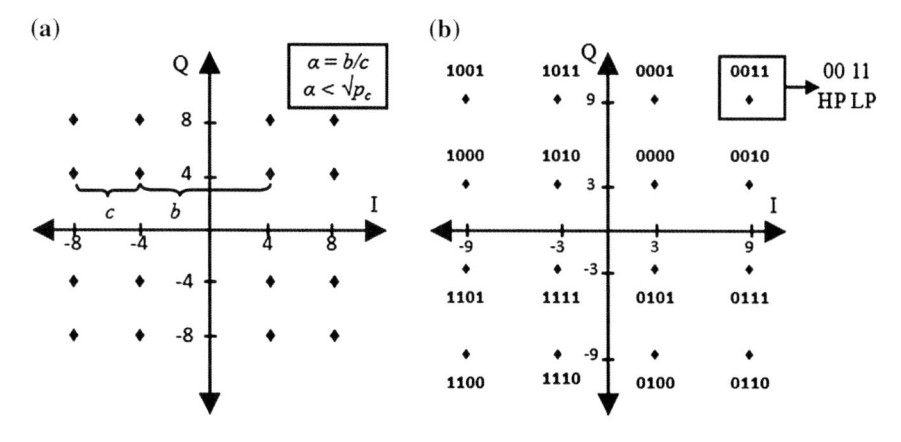

Fig. 17.3 Constellation diagram of 16-HQAM for **a** α = 2 and **b** α = 1

not exceed the square root of the carrier power pc. Otherwise, the constellation points of the same quadrant will overlap. When α = 1, i.e. b = c then HQAM results in QAM as can be seen from Fig. 17.3b.

Referring to Fig. 17.3b, the two MSB represents the HP bits which have lower BER than the two LSB bits. LSB bits are representing LP bits. It can be seen that the four symbols in every quadrant have the same HP bits but different LP bits; this is also called constellation overlapping and ensure that the HP bits to be transmitted correctly [5].

17.5 Simulations and Results

A simulation for gray scale image transmission and reception was carried out using 16-HQAM and Median filter is used for suppression of impulsive noise in the image to improve the PSNR. The flow diagram of the proposed simulation is shown in Fig. 17.4 and simulated using MATLAB 7.8 (R2009a). This simulation transmits and receives grayscale images through the wireless erroneous channel and calculates the PSNR of the image.

Figure 17.5 shows the PSNR versus modulation parameter curves for before and after median filtering (over AWGN channel) using hierarchical 16-QAM for α = 1–5. There are two different images (Peppers and Lena) used to be transmitted. It is seen that as expected by increasing the value of α, PSNR of the received images increase when the median filtering is used. However, HP (significant) data of the images are highly protected for the larger value of α for lower channel SNR and also improve the image quality (higher PSNR).

Gray scale test images shown in Fig. 17.6a and d were used to obtain the performance of 16-HQAM using MATLAB. Different values of the modulation

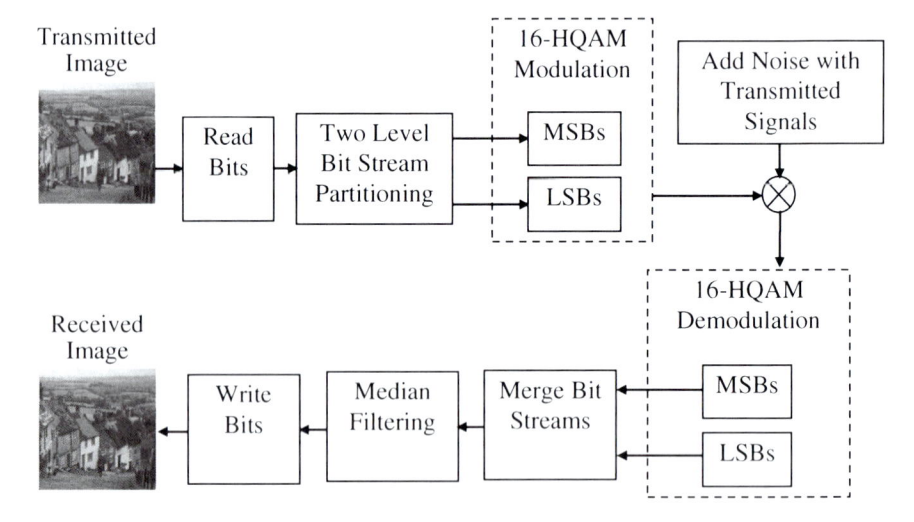

Fig. 17.4 Simulation of image transmission and reception using 16-HQAM and median filtering

Fig. 17.5 PSNR versus modulation parameter, $\alpha = 1\text{--}5$

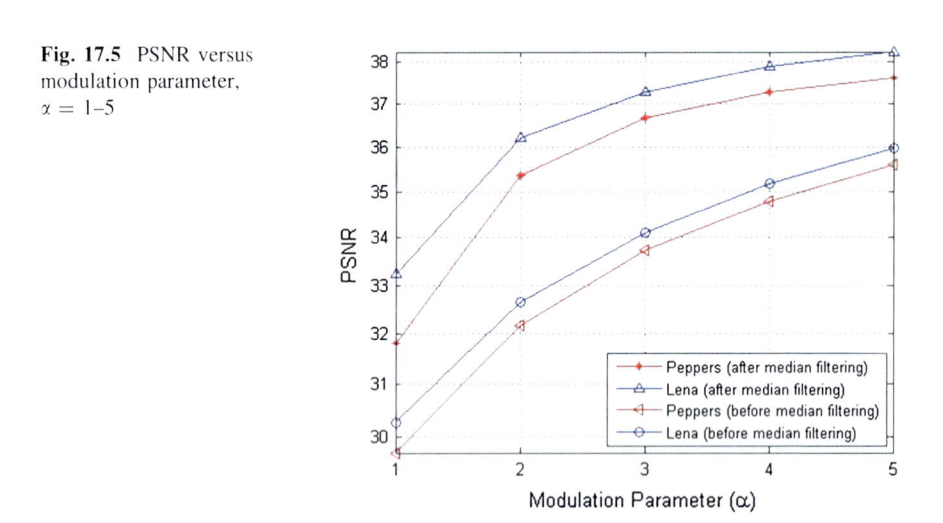

parameter α were used to see the improvement in the distortion of the received image with α before and after median filtering. The SNR was kept at a fixed value of 14 dB. The results are shown in Fig. 17.6b and e before filtering and Fig. 17.6c and f after filtering when $\alpha = 5$. It has been seen that the best-quality images are received by proposed simulation when $\alpha = 5$ and SNR = 14 dB by using the median filter.

Fig. 17.6 a Original image (*Peppers*). **b** Received image (*Peppers-before filtering*). **c** Received image (*Peppers-after filtering*). **d** Original image (*Lena*). **e** Received image (*Lena-before filtering*). **f** Received image (*Lena-after filtering*)

17.6 Conclusion

In image transmission over erroneous wireless channels, if the sensitive data of transmitted image is corrupted then it is difficult to recover that image for low channel SNR. Hierarchical QAM (HQAM) overcomes this disadvantage by providing more protection to the higher-priority bits and less protection to the lower-priority bits of the image data. To increase the PSNR of reconstructed median filter is used. Thus HQAM being a simple technique provides a more efficient means of image transmission over the erroneous wireless channel with low SNR.

References

1. El-Khamy SE, Qabel IMA (2008) Robust image transmission through wireless communication channels using unequal error protection based on adaptive image segmentation. In: National radio science conference 2008 (NRSC 2008), 18–20 March 2008, pp 1–9
2. Barmada B, Gandhi MM, Jones EV, Ghanbari M (2005) Prioritized transmission of data partitioned H. 264 video with HQAM. IEEE Sig Process Lett 12(8):577–580
3. Ghandi MM, Ghanbari M (2006) Layered H. 264 video transmission with hierarchical QAM. Elsevier J Visual Commun Image Represent, Special issue on H.264/AVC 17(2):451–466
4. Toh KKV, Ibrahim H, Mahyuddin MN (2008) Salt-and-pepper noise detection and reduction using fuzzy switching median filter. IEEE Trans Consumer Electron 54(4):1956–1961
5. Hanzo L, Webb W, Keller T (2000) Basic QAM techniques. In: Single and multi carrier quadrature amplitude modulation: principles and applications for personal communications, WLANs and broadcasting. Wiley, London
6. Zhang S, Karim MA (2002) A new impulse detector for switching median filter. IEEE Signal Process Lett 9(11):360–363
7. Pratt WK (1978) "Image enhancement" in digital image processing, 4th edn. Wiley, New York
8. Barmada B, Jones EV (2002) Adaptive mapping and priority assignment for OFDM. In: Third international conference on 3G mobile communication technologies, 2002. (Conf. Publ. No. 489), May 2002, pp 495–499
9. Mirabbasi S, Martin K (2000) Hierarchical QAM: a spectrally efficient dc-free modulation scheme. Commun Mag IEEE 38(11):140–146
10. Kader MA, Farid G, Ahmad RB (2011) Development and performance of hierarchical quadrature amplitude modulation (HQAM) for image transmission over wireless channels. In: 3rd international conference on computational intelligence, modeling and simulation (CIMSim-2011), Malaysia, 20–22 Sept 2011, pp 272–232

Chapter 18
Frog Identification System Based on Local Means K-Nearest Neighbors with Fuzzy Distance Weighting

Haryati Jaafar, Dzati Athiar Ramli, Bakhtiar Affendi Rosdi and Shahriza Shahrudin

Abstract Frog identification based on the vocalization becomes important for biological research and environmental monitoring. As a result, different types of feature extractions and classifiers have been employed. Yet, the k-nearest neighbor (kNN) is one of the popular classifiers and has been applied in various applications. This paper proposes an improvement of kNN in order to evaluate the accuracy of frog sound identification. The recorded sounds of 12 frog species obtained in Malaysia forest have been segmented using short time energy and short time average zero crossing rate while the features are extracted by mel frequency cepstrum coefficient. Finally, a proposed classifier based on local means kNN and fuzzy distance weighting have been employed to identify the frog species. Comparison of the system performances based on kNN, local means kNN and the proposed classifier i.e. fuzzy kNN with manual segmentation and automatic segmentation is evaluated. The results show the proposed classifier outperforms the baseline classifier with accuracy of 94.67 % and 98.33 % for manual and automatic segmentation, respectively.

Keywords Frog identification · kNN · Local means KNN · Fuzzy kNN · Distance weighting

---

H. Jaafar (✉) · D. A. Ramli · B. A. Rosdi
School of Electrical and Electronic Engineering, USM Engineering Campus, 14300, Nibong Tebal, Pulau Pinang, Malaysia
e-mail: haryati_jaafar@yahoo.com

D. A. Ramli
e-mail: dzati@eng.usm.my

B. A. Rosdi
e-mail: eebakhtiar@eng.usm.my

S. Shahrudin
School of Pharmacy Sciences, USM, 11800, Minden, Pulau Pinang, Malaysia
e-mail: shahriza18@usm.my

18.1 Introduction

Frogs are unique creatures that have been living in this planet for more than 250 million years. Over a decade, these amphibians become crucial since their impact of whole ecosystem is great as bio indicators. In addition, their bodies may keep the important key for new discoveries in medical research. The chemical compounds in their skin may provide antimicrobial peptides that used to treat pain and block infections [1]. Commonly, frog relies on their sound to present the presence, behaviors and species. This is because their sound can be received over varying distance that allow and obstructive detection of their existence [2]. Different techniques which involved feature extractions and classifiers have been studied and proposed in order to identify the frog species based on their vocalization automatically [3, 4]. Among of the classifiers, k nearest neighbor (kNN) becomes the most popular nonparametric classifier which has widely been used in pattern classification application and generally archives good result. Nonetheless, this classifier required a large number of training samples to determine desired values of probability of correct classification [5]. Moreover, this classifier suffers from existing outliers particularly in small training sample size situation [6]. Hence, the improvement of kNN has been investigated actively [5, 7–10]. This paper proposes an improvement of kNN by employing local means kNN with fuzzy distance-weighting (LMKNN-FDW). As compared with the previous papers, the distance between query pattern or testing sample and local means vector is assigned using fuzzy algorithm. In addition, the comparative studies with kNN, FKNN and LMKNN are discussed. The various frog sounds in manual segmentation and automatic segmentation based on short time energy (STE) and short time average zero crossing rate (STAZCR) are conducted in this experiments. Consequently, a standard mel frequency ceptrum coefficient (MFCC) is executed as feature extraction in this study. The first objective is to improve kNN classifier by proposing LMKNN-FDW. The second is to compare the performance results between proposed classifier with the baseline classifiers. This paper is outlined as follows. In Sect. 18.2, the methodology of this study is discussed. Section 18.3 describes the proposed classifier in detail and the experimental results are presented in Sect. 18.4 and the conclusion are summarize in Sect. 18.5.

18.2 Methodology

18.2.1 Data Acquisition

All of 12 frogs sound were recorded from locations around Baling and Kulim, Kedah, Malaysia using Sony Stereo IC Recorder ICD-AX412F supported with Sony electret condenser microphone in 32-bit wav files at a sampling frequency of

48 kHz. Consequentially locations were selected based on frog's potential habitat such as next to a swamp, running stream and ponds from 8.00 pm to 12.00 pm.

18.2.2 Syllables Segmentation

The syllables segmentation based on STE and STAZCR were applied where the principle of the techniques is to determine the endpoint of syllable boundaries accurately to detect the syllable signal that has been segmented [11].

18.2.2.1 Short Time Energy

This technique is used to classify voiced and unvoiced parts. The voice part has high energy than unvoiced part due to the periodicity. The STE function is defined by the following expression;

$$E_n = \frac{1}{N} \sum_{m=1}^{N} [x(m)w(n-m)]^2 \qquad (18.1)$$

where E_n is the energy of the sample n of the signal, $x(m)$ is the discrete-time signal and w[m] is a hamming window of size N.

18.2.2.2 Short Time Average Zero Crossing Rate

On the other hands, STAZCR is often used as a part of the front-end processing in automatic speech recognition system. During the frog signal processing, the amplitudes of the unvoiced part normally have higher values and vice-versa. The ZCR is the rate at which signal changes from positive to negative and back and defined as;

$$Z_n = \frac{1}{2N} \sum_{m=1}^{N} |\text{sgn}x(m) - \text{sgn}[x(m-1)]| w(n-m) \qquad (18.2)$$

where

$$\text{sgn}[x(m)] = \begin{cases} 1, & x(m) \geq 0 \\ -1, & x(m) < 0 \end{cases} \qquad (18.3)$$

18.2.3 Feature Extraction

MFCC is selected due to the features are robust to noise which is suitable to be implemented in outdoor environment that contains interference of background

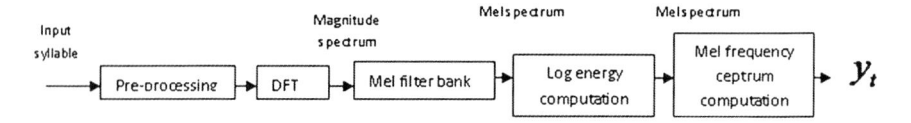

Fig. 18.1 Typical MFCC process

noises such as sound of wind, running water and other animal calls where MFCC processing is shown in Fig. 18.1. There are 12 mel cepstrum coefficients, one log energy coefficient and three delta coefficients per frame have been set in the experiments [12].

18.3 Propose Classifier, LMKNN-FDW

In order to design a simple classifier based on kNN, the following steps are executed. Let $X_i = \{x_n \in R^m\}_{n=1}^{N}$ be a training sample where m is the number of dimensional in feature space, N is the total number of training sample and $y_n \in \{c_1, c_2, \ldots, c_M\}$ denotes the class label for x_n. A query pattern is first determined;

1. Determine the k nearest neighbor from the set X_i for each class y_n by Euclidean distance where $k \leq N$;

$$d\left(x, x_{ij}^{N}\right) = \sqrt{\left(x - x_{ij}^{N}\right)^{T}\left(x - x_{ij}^{N}\right)} \qquad (18.4)$$

2. Search the local mean vector Y_{ik} by applying k nearest neighbor of training sample such that;

$$Y_{ik} = \frac{1}{k}\sum_{j=1}^{k} x_{ij}^{N} \qquad (18.5)$$

3. In the fuzzy method, the testing data value or query pattern is classified by assigning membership values, $U_{ij}(k)$ in particular class based on percentage of neighbors in that class weighted. Hence, by applying Eq. (18.5) in the fuzzy method, the query pattern, x is classified as follows;

$$u_{ij} = \frac{\sum_{j=1}^{k} u_{ij}\left[\frac{1}{\|x-Y_{ik}\|^{2/(m-1)}}\right]}{\sum_{j=1}^{k}\left[\frac{1}{\|x-Y_{ik}\|^{2/(m-1)}}\right]} \qquad (18.6)$$

where m is the scaling factor for fuzzy weight. Note the notation m denote the fuzzy weight of the distance or fuzzy relationship. If value m increases, the neighbors are more evenly weighted. This caused the distance between training and query pattern have less effect on each other and vice versa. In this paper, the value of $m = 2$ is used for the proposed classifier.

Table 18.1 Manual segmentation results

Scientific name	kNN	LMKNN	FKNN	LMKNN-FDW
Hylarana glandulosa	25	24	25	24
Kaloula pulchra	25	22	15	23
Odorrana hossi	11	19	22	22
Polypedates leucomystax	25	25	25	25
Kaloula baleata	25	23	25	24
Philautus mjobergi	24	23	24	22
Phrynoidis aspera	19	25	23	25
Microhyla heymonsi	13	20	18	23
Microhyla butleri	23	21	25	25
Rhacophorus appendiculatus	25	23	25	25
Hylarana labialis	25	25	25	25
Philautus petersi	7	14	16	21
Total	**247**	**264**	**268**	**284**
Percentage (%)	**82.33**	**88**	**89.33**	**94.67**

18.4 Experimental Result

The experiments are implemented in Intel Core i5, 2.1 GHz CPU, 2G RAM and Window 7 operating system. In this experiment, 540 syllables in total have been extracted with 20 syllables are used for training and 25 for testing while the value of $k = 3$ is used for all classifiers. The experiments have then been divided into two techniques of segmentation i.e. manual and automatic. Gold Wave software has been used to segment the samples manually while endpoint detection techniques have been employed for automatic segmentation. The classification accuracy (C_A) is defined as;

$$C_A = \frac{N_C}{N_T} \times 100\,\% \qquad (18.7)$$

where N_c is the number of syllables which are recognized correctly and N_T is the total number of test syllables.

Table 18.1 lists the analytical results of the manual segmentation. The results show that all of the classifiers have been able to identify with more than 80 % of accuracies. By improving the kNN classifier, all of modified kNN classifiers show the improvement in performances compared to baseline kNN with 88, 89.33 % for LMKNN, FKNN, respectively and the proposed classifier, LMKNN-FDW gives the best performance i.e. 94.67 %. Table 18.2 lists the analytical results of the automatic segmentation. After applying the automatic segmentation, improvement of the results are observed. However, the percentage of accuracy for FKNN is slightly less than basic kNN with 96 % compared than 96.67 % to kNN. Nevertheless, the proposed classifier is the most outstanding classifiers compared to the other classifiers with 98.33 % of accuracy with 8 species can be identified 100 % accurately.

Table 18.2 Automatic segmentation results

Scientific name	kNN	LMKNN	FKNN	LMKNN-FDW
Hylarana glandulosa	24	24	24	24
Kaloula pulchra	25	25	25	25
Odorrana hossi	25	25	25	25
Polypedates leucomystax	24	25	24	24
Kaloula baleata	25	25	25	25
Philautus mjobergi	23	23	23	25
Phrynoidis aspera	25	24	25	25
Microhyla heymonsi	21	22	20	23
Microhyla butleri	25	23	22	24
Rhacophorus appendiculatus	25	25	25	25
Hylarana labialis	24	25	25	25
Philautus petersi	24	25	25	25
Total	290	291	288	295
Percentage (%)	96.67	97	96	98.33

18.5 Conclusion

In this paper, an improvement classifier based on kNN is proposed to overcome the problem of existing outliers particularly in small training sample size situation. The overall accuracy shows that the proposed classifier outperforms the other classifiers with the most outstanding result using the automatic segmentation. By using automatic segmentation, their rates were further improved remarkable. From this experiment, it may be inferred that proposed classifier is effective for frog identification system and is comparable to several state of-the-art methods regardless of their training sample size and future space dimension.

Acknowledgments The authors would like to thank the financial support provided by Universiti Sains Malaysia Short Term Grant, 304/PELECT/60311048, Research University Grant 814161 and Research University Grant 814098 for this project.

References

1. Bevier CR, Sonnevend A, Kolodziejek J, Nowotny N, Nielsen PF, Conlon JM (2004) Purification and characterization of antimicrobial peptides from the skin secretions of the mink frog Rana septentrionalis. Comp Biochem Physiol 139(1–3):31–38
2. Obrist MK, Pavan G, Sueur J, Riede K, Llusia D, Márquez R (2010) Bioacoustic approaches in biodiversity inventories. In: Manual on field recording techniques and protocols for all taxa biodiversity inventories. Abc taxa, vol 8. pp 68–99
3. Huang CJ, Yang YJ, Yang DX, Chen YJ (2009) Frog classification using machine learning techniques. Expert Syst Appl 36:3737–3743
4. Han NC, Muniandy SV, Dayou J (2011) Acoustic classification of Australian anurans based on hybrid spectral-entropy approach. J Appl Acoust 72:639–645

5. Mitani Y, Hamamoto Y (2006) A local mean-based nonparametric classifier. Pattern Recogn Lett 27:1151–1159
6. Fukunaga K (1990) Introduction to statistical pattern recognition, 2nd edn. Academic, London
7. Zeng Y, Yang Y, Zhao L (2009) Nonparametric classification based on local mean and class statistics. Expert Syst Appl 36:8443–8448
8. Gou J, Yi Z, Du L, Xiong T (2012) A local mean-based k-nearest centroid neighbor classifier. Comput J 55(9):1058–1071
9. Zuo W, Wang K, Zhang H, Zhang D (2007) Kernel difference-weighted k-nearest neighbors classification. ICIC 2:861–870
10. Jena PK, Chattopadhya S (2012) Comparative study of fuzzy k-nearest neighbor and fuzzy c-means algorithms. Int J Comput Appl 57(7):22–32
11. Jaafar H, Ramli DA (2013) Automatic syllables segmentation for frog identification system. In: 2013 IEEE international colloquium on signal processing and its application, vol 9
12. Hasan MH, Jaafar H, Ramli DA (2012) Evaluation on score reliability for biometric speaker authentication system. J Comput Sci 8(9):1554–1563

Chapter 19
A New Robust Reversible Watermarking Method in the Transform Domain

Rasha Thabit and Bee Ee Khoo

Abstract Robust (or semi-fragile) reversible watermarking methods have been proposed to provide robustness against unintentional attacks (e.g., noise addition, JPEG compression). This kind of watermarking schemes has recently attracted more attention. This paper presents a new robust reversible watermarking scheme in the transform domain. In the proposed method, the Slantlet transform (SLT) matrix has been used to transform small blocks of the original image and the mean values of the SLT coefficients in the high frequency subbands have been shifted to carry the watermark bits. The problem of overflow/underflow has been avoided by using the histogram modification according to specific rules. The results prove that the proposed method is completely reversible with improved capacity, robustness, and invisibility.

Keywords Robust reversible watermarking (RRW) · Histogram modification · Reversibility · Slantlet transform (SLT) · Matrix multiplication

19.1 Introduction

Reversible watermarking techniques [1–3] have been applied in many fields, especially those which require recovering the original image after the extraction of the watermark at the receiver side. Medical imaging and military imaging are examples of the fields that give attention to the original image and the watermark

---

R. Thabit (✉) · B. E. Khoo
School of Electrical and Electronic Engineering, University Sains Malaysia,
Penang, Malaysia
e-mail: rtm11_eee086@student.usm.my; rashathabit@yahoo.com

B. E. Khoo
e-mail: beekhoo@eng.usm.my

simultaneously. In most of the reversible watermarking methods, the watermark cannot be recovered in case of unintentional attacks (e.g., noise addition, JPEG compression) and intentional attacks (e.g., active/passive attacks, collusion attacks). It is usually a regular way to use the lossy compression for saving the storage space required for images. Thus, the watermarking methods that can withstand incidental distortion and can be destroyed by deliberate tampering became necessary.

Recently, a number of robust (or semi-fragile) reversible watermarking methods have been proposed. These methods can be classified into two categories: the spatial domain methods [4–6] and the transform domain method [7]. In [4], a grayscale histogram rotation based scheme has been proposed. This method has robustness against JPEG compression but the salt-and-pepper noise in the watermarked images is the main problem. To prevent the salt-and-pepper noise effect, in [5], a semi-fragile image authentication scheme has been designed. The embedding process depends on applying the histogram distribution constrained. In [6], two robust reversible watermarking schemes have been proposed. In these methods different strategies of histogram rotation have been applied. The aim of those methods is to enhance the performance of the method used in [4]. In [7], a semi-fragile lossless watermarking scheme based on shifting the absolute mean values of the integer wavelet transform (IWT) coefficients has been proposed. According to the study made in [8], the methods in [5, 6] suffer from unstable reversibility and robustness.

In this paper, we develop a new robust reversible watermarking method in the transform domain. In the proposed method the Slantlet Transform (SLT) matrix has been used to transform the image blocks and then the data bits are embedded into the SLT coefficients of a selected high frequency subband by shifting their mean values. To avoid the overflow and/or underflow, a histogram modification process has been utilized before embedding the watermark. The purpose of the proposed method is to ensure the reversibility and to improve the robustness, capacity, and invisibility.

The rest of the paper is organized as follows. In Sect. 19.2, the SLT matrix is presented. The data embedding method that has been used in [7] is explained in Sect. 19.3. Section 19.4 contains the details of the proposed method. The experimental results and discussion are in Sect. 19.5. Section 19.6 contains the conclusions.

19.2 Slantlet Transform Matrix

The Slantlet transform has been proposed as a wavelet-like transform [9]. The filters that were used to construct the Slantlet filter bank are $g_i(n)$, $f_i(n)$, and $h_i(n)$ [9]. For L-scale filter bank the orthogonal matrix dimension will be 2^L. The first row of the matrix will be corresponded to $h_L(n)$ and the second row will be corresponded to $f_L(n)$. Each of the remaining rows will be constructed by the sequences $g_i(n)$, its time reverse and their shifts by 2^{i+1}, for $i = 1,..., L - 1$. As an example, for the 3-scale SLT filter bank the SLT matrix dimension will be $2^3 = 8$, thus the size of the matrix will be 8×8. For more details about the SLT matrix the reader can refer to [9].

As explained by Mulcahy [10], the image can be transformed by applying rows transformation and columns transformation. The complete Slantlet transform can be represented in matrix format by:

$$S = SLT_N \; s \; SLT_N^T.$$

where, (s) is the original two-dimensional signal, (S) is the Slantlet transform of the original signal, and (SLT_N) is an $N \times N$ Slantlet matrix. Note that s, S, and SLT_N have the same size ($N \times N$).

The inverse SLT transform can be obtained by:

$$s = SLT_N^T \; S \; SLT_N.$$

19.3 The Data Embedding Method

In [7], the original image is transformed using IWT (wavelet family 5/3); then one of the high frequency subbands was chosen to carry the data. The carrier subband was divided into non-overlapping blocks. The algorithm scans all the blocks to find the maximum absolute mean value (m_{max}) of the coefficients. Then a threshold value T adjusted to be the smallest integer number which is greater than m_{max}. In this way, the absolute mean values of all the blocks will be less than the threshold T. A binary bit '1' can be embedded into one block by shifting the mean value of this block by S, which is equal to or greater than T. Because of the shifting process, some of the pixel values in the spatial domain will suffer from overflow/underflow. To avoid this problem, Zou et al. [7] investigated the effect of changing the coefficients on the spatial domain values. According to their investigation, they found the corresponding blocks in the spatial domain and these blocks were classified into four types. In some cases the block in the transform domain was left without any change, even though the data bit is '1', this causes error in the extracted watermark, to correct those errors they used the error correction coding (ECC) [BCH (15,11)] to extract the watermark correctly. The use of ECC reduced the capacity. Also the method of finding the spatial domain blocks is not precise [8] and thus the method is not completely reversible.

19.4 The Proposed Robust Reversible Watermarking Method

19.4.1 Overflow/Underflow Problem

In the proposed method, we suggested the use of the SLT instead of the IWT and the original image will be divided before the transformation. To embed a watermark bit, the main idea that has been proposed in [7] (i.e., mean value shifting) will be employed. In order to prevent the overflow/underflow problem, the histogram of the spatial domain blocks has been modified before embedding the data. We investigate the effect of shifting the SLT coefficients on the pixel values. Two main factors can decide the effect of shifting the subband coefficients: (1) the location of the pixel values that were changed and (2) the scale of this change. An extensive study was made to decide these two factors; we found that when the mean value of the SLT subband is shifted by a shift value (β), in the worst case all the pixel values in the spatial domain block will be affected. The scale of the change in the spatial domain had been always less than the shift value (β). Thus, we can check each spatial domain block to find out if the block will suffer from the problem of overflow and/or underflow by comparison between the pixel value and the shift value (β). The checking process depends on the shift value (β) and the pixel values in the spatial domain block. The histogram modification process will be applied only when it is necessary and it can be performed as follows:

$$I'_m(i,j) = \begin{cases} I_m(i,j) - (\beta+1), & \text{if } I_m(i,j) > 255 - \beta \\ I_m(i,j) + (\beta+1), & \text{if } I_m(i,j) < \beta \end{cases}.$$

where $I_m(i,j)$ is the grayscale value of the pixel at the coordinates (i,j) in the image I_m, $I'_m(i,j)$ is the modified pixel value. The pixel values that were changed and their locations must be saved as bookkeeping information. The bookkeeping information together with the name of the carrier subband and the shift value must be transmitted to the receiver side to recover the original image.

19.4.2 Watermark Embedding Process

The watermark embedding method can be summarized in the following steps:

Step 1: The original image is divided into non-overlapping blocks.
Step 2: Each block in the spatial domain will be transformed using the SLT matrix. The spatial domain block and the SLT matrix have the same size. The coefficients are divided into 4-subbands (LL, LH, HL, and HH). Then the mean value of the coefficients in the carrier subband must be calculated (only HL subband or LH subband will be used). This step will

be repeated for all the blocks. Then the maximum absolute mean value (m_{max}) must be found to set the threshold, which is the smallest integer number greater than m_{max}.

Step 3: The histogram modification process is applied to prevent the overflow/underflow. Thus all the blocks can be used for the watermark embedding process.

Step 4: A binary bit '1' can be embedded into a block by shifting the mean value of the carrier subband of this block by using the threshold value that has been obtained in step 2.

Step 5: After the watermark embedding process is completed for all the watermark bits, the original subbands in the SLT matrix must be replaced by the modified subbands. Then the inverse SLT must be applied. Thereafter, the result must be rounded to integer numbers.

19.4.3 Watermark Extraction Process

The watermark extraction process can be explained in the following steps:

Step 1: The watermarked image will be divided into non-overlapping blocks.
Step 2: Transform each block using SLT matrix.
Step 3: Calculate the absolute mean value of the carrier subband in each block. Then compare the absolute mean value with the shift value to decide watermark bit. The extraction process can be explained as follows:

$$b = \begin{cases} 1, & \text{if } m \geq \beta \\ 0, & \text{if } m < \beta \end{cases}.$$

where b is the extracted bit, m is the absolute mean value and β is the shift value. To recover the original image, when the embedded watermark is '1' shift back the mean value of the carrier subband. Then apply inverse SLT transform and round the values to be integer. Check the bookkeeping information and replace the pixel values that were already changed in the histogram modification process. Thus, the original image recovering process will be lossless.

19.5 Experimental Results and Discussion

19.5.1 Comparison with Zou et al. Method

In all the experiments, the proposed method was completely reversible. Thus the proposed method is better than the method in [7] in terms of reversibility. The capacity of the proposed method is higher than the capacity of the previous

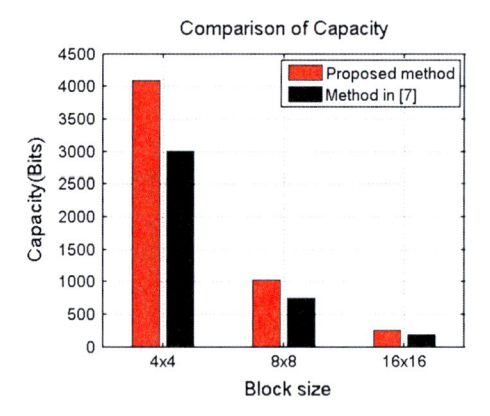

Fig. 19.1 Comparison between the capacity at different block sizes

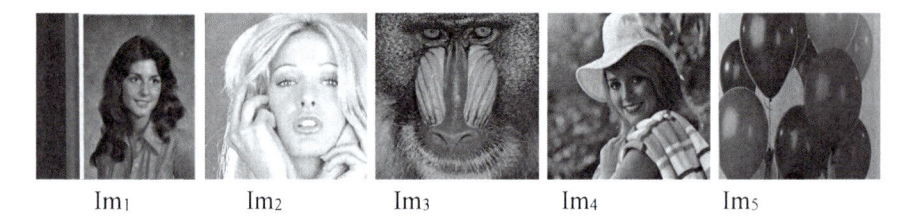

Fig. 19.2 Example test images

Table 19.1 Comparison of invisibility [$PSNR(dB)$]

Original image	Watermark size = 3,003 bits		Watermark size = 750 bits	
	Method in [7]	Proposed method	Method in [7]	Proposed method
Im_1	42.4635	**42.8047**	47.5490	**51.5112**
Im_2	40.6735	**42.1149**	47.8712	**48.0126**
Im_3	**39.0605**	38.1759	43.2399	**44.2695**
Im_4	42.3077	**42.8308**	47.7013	**51.5128**
Im_5	42.5180	**42.8308**	43.034	**48.5025**

transform domain method in [7] because all the blocks will be used to carry the watermark bits and in addition, ECC is not required. As an example, Fig. 19.1 shows the comparison results between the capacity for the grayscale images with size (512 × 512).

To compare the watermarked image quality, five test images of size (512 × 512) have been used and a randomly generated watermark has been embedded in each image. The peak signal-to-noise ratio (*PSNR*) between the original image and the watermarked image has been calculated. The test images are shown in Fig. 19.2; and Table 19.1 contains the comparison results of the invisibility test.

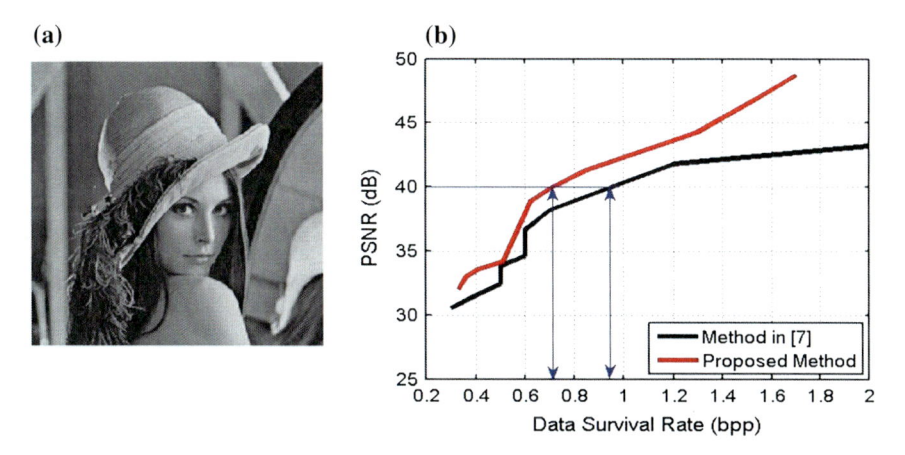

19 A New Robust Reversible Watermarking Method

Fig. 19.3 Robustness evaluation. **a** Lena image. **b** Data survival rate

Table 19.2 The invisibility and robustness test for two different watermarks

Original image	Original watermark	PSNR	AGN (0.0, 0.0005)	AGN (0.0, 0.001)	AGN (0.0, 0.002)	JPEG 60 %
Im_1	USM	45.3114	USM NC= 0.7279	USM NC= 0.7098	USM NC= 0.5734	USM NC= 0.9822
Im_1	I	46.3645	I NC= 0.7481	I NC= 0.7145	I NC= 0.5655	I NC= 0.9901
Im_3	USM	40.6990	USM NC= 0.8565	USM NC= 0.8206	USM NC= 0.7711	USM NC= 0.9970

In order to compare the robustness against JPEG2000 compression, the surviving level must be calculated at different threshold values as implemented in the experiments of [7]. The surviving rate means the watermark can be extracted correctly when the rate of the JPEG2000 compression is more than or equal to that surviving rate. The robustness tested at block size 8×8 (Lena image size $512 \times 512 \times 8$ bit). Figure 19.3 shows Lena image and the robustness test results. It is clear that the robustness of the proposed method is better because at the same *PSNR*, the data survival rate is less which means more robust. In other words at the same rate the *PSNR* is higher.

19.5.2 Studying the Effect of Embedding Different Watermarks

The effect of embedding two different binary watermarks of size (64×64) using the proposed method is shown in Table 19.2. For robustness evaluation, the normalized correlation (NC) between the original watermark and the extracted watermark after an attack [e.g., Additive Gaussian Noise (AGN) and JPEG compression] has been calculated. From the results, when two different watermarks are embedded in the same image they gives different *PSNR* and when the same watermark is embedded in different images the robustness of the watermark (i.e., NC) will be different.

19.6 Conclusions

The method proposed in this paper includes: (1) dividing image into non-overlapping blocks and transforming these blocks using SLT matrix, (2) modifying the histogram of the spatial domain blocks in order to avoid the problem of overflow and/or underflow, and (3) shifting the mean value of the carrier subband when a watermark bit is one. We concluded that dividing the image into blocks before transformation gives better control on the overflow/underflow problem. And the use of the SLT increases the robustness. Better watermarked image quality has been obtained in comparison with the method in [7] for the same watermark size. Improving the capacity and the visual quality will be the future work.

References

1. Kuo W, Jiang D, Huang Y (2007) Reversible data hiding based on histogram. In: Huang D.-S, Heutte L, Loog M (eds) ICIC 2007, LNCS(LNAI) 4682. Springer, Berlin, pp 1152–1161
2. An L, Gao X, Deng C, Ji F (2009) Reversible watermarking based on statistical quantity histogram. In: Muneesawang P et al (eds) PCM 2009, LNCS 5879. Springer, Berlin pp 1300–1305
3. Xuan G, Shi YQ, Chai P, Cui X, Ni Z, Tang X (2008) Optimum histogram pair based image lossless data embedding. In: Shi YQ, Kim H-j, Katzebeisser S (eds) IWDW 2007, LNCS 5041. Springer, Berlin, pp 264–278
4. De Vleeschouwer C, Delaigle J, Macq B (2003) Circular interpretation of bijective transformations in lossless watermarking for media asset management. IEEE Trans Multimedia 5(1):97–105
5. Ni Z, Shi YQ, Ansari N, Su W, Sun Q, Lin X (2008) Robust lossless image data hiding designed for semi-fragile image authentication. IEEE Trans Circuits Syst Video Technol 18(4):497–509
6. An L, Gao X, Deng C (2010) Reliable embedding for robust reversible watermarking. In: Proceedings of the second international conference on internet multimedia computing and service (ICIMCS'10), Harbin, China, pp 57–60
7. Zou D, Shi Y, Ni Z, Su W (2006) A semi-fragile lossless digital watermarking scheme based on integer wavelet transform. IEEE Trans Circuits Syst Video Technol 16(10):1294–1300
8. An L, Gao X, Deng C, Ji F (2010) Robust lossless data hiding: analysis and evaluation. In: Proceedings of international conference on high performance computing and simulation (HPCS 2010), Caen, France, pp 512–516
9. Selesnick IW (1999) The slantlet transform. IEEE Trans Signal Process 47(2):1304–1313
10. Mulcahy C (1997) Image compression using the haar wavelet transform. Spelman Sci Math J 1:22–31

Chapter 20
A Review for Image Segmentation Approaches Using Module-Based Framework

Guannan Jiang, Chin Yeow Wong, Stephen Ching-Feng Lin and Ngaiming Kwok

Abstract Image segmentation assists image understanding by perceptually partitioning an image into several homogeneous regions. This topic has been reported in a vast amount of literature. The techniques covered in the literature, however, are often concealed in specific applications with no explicit categorization. In order to cater the needs for both reasonably choosing methods under various circumstances and guiding further algorithm development, a review that can reveal the characteristics of typical segmentation approaches under detailed categorization is in demand. A natural categorization of the segmentation techniques with respect to their modeling methods is advocated here. Using such categorization as basis, different segmentation methods are presented within a general module-based framework proposed in this paper, where investigation for each module is carried out in a coherent manner.

Keywords Image segmentation · Categorization · Module-based framework

20.1 Introduction

Image segmentation can be intuitively perceived as a pixel clustering or graph partitioning process where assignments of every pixel to different object labels are determined. There are two main challenging issues needing to be addressed in

---

G. Jiang (✉) · C. Y. Wong · S. C.-F. Lin · N. Kwok
School of Mechanical and Manufacturing Engineering, The University of New South Wales, Sydney, NSW 2052, Australia
e-mail: guannan.jiang@unsw.edu.au; jstaind@hotmail.com

C. Y. Wong
e-mail: chin.wong@unsw.edu.au

S. C.-F. Lin
e-mail: stephen.lin@unsw.edu.au

N. Kwok
e-mail: nmkwok@unsw.edu.au

practical segmentation applications. On one hand, robustness of the segmentation needs to be considered. Several factors [1] that commonly influence the performance of segmentation include noise, luminance variation, moving background, cluttered background, multimodal background, camera motion and shadows. Numerous methods and algorithms have been proposed aiming to resolve these issues, which mainly distinguish each other by exploiting different visual cues such as color, texture, edge and spatial-temporal; model representations including hard assignment, single modal or multimodal; modeling methods encompassing simple merge and split, mode finding, spectral partitioning and energy minimization; and model initialization requirements on initial parameterization and other human involvements. On the other hand, the computational efficiency of the algorithm also needs to be addressed. The output of image segmentation is often input for higher level computer vision applications such as video surveillance, human-computer interface, vehicle navigation and robotic vision, where only limited computation resource could be allocated for the segmentation process. In practice, the diminishing of one issue is often at cost of aggravating the other. It thus calls for a rational balance between these two issues with the assistance of detailed analysis of available techniques.

In this review, it is inspired to group up and evaluate strategies underlying representative methods in different categories by means of a structured framework, so as to form a basis for both constructing appropriate segmentation systems for specific application and further improvement of available techniques. The rest of the paper will be organized as follows: a module-based framework that can adapt different segmentation methods is proposed in Sect. 20.2. In Sect. 20.3, representative approaches will be broken down into modules according to the proposed structure. Evaluations and improvements of these components for each segmentation technique will also be discussed in this section. In Sect. 20.4, further discussion will be carried out based on recent developments of representative techniques. A conclusion will be drawn in Sect. 20.5.

20.2 A Module-Based Framework

A generalized module-based framework shown in Fig. 20.1 is proposed here to efficiently evaluate the influence of each aspect of segmentation techniques through enormous literature. Components in the segmentation procedure include preprocessing, parameter initialization and image modeling.

Most input images need to be preprocessed to convey meaningful information. Common preprocessing includes noise deduction by convolving the input images with certain smooth kernel, frame size and frame rate deduction of input video for computing efficiency and image registration for multi-view or multi-camera situations.

Parameter initialization is often required to generate the starting sets of parameters in different segmentation methods, of which some might require human

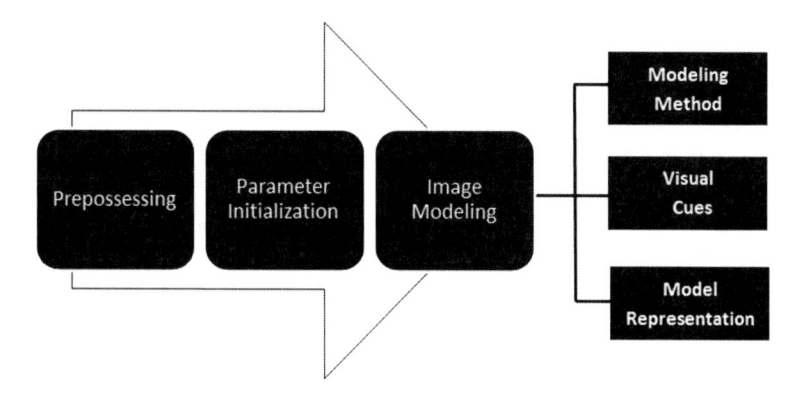

Fig. 20.1 Common procedure for segmentation with components and sub-components

involvement for the supervision of parameters as segmentation progress. While useful in certain applications [2], we will not include here methods that require human supervision since we want to focus our discussion on generalized methods that do not depend on ad-hoc techniques.

Image modeling includes modeling methods, choice of visual cues and model representation. It is the central component for a segmentation system, which the categorization of our review is based on. Detailed discussion of this component will be found for all representative segmentation techniques in the following section.

20.3 Representative Segmentation Methods

While different criteria can be used to categorize available algorithms, a natural approach is to do so using modeling methods. The rationale behind such categorization is that the parameter initialization, choices of visual features and model representations are often determined in a way to optimize the performance for certain modeling method. Aside from the basic region splitting or merging methods [3, 4] that utilize simple thresholding, there are three major categories identified in the literature, which are mode-finding clustering methods, spectral partitioning methods and energy minimizing methods. In the following subsections, each representative modeling method will be presented in accordance to the proposed framework.

20.3.1 Mode-Finding Clustering

K-means and other mode-finding clustering techniques have been explored extensively for image segmentation. The basic idea is to model each pixel in the scene as a vector of low level visual cues, and then try to cluster the pixels in a recursive updating manner [5]. Methods in this field can be further divided into parametric methods and non-parametric methods.

K-means Clustering. K-means clustering is an iterative region center updating method which is famous for its efficiency andeasy implementation.

Parameter Initialization. Number of clusters and seed locations have to be specified when starting the algorithm.

Modeling Method. Classic K-means methods tackle the pixel clustering problem by hard or soft assignment of pixels into K segments of image parametrically. The most popular method based on Lloyds algorithm [6] mainly contains three steps for each update, namely, parameter initialization, pixel membership assignment and parameter updating. Firstly, cluster number K is given by the user and the initial cluster centers are randomly chosen from all possible values in the visual cue vectors. Then, each pixel is assigned to clusters based on the minimum Euclidean distance from its cluster center. At last, cluster centers are re-computed based on the membership shift of each pixel in the updating step. This updating cycle stops when pixel membership does not change anymore with further parameter updating.

Choice of Visual Cues. Single low level visual cue including intensity, color, and texture are usually used for efficiency. More than one visual cue with associated weights are employed to enhance the robustness [7].

Model Representation. Although hard assignment of pixels without probability density distribution is possible, Mixture of Gaussians [8] has been favored in recent literature to represent each pixel model as a probability density function with several Gaussians to account for multi-modality background in realistic situations.

K-means clustering methods utilizing heuristic algorithm converge very fast to local extrema. The parametric natural of K-means clustering method decides one intrinsic drawback in this strategy, namely, the high dependence of performance on careful choice of initializing parameters [9]. This drawback limits the generalization of this method although alleviation of its effect can still be found in resent literature [10, 11]. Apart from that, the Euclidean distance metric is not a very robust measurement of affinity. For larger value of K, clusters could be found in isolated locations. These features ensure that the application of K-means clustering is most suitable for images with fewer objects but require high computation efficiency.

Mean Shift. Mean Shift [12] is a non-parametric mode-finding clustering method where pixel distribution is implicitly modeled by computing density gradient ascent coupled with appropriate kernel function. In contrast to K-means clustering techniques, arbitrary assumptions of the distribution shape and cluster number are not needed.

Model Initialization. Kernel function and size need to be specified according to image.

Modeling Method. Analogous to K-means mode finding methods, Mean Shift updates mean in each cluster using a Mean Shift vector:

$$M_h(y_0) = \frac{\sum_{i=1}^{n_i} w_i(y_0) x_i}{\sum_{i=1}^{n_i} w_i(y_0)} - y_0, \qquad (20.1)$$

$$\hat{y} = y_0 + M_h(y_0), \qquad (20.2)$$

where x_i are the data points in the quest neighborhood, w_i stands for weights of all the data points determined by the kernel function and size. n_i is the number of points in the kernel, y_0 is the initial mean location, and $\hat{y}$ is the new target center for next shift.

The convergence of the Mean Shift algorithm is achieved when the Eq. 20.1 is minimized. To avoid being trapped in false local extrema, Comaniciu and Meer [13] apply Mean Shift in various starting positions and achieve convergence of local extrema under rational weak kernel assumption.

Choice of Visual Cues. Mean shift gains its popularity in the segmentation field by the introduction from Comaniciu etã CPÖal. using color cues [12]. But the deterministic nature of the method tends to fail when nearby regions demonstrate similar color distribution of the object. Joint domain of color and other visual cues can generally produce better results.

Model Representation. Selection of kernel type and size represents the configuration of segmentation model. Epanechnikov kernel is a popular choice for its guarantee of convergence [12, 13].

As a well-established method, Mean Shift clustering is widely used in segmentation applications where little prior knowledge of images is available. Various modifications such as smart kernel manipulation [14], novel kernel sampling methods [15], and Particle Filter based sampling methods [16] can be seen from the literature trying to speed up the shifting procedure. Furthermore, choice of the kernel remains an unsolved issue to date. Many Mean Shift based algorithms try to address the kernel issue within specific applications. The scale and shape of the kernel utilized determine the tradeoff between sampling artifacts and accuracy loss. The high computation cost associated with it hinders its applications in real-time situations.

20.3.2 Spectral Partitioning Methods

Spectral partitioning methods deem the quest of global energy minimum as a generalized eigenvalue problem [17]. Eigenvectors of an affinity matrix are used to determine partitions of the image where recursive partitioning process can be continued. Introduced by Shi and Malik [18] for the formulation and minimization of a normalized cut criterion, a segmentation system based on spectral graph theory with promising result can be expressed using our framework as below.

Model Initialization. Definition of affinity matrix with respect to chosen similarity measure is initialized. Different graph Laplacians are defined in different spectral cut function.

Modeling Method. Treating the segmentation of an image in a hierarchical divisive manner, the neighboring pixels are examined by their affinity measurement to separate groups with weak similarity in a normalized way. A minimum normalized cut is formed where relative weak affinity measure happens, which is defined as:

$$Nassoc(A, B) = \frac{assoc(A, A)}{assoc(A, V)} + \frac{assoc(B, B)}{assoc(B, V)}, \quad (20.3)$$

where $assoc(A, A)$ and $assoc(B, B)$ stand for total energy of edges connecting nodes within A and B respectively, and $assoc(A, V)$ and $assoc(B, V)$ is the sum of all the energy associate with nodes in A and B respectively.

To solve this NP-complicate problem, an approximate discrete solution is suggested in Shis work [18], which ends up measuring an equivalent equation:

$$min_x\{Ncut(x)\} = min_y \left\{ \frac{y^T(D-W)y}{y^T Dy} \right\}, \quad (20.4)$$

by solving an eigenvalue equation:

$$(D - W)y = \gamma Dy. \quad (20.5)$$

Utilizing the simple fact in the Rayleigh quotient, the second smallest eigenvector of Eq. 20.5 is the solution to minimize the normalized cut problem, which gives a global minimum to partition the image into two segments. Repeating construction of the eigenvalue equation can be used to further dividing the image hierarchically.

Choice of Visual Cues. In [18], various visual cues have been exploited to justify their modeling method. It is clearly indicated that all the visual cues including luminance intensity, color cues, spatial-temporal cues and texture cues captured in various scales and orientation can be used to model the affinity of pixels. For rational choice of visual cues to balance robustness and speed, an investigation such as [7] on the comparison of using different visual cues for segmentation performance is necessary.

Model Representation. Eigenvectors from the approximate discrete solution are used to represent model. In [19], real-value assignment that takes +1 for grouping and −1 for separating and distance for weights is used for the formulation of Eq. 20.4. Other model representation modifications had been reported such as in [20], where intervening contours are used as contour weights.

Spectral partitioning framework utilizes generalized eigenvalue equation to provide a real value solution to the minimization criterion. Normalize cut is one of the most popular application of this theory seeking to find global minimum of affinity in the image, which confirms the segmentation perceptually in high level. But the process of finding a cut under such framework is NP-hard, which still calls for more efficient approximation. Extensions and modifications for the basic normalized cut are plentiful in terms of using new affinity matrix [21] and using special treatment to accelerate eigenvalue computation [22].

20.3.3 Energy Minimizing Methods

Energy minimizing methods can be divided into contour-based ones that use energy-minimizing splines to localize continuous object boundaries and others utilizing region-based properties defined on discrete set of variables. The essential modeling method is to minimize the cost function to grow towards the true segmentation of objects.

Contour-based Method. Contour-based methods, also known as active contour, evolves the curve in the direction of negative energy gradient by minimizing the combinatory costs of internal energy that characterized by smoothness and external energy characterized by image-based potentials to grow towards the object boundary. Kass etã CPÖal. [23] raised a revolutionary model called Snake which precedes many contour-based methods.

Model Initialization. There are multiple degrees of freedom in the formulation of cost equation, which renders the contour-based methods very sensitive to the initial parameterization.

Modeling Method. An explicit curve energy function is normally adopted in this category:

$$-\int |\nabla I(C)|^2 ds + v_1 \int |C_s| ds + v_2 \int |C_{ss}|^2 ds, \qquad (20.6)$$

where the first term accounts for the external energy that favors large image gradient, the second and third term account for the internal energy that addresses smoothness of the contour. Expectation maximization method is integrated in many variants of the original snake, e.g. Kalman snakes [24] and CONDENSATION [25] based on particle filtering.

Choice of Visual Cue. Edge feature of the image is normally used. Because an intuitive probabilistic interpretation does not exist, integration of other visual cues such as color, texture, motion or intensity is hard to be implemented.

Model Representation. Probability function with respect to visual cue is normally used to account for the image degradation from noise.

Contour-based segmentation methods can be very good in capturing fine details in real applications. The explicit parameterization method it depends on, at the meantime, causes it heavily relying on accurate model initialization, which could cause the contour evolution stuck in local minima and not robust against noise.

Region-based Method. Region-based methods utilize implicit modeling of the embedding function to measure the consistency of the image statistics inside and outside a segmented region. The development of these methods is based on the Mumford-Shah model [26]. Level set methods [27] and graph cut methods [28, 29] are popular applications of this established model. The graph cut method [29] will be elaborated here for its popularity in resent extensions and improvements.

Model Initialization. Graph cut does not require any initial contour, but it can benefit from some shape prior to shorten the time needed for finding appropriate seeds.

Modeling Method. The energy minimization problems have been formulated as a combinatorial max-flow/min-cut optimization process [29]. Graph cut concerns the segmentation as a binary problem based on $s-t$ graph cut algorithm which utilizes both region information and boundary information.

Choice of Visual Cues. Standard graph cut method uses intensity cue for region-based similarity. However, by using the *max-flow/min-cut* algorithm there would not be convergence issues when multiple visual cues are used.

Model Representation. Affinity measure in feature space between all the nodes are used to represent the cost in the min-cut algorithm. Mixture of Gaussian is exploited in the work of Blake etãCPÖal. [30] to emphasize regional multi-modality.

Based on the heuristics of finding global minima by implicit modeling region and boundary properties and using only intuitive binary variables assigning pixels to either inside or outside segments, energy minimization based graph cut is widely applicable to many types of images. One drawback is the limitation in constraints to guide segmentation process. Because of its high demand on memory, graph cut is not practical to use in large scale problems. Many extensions and improvement have been proposed in recent years for specific applications [31, 32].

20.4 Recent Developments

It is evident that all the representative methods suffer from different shortcomings respectively. Recent development based on these representative image segmentation techniques has shown several new trends for the purpose of enhanced performance.

Segmentation techniques formulated from integrated color-texture visual cues have attracted substantial interest because of their strong similarity to human perception in dealing with natural images. Wide spectrums of schemes for effective color-texture feature integration are prompted in resent literature, among which three types could be summarized. Texture feature is explored [33], in different color channels in a hierarchically refining manner. A sequential integrating approach is noted [34], where color and texture based segmentation processes are conducted separately in order. Finally, [35, 36] belong to the approach that independently segments images in different channels and then seeks appropriate integration scheme. The latter two approaches have shown popularity in resent researches.

As discussed above, different methods have their own desired attributes, e.g. low computation burden of K-means clustering methods and global optimizing ability of graph cut based methods. Organic Integration of different modeling methods can lead to promising result in terms of accuracy and speed towards natural images. Segmenting images in a multistage manner with K-means clustering and normalized cut is presented in [37], which claims to alleviate heavy computation demand of the original normalized cut algorithm by firstly dividing the image into same size cells and then performing normalize cut in the reduced areas. A fast segmentation algorithm with high accuracy is presented by Tan and Wang [38], where pre-segmented image by Mean Shift is used as prior in a Bayesian framework to guide the further image partition using Kway-Ncut. The smart combination of regional clustering method and graph partitioning method are pervasive in resent development, however, attention needs to be paid to address the sensitivity of final segmentation performance to the region segmentation result.

20.5 Conclusion

Wide variety of segmentation schemes have been proposed to accommodate particular issues aroused from applications where the method categorization is commonly blurred. In order to determine a clear categorization that can guide technique selection for various applications and further research needs in this field, a module-based framework is developed in this paper. We argue that distinction among different methods can be mainly divided into three components along the segmentation procedure. The segmentation techniques are categorized by their modeling methods, which coincides with the technique formation process perceptually. Representative methods and their characteristics in each category are then presented in a module by module manner according to the proposed framework. Not only comparison for segmentation techniques is made direct and convenient by adapting methods into this module-based framework, but also several trends for improvements in each component have emerged. Overall, Trade-off between robustness and speed is a main concern in the development of a segmentation system, which newly developed algorithms are prompted to resolve.

References

1. Toyama K, Krumm J, Brumitt B, Meyers B (1999) Wallflower: principles and practice of background maintenance. In: The Proceedings of the seventh IEEE international conference on computer vision, vol 1. pp 255–261
2. Mortensen EN, Barrett WA (1998) Interactive segmentation with intelligent scissors. Graph Models Image Process 60:349–384
3. Ohlander R, Price K, Reddy DR (1978) Picture segmentation using a recursive region splitting method. Comput Graph Image Process 8(3):313–333
4. Brice CR, Fennema CL (1970) Scene analysis using regions. Artif Intell 1(3):205–226
5. Bishop CM, Nasrabadi NM (2006) Pattern recognition and machine learning, vol 1. Springer, New York
6. Lloyd S (1982) Least squares quantization in PCM. IEEE Trans Inf Theory 28(2):129–137
7. Barnard K, Duygulu P, Guru R, Gabbur P, Forsyth D (2003) The effects of segmentation and feature choice in a translation model of object recognition. In: Proceedings 2003 IEEE computer society conference on computer vision and pattern recognition, vol 2. IEEE, II–675
8. Stauffer C, Grimson WEL (1999) Adaptive background mixture models for real-time tracking. In: IEEE computer society conference on computer vision and pattern recognition, vol 2. IEEE

9. Kanungo T, Mount DM, Netanyahu NS, Piatko CD, Silverman R, Wu AY (2002) An efficient k-means clustering algorithm: analysis and implementation. IEEE Trans Pattern Anal Mach Intell 24(7):881–892
10. Isa NAM, Salamah SA, Ngah UK (2009) Adaptive fuzzy moving k-means clustering algorithm for image segmentation. IEEE Trans Consum Electron 55(4):2145–2153
11. Siddiqui FU, Mat Isa NA (2011) Enhanced moving k-means (EMKM) algorithm for image segmentation. IEEE Trans Consum Electron 57(2):833–841
12. Comaniciu D, Ramesh V, Meer P (2000) Real-time tracking of non-rigid objects using mean shift. In: Proceedings of the IEEE conference on computer vision and pattern recognition, vol 2. IEEE, pp 142–149
13. Comaniciu D, Meer P (2002) Mean shift: a robust approach toward feature space analysis. IEEE Trans Pattern Anal Mach Intell 24(5):603–619
14. Freedman D, Kisilev P (2009) Fast mean shift by compact density representation. In: IEEE conference on computer vision and pattern recognition (CVPR 2009). IEEE, pp 1818–1825
15. Paris S, Durand F (2007) A topological approach to hierarchical segmentation using mean shift. In: IEEE conference on computer vision and pattern recognition (CVPR'07). IEEE, pp 1–8
16. Shan C, Wei Y, Tan T, Ojardias F. (2004) Real time hand tracking by combining particle filtering and mean shift. In: Proceedings of the sixth IEEE international conference on automatic face and gesture recognition. IEEE, pp 669–674
17. Oliver NM, Rosario B, Pentland AP (2000) A bayesian computer vision system for modeling human interactions. IEEE Trans Pattern Anal Mach Intell 22(8):831–843
18. Shi J, Malik J (2000) Normalized cuts and image segmentation. IEEE Trans Pattern Anal Mach Intel 22(8):888–905
19. Golub GH, Van Loan CF (2012) Matrix computations, vol 3. JHU Press, Baltimore
20. Malik J, Belongie S, Leung T, Shi J (2001) Contour and texture analysis for image segmentation. Int J Comput Vis 43(1):7–27
21. Stein AN, Stepleton TS, Hebert M (2008) Towards unsupervised whole-object segmentation: combining automated matting with boundary detection. In: IEEE conference on computer vision and pattern recognition (CVPR 2008). IEEE, pp 1–8
22. Sharon E, Galun M, Sharon D, Basri R, Brandt A (2006) Hierarchy and adaptivity in segmenting visual scenes. Nature 442(7104):810–813
23. Kass M, Witkin A, Terzopoulos D (1988) Snakes: active contour models. Int J Comput Vis 1(4):321–331
24. Blake A, Curwen R, Zisserman A (1993) A framework for spatiotemporal control in the tracking of visual contours. Int J Comput Vis 11(2):127–145
25. Isard M, Blake A (1998) Condensationconditional density propagation for visual tracking. Int J Comput Vis 29(1):5–28
26. Mumford D, Shah J (1989) Optimal approximations by piecewise smooth functions and associated variational problems. Commun Pure Appl Math 42(5):577–685
27. Osher S, Paragios N (2003) Geometric level set methods in imaging, vision, and graphics. Springer, Berlin
28. Greig D, Porteous B, Seheult AH (1989) Exact maximum a posteriori estimation for binary images. J Roy Stat Soc B (Methodological), 271–279
29. Boykov Y, Kolmogorov V (2004) An experimental comparison of min-cut/max-flow algorithms for energy minimization in vision. IEEE Trans Pattern Anal Mach Intell 26(9):1124–1137
30. Blake A, Rother C, Brown M, Perez P, Torr P (2004) Interactive image segmentation using an adaptive GMMRF model. In: Computer vision-ECCV 2004, Springer, Berlin, pp 428–441
31. Lermé N, Malgouyres F, Létocart L (2010) Reducing graphs in graph cut segmentation. In: 17th IEEE international conference on image processing (ICIP) 2010. IEEE, pp 3045–3048
32. Vicente S, Kolmogorov V, Rother C (2008) Graph cut based image segmentation with connectivity priors. In: IEEE conference on computer vision and pattern recognition (CVPR 2008). IEEE, pp 1–8

33. Hoang MA, Geusebroek JM, Smeulders AW (2005) Color texture measurement and segmentation. Signal Process 85(2):265–275
34. Hedjam R, Mignotte M (2009) A hierarchical graph-based markovian clustering approach for the unsupervised segmentation of textured color images. In: 16th IEEE international conference on image processing (ICIP 2009). IEEE, pp 1365–1368
35. Kim JS, Hong KS (2009) Color–texture segmentation using unsupervised graph cuts. Pattern Recogn 42(5):735–750
36. Serrano C, Acha B (2009) Pattern analysis of dermoscopic images based on markov random fields. Pattern Recogn 42(6):1052–1057
37. Choong MY, Kow WY, Khong WL, Liau CF, Teo KTK An image segmentation using normalised cuts in multistage approach. Image, 5(6):7
38. Tan LY, Wang SJ (2013) Multilevel and mean shift based image segmentation using kway-ncut. Moshi Shibie yu Rengong Zhineng/Pattern Recogn Artif Intell 26(4):328–336 cited By (since 1996)

Chapter 21
Adaptive Image Super-Resolution with Neural Networks

Kah Keong Chua and Yong Haur Tay

Abstract A framework for image super resolution using Multilayer Perceptron (MLP) was developed. In this paper, we focus on verifying the influence of training images to the performance of Multilayer Perceptron. The training images were collected from various categories, i.e. flowers, buildings, animals, vehicles, human and cuisine. The neural network trained with single image category has better performance compared to other methods. Multiple categories of training images made the MLP more robust towards different test scenarios. Common image degradation, i.e. motion blur and noise, can be reduced when the MLP is provided with proper training samples.

Keywords Image super resolution · MLP · Motion blur · Image denoising

21.1 Introduction

Super-Resolution, generally speaking, is the process of recovering a high resolution (HR) image from single or multiple low resolution (LR) images.

Different approaches have been proposed with foremost attempt worked in the frequency domain by Tsai and Huang [1]. In order to overcome the image artefacts e.g. jagged edges and blurred contours, new methods have been proposed to upscale the images [2–4]. Interpolation-based approach reconstructs high-resolution (HR) image by projecting the low-resolution (LR) images obtained to the respective

---

K. K. Chua (✉) · Y. H. Tay
Centre for Computing and Intelligent Systems, Universiti Tunku Abdul Rahman, Kuala Lumpur, Malaysia
e-mail: chua.kah.keong@gmail.com

Y. H. Tay
e-mail: tayyh@utar.edu.my

referencing image, then followed by blending all the accessible information from each image to form the result [5, 6]. A neural network can be trained to estimate non-linear functions and perform parallel computation, minimizing the processing time of super-resolution systems [2, 3]. Throughout the time consuming training process, the weight factors are adjusted and every epoch fine-tuned [3, 7, 8]. Over the years, neural network based methods gained their popularity due to the network flexibility.

In a nutshell, the contribution of our work in this paper is to perform low level image processing using neural network, i.e. image super resolution, by creating a framework to test the behaviour of neural network based on different types of training samples provided, in contrast with several methods in image super-resolution.

21.2 Multilayer Perceptrons

Multilayer Perceptron (MLP) is a feed-forward back-propagation artificial neural network model. MLP works on a supervised back-propagation learning technique during the training phase of the network. MLP generally consists of input layer, hidden layer and output layer. Each layer has multiple numbers of neurons, which vary according to the complexity of the network.

The learning process takes place in the perceptron by altering the weight factors after each training epoch. Weight factors are adjusted accordingly by calculating the mean error of the expected result in contrary with the output result. The aim of training a neural network is to search for a set of weight factors which links the provided input with the expected output. The process of optimizing the number of hidden layers and amount of neurons used in each layer greatly affects the performance of the entire network. Moreover, validation is required throughout the training to prevent over-fitting.

21.3 System Overview

The MLP used for this research has been configured to match the requirement of the network. The MLP is a typical one hidden layer MLP of 20 hidden neurons with hyperbolic tangent (tanh) activation function, while the output is logistic regression layer. The network considers the desired pixel alongside with eight neighboring pixel as reference, then outputs four up-scaled pixels. The process is repeated on the adjacent pixel, throughout the entire image. The Fig. 21.1 shows the architecture of the proposed MLP system.

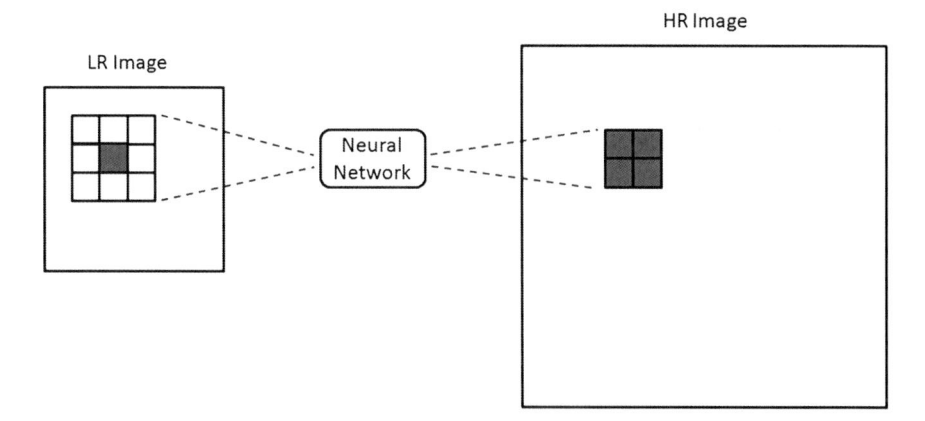

Fig. 21.1 Architecture of the proposed MLP system

21.4 Experiment Framework and Result

To test the framework, all the methods will undergo a 2x image upscale process. The enlarged image will then be compared with the original high resolution image to yield quantitative results. The difference of the images will be measured using Mean Square Error, Peak Signal-to-Noise Ratio and Structural Similarity Index.

For the purpose of benchmarking, the algorithms tested include Nearest Neighbour, an improved version of New Edge-Directed Interpolation (NEDI) [9], as well as the Fast Curvature Based Interpolation (FCBI) [6] and Iterative Curvature Based Interpolation (ICBI) [6] proposed by Giachetti and Asuni. These methods are regenerated using the evaluation scripts available at the website: http://www.andreagiachetti.it/icbi.

MLP will be trained using different sets of samples to tackle the problem effectively. The first training set (MLP_gen) consists of sample from multiple sources, customized at such that, in order to provide a more robust MLP network which will not suffer from performance loss under different scenarios. Several training sets (MLP_sp) are created solely on image samples from similar categories.

In addition, another two special training sets are created to tackle common problems encountered in image capturing, namely image noise and motion blur.

Table 21.1 is the PSNR and SSIM of MLP and other image up-scaling methods. Tables 21.2 and 21.3 are the experiment results for noisy and motion blurred images respectively. Typical values for PSNR are between 30 and 50 dB, where higher value indicates better result. SSIM values are from 0 to 1, where higher resemblance of output and original image provides higher values.

Table 21.1 Test results on zebra and buildings

Methods	Zebra		Buildings		Flowers	
	PSNR	SSIM	PSNR	SSIM	PSNR	SSIM
N. Neighbour	31.85	0.70	33.26	0.74	34.02	0.81
NEDI	32.00	0.72	33.83	0.78	34.18	0.84
FCBI	31.76	0.71	32.97	0.74	33.80	0.84
ICBI	32.04	0.73	33.17	0.76	34.17	0.86
MLP_gen	35.38	0.87	**35.30**	**0.84**	36.22	0.90
MLP_sp	**35.92**	**0.90**	34.55	0.72	**37.38**	**0.92**

Table 21.2 Test results on noisy images (MLP_noise)

Methods	PSNR	SSIM
Buildings	33.39	0.71
Flowers	33.50	0.75
Random textures	31.60	0.69
Tiger	33.03	0.69
Zebra	33.59	0.76

Table 21.3 Test results on motion blurred images (MLP_motion)

Methods	PSNR	SSIM
Buildings	32.98	0.73
Flowers	32.55	0.64
Random textures	31.28	0.74
Tiger	32.54	0.73
Zebra	32.34	0.79

21.5 Discussion

21.5.1 Comparison of Different Up-Scaling Method Results

The simulation results show that MLP generally perform better than ordinary image up-scaling methods. In most of the cases, MLP yields very low MSE in contrast with other methods, while PSNR are about the same. SSIM index shows significant resemblance of original reference images and the up-scaled images from MLP (Fig. 21.2).

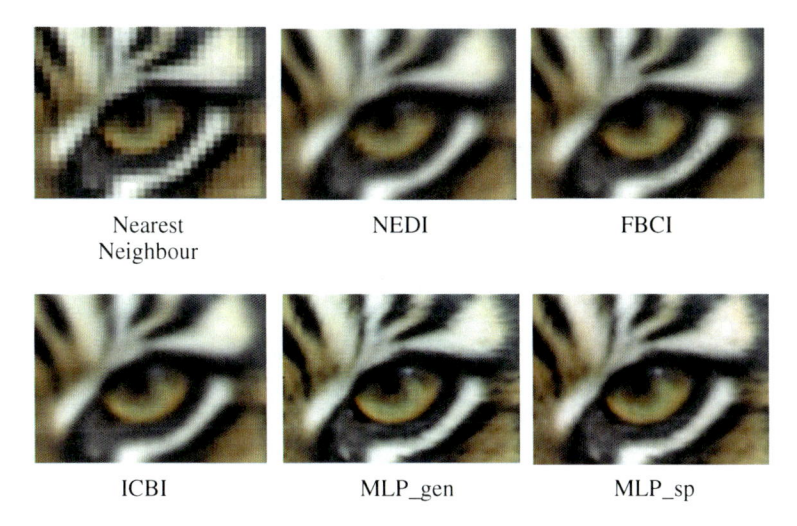

Fig. 21.2 Comparison between different types of image up-scaling methods

Fig. 21.3 Comparison of the MLP results trained using different samples

21.5.2 Comparison Between MLP Trained with Different Samples

When provided with sufficient samples, MLP_sp trained with similar images can perform better than MLP_gen trained with samples from different categories. However, MLP_sp is not as robust as MLP_gen handling colours beyond its training sets (Figs. 21.3 and 21.4).

By comparing the two MLPs trained with different samples, MLP can upscale the image provided accurately for most of the regions, but suffers performance loss at most of the edges in the image.

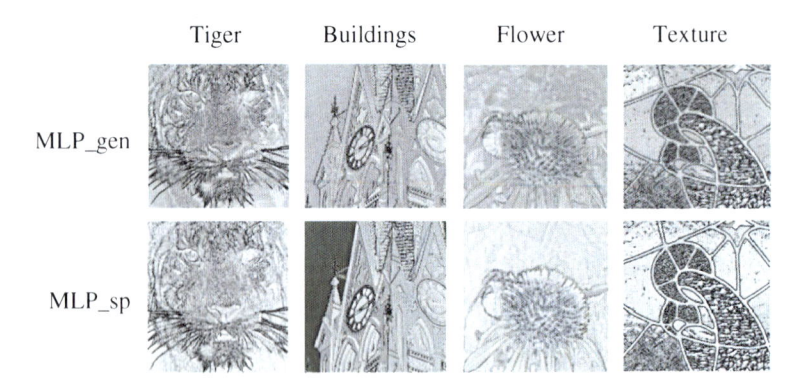

Fig. 21.4 The difference of original reference image and up-scaled images

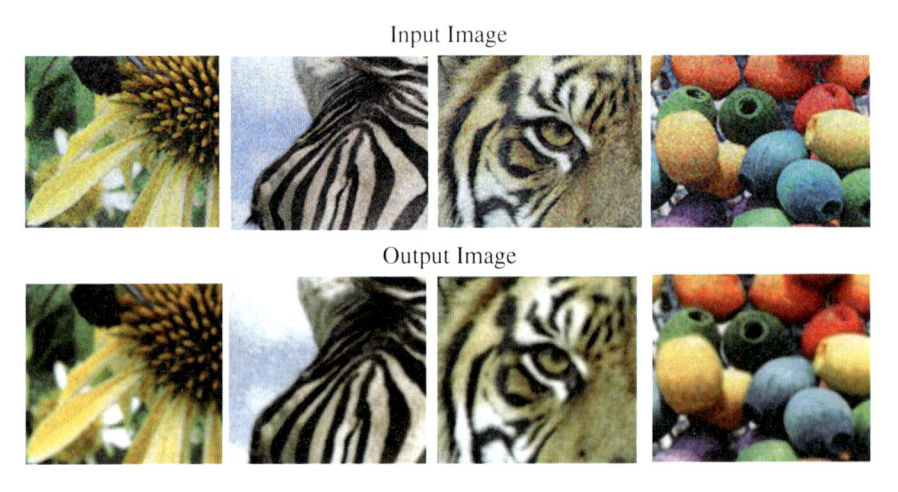

Fig. 21.5 Comparison of noisy input and up-scaled output images

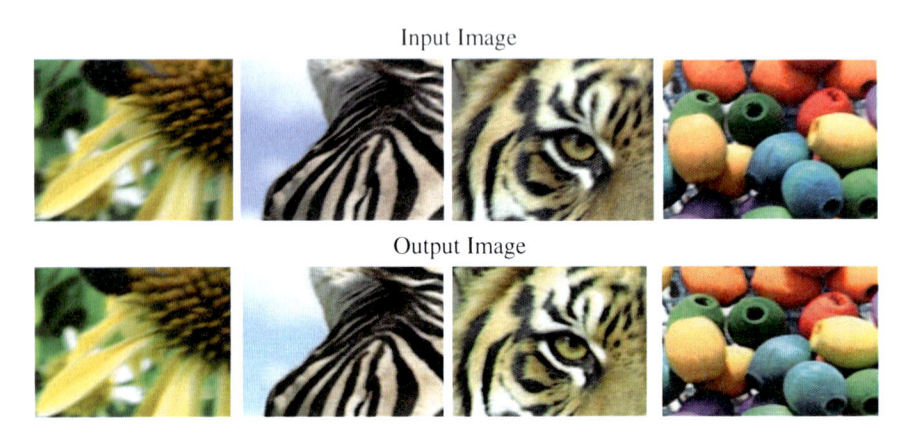

Fig. 21.6 Comparison of motion blurred input images and up-scaled output images

21.5.3 Motion Blurred and Noisy Images

In MLP_noise, we can clearly notice that noise in the test image is greatly reduced, however not fully eliminated. Meanwhile, MLP_motion managed to minimize the blurry edges caused motion blur. Noise and motion blur often cause image degradation and the loss of crucial information, especially in surveillance images (Figs. 21.5 and 21.6).

21.6 Conclusions and Future Works

To sum it all, a neural network trained with single type of samples performs splendidly on images of similar type; however it is not robust against images beyond the training samples or category. Meanwhile, a more robust neural network, trained with samples from various categories, can handle different images accurately without introducing artefacts. The adaptation ability is a major advantage of neural network, utilizing this feature by pre-training some filters greatly help in artefacts reduction. A well-trained neural network with filters applied can be used to reduce motion blur and noise when performing image super resolution.

References

1. Asuni N, Giachetti A (2008) Accuracy improvements and artifacts removal in edge based image interpolation. In: Proceedings of 3rd international conference on computer vision theory and application (VISAPP), pp 58–65
2. Tian J, Ma K-K (2011) A survey on super-resolution imaging. Springer, London
3. Torrieri D, Bakhru K (1997) Neural network super resolution. IEEE, pp 1594–1598
4. Egmont-Petersen M, Ridder D, Handels H (2002) Image processing with neural networks—a review. Pattern Recogn 35:2279–2301
5. Giachetti A, Asuni N (2011) Real-time artifact-free image upscaling. IEEE Trans Image Process 20(10):2760–2768
6. Tsai RY, Huang TS (1984) Multiframe image restoration and registration. JAI Press Inc., London
7. Sudheer Babu R, Sreenivasa Murthy KE (2011) A survey on the methods of super-resolution image reconstruction. Int J Comput Appl 15(2)
8. Liu HY, Zhang YS, Ji S (2008) Study in the methods of super-resolution image reconstruction. Int Arch Photogrammetry, Remote Sens Spat Inf Sci 37(B2)
9. Chen MJ, Huang CH, Lee WL (2002) A fast edge-oriented algorithm for digital images. Image Vis Comput 20:805–812
10. Elad M, Fueur A (1997) Restoration of a single super-resolution image from several blurred, noise and undersampled measured images. IEEE Trans Image Process 6(12):1646–1658
11. Freeman WT, Jones TR, Pasztor EC (2002) Example-based super-resolution. IEEE Comput Graphics Appl 22(2):56–65
12. Baker S, Kanade T (2002) Limits on super resolution and how to break them. IEEE TPAMI 24(9):1167–1183

Chapter 22
Fingerprint Recognition Based on Multi-Resolution Histogram of Gradient Descriptors

Munalih Ahmad Syarif, Thian Song Ong and Connie Tee

Abstract This paper proposes a method to extract the fingerprint feature by using multi-resolution Histogram of Oriented Gradient representation. In this work, fingerprint is first enhanced for better ridge appearance. Next, Histogram of Oriented Gradient (HOG) is applied to model ridge valley structure as the occurrence of gradient information into a histogram bin size to obtain the fingerprint descriptor. Specifically, Multi-resolution fingerprint representation with HOG descriptor is used to isolate and analyse the fingerprint ridge structures in different resolution for better recognition performance. Experimental analysis shows that the proposed method is feasible in both performance accuracy and computational time as compared to conventional methods.

Keywords Fingerprint recognition · Histogram of oriented gradient · Multi-resolution representation

22.1 Introduction

Biometrics has been widely accepted to replace traditional password and PIN for more reliable user authentication. Among the different biometric modalities, fingerprint is one of the oldest biometrics systems being used with promising recognition performance. Fingerprint biometrics uses the impressions made by the minutiae formations or ridge patterns found on the fingertip to identify a person. Fingerprint recognition is widely accepted by public for user authentication and access control due to its reliable and stable performance.

Generally, fingerprint recognition methods can be categorized into two different approaches, minutiae-based and texture-based. Specifically, the minutiae approach focuses on the point extraction of discontinuity of the ridge structure such as ridge termination (the ridge comes to an end) and ridge bifurcation (the ridge divides to form two ridges) [1]. The minutiae-based fingerprint recognition system may also include one or more global attributes such as finger orientation, core point location and fingerprint class information [2]. Matching is done by comparing the reference template and input minutiae sets. However, the minutiae might not be detected completely sometimes due to poor image quality, and this might lead to low matching performance [3].

On the other hand, the overall pattern or texture information of the fingerprint could be extracted more reliably than minutiae approach, especially in low quality images. The matching of the texture-based method is performed by comparing the pattern information such as frequency and orientation features between the input and template images. Gabor filters is the most popular fingerprint feature extraction method. Jain et al. [4] first proposed a filter-based algorithm to obtain both global and local pattern of the ridges and valleys information of a fingerprint and form a compact feature vector, ψ with length d by using a bank of Gabor filters. For this purpose, a region of interest and center point for the fingerprint image, called core point was determined. Then, the region of interest (ROI) was normalized and tessellated around the center point into 80 sectors and the filter was tuned with a preset mask size on the ROI in eight different directions, $\theta \in \{0°, 22.5°, 45°, 67.5°, 90°, 112.5°, 135°, 157.5°\}$ in order to filter out the noise and capture the genuine local ridge and valley details and thus preserved the distinctive information contained in a particular orientation in the image. The method showed that the result was better than minutiae-based fingerprint recognition. However, the computation load was high due to the need to achieve rotation-invariance for better performance accuracy. Moreover, Ross et al. [5] found some problem with Jain et al. method [4] in 2003. Their study in [5] explained that the circular tessellation method that [4] used suffers from following problems: (1) a core point is not always present and detectable in small-size image. (2) The method aligns the fingerprint based on a single reference point, and is prone to errors in locating/detecting the core point accurately. (3) The circular tessellation used in the method does not cover the entire image. Moreover, if the core point were detected to be close to the image boundary, the tessellation could cover only a very small portion of the image. Due to the mentioned problem, Ross et al. [5] suggested to use rectangular tessellation which covered the entire image with all the tessellated cells having the same size. Ross et al. [5] believed this can be more robust than circular tessellation as the fingerprint images were aligned using the overall minutiae information as compared to [4] which only used one core point for aligning the fingerprint images.

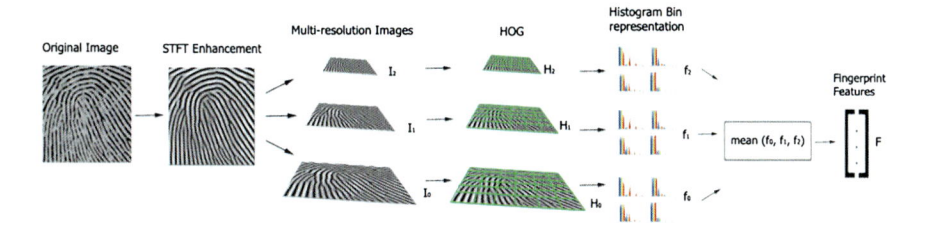

Fig. 22.1 Block diagram of the proposed method

In addition, Lee and Wang [6] proposed a scale and shift invariant representation method that made use of block wise Gabor filter response to directly extract the fingerprint magnitude features from grey scale fingerprint image. The proposed method achieved very low Equal Error Rate (EER) using k-Nearest Neighbour (K-NN) classifier, even without any pre-processing [7]. However, it was only tested on a small dataset with 192 inked fingerprints from 16 persons.

22.2 Proposed Method

In this research, we propose a method to extract fingerprint features by using multi-resolution Histogram of Oriented Gradient (MHOG). Traditionally, Histogram of Oriented Gradient (HOG) [8] has been used for object detection and recognition purpose. The HOG descriptor captures gradient information that characterizes the local spatial representation of a fingerprint image which takes into account the local contrast and illumination changes. The study in [9] mentioned that HOG with proper pre-alignment of fingerprint outperformed Locally Binary Pattern (LBP) and Gabor Filter.

In this paper, the fingerprint image is first enhanced using Short Time Fourier transform (STFT) [10] to enhance appearance of the ridge quality of fingerprint. On every fingerprint image, we extract the feature set from different resolution of fingerprint representation. Specifically, MHOG is designed to isolate, analyze and interpret the fingerprint ridge structure at different level of resolution, which we believe will be able to improve the fingerprint recognition performance. Next, the mean of the extracted feature set at different resolution levels will be determined to form the final fingerprint feature vector for matching purpose. The detail of the process is shown in Fig. 22.1.

22.2.1 Feature Extraction

In this work, we propose a multi resolution Histogram of Oriented Gradient (MHOG) method to extract the fingerprint feature. The multi-resolution image extraction using HOG can extract the feature in finer details and thus improve the fingerprint recognition performance. The detailed implementation of the proposed method is as below:

1. To obtain multi-resolution of the fingerprint representation, the original resolution of the images is decomposed two times, which means we have 3 different resolutions of a fingerprint image. The details of obtaining multi-resolution image are as follow:

   - Level 0 is the original resolution of the image, I_0.
   - Level 1 (I_1) and Level 2 (I_2) are $(100 - (n \times 30))\%$ resolution size of the I_{n-1}.

   where n is the number of level.

2. HOG is implemented on each level of the image, I_0, I_1 and I_2. I_0, which is then filtered using the following filter kernel $D_x = [-1, 0, 1]$ and $D_y = [-1, 0, 1]^T$. The results of the filtering process are $I_{0_x} = I_0 \times D_x$ and $I_{0_y} = I_0 \times D_y$. Then, the filtered image I_{0_x} and I_{0_y} are divided into m number of overlapping cell. In each cell, the calculation of the magnitude and angles (orientation) is performed. The magnitude and angles (orientation) calculation is described as follows,

   - The magnitude of the gradient is calculated as

   $$|G_0| = \sqrt{I_{0_x}^2 + I_{0_y}^2} \qquad (22.1)$$

   - The orientation, θ of the gradient computed as,

   $$\theta_0 = \arctan\left(\frac{I_{0_y}}{I_{0_x}}\right) \qquad (22.2)$$

   In this paper, the Rectangular HOG (R-HOG) is used to calculate the histogram channels. Each fingerprint is divided into a set of rectangular cells to preserve ridge continuity. The rectangular cells overlap half of their area, meaning that each rectangular cell contributes more than once to the final feature vector [11]. The magnitude G_0 and orientation θ_0 are converted and represented by h number of histogram bins to form feature vector, H_0. This process is repeated 3 times for I_0, I_1 and I_2 to obtain H_0, H_1 and H_2.

3. Next, H_0, H_1 and H_2 are filtered using Gaussian function in order to suppress high spatial frequency components of the image s, the detail of the implementation is described as follows,

$$f_n(x, y : s) = H_n(x, y) \times G(x, y : s) \qquad (22.3)$$

$$G(x, y : s) = \frac{1}{2\pi s^2} e^{-\frac{x^2+y^2}{2s^2}} \tag{22.4}$$

where G is a Gaussian low filter kernel, s is a scale. In this work we set the value of s into 1.2 and x and y are the coordinates position.

4. Then, we take the mean of f_0, f_1 and f_2 to obtain the final fingerprint feature, F.

$$F = \text{mean}(f_0, f_1, f_2). \tag{22.5}$$

22.3 Experimental Analysis

The experiment was performed using the standard FVC2002 DB1 [12] database. The database consists of sample images from 100 different persons with 8 different impressions per person. The total number of sample images used is $100 \times 8 = 800$ images. The size of each image is 176×176 pixels. Euclidean distance and Support Vector Machine (SVM) were used as the matching methods to evaluate the performance of the proposed method. The experiment was performed on a PC with 2.5 GHz Intel I5 processor and 4 GB RAM and the programming language used was MATLAB for simulation. The performance is indicated by Equal Error Rate (EER) and Receiver Operating Curve (ROC).

22.3.1 Performance Comparison with Different Size of HOG Rectangular Cell and Number of Histogram Bin

The experiments were conducted by examining the combinations of different number of cells, m and number of histogram bins, h. Specifically, m = 36, 25 and 16, h = 9 and h = 12 were used or performance comparison and the result is shown in Fig. 22.2. In this experiment, the Euclidean Distance was used as the classifier method. The experiment reveals that the best and consistent result can be obtained by setting m = 36 and h = 12 which yields 86.96 % of accuracy. This setting, m = 36 and h = 12 is used for the rest of the experiment conducted in this paper.

22.3.2 Performance Comparison of MHOG and Gabor Filter

Performance comparison was conducted to compare the proposed MHOG method to the other two well-known methods in the literature, namely FingerCode approach by Jain et al. in [2] and block Gabor Filter method in Lee and Wang [6].

Fig. 22.2 Result of experiment using the combinations of different number of cells, m and number of histogram bins, h

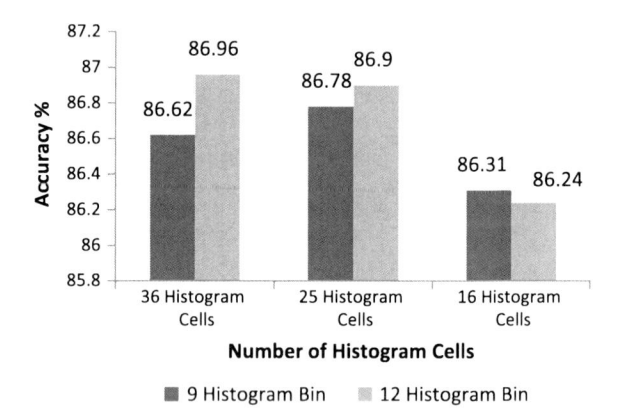

Fig. 22.3 Result of experiment using SVM as classifier with different number of training set

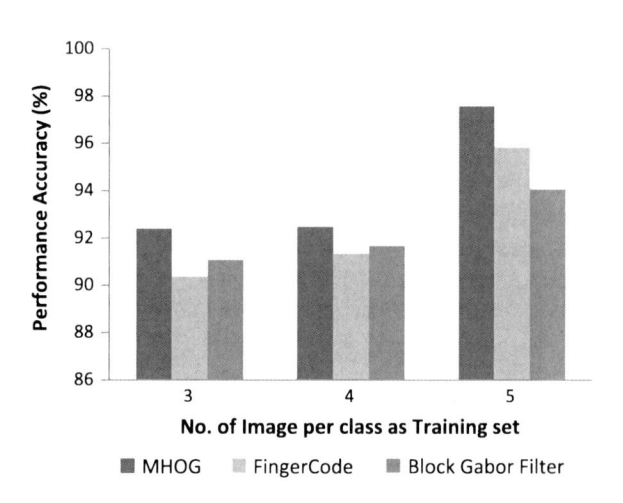

We validate that the proposed method outperform both Gabor Filter methods based on Support Vector Machine (SVM) classifier. In this experiment, the cross validation was applied. The experiment for each set number of training was conducted three times (3 observations). Each observation used different images as the training sample. For example, first observed sample image number 1 and 2 as training, the second observed image number 3 and 4 as training and the third observation used image number 5 and 6 as training. In this experiment, we used linear SVM. Figure 22.3 shows the result of the experiment using different number of training set. From Fig. 22.3 we can conclude that the best result can be obtained by setting the number of training set to 5 images per class, as we can obtain performance accuracy of 97.57 % using MHOG method. Table 22.1 shows the detailed result of the experiment using 5 images per class as the SVM training set. Figure 22.4 shows the ROC of the experiment.

Table 22.1 Result of experiment fingerprint feature extraction using MHOG, FingerCode and Gabor filter

Feature extraction method	Classification method	Recognition rate (%)	Feature extraction and matching time (s)
MHOG	SVM	97.57	0.17
FingerCode [2]		95.83	0.22
Block Gabor filter [6]		94.06	0.28

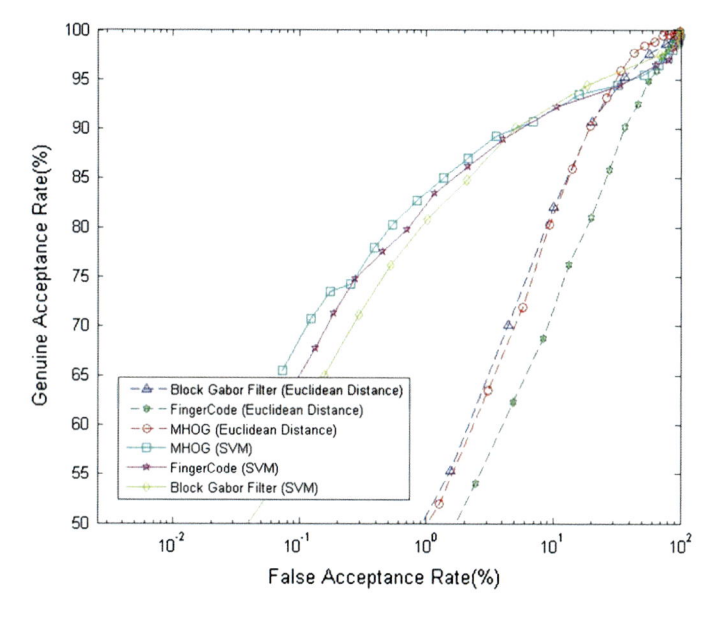

Fig. 22.4 ROC curves experiment fingerprint feature extraction using MHOG, FingerCode and Gabor filter

22.4 Conclusion

In this paper, we propose a method for fingerprint recognition by using multi-resolution Histogram of Oriented Gradient (MHOG) to isolate and interpret structures of fingerprint image in different resolutions for better performance accuracy. The viability of the proposed method has been evaluated through the experimental analysis on FVC2002 fingerprint databases and we find that the proposed algorithm provides notable SVM classification accuracy in efficient time.

Acknowledgments This research was supported by Fundamental Research Grant Scheme (FRGS) funded by the Ministry of Education Malaysia.

References

1. Maltoni D, Cappelli R (2008) Fingerprint recognition. Handbook of biometrics. Springer, Berlin, pp 23–42
2. Jain AK, Hong L, Pankanti S, Bolle R (1997) An identity-authentication system using fingerprints. Proc IEEE 85:1365–1388
3. Yang J (2011) Non-minutiae based fingerprint descriptor. InTech, Shanghai
4. Jain AK, Prabhakar S, Hong L, Pankanti S (2000) Filter-bank-based fingerprint matching. IEEE Trans Image Process 9:846–859
5. Ross A, Jain AK, Reisman J (2003) A hybrid fingerprint matcher. Pattern Recogn 36:1661–1673
6. Lee CJ, Wang SD (1999) Fingerprint feature extraction using gabor filters. Electron Lett 35:288–290
7. Nemati RJ, Javed MY (2008) Fingerprint verification using filter-bank of gabor and log gabor filters. In: Systems, signals and image processing, 208. IWSSIP 2008
8. Dalal N, Triggs B (2005) Histograms of oriented gradients for human detection. In: Proceedings of the 2005 computer society conference on computer vision and pattern recognition, Montbonnot
9. Nanni L, Lumini A (2009) Descriptors for image-based fingerprint matchers. Expert Syst Appl Int J 36(10):12414–12422
10. Chikkerur S, Cartwright AN, Govindaraju V (2007) Fingerprint enhancement using STFT analysis. Pattern Recogn 40:198–211
11. Ludwig O, Delgado D, Goncalves V, Nunes U (2009) Trainable classifier-fusion schemes: an applications to pedestrian detection. In: 12th international IEEE conference on intelligent transportation systems, vol 1. pp 4–7
12. FVC (2002). Available at http://bias.csr.unibo.it/fvc2002/

Chapter 23
A Phase Linearization Method for IF Sampling Digitizer

Eric Chan

Abstract A digitally modulated signal has phase of the carrier modulated in addition to amplitude modulation. If time domain measurement or demodulation on the modulated signal is to be performed, the measurement system or receiver must not introduce phase distortion onto the signal prior to measurement process. Traditionally, receiver IF bandwidth is made sufficiently wide compared to the signal bandwidth such that the desired signal spectrum falls within the linear phase portion of the receiver's IF response. This approach becomes impractical for new multi-carrier modulation such as various OFDM schemes with signal bandwidth in tens or even hundreds of megahertz. With limited digitizer sampling rate, the digitizer front end response (RF + IF + anti-aliasing) cannot be made sufficiently wide band to be presented as a linear phase channel to the received modulated signal. As such, a post-ADC phase linearizing scheme is required before accurate measurement of time-domain parameters of the signal can be performed. A design procedure and algorithm is presented with simulated results of a derived phase linearizing FIR filter compensating the nonlinear phase response of a 9th order Chebyshev BPF.

Keywords Phase linearization · Phase compensation

23.1 Introduction

Measuring time domain characteristics of a modulated RF/IF signal is often needed for troubleshooting transmitter and receiver. Such time domain characteristics are RF pulse rise time, fall time, pulse droop rate, pulse width etc.

E. Chan (✉)
Basic Instruments Division, Agilent Technologies Malaysia, Bayan Lepas Free Trade Zone, Bayan Lepas, Penang, Malaysia
e-mail: eric-th_chan@agilent.com; thchan2005@gmail.com

Demodulation of such a signal in a receiver is also needed if measurement such as EVM is desired. To perform measurements and demodulation of such signals, the receiver must not introduce significant phase distortion or frequency varying group delay onto the signal prior to measurement process. Unfortunately, such expectation cannot be realized for signal bandwidth occupying significant portion of the receiver IF bandwidth. A wide-band analogue front end with linear phase throughout the entire band of interest is difficult to realize if not impossible. A post-ADC, digital correction for phase distortion introduced to desired signal is now made possible with high speed real-time DSP operation in FPGAs or ASICs. What's left is to devise a procedure to measure the phase distortion introduced by RF front end and to design a digital compensation algorithm to remove the phase distortion introduced onto the signal. This paper discusses a compensation algorithm for phase distortion removal and its results, making assumption that the function of phase distortion versus frequency of the digitizer front end is known prior to correction.

23.2 Deriving Desired Compensation Response

23.2.1 Modeling a Front End Response for Phase Distortion

An IF sampling digitizer will have a band pass filter for IF filtering and anti-aliasing filtering. In this study a theoretical 9th order Chebyshev band pass response centered around 62.5 MHz is assumed to be the front end response of an IF digitizer and is to be digitally compensated, serving as an example for all the calculation in this paper. The filter transfer function can be written as:

$$T(s) = \frac{b_0}{\sum_{n=0}^{9} \left(\frac{s^2 + \omega_0^2}{Bs}\right)^n} \tag{23.1}$$

where $b_0....b_8$ are the coefficients calculated via standard filter table and filter transformation procedure. The values of the coefficients are given in Table 23.1.

With aid of a computer program or software packages, the wrapped and unwrapped phase function of the modeled front end can be calculated from Eq. (23.1). What is left is to devise a mathematical procedure to linearize the phase function via a digital FIR filter.

23.2.2 Nonlinear Phase FIR with Real Coefficients

Once the nonlinear phase of the front end is known (modeled), a FIR with phase response which corrects for the phase nonlinearities of the front end is to be obtained. We seek such a FIR with real coefficients for ease of implementations in

23 A Phase Linearization Method

Table 23.1 Coefficients values

$b_0 = 0.00767667$	$b_5 = 2.37811881$	$\omega_0 = 6.250000e8$
$b_1 = 0.07060479$	$b_6 = 1.88147976$	
$b_2 = 0.24418637$	$b_7 = 2.67094683$	
$b_3 = 0.78631094$	$b_8 = 0.91754763$	
$b_4 = 1.20160717$	$B = 5.3407075e8$	

FPGAs. A FIR with complex coefficients would require significantly more resources to implement in FPGAs as compared to FIR with real coefficients. The frequency response of a FIR is given by,

$$H(\omega) = \sum_{n=0}^{N-1} h(n) e^{-j\omega n} \qquad (23.2)$$

which can be re-written as,

$$H(\omega) = e^{-j\omega M} \sum_{n=0}^{N-1} h(n) e^{-j\omega(n-M)} \qquad (23.3)$$

where $M = (N-1)/2$.

For $h(n)$ to have linear phase it must be either symmetric or anti-symmetric and must be fully defined for $t \geq 0$ (causality). The need for linear phase $h(n)$ to be symmetric around $(N-1)/2$ becomes obvious. For a FIR that corrects for phase distortion, causality is still required desired $h(n)$ must be neither symmetric nor anti-symmetric in order for it to exhibit non-linear phase response. Exactly how non-linear phase can be achieved by giving up symmetric (or anti-symmetric) properties of $h(n)$ can be made clear by realizing that $h(n)$ can be thought of as a combination of a symmetric sequence $h_s(n)$ and another anti-symmetric sequence $h_a(n)$.

$$h(n) = h_s(n) + h_a(n) \qquad (23.4)$$

Substituting $h_s(n) + h_a(n)$ into $h(n)$ gives,

$$H(\omega) = e^{-j\omega M} \left[\sum_{n=0}^{N-1} h_s(n) e^{-j\omega(n-M)} + \sum_{n=0}^{N-1} h_a(n) e^{-j\omega(n-M)} \right] \qquad (23.5)$$

By symmetric properties of the sequences and by Euler's equation,

$$H(\omega) = e^{-j\omega M} \left[\sum_{n=0}^{M-1} 2h_s(n) \cos(\omega(n-M)) + \sum_{n=0}^{M-1} j2h_a(n) \sin(\omega(n-M)) \right] \qquad (23.6)$$

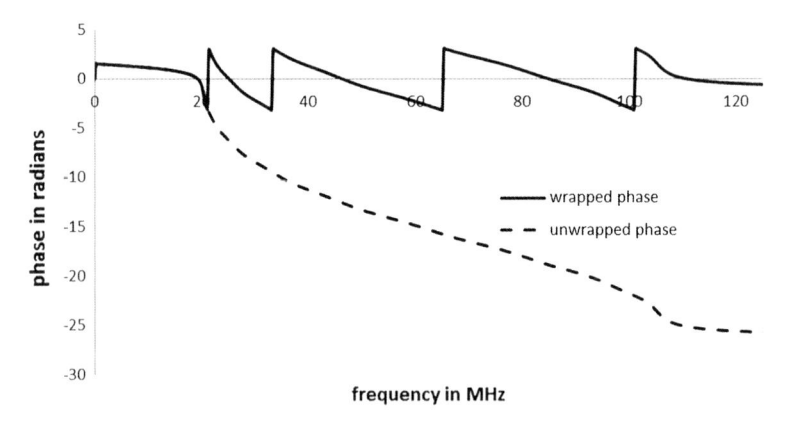

[Fig. 23.1 chart]

Fig. 23.1 Wrapped phase and unwrapped phase of the front-end

Stating $\emptyset(\omega)$ as the phase function we need for correcting the front-end phase distortion, it is obvious that,

$$\cos(\emptyset(\omega)) = \sum_{n=0}^{M-1} 2h_s(n)\cos(\omega(n-M)) \qquad (23.7)$$

$$\sin(\emptyset(\omega)) = -\sum_{n=0}^{M-1} 2h_a(n)\sin(\omega(n-M)) \qquad (23.8)$$

Depending on how $\emptyset(\omega)$ is known (numerically or analytically), various methods can be used to obtain $h_s(n)$ and $h_a(n)$. Frequency sampling [1], Remez exchange [2] and multi-dimensional optimization are few options available. Complexity and difficulty in implementing Remez exchange and optimization methods are challenges which test and calibration engineers encounter when developing test and calibration software and setup. Frequency sampling design method while not the most efficient in terms of resulting filter length, is being used here for be ease of implementation in a test and calibration system.

23.2.3 Deriving the Linearizing Function $\emptyset(\omega)$

The unwrapped phase function shown in Fig. 23.1, has points with fast changing slopes at both regions close to the ends. These regions are outside the pass band where signal components are attenuated by the front-end. Points with fast changing slopes can be replaced with points with slow changing slope, as shown in Fig. 23.2, to minimize rippling in the synthesized linearizing phase response. The resulting curve is actually a sum of linear function and non-linear function. The non-linear portion of the function represents the phase to be subtracted during

23 A Phase Linearization Method

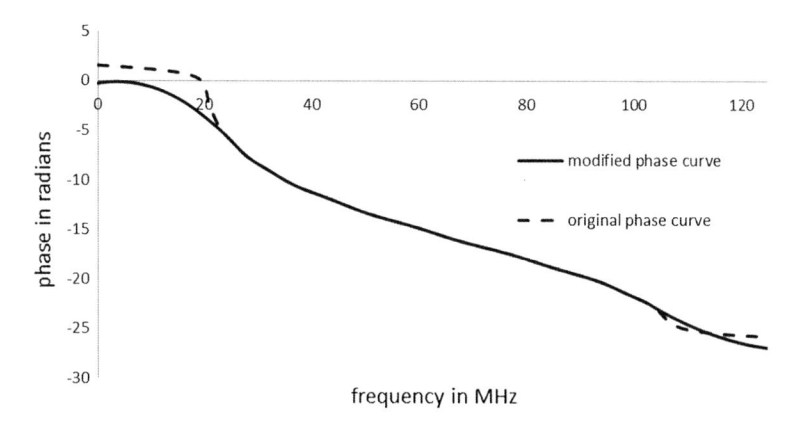

Fig. 23.2 Modifying phase function outside the frequency band of interest

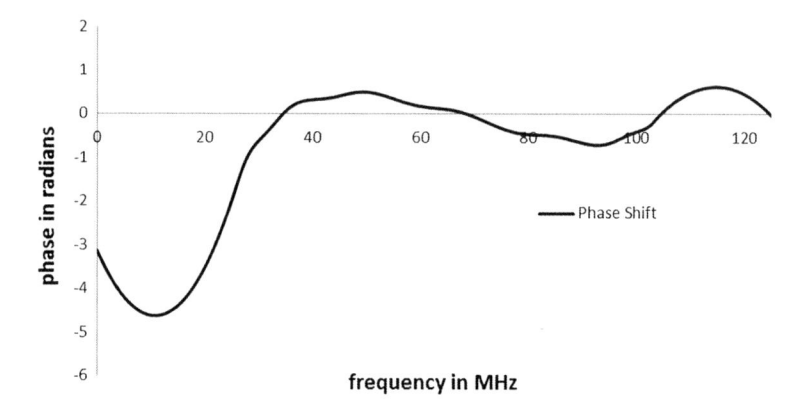

Fig. 23.3 Target phase response of the phase linearizing filter

digital compensation. Linear regression can be carried out on points of the curve to extract the value of slope and intercept of the linear phase function. In this example, the linear phase function is calculated to be $-1.878f - 3.377$. After the linear phase portion is being subtracted from the modified phase curve, whatever left is the non-linear portion which is to be further removed by the linearizing filter. Hence $\varnothing(\omega)$ is found and in this example $\varnothing(\omega)$ is shown in Fig. 23.3.

23.2.4 Calculating Phase Linearizing FIR Coefficients

In this example, 101-tap, odd length filter is assumed. As such, the filter coefficients can be calculated via frequency sampling formulas below,

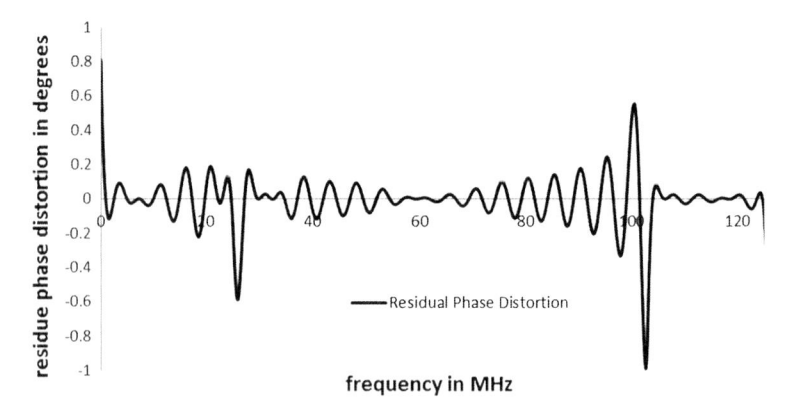

Fig. 23.4 Residual phase distortion after application of phase linearizing filter

$$h_s(n) = \frac{1}{N}\left[\sum_{k=1}^{L}\left(2\cos(PhaseShift_k)\cos\left(\frac{2\pi k}{N}(n-L)\right)\right) + \cos(PhaseShift_0)\right] \quad (23.9)$$

$$h_a(n) = \frac{1}{N}\left[\sum_{k=0}^{L}2\sin(PhaseShift_k)\sin\left(\frac{2\pi k}{N}(L-n)\right)\right] \quad (23.10)$$

The frequency response of the phase linearizing filter is given by.

$$J(\omega) = \sum_{k=0}^{N-1}(h_s(n) + h_a(n))e^{-j\omega k} \quad (23.11)$$

The residual phase distortion after the application of the synthesized phase linearing filter can then be calculated and shown in Fig. 23.4. In this example, the residual phase error within the band of interest is less than 1°.

23.3 Conclusion

Synthesizing a non-linear phase real FIR requires a set of non-symmetric filter coefficients. Breaking up the coefficients into of symmetric and anti-symmetric coefficients allows separate synthesis of real and imaginary parts of the linearizing filter frequency response. By careful shaping of front-end phase function at the stop bands, the synthesized phase linearizing filter can have phase response that minimizes the overall system phase distortion.

References

1. Parks TW, Burrus CS (1987) Digital filter design. Wiley-Interscience, Hoboken
2. McClellan JH, Parks TW, Rabiner LR (1973) A computer program for designing optimum FIR linear phase digital filters. IEEE Trans Audio Electroacoust AU-21(6):506–526
3. Proakis JG, Manolakis DG (2007) Digital signal processing principles, algorithms and applications, 4th edn. Pearson Prentice Hall, Upper Saddle River

Chapter 24
Driving Circuitry of Complementary Metal Oxide Semiconductor (CMOS) Area Image Sensor for Optical Tomography Instrumentation System

Suhaila Mohd Najib, Mariani Idroas and Muhammad Nasir Ibrahim

Abstract Process tomography is a system that can be used to visualize the exact behavior of internal flow in any process. An optical tomography system can safely reconstruct images of the internal flow in non-intrusive and non-invasive sensor. The optical sensor used in the optical tomography system is the most critical part that determines the accuracy and reliability of the system. Therefore, a complementarymetal oxide semiconductor (CMOS) area image sensor is used in the developed optical tomography system, in order to produce a high resolution image without disturbing the process flow. This paper describes the development of the driving circuitry system to test the performance of the monochrome CMOS area image sensor.

Keywords CMOS area image sensor · Optical tomography

24.1 Introduction

Various types of optical tomography systems have been designed for process tomography. These systems involve the implementation of optical waveguide detector [1], photodiode [2], optical fiber sensor [3], and charged couple device [4].

M. Idroas
Faculty of Petroleum and Renewable Energy Engineering, Universiti Teknologi Malaysia, 81310 Skudai, Johor, Malaysia
e-mail: mariani@petroleum.utm.my

S. Mohd Najib (✉) · M. N. Ibrahim
Faculty of Electrical Engineering, Universiti Teknologi Malaysia, 81310 Skudai, Johor, Malaysia
e-mail: suhailamohdnajib@yahoo.com

M. N. Ibrahim
e-mail: mnasir@fke.utm.my

Nowadays, with the existence of the advance technology, it can provide an optical tomography system with higher capability to produce a simple and smaller circuit size, high resolution image, high performance system, all at a low cost.

The main purpose of this project is to minimize the complexity of optical tomography system as well as increasing the resolution of the reconstructed image by implementing the new optical sensor which is the Complementary Metal Oxide Semiconductor (CMOS) area image sensor. Other than that, advanced technology provided by the microcontroller allows us to develop a simple but complete optical tomography system with a reasonable cost. Referring to previous researches, the usage of incredibly fast and powerful a well-known microprocessor, which is the Digital Signal Processor (DSP), made the optical tomography system simpler, with lower cost and competitive in performance where the microprocessor itself acts as a Data Acquisition System (DAS) [3]. The research is similar to Pearson (2009) where the author developed a system to detect and separate defected grains with the combination of a CMOS colour image sensor and a Field-Programmable Gate Array (FPGA) without the need of an external personal computer to execute image processing in real-time [5]. Additionally, Idroas (2004) investigated the advantages of Charge Coupled Device (CCD) to replace the previous optical sensor where it produces a high resolution image [4].

The overall aim is to assist in understanding the system in order to realize the potential of using CMOS area image sensor together with microcontroller, in order to produce a non-invasive and non-intrusive tomographic imaging of process industry.

24.2 Hardware Design

A typical process tomography system consists of three basic parts, which are the sensor, the conditioning circuit and the display unit [6]. The sensor is the most critical part as it will examine closely the process where it will affect the accuracy of the system.

Figure 24.1 shows the block diagram of the developed optical tomography system. There are four main elements in the system—the lighting system, the measurement section, the microcontroller module and the image reconstruction system. When the collimated laser light propagates through the measurement section, it will strike the CMOS area image sensor. The outputs from the CMOS area image sensor are formed where each pixel of the sensor contains its own photodiode, an Analog-to-Digital converter (ADC) and an active amplifier. The output is captured, sampled and transferred to a personal computer through the microcontroller based DAS where SD card as a temporary data storage. Finally, the output sensor data will be used in MATLAB software to obtain the internal properties inside the process vessel and the image will be constructed using the Inverse Problem method. In this paper, we will only focused on developing the driving circuit for CMOS area image sensor before it is being captured by the microcontroller.

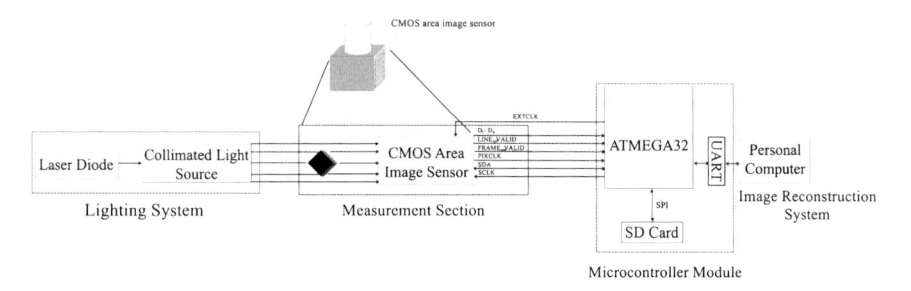

Fig. 24.1 Block diagram of the optical tomography system

Figure 24.2 shows one projection of the optical tomography system, where the collimated laser illuminated an object in a pipe and the shadow of the object is casted in front of the CMOS area image sensor. Figure 24.3 shows the monochrome CMOS area image sensor Model MT9M001C12STM used in the system. The CMOS area image sensor generates a stream of pixel data activated by LINE_VALID and FRAME_VALID signals. The active pixel array is configured as 1,280 columns by 1,024 rows which produced 1.3 Megapixels area image. One pixel of the image sensor has a size of 5.2 micron by 5.2 micron, and it is formed in 10 bits digital output. When the image is read out from the sensor, it is being read one row at a time starting with the first pixel read out-pixel (16, 8) which is in the top left corner of the image sensor (see from FRONT).

Figure 24.4 illustrates the driving circuit for the CMOS area image sensor which oscillates at 1.5 kHz. The 10 bits output data from the CMOS area image sensor are connected to the 10 bits Digital to Analog converter (DAC) which is designed based on R/2R resistors network. The output of the DAC is amplified using operational amplifier LM741. The DAC is used in order to get the general view of the monochrome CMOS area image sensor output. The output of the op-amp will be equivalent to the weight of the binary number on lines D_0-D_9.

24.3 Result and Discussions

Several experiments with various conditions were conducted to verify the performance of the circuit. The verification was done by testing the performance of the CMOS area image sensor with two (2) modes, which are the row mode and frame mode. For row mode, it involves five different conditions of obstacles which are (a) fully opened, (b) fully blocked, (c) partially right blocked, (d) partially left blocked, and (e) vertically center blocked. For the frame mode, the experiment was conducted with six different conditions: (a) fully opened with void area, (b) centrally blocked with void area, (c) partially left blocked with void area, (d) partially lower blocked, (e) partially upper blocked, and (f) horizontally center blocked. The results for each experiment are shown in Figs. 24.5 and 24.6.

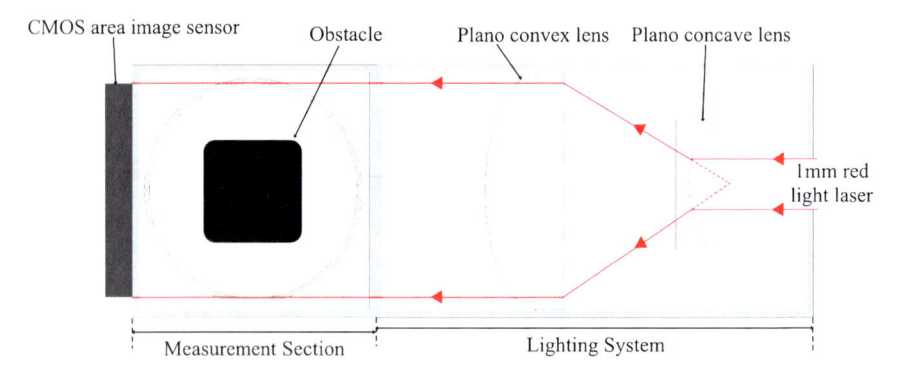

Fig. 24.2 The arrangement inside of the one projection

Fig. 24.3 Back and front image of MT9M001C12STM

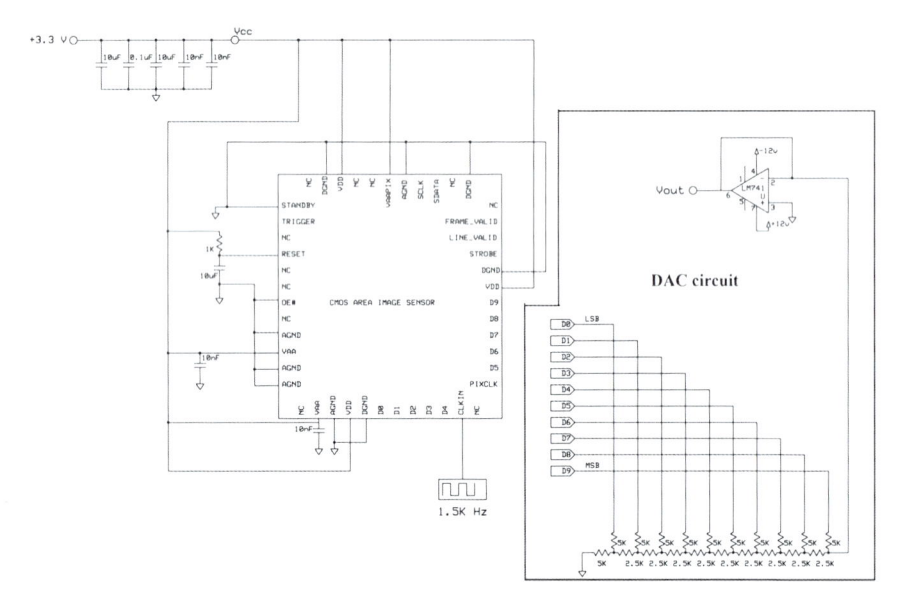

Fig. 24.4 Driving circuit of the CMOS area image sensor

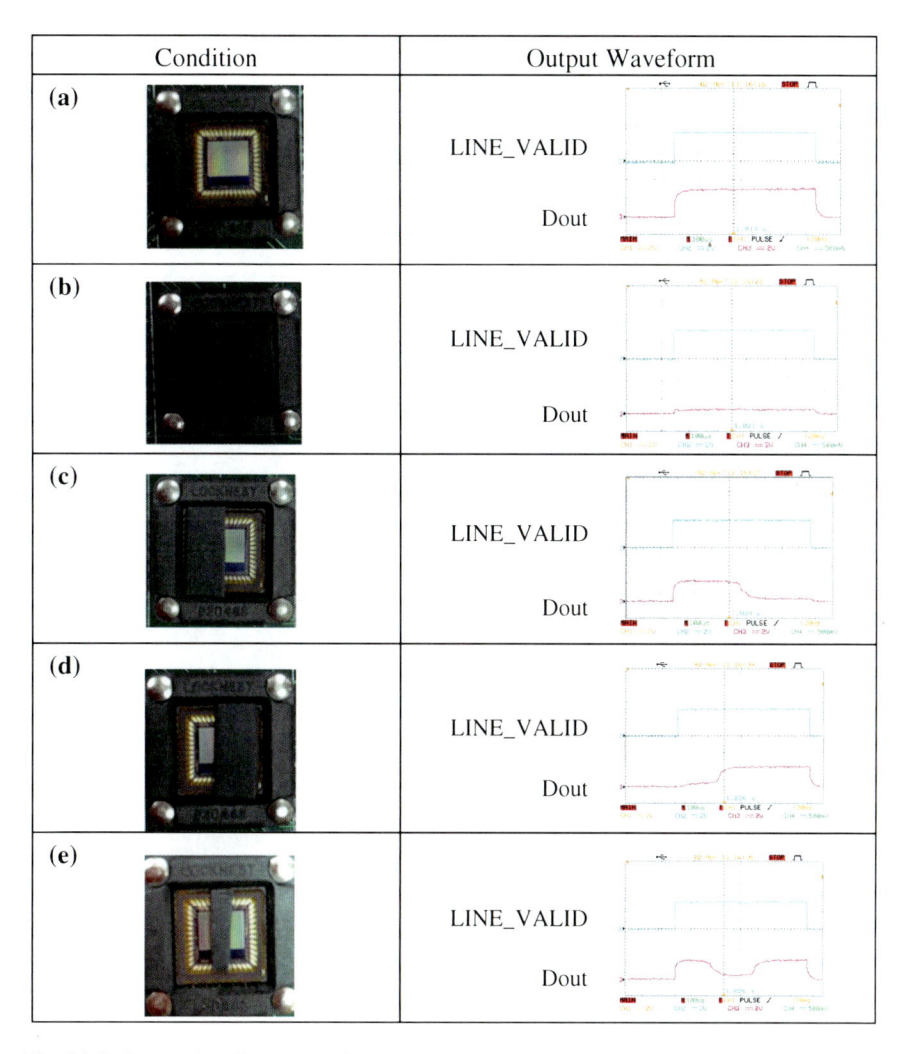

24 Driving Circuitry of Complementary Metal Oxide Semiconductor 209

	Condition	Output Waveform
(a)		LINE_VALID Dout
(b)		LINE_VALID Dout
(c)		LINE_VALID Dout
(d)		LINE_VALID Dout
(e)		LINE_VALID Dout

Fig. 24.5 Output data for row mode

All experiments were done in controlled ambient light to allow high performance of the sensor and to avoid it being saturated. The experiment pictures show the back of the image sensor which was covered by the obstacle. The output data is synchronized with the PIXCLK and it is used to sample the data. When the LINE_VALID is HIGH, one pixel datum is output at every PIXCLK period. When FRAME_VALID is HIGH, one frame of data was completely sampled.

The monochrome CMOS area image sensor produces an output data with periodic pattern of nearly ground potential when it was blocked from light. When the image sensor is fully exposed to the light, the output pattern will be high. By referring to Fig. 24.5c, the CMOS image sensor was partially left blocked (seen from back). The waveform consists of the LINE_VALID signal that showed a

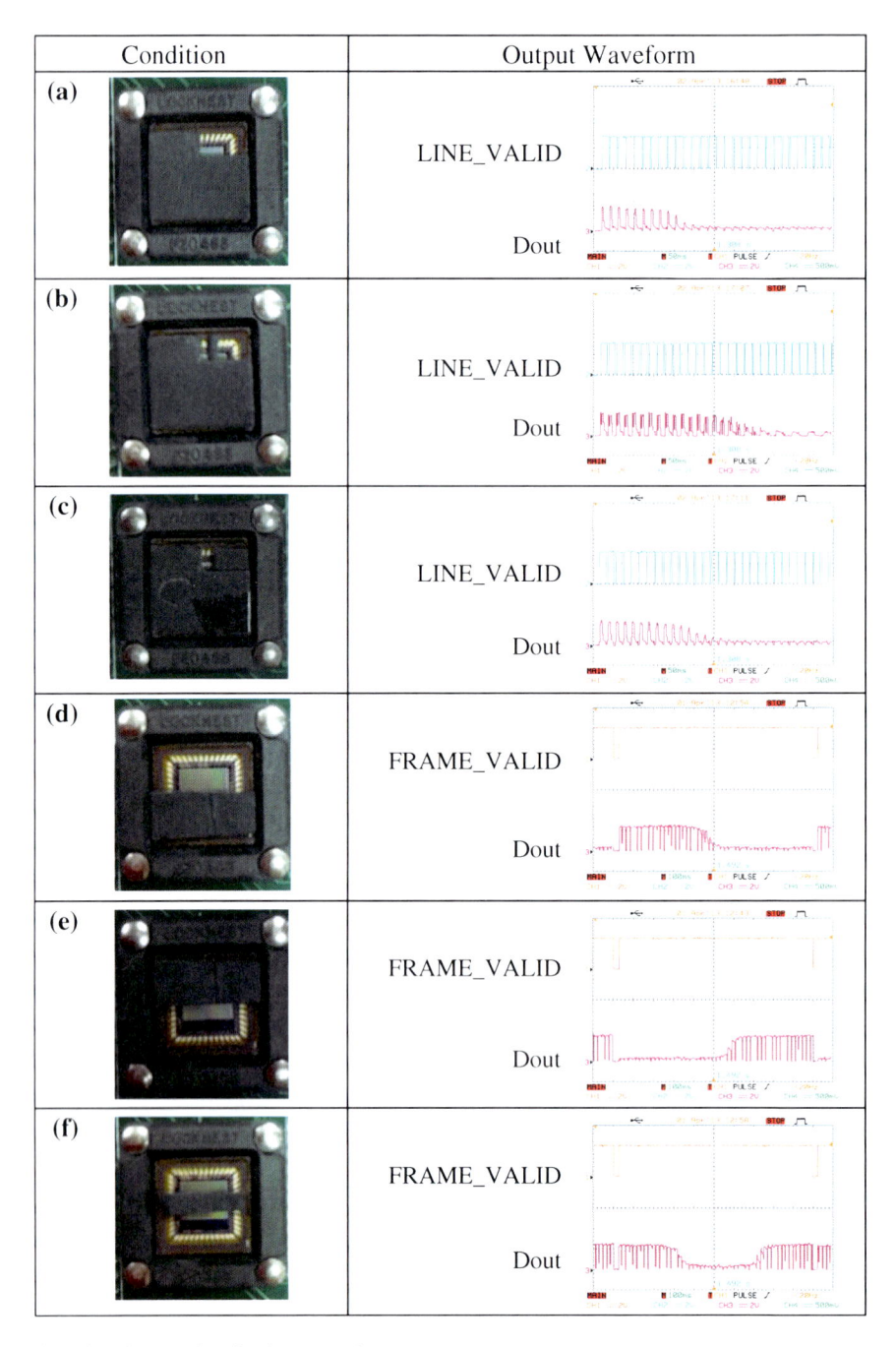

Condition	Output Waveform
(a)	LINE_VALID Dout
(b)	LINE_VALID Dout
(c)	LINE_VALID Dout
(d)	FRAME_VALID Dout
(e)	FRAME_VALID Dout
(f)	FRAME_VALID Dout

Fig. 24.6 Output data for frame mode

complete one output row, where the output data produced high voltage for the first half of the row.

In Fig. 24.6a–c, the area covered by the black card is considered as a void area. The purpose of these cases is to identify the output data when the CMOS area image sensor is exposed to a small area. For Fig. 24.6d–f, the cases show the output voltage attribute for the complete frame by referring to the FRAME_VALID signal. For example in Fig. 24.6d, when the CMOS area image sensor was partially blocked, the output image sensor produce almost ground voltage at the last half frame.

24.4 Conclusion

Then experiment was done to ease our understanding on the CMOS area image sensor outputs were being sampled. For this reason, in future system development, we will use the digital output from the CMOS area image sensor directly to the microcontroller. A CMOS area image sensor has a high potential of producing a high resolution reconstructed image as it has 1.3 Megapixels resolution with a pixel size of 5.2 μm^2. Moreover, the cost of an optical tomography system can be reduced since the CMOS area image sensor provides a digital output which is ready for image processing without extra circuitry needed.

24.5 Acknowledgment

The authors would like to acknowledge the Malaysian Ministry of Education and Universiti Teknologi Malaysia for research grant (PRGS Vot number 04L611) for their support given.

References

1. Mariusz RR, Andrzej P (2003) Application of optical tomography for measurements of aeration parameters in large water tanks. Meas Sci Technol 14:199–204
2. Margi S, Hariyadi S (2010) A design of simple, portable optical tomography apparatus using 904 nm NIR laserdiode. Optik 121:1418–1422
3. Ruzairi AR, Chiam KT, MHF Rahiman (2008) An optical tomography system using a digital signal processor. Sensors 8:2082–2103
4. Idroas M (2004) A charge couple device based optical tomographic instrumentation system for particle sizing, Ph.D. Thesis
5. Tom P (2009) Hardware-based image processing for high-speed inspection of grains. Comput Electron Agric 69:12–18
6. Dickin FJ, Hoyle BS, Hunt A, Huang SM, Ilyas O, Lenn C, Waterfall RC, Williams RA, Xie CG, Beck MS (1992) Tomographic imaging of industrial process equipment: techniques and applications. In: IEE proceedings-G, vol 139, no I. pp 72–82

Chapter 25
A Novel Technique for Mammogram Mass Segmentation Using Fractal Adaptive Thresholding

P. Shanmugavadivu and V. Sivakumar

Abstract Digital mammogram is emerged as a most reliable screening technique for the early diagnosis of breast cancer and it paves an opportunity for researchers to develop novel algorithms for computer aided detection. Presence of clusters of microcalcifications as masses in mammograms is an important early indication of breast cancer. Fractal geometry is an efficient mathematical approach that deals with self-similar, irregular geometric objects called fractals. As the breast background tissues have high local self-similarity, which is the basic property of fractals, a new fractal method is proposed in this paper for the detection and segmentation of circumscribed masses from mammograms. The median filtering, label removal and contrast enhancement are done as pre-processing measures which makes the process of segmentation of masses, easier. The proposed technique then segments the circumscribed masses using Fractal adaptive thresholding with the application of morphological operations. This Fractal based mammogram mass segmentation is able to produce encouraging results that substantiate the merit of the proposed technique.

Keywords Fractal dimension · Median filtering · Mammogram · Thresholding · Image segmentation

25.1 Introduction

The application of Digital Image Processing (DIP) in Medical Image Processing has gained a popular momentum in the diagnosis of medical images. Image segmentation, an element of DIP that deals with subdividing an image into its

---

P. Shanmugavadivu · V. Sivakumar (✉)
Department of Computer Science and Applications, Gandhigram Rural Institute,
Deemed University, Gandhigram 624302 Tamil Nadu, India
e-mail: sivakumar.vengusamy@gmail.com

P. Shanmugavadivu
e-mail: psvadivu67@gmail.com

constituent regions or objects, is a highly demanding goal in health diagnosis. Accurate and simple segmentation results help to capture the necessary vital objects/regions of interest with respect to a given property and ignore the insignificant details of an image [1, 2].

Breast Cancer is one of the most significant causes of fatality among all cancers for women in both developed and developing countries. In India, a death rate of one in eight women has been reported due to breast cancer [3–6]. Recently mammography has emerged as the most effective, low cost and reliable method used for the early detection of breast cancer and other abnormalities in the breast tissues. According to the radiologists' version, a patient is declared to have breast cancer if certain types of masses or calcifications which appear as small white specks are identified in the mammogram [7–10]. Fractal-based techniques are applied in mammogram segmentation as the breast tissues are found to possess self-similar property of fractal objects and in the proposed technique it is used for the segmentation of masses from mammograms [11–14]. Fractal dimension, an important characteristic of fractal objects, has got wider applications in the fields of image segmentation and shape recognition [15–22].

In this paper, Sect. 25.2 describes the basics of fractal dimension, Median filtering and morphological operators. Section 25.3 presents the computational methodology of the proposed mass segmentation. The results and discussion are presented in Sect. 25.4 and the conclusions are drawn in Sect. 25.5.

25.2 Techniques Adopted in the Present Work

25.2.1 Fractal Dimension

Self-similarity is defined as a property, where a subset is indistinguishable from the whole, when magnified to the size of the whole. Fractals are of rough geometric self-similar and irregular shapes which can be subdivided in parts, each of which is reduced to similar of the whole [15, 16]. Fractal objects are characterized by their Fractal dimension (defined as D) which is an important characteristic of fractals as it has got information about their geometric structure. In Euclidean n-space, the bounded set X is said to be self-similar when X is the union of N_r distinct non-overlapping copies of itself, each of which is similar to X scaled down by a ratio r. Fractal Dimension D of X can be derived from the relation [15], as

$$D = \frac{\log(N_r)}{\log\left(\frac{1}{r}\right)} \qquad (25.1)$$

The application of fractals in the present work facilitates for an effective mass segmentation because the masses are well separated in the image using fractal adaptive thresholding and therefore promises an easier and more robust segmentation.

25.2.2 Median Filtering

Order statistic filters are non-linear spatial filters whose response is based on ordering the pixels contained in an image neighbourhood and then replacing the value of the center pixel in the neighbourhood with the value determined by the ranking result. The best-known order statistic filter in digital image processing is the median filtering technique which corresponds to the 50th percentile [1, 2]. The median filter compares each pixel in the image to its surrounding neighbour pixels and classifies pixels as noise. Then these noise pixels are replaced by median pixel value of the neighbourhood pixels. The approach of Median filtering is a useful tool to smoothen the image and to reduce the excessive distortions such as thinning or thickening of object boundaries. Median filtering helps to remove the digitization noise and high frequency components from the mammography images.

25.2.3 Morphological Operators

The mathematical morphology is used as a tool for extracting image components that are useful in the description and representation of region shapes in an image [1]. In digital morphology, a small pattern or shape, which is known as structuring element, probes the image. The morphological operations used in the present paper are dilation, erosion, opening and reconstruction. For the dilation operation, the area around a pixel is set as the structuring element and the original object is allowed to grow larger. Erosion is an operation on the image in which the pixels matching the structuring element are deleted. Opening is an important morphological transformation created as a result of erosion followed by dilation. Opening with the structuring element is used to eliminate specific image details smaller than the structuring element. Reconstruction is a morphological transformation which makes use of one image that contains the starting points for transformation and other image to constrain the transformation with the help of structuring element that defines connectivity. The definitions of these operations are dependent on the image types, such as binary, gray level or color of the image being processed [1, 2].

25.3 Methodology

The proposed technique initially finds the fractal dimension, D by differential box counting method, for an input mammogram image I, using Eq. (25.1). Then fractal Hurst H is calculated by finding difference between D and the dimension value of the image. Also statistical measure standard deviation S is found out for the image. Now two fractal thresholding values T1 and T2 are formulated by calculating $T1 = (H*H)/S$ and $T2 = (H/S)*D$. Filtering is then applied on the image I using median filtering technique to acquire I_{FM}. Then the film artifacts such as labels and x-ray marks are removed from I_{FM} by making the pixels as ones whose intensity values are greater than fractal thresholding value T1 and else pixels as zeros along with the usage of morphological operations to produce the image I_{LABEL}. Further, the label removed image I_{LABEL} is enhanced, in order to increase the variation in brightness and to improve the computational consistency, by framing a look up table with respect to intensity value ranges and converting the values in I_{LABEL} based on the look up table to acquire I_E, thus increasing the contrast of the filtered label removed image. Now, the mass is detected from I_E by considering only those pixels whose intensity values are greater than the Fractal thresholding value T2 which is represented by the image I_{DETECT}. Application of morphological operations dilation, erosion and reconstruction again on the image I_{DETECT} suppresses the pectoral muscle and results in the final segmented mass image I_{MASS} from the input mammogram image.

The algorithm for the above methodology is given as follows.

Algorithm: Segmentation of masses from mammogram
Aim: To segment the masses from mammogram
Input: A 2-Dimensional mammogram image, I
Output: Segmented mass from mammogram

Stage 1: Computation of Fractal Thresholding
1: Read a 2-D input mammogram image I.
2: Cover the image with boxes of size r.
3: [X, Y] ← IMSIZE [I].
4: If X > Y then r ← X; Else r ← Y.
5: Let min. and max. gray levels of the image fall in k and l box respectively.
6: Calculate $n_r(i, j) = l - k + 1$ where $n_r(i, j)$ is the contribution at the (i,j)th grid.
7: Find $N_r = \sum n_r(i, j)$, where N_r is the summation of n_r with respect to r.
8: Compute fractal dimension D using Eq. (25.1).
9: Calculate standard deviation S for I.
10: Calculate fractal Hurst H = D-2.
11: Formulate Fractal thresholding values using H and S, by (i) T1 = (H*H)/S and (ii) T2 = (H/S)*D.

Stage 2: Preprocessing of I (Filtering, Label removal and Enhancement)
12: Filter the input image I using median filtering to acquire I_{FM}.
13: Mark all the intensities of I_{FM} such that $I_{FM}(i, j) > T1$ into I_{LABEL} to acquire the label removed image.
14: Enhance the image I_{LABEL} to I_E by framing a look up table.

Stage 3: Detection and Segmentation of Masses in I
15: Mark all the intensities of I_E such that $I_E(i, j) > T2$ to I_{DETECT}.
16: Dilate, Erode and reconstruct the image I_{DETECT} to suppress the pectoral muscle and to produce the final segmented mass image I_{MASS}.
17: Stop.

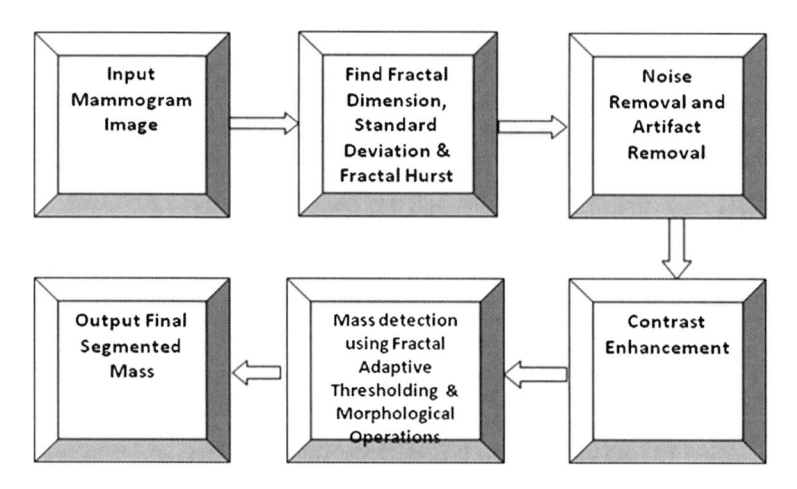

Fig. 25.1 Block diagram for computational methodology of mass segmentation

The above methodology is represented diagrammatically in block diagram in Fig. 25.1.

25.4 Results and Discussion

The presented procedure is implemented using Matlab 7.8. To ascertain the merits of the proposed method, experiments are conducted on different mammogram images collected from the Mini-Mammographic database of the Mammographic Image Analysis Society from the Pilot European Image Processing Archive (PEIPA) [23] at the University of Essex and the proposed algorithm is tested on over thirty mammograms with circumscribed masses. Since mammograms are medical images that are difficult to interpret, a preprocessing stage for mammograms is required to improve the image quality in order to make the segmentation results more accurate. Hence, according to the principle of the proposed methodology, as a preprocessing step, initially the input image has been filtered using median filtering. Then, the labels are removed from filtered image by considering only those pixels whose intensity values are greater than the fractal thresholding value T1 and further by applying morphological erosion operation. Further, contrast enhancement is done for the label removed image by forming a look up table with respect to intensity value ranges and converting those values based on the look up table.

For illustrative purpose, the results of 4 mammograms (mdb010.pgm, mdb025.pgm, mdb028.pgm and mdb132.pgm) with circumscribed masses are depicted in Fig. 25.2a1–f1, a2–f2, a3–f3, a4–f4, a5–f5 and a6–f6 respectively. The original image, median filtered image, label removed image and the contrast

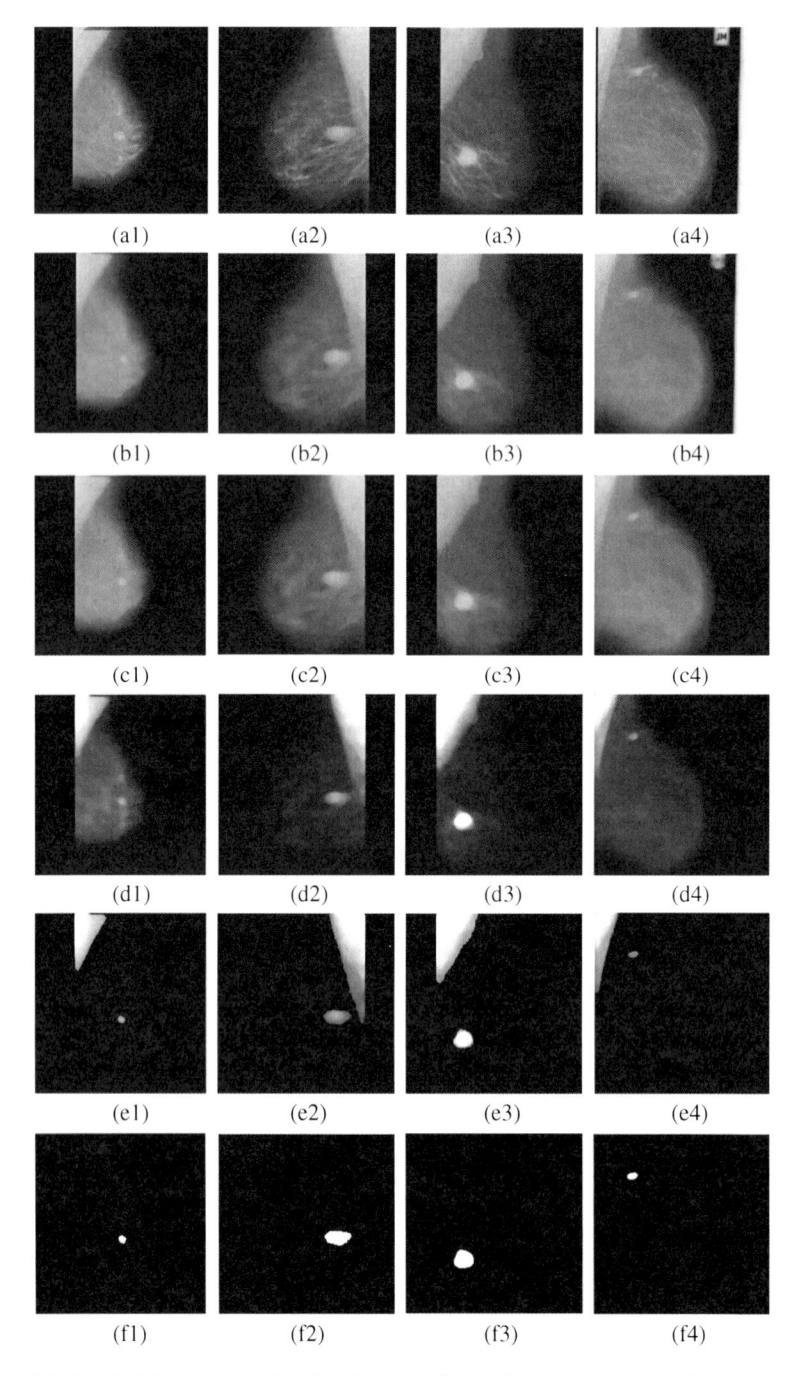

Fig. 25.2 Original image of **a1** mdb010.pgm, **a2** mdb025.pgm, **a3** mdb028.pgm and **a4** mdb132.pgm. **b1–b4** Median filtered image of **a1–a4**. **c1–c4** Label removed image of **b1–b4**. **d1–d4** Contrast enhanced image of **c1–c4**. **e1–e4** Detected image of **d1–d4**; **f1–f4** Final segmented mass from **e1–e4**

enhanced image for the input mammograms are depicted in Fig. 25.2a1–a4, b1–b4, c1–c4 and d1–d4 respectively. Now, the mass is detected from the contrast enhanced image by taking into account only those pixels whose intensity values are greater than the Fractal thresholding value T2 and the final mass segmentation is achieved by suppressing the pectoral muscle in the detected image with the application of morphological dilation and erosion operations along with morphological reconstruction operation. The detected image and the final mass segmentation results are shown in Fig. 25.2e1–e4 and f1–f4 respectively. The segmentation results clearly indicate that this proposed technique using Fractal dimension based adaptive thresholding is proved to be accurate in the segmentation of mass from a digital mammogram.

25.5 Conclusion

A novel fractal oriented mammogram mass segmentation method is designed and presented in this paper that enables the early detection of breast tumor. The obtained results over various mammograms with circumscribed masses from miniMIAS database have justified the merit of this method. Moreover, the results indicate that the presented technique can facilitate and assist the radiologists in the early diagnosis of breast cancer. As a future scope the proposed fractal based method can also be used for the classification of those detected masses into either as benign or malignant.

References

1. Gonzalez RC, Woods RE (2009) Digital image processing, 2nd edn. Prentice Hall, Englewood Cliffs
2. Sonka M et al (2008) Digital image processing and computer vision. Cengage Learning, Stamford
3. Cheng HD, Cai X et al (2003) Computer-aided detection and classification of microcalcifications in mammograms: a survey. Pattern Recognit 36:2967–2991
4. Marrocco C, Molinara M et al (2010) A computer-aided detection system for clustered microcalcifications. Artif Intell Med 50:23–32
5. Kom G, Tiedeu A, Kom M (2007) Automated detection of masses in mammogram by local adaptive thresholding. Comput Biol Med 37:37–48
6. Maitra IK et al (2012) Technique for preprocessing of digital mammogram. Comput Methods Programs Biomed 107:175–188
7. Stojic T et al (2006) Adaptation of multifractal analysis to segmentation of microcalcifications in digital mammograms. Phys A 367:494–508
8. Vuduc R (1997) Image segmentation using fractal dimension. Report on GEOL 634, CU
9. Ferrari RJ, Rangayyan RM et al (2004) Automatic detection of the pectoral muscle in mammograms. IEEE Trans Med Imaging 23(2):232–245
10. Raba D, Oliver A et al (2005) Breast segmentation with pectoral muscle suppression on digital mammograms. Lecture notes on computer science, Springer Series

11. Bhadoria S, Bharwani Y, Pati A (2012) Removal of pectoral muscle in mammograms using statistical parameters. Int J Comput Appl 43(6):0975–8887
12. Shanmugavadivu P, Sivakumar V (2012) Fractal approach in digital mammograms: a survey. NCSIP-2012, pp 141–143, ISBN: 93-81361-90-8
13. Shanmugavadivu P, Sivakumar V (2012) Fractal dimension based texture analysis of digital images. Procedia Eng (Elsevier-Science Direct) 38:2981–2986, ISSN: 1877-7058
14. Mohamed WA, Alolfe MA, Kadah YM (2009) Fast fractal modeling of mammograms for microcalcifications detection. In: 26th National radio science conference, Egypt
15. Addison PS (2005) Fractals and Chaos. IOP Publishing, Bristol
16. Welstead ST (1999) Fractal and wavelet image compression techniques. Tutorial texts in optical engineering, vol TT40
17. Lopes R (2009) Fractal and multifractal analysis: a review. Med Image Analysis 13:634–649
18. Chen DR et al (2005) Classification of breast ultrasound images using fractal feature. Clin Imaging 29:235–245
19. Shanmugavadivu P, Sivakumar V (2012) Fractal based detection of microcalcification clusters in digital mammograms. In: Proceedings of ICECIT-2012, Elesevier India. ISBN: 978-81-312-3411-2, pp 58–63
20. Biswas MK et al (1998) Fractal dimension estimation for texture images: a parallel approach. Pattern Recogn Lett 19:309–313
21. Shanmugavadivu P, Sivakumar V (2012) Comparative analysis of microcalcifications detected in mammogram images by edge detection using fractal hurst co-efficient and fudge factor. In: INCOSET-2012, IEEE Xplore. ISBN: 978-1-4673-5141-6, pp 174–179
22. Shanmugavadivu P, Sivakumar V (2013) Segmentation of pectoral muscle in mammograms using Fractal method. In: ICCCI-2013, IEEE Xplore. ISBN: 978-1-4673-2906-4, pp 1–6
23. ftp://peipa.essex.ac.uk/ipa/pix/mias/
24. MIAS database, UK

Chapter 26
Vehicle Classification Using Visual Background Extractor and Multi-class Support Vector Machines

Lee Teng Ng, Shahrel Azmin Suandi and Soo Siang Teoh

Abstract This paper describes a method to classify vehicle type using computer vision technology. In this study, Visual Background Extractor (ViBe) was used to extract the vehicles from the captured videos. The features of the detected vehicles were extracted using Histogram of Oriented Gradient (HOG). Multi-class Support Vector Machine (SVM) was used to recognise four classes of images: motorcycle, car, lorry and background (without vehicles). The results show that the proposed classifier was able to achieve an average accuracy of 92.3 %.

Keywords Vehicle classification · Visual background extractor · Support vector machines · Histogram of oriented gradient

26.1 Introduction

Vehicle classification system has become more important for several purposes such as transportation planning for traffic operation and pavement design [1]. A system that able to classify and keep track of vehicle patterns and volumes is required, due to considerable differences in size, weight, and performance between light and heavy vehicles [2]. It is also important to have vehicle data collection for safety purposes due to the high percentage of fatal accidents involving heavy

L. T. Ng (✉) · S. A. Suandi · S. S. Teoh
Intelligent Biometric Group, School of Electrical and Electronic Engineering, USM Engineering Campus, University Sains Malaysia, Seri Ampangan, Nibong Tebal 14300 Pulau Pinang, Malaysia
e-mail: leeteng_87@hotmail.com

S. A. Suandi
e-mail: shahrel@eng.usm.my

S. S. Teoh
e-mail: eeteoh@eng.usm.my

vehicles [1]. Moreover, road wear is affected by loading or weight of the vehicles and traffic flow [3]. Hence, varieties of traffic monitoring systems are appearing to solve these problems.

There are many methods have been developed to detect vehicles. These methods can be classified into two: intrusive and non-intrusive [4]. Sensors of intrusive methods need to be installed directly to the pavement surface [4]. Among these methods, one of the most conventional intrusive methods in determining the type of vehicle is through the counting of vehicles' axles number using inductive loop detectors [1]. However, the pavement life decreases when the road is dug for installation. These will cause delay and further disruption to the traffic.

On the other hand, there is no hard installation of sensors on the road surface for non-intrusive methods [4]. These sensors normally mounted overhead or on the roadside. These methods can be further classified into two: imaging and non-imaging methods. Non-imaging method includes magnetic field sensors, active and passive infrared sensors, microwave radar, passive acoustic array sensors and ultrasonic sensors [4]. However, these methods have low spatial resolution and slow scanning speed compared to imaging method. Moreover, the cost of the imaging method is the lowest among non-intrusive methods [5].

This paper proposes a vehicle classifier based on the Multi-class Support Vector Machines (SVM). The classifier uses uncalibrated video images. It classifies an image into one of the following four categories: motorcycle, car, lorry and non-vehicle. Visual Background Extractor (ViBe) is used to detect moving object in the captured video, while Histogram of Oriented Gradient (HOG) technique is used to extract the features of the collected images for the classification.

The paper starts by the description of the related works in Sect. 26.2, while Sect. 26.3 describes the methodology of our approach. Then in Sect. 26.4, the experimental results are presented and discussed. Finally, a conclusion is given in the last section.

26.2 Related Works

26.2.1 Detection of Moving Object

In surveillance system, background subtraction is often used to extract the moving object from the video frames [6]. Several methods have been developed in this matter, where the current frames will be compared to at least one reference frame (background) to detect the moving object [6]. Simple methods, such as frame differencing, mean filter, median filter, used a single background model to extract the moving object. However, these methods would fail if the background is bimodal or contained slow moving object. These problems could be solved using multi-modal background model as the reference frame. A multi-modal background modelling method that frequently used in extracting moving objects, namely

Gaussian Mixture Model (GMM) [7], used three to five modals to represent the background and foreground model. Recently, Barnich and Van Droogenbroeck [8] proposed a more robust multi-modal background subtraction method compared to the other state-of-the-art methods, namely visual background extractor (ViBe). It was proven to outperform other tested background subtraction algorithms in term of classification score and speed.

26.2.2 Object Classification Methods

Once the moving objects are detected using background subtraction methods, it is important to identify the detected objects. There are a lot of techniques for object classification. One of the earliest approaches used for vehicle classification is template matching [9]. Generally, this approach uses a metric distance as the similarity measure. Neural networks are a popular classifier. It consists of a massive amount of simple processors with many connections. In neural networks system, sequential training procedures are used to learn complex nonlinear input–output relationships and it is able to adapt itself to the data [9].

26.3 Proposed Method

The proposed vehicle classification process consists of three successive steps: moving object detection (ViBe), feature extraction (HOG) and pattern classification (SVM). Before the classification process, a large number of positive training images are collected using ViBe and labelled as either motorcycle, car or lorry, while the negative or background images were manually collected from the captured videos. The HOG features of the labelled images are extracted and used to train the multi-class SVM classifier. The trained multi-class SVM model will be used as the classifier in this system.

26.3.1 Visual Background Extractor

ViBe is used as the moving object detection algorithm in this study due to its robustness and speed [8]. This method used N background samples for each pixel and these samples are randomly selected to be updated in $1/\phi$ chances when the corresponding pixel is detected as background. Besides, the neighbouring pixels of the detected background will also be randomly selected for update with the detected background's value. This is to recover from the deadlock cases when foreground pixels were wrongly classified as background. Euclidean distance algorithm with R threshold on the current pixel to background models' pixels was performed to determine whether a pixel should be classified as a foreground or background. Once #$_{min}$ number or more of the background models' pixel values were matched with the current pixel, the corresponding pixel is determined as the background. The sensitivity of this system is determined by #$_{min}$/N value (the higher this value, the more sensitive the system will classify a pixel as foreground). In this study, the values of parameters N, ϕ, R and #$_{min}$ used were 20, 16, 20 and 2, respectively. In literature, these parameters setting yield excellent results in every situation [8].

26.3.2 Histogram of Oriented Gradient

Dalal and Triggs [10] proposed HOG features for human detection. The implementation of these features extraction is achieved by scaling the segmented images into 32 × 32 pixels images. Then, each image is divided into 16 cells (8 × 8 pixels each) and 9 overlapping are blocks formed with 2 × 2 neighbouring cells [11]. Gradient's orientations for each pixel in the blocks are calculated as the features. The horizontal and vertical gradients are calculated using Eqs. (26.1) and (26.2) respectively, while the gradient orientation, $\theta(x, y)$, is calculated using the Eq. (26.3):

$$dx = I(x+1, y) - I(x-1, y) \tag{26.1}$$

$$dy = I(x, y+1) - I(x, y-1) \tag{26.2}$$

where dx and dy are horizontal and vertical gradients respectively; $I(x, y)$ is the pixel intensity value.

$$\theta(x, y) = tan^{-1}\left(\frac{dy}{dx}\right) \tag{26.3}$$

These orientations are quantised into a predefined number of bins. A histogram, which represents the number of pixels that fall on the specific bin's range, for each block is calculated and used as the feature vector for classification. Increasing the number of bins would increase the accuracy but this would also cause the processing time to increase drastically [12]. According to [12], vehicles classification using SVM on HOG with 8 bin features provides good results (90.8 %) and processing time at 22.61 ms. Hence, this configuration of HOG will be used in this study.

Fig. 26.1 Examples of images used in classification (ViBe's output images for three different vehicles' classes and manually segmented background images): from *top left* motorcycle, lorry, car and background

26.3.3 Multi-class Support Vector Machines

SVM is frequently used in pattern recognition with good accuracy [9]. This classifier finds a hyperplane (in the higher dimensional space) to classify the data. It is originally used for binary classification (two-class classifier). Radial Basis Function is used as kernel function since it has fewer numerical difficulties and can handle nonlinear relationship between class labels and its attributes.

There are a few methods for multi-class classification using SVM such as "one-against-one" and "one-against-all" methods proposed by Hsu and Lin [13]. It was reported that the former is more suitable for practical used as this method is faster than the latter in training and testing time. Hence, in this study, multi-class SVM using "one-against-one" is used to classify the four targeted classes. For this method, each pair of the classes is trained and forms $n(n-1)/2$ classifiers, where n is the number of classes. Voting with "Max Wins" strategy is used for classification, where the most voted class will be selected. This multi-class SVM was run using Support Vector Machine software library (LIBSVM) developed by Chang and Lin [14].

26.4 Experimental Results

In order to get some pre-recorded video streams, a digital camera was set on a flyover by using a tripod. The training and testing images, which consist of motorcycle, car, lorry and background, are segmented using ViBe from the recorded video streams and used for vehicle classification. 2,400 images (600 images for each class) are used for training purpose. For classifier testing, 100 images for motorcycle, 100 images for car, 100 images for lorry and 100 images for background are used. Figure 26.1 shows some of the ViBe's output images used for vehicle classification. From Table 26.1, the results show that the classification of lorry has lowest accuracy (82.0 %) compared to motorcycle (99.0 %),

Table 26.1 Results of vehicle classification

Type of classes	Rate for classification		
	Total number of test images	Number of correct classification	Accuracy (%)
Motorcycle	100	99	99.0
Car	100	97	97.0
Lorry	100	82	82.0
Background	100	91	91.0
Total	400	369	92.3

Fig. 26.2 False classification of SVM due to shadow

while car and background have 97.0 and 91.0 %, respectively. Lorry has lowest accuracy as there were more variations of lorry's appearance compared to other tested classes, while the similarity of appearance for all motorcycle yields almost perfect results in this experiment.

However, there is also another factor that caused false classification in the proposed system. Figure 26.2 shows the images that were falsely classified. The first two images from left are classified incorrectly as motorcycle, while the others are classified as lorry. These images were cropped together with the shadow as ViBe detects the moving objects without excluding the shadow. This caused confusion to the classifier in this system. Nevertheless, the proposed vehicle classification system is still able to achieve 92.3 % detection rate even without shadow removal algorithm as the training and testing data are collected during the same time of day. Performance of the system is expected to improve by adding shadow removal stage before the feature extraction.

26.5 Conclusion

In this paper, a vehicle classification system based on ViBe (moving object detector), HOG (feature extraction) and Multi-class SVM (classifier) has been implemented and evaluated for multiclass vehicle classification. The average classification accuracy of the targeted four object classes (motorcycle, car, lorry and background) is 92.3 %. As future works, n-fold cross validation method could be implemented to improve the classification accuracy. Furthermore, the effect of shadow removal (before feature extraction) on the classification accuracy will also be investigated. In fact, there are more types of vehicle classes found on road that have been ignored in this study, such as van, multi-purpose vehicle, sport utility vehicle and bus. If these vehicle classes are to be included, a more complex classifier is needed.

Acknowledgments Specially thanks to Chih-Chung Chang and Chih-Jen Lin for sharing LIBSVM, which contributed greatly to this study.

References

1. Avely RP, Wang Y, Rutherford GS (2004) Length-based vehicle classification using images from uncalibrated video cameras. IEEE intelligent transportation systems conference, pp 737–742
2. Zhang G, Avery R, Wang Y (2007) A video-based vehicle detection and classification system for real-time traffic data collection using uncalibrated video cameras. Transp Res Rec: J Transp Res Board 1993:138–147
3. Hjort M, Haraldsson M, Jansen JM (2008) Road wear from heavy vehicles. NVF committee vehicles and transports
4. Daubaras A, Zilys M (2012) Vehicle detection based on magneto-resistive magnetic field sensor. Electron Electr Eng 118:27–32
5. Sun Z, Bebis G, Miller R (2004) On-road vehicle detection using optical sensors: a review. In: Proceedings of the 7th international IEEE conference on intelligent transportation systems, pp 585–590
6. Bouwmans T, El Baf F, Vachon B (2008) Background modeling using mixture of Gaussians for foreground detection—a survey. Recent Pat Comput Sci 3:219–237
7. Stauffer C, Grimson W (1999) Adaptive background mixture models for real-time tracking. In: IEEE computer society conference on computer vision and pattern recognition
8. Barnich O, Van Droogenbroeck M (2011) ViBe: a universal background subtraction algorithm for video sequences. IEEE Trans Image Process: Publ IEEE Signal Process Soc 20:1709–1724
9. Razali MT, Jantan A (2008) Support vector machine for classify dynamic human/vehicle shapes. In: International conference on electronic design, pp 1–6
10. Dalal N, Triggs B (2005) Histograms of oriented gradients for human detection. In: IEEE computer society conference on computer vision and pattern recognition
11. Teoh SS, Braunl T (2012) Symmetry-based monocular vehicle detection system. Mach Vis Appl 23:831–842
12. Teoh SS (2011) Development of a robust monocular-based vehicle detection and tracking system. Ph.D. Thesis, The University of Western Australia
13. Hsu CW, Lin CJ (2002) A comparison of methods for multiclass support vector machines. IEEE Trans Neural Networks 13:415–425
14. Chang CC, Lin CJ (2013) LIBSVM : a library for support vector machines. ACM Trans Intell Syst Technol 2(3):1–39

Chapter 27
A Robust Fuzzy-Based Modified Median Filter for Fixed-Value Impulse Noise

P. Shanmugavadivu and P. S. Eliahim Jeevaraj

Abstract This paper describes the development of Fuzzy-based Modified Median (FMM) filter which uses a simple membership function for fuzzification of intensity values of the corrupted image using which noise detection is performed in the order of a 3×3 overlapping sliding window. If the central pixel of the window is found corrupted, this filter computes four different medians on its neighbourhood pixels using which the filter estimates the original value of the corrupted pixel. This filter has produced higher Peak Signal-to-Noise Ratio (PSNR) and Mean Structural Similarity Index Matrix (MSSIM) values comparable with other high performing median-based fixed-value impulse noise filters. The added advantage of the filter is its reduced computational speed and complexity. This filter provides denoising solutions to the application domains like Document Imaging, SEM and TEM images wherein the images are often corrupted with fixed-value impulse noise.

Keywords Image restoration · Median filter · Noise detection · Noise correction · Highly corrupted image · Fuzzy systems

27.1 Introduction

Image Restoration aims to restore the quality of the images corrupted due to bluring and noises. Noise, the unwanted/undesirable information distributed across an image is categorized based on mathematical/statistical model. Impulse noise

P. Shanmugavadivu · P. S. Eliahim Jeevaraj (✉)
Department of Computer Science and Applications, Gandhigram Rural Institute-Deemed University, Gandhigram 624302 Tamil Nadu, India
e-mail: eliahimps@gmail.com

P. Shanmugavadivu
e-mail: psvadivu67@gmail.com

commonly creeps into an image due to the limitations of the capturing/transmission devices and undesirable atmospheric conditions during acquisition and transmission [1, 2].

As per the recent literature, noise filters are being developed to adaptively suppress noise either linearly or non-linearly. The Standard Median (SM) filter that works on the noisy images replaces the central pixel of a subimage which is assumed to be corrupted, by the median of its neigbourhood pixels [2]. This principle forms the basis for all median-based filters. The Center Weighted Median (CWM) filter assigns weightage to central pixel while finding median and replaces the center pixel with the computed median [3]. Iterative Median Filter applies median filtering technique iteratively which is more powerful than SM and CWM [4]. Signal Dependent Rank Order Median (SDROM) Filter focuses on the false signals and replaces those signals, based on their neighbouring true information, [5]. Researchers move towards the Switching median filters like Progressive Switching Median (PSM) filter and Tri State Median (TSM) filter which use the optimum thresholds for noise detection as well as correction [6–10]. Adaptive median based denoising techniques and PDE-based Filters are proved to be effective than classical linear and non-linear filters [11, 12]. Fuzzy systems provide freedom to frame the problem-specific rules and inferences, which are computationally simple.

The FMM filter presented in this paper uses fuzzy rules for noise detection and Median-based intensity estimation for noise correction. The fixed-value impulse noise model is explained in Sect. 27.2 and the algorithmic description of Fuzzy-based Modified Median Filter is detailed in Sect. 27.3. The results and discussion is presented in Sect. 27.4 and the conclusion in Sect. 27.5.

27.2 Noise Model

Fixed-value impulse noise is a type of impulse noise that assumes either minimum or maximum intensity value of the dynamic range of image and the percentage of distribution is evenly divided by those two intensities. The probability distribution of fixed-value impulse noise of p % in a corrupted image I' is described as:

$$I' = \begin{cases} 0 \text{ with probability } (p/2)\,\% \\ 255 \text{ with probability } (p/2)\,\% \\ U_{i,j} \text{ with probability } (1-p)\,\% \end{cases} \quad (27.1)$$

where $U_{i,j}$s are the uncorrupted pixels in I' [11].

27.3 Fuzzy-Based Modified Median Filter

The two-phase FMM filter performs noise detection in Phase I and noise correction in Phase II.

In noise detection phase, the pixel values of the noisy image are transformed into fuzzy member values using a fuzzy membership function given in Eq. (27.2). The membership function of this filter falls under the category of bell-shaped membership function. The special feature of this membership is non-zero values for all the points. For the pixel of a given input image, the membership values computed using bell-shaped membership function are found to lie between 0.5 and 1.0. The fuzzy membership function of the FMM filter is defined as:

$$\mu(i,j) = \frac{1}{1 + \frac{I'(i,j)}{Max}} \quad (27.2)$$

where $I'(i, j)$ represents the intensity of a pixel at (i,j) and $\mu(i,j)$ is the respective computed fuzzy member value of the pixel at (i,j). The Max denotes the maximum intensity value of the noisy image.

The membership values of the noisy image are divided into subimages using an overlapping sliding window (W) size of 3×3. The pixels of each window are processed by Fuzzy inference and fuzzy normalization for noise detection and suppression respectively.

The following two fuzzy rules are framed to classify the corrupted and uncorrupted pixels.

Rule 1: If $\mu(i, j) = 0.5$ or $\mu(i, j) = 1$, then $\mu(i, j)$ is corrupted
Rule 2: If $0.5 < \mu(i, j) < 1$, then uncorrupted

As per the fuzzy rules, if the central pixel (C) of a window is found to be corrupted, it undergoes fuzzy normalization in order to estimate the intensity of the corrupted pixels.

The fuzzy normalization procedure uses the four different neighbourhood medians of W as, median of 4-Neighubourhood ($N_4(C)$), median of Diagonal elements ($N_D(C)$), median of 8-Neighubourhood elements ($N_8(C)$) and M_U, the median of uncorrupted pixels in W. Then, the median (MED) of those four medians is computed. The central value of W is replaced by MED. This process is carried over in the entire image.

After the fuzzy normalization, the fuzzy member value is converted into image intensity values using the defuzzification function. The defuzzification function is defined as given below:

$$r(i, j) = \frac{Max}{\mu(i, j)} - Max \quad (27.3)$$

where $r(i, j)$ represents the intensity values of restored image, $\mu(i, j)$ is normalized fuzzy values and Max is the maximum intensity of the image.

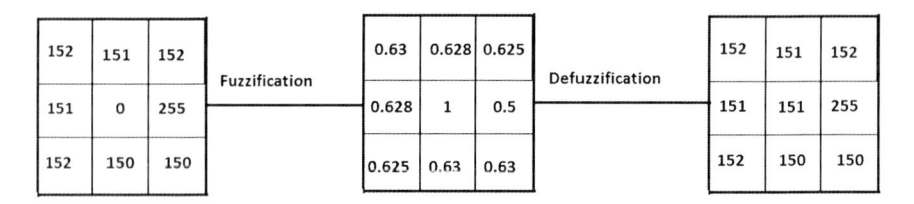

Fig. 27.1 FMM Filter on 3 × 3 subimage

Figure 27.1 illustrates the mechanism of FMM filter on a subimage of size 3 × 3. The intensities of the image are converted into fuzzfiied values and the image is divided as subimage. The central pixel (C) of W is to be checked whether it is corrupted or not, using fuzzy rules. In the normalization process, the medians of $N_4(C)$, $N_D(C)$, $N_8(C)$ and M_U are 0.628, 0.628, 0.628 and 0.628 respectively. Then, the median (MED) of medians is calculated, 0.628. Finally, the fuzzy member values are converted into intensity values using Eq. (27.3).

27.3.1 The Algorithm of FMM Filter

The algorithmic description of the FMM filter is given herein-under:
  Input : Noisy Image, I'
  Output : Restored Image, R
Step 1: Read the corrupted image (I').
Step 2: Noisy image is converted into fuzzy members set (μ) using membership function as in Eq. (27.2).
Step 3: Divide μ into subsets using an overlapping sliding window (W) of size n × n.
  For each window, check whether the central member value (C) is corrupted.
Step 4: If C = 0.5 or C = 1, then C is corrupted otherwise uncorrupted.
Step 5: If C is uncorrupted, go to step 3
Step 6: Compute MED:

  (a) $M_1 \leftarrow$ Median($N_4(C)$ of W)
  (b) $M_2 \leftarrow$ Median($N_D(C)$ of W)
  (c) $M_3 \leftarrow$ Median($N_8(C)$ of W)
  (d) $M_4 \leftarrow$ Median(Uncorrupted values in W)
  (e) MED $\leftarrow$ Median(M_1, M_2, M_3, M_4)

Step 7: Replace C by MED.
Step 8: Goto Step 3.
Step 9: Defuzzify the values using Eq. (27.3) and construct restored image (R).

Table 27.1 MSSIM values for lena image

Noise density (%)	TSM	FMM
10	0.9548	0.9814
20	0.8511	0.9184
30	0.6477	0.7886
40	0.4024	0.5982
50	0.2270	0.4174
60	0.1250	0.2731
70	0.0680	0.1648
80	0.0372	0.0977
90	0.0180	0.0452

Table 27.2 PSNR values for lena and mandrill image

Method	Noise density									
	10 %		30 %		50 %		70 %		90 %	
	L	M	L	M	L	M	L	M	L	M
Corrupted	15.5	15.7	10.7	10.9	8.5	8.6	7.0	7.2	5.9	6.1
SM	28.7	23.8	22.6	20.7	15.0	14.7	9.8	10.9	6.5	6.8
CWM	29.7	25.5	19.5	18.9	13.0	12.9	8.9	9.1	6.3	6.6
PSM	30.7	27.8	26.9	25.0	20.0	19.5	11.1	11.0	6.4	6.5
IMF	27.2	23.0	26.1	22.6	23.9	21.5	16.6	16.0	8.0	8.3
SDROM	30.3	26.0	22.0	20.8	14.4	14.3	9.4	9.6	6.4	6.6
SDROMR	30.5	25.8	25.7	23.2	20.7	19.7	14.2	14.2	6.9	7.1
ACWM	30.9	27.1	22.4	21.4	14.8	14.8	9.7	9.9	6.5	6.8
ACWMR	31.4	27.5	25.8	23.9	20.8	20.2	15.0	15.4	8.1	8.8
RUSSO	31.0	29.5	24.9	24.3	20.3	20.2	14.7	15.0	8.6	8.9
SUN	31.0	26.1	23.0	21.3	15.0	14.8	9.8	9.9	6.5	6.7
IRF	30.2	25.8	22.5	21.2	14.9	14.7	9.7	9.9	6.5	6.8
TSM	30.3	25.4	19.6	18.9	12.7	12.6	8.4	8.6	6.0	6.2
FMM (Proposed filter)	32.5	30.0	25.1	23.6	19.1	18.5	13.9	14.0	9.6	9.8

27.4 Results and Discussion

The proposed FMM filter is developed in Matlab 7.8. This newly devised filter is tested on the standard images, Lena, Mandrill, Saturn, Cameraman and Peppers and also tested on a few real images size of 256×256 and the optimal window size is chosen as 3×3.

The FMM filter is proved to exhibit greater details and edges preservation in the restored image which is ascertained by Mean Structural Similarity Index Matrix (MSSIM) values given in Table 27.1. This metric quantifies the structural similarity between the original image and the restored image which lies between 0 and 1. MSSIM values closer to 1 signify higher degree of similarity and vice versa

Fig. 27.2 **a** Lena original image; **b** 10 % noise image **c** restored image of (**b**); **d** 30 % noise image; **e** restored image of (**d**); **f** 50 % noise image **g** restored image of (**f**)

otherwise. It is obvious from Table 27.1 that the MSSIM of FMM filter is greater than its competitive filter, TSM.

The performance of the proposed filter is compared with other high performing noise filters in terms of PSNR values listed in Table 27.2 for Lena (L) and Mandrill(M) images. Table 27.2 shows that the FMM filter produces the highest PSNR values for Lena and Mandrill images than other high performing filters for the 10–20 and 80–90 % noise densities. For the rest of the noise densities, FMM filter generates PSNR values which are comparable with PSM and IM filters. In addition, the human visual perception of the restored images serves as qualitative indicator for FMM filter. Figures 27.2 and 27.3 visually reiterates the edges and details preserving ability of FMM filter.

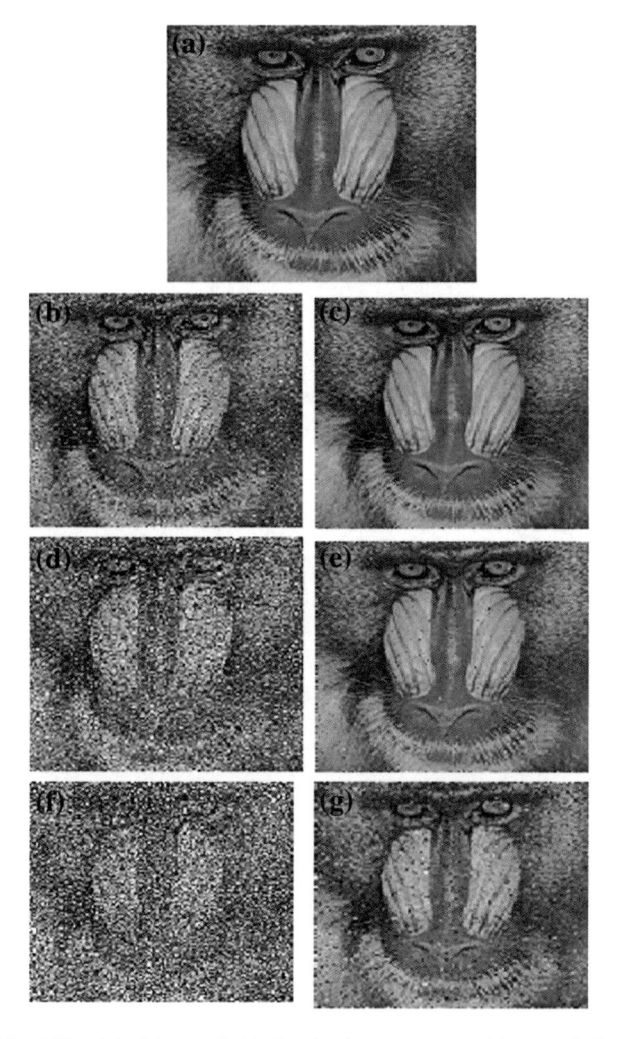

Fig. 27.3 a Mandrill original image; **b** 10 % noise image **c** restored image of (**b**); **d** 30 % noise image; **e** restored image of (**d**); **f** 50 % noise image **g** restored image of (**f**)

27.5 Conclusion

The FMM filter effectively denoises the images corrupted with fixed-value impulse noise of probabilities 10–90 % using the principle of fuzzy systems and median filters. It performs noise detection prior to noise correction. The accuracy of noise detection of FMM filter assures high degree of edges and details preservation. This filter is found to be more efficient than other high performing fixed-value impulse noise filters in terms of PSNR and MSSIM values. The human visual perceptions endorse the merit of the proposed filter.

Acknowledgments Authors wish to place on record the financial assistance received in the form of a Major Research Project from UGC, New Delhi.

References

1. Satpathy SK, Panda S, Nagwanshi KK, Ardil C (2010) Image restoration in non-linearing domain using MDB approach. IJICE J 6(1):45–49
2. Sung-Jea Jea Ko and Yong-Hoon Hoon Lee (1991) Center weighted median filters and their applications to image enhancement. IEEE Trans Circuits Syst 38(9):984–993
3. Forouzan AR, Araabi BN (2003) Iterative median filtering for restoration of images with impulsive noise. In: Proceedings of ICECS, vol 1. pp. 232–235
4. Abreu E, Mitra SK.(1995) A signal-dependent rank ordered mean (SD-ROM) filter-a new approach for removal of impulses from highly corrupted images. In: Proceedings of the ICASSP, vol 4. pp 2371–2374
5. Lin Tzu-Chao (2007) A new adaptive center weighted median filter for suppressing impulsive noise in images. Inf Sci 177(4):1073–1087
6. Chen T, Ma KK, Chen LH (1999) Tri state median filter. IEEE Trans Image Process 8(12):1834–1838
7. Wang Z, Zhang D (1999) progressive switching median filter for the removal of impulse noise from highly corrupted images. IEEE Trans Circuits Syst: Analog Digital Signal Process 46(1):78–80
8. Shanmugavadivu P Eliahim Jeevaraj PS (2011) Fixed—value impulse noise suppression for images using PDE based adaptive two-stage median filter, ICCCET-11 (IEEE Explore), pp 290–295
9. Shanmugavadivu P, Eliahim Jeevaraj PS (2012) Laplace equation-based adaptive median filter for highly corrupted images. In: International Conference on Computer Communication and Informatics (ICCCI-2012), pp. 47–51
10. Shanmugavadivu P, Eliahim Jeevaraj PS (2011) Fixed-value impulse noise suppression for images using pde based adaptive two stage median filter. International Conference Computer, Communication and Electrical Technology, pp 290–295
11. Russo F, Ramponi G (1996) A Fuzzy filter for images corrupted by impulse noise. IEEE Signal Process Lett 3(6):168–170
12. Zhong-gui S, Jie C Guang-wu M (2008) An impulse noise image filter using fuzzy sets. In: International Symposiums on Information Processing (ISIP), pp 183–186

Part III
Artificial Intelligence and Computer Applications

Chapter 28
Vision-Based Human Gesture Recognition Using Kinect Sensor

Huong Yong Ting, Kok Swee Sim, Fazly Salleh Abas and Rosli Besar

Abstract Gestures are indeed important in our daily life as they serve as one of the communication platform by using body motions in order to deliver information or effectively interact. This paper proposes to leverage the Kinect sensor for close-range human gesture recognition. The orientation details of human arms are extracted from the skeleton map sequences in order to form a bag of quaternions feature vectors. After the conversion to log-covariance matrix, the system is trained and the gestures are classified by multi-class SVM classifier. An experimental dataset of skeleton map sequences for 5 subjects with 6 gestures was collected and tested. The proposed system obtained remarkably accurate result with nearly 99 % of average correct classification rate (ACCR) compared to state of the art method with ACCR of 95 %.

Keywords Kinect sensor · SVM · Gesture recognition · Quaternions

28.1 Introduction

Gestures play vital role in our daily framework of communication where these expressive and meaningful body motions mean to convey information or to more effectively with the environment. Gesture recognition is a complex, yet interesting research topic in computer vision trying to detect and interpret human movement, postures, and gestures via computerized algorithms. Generally, vision-based gesture recognition consists of four main steps, such as model initialization, tracking, pose estimation, and gesture classification [1].

H. Y. Ting (✉) · K. S. Sim · F. S. Abas · R. Besar
Faculty of Engineering and Technology, Multimedia University, Jalan Ayer Keroh Lama 75450 Bukit Beruang, Melaka, Malaysia
e-mail: hyting@mmu.edu.my

Human gesture and action recognition have been extensively researched and numerous literatures have been reported. Due to the emergence of inexpensive, reliable, and robust algorithms to capture the depth information, gesture recognition using Microsoft Kinect sensor is becoming more prevalent. Shotton et al. [2] proposed a new method to predict 3D positions of body joints from a single depth image rapidly and accurately. In addition, Reyes et al. [3] presented an action recognition system using dynamic time wrapping, where weights are assigned to features based on inter-intra class action variability. Furthermore, real-time dance gesture classification system was proposed by Raptis et al. [4] where a large number of dance gestures classes were used. Moreover, there are numerous of researches in human gesture recognition by adopting Microsoft Kinect sensor [5–7].

However, the employment of coordinates and joint angle features in human gesture recognition are lacking of orientation information. Hence, we propose to represent the orientation of skeleton link by using quaternions in this paper. The extraction of quaternions for each link will be treated as "bag of quaternions feature vectors" and convert to log-covariance matrix subsequently. Finally, the system is trained and tested with multi-class support vector machine.

28.2 Proposed Method

28.2.1 Skeleton Model

The depth map with spatial resolution 640×480 operating at 30 frames per second which is generated by Microsoft Kinect sensor provides body silhouette of human in real-time. From the body silhouette, the body joints are estimated by using method from [2] and the skeleton model is constructed. The skeleton model also known as "stick model" encompasses 20 body joints and the joint axis are defined as illustrated in Fig. 28.1. The 3D coordinates of these 20 body joints will be tracked in real-time. Moreover, the skeleton model is quite robust to variations in shape and size of human body, the color and texture of clothing, and background. In this research, we focus on upper-body gestures and therefore, the orientations of arms are then of greatest interest. We have discovered that 6 out of 19 links (bones) contribute significantly to the upper-body gesture recognition system, i.e., left shoulder (ls), left upper-arm (lu), left forearm (lf), right shoulder (rs), right upper- arm (ru), and right forearm (rf).

28.2.2 Gesture Recognition Algorithm

In this research, the orientations of the links are represented using quaternions. Quaternions indeed a more compact and complete rotational representation in 3D space as compared with rotation matrix where the matrix suffer from representation

Fig. 28.1 Skeleton model and the joint axes

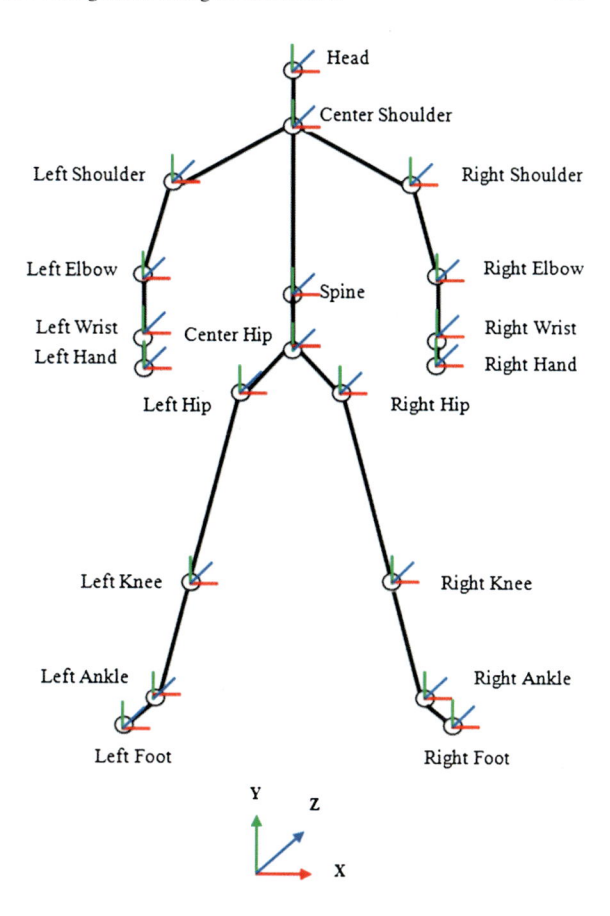

problem namely gimbal lock [8]. Furthermore, rotational matrix requires more storage, 9 storage units as compared with quaternions 4 storage units. Quaternions are constructed with 4D tuples (W, X, Y, Z), has a norm of one and is generally represented by one real dimension and three imaginary dimensions. In each skeleton frame, the quaternions of the link are extracted with respect to the joint, e.g., the orientation of ls is relative to the joint of left shoulder. Then, 24-dimensional quaternions feature vector (for each skeleton frame) is defined as follows:

$$g_n = \left[q_{ls}, q_{lu}, q_{lf}, q_{rs}, q_{ru}, q_{rf}\right]$$

where $q_{ls}, q_{lu}, q_{lf}, q_{rs}, q_{ru}$, and q_{rf} are the quaternions for each part of left and right arms respectively. We use $N = 60$ vector g_n in our gesture recognition algorithm. The covariance matrix of carefully selected features can provide a remarkably discriminative representation for human gesture recognition [9]. Moreover, the computation of the matrix is simple, effective, requires low storage, and low processing requirements.

Let $\mathbb{G} = \{g_n\}_{n=1}^{N}$ denotes a bag of quaternions feature vectors. The covariance matrix of $\mathbb{G}$ is shown in Eq. 28.1:

$$\mathbb{C} = \frac{1}{N-1}(\mathbb{G} - \mu)^T(\mathbb{G} - \mu) \tag{28.1}$$

where $\mu = \frac{1}{N}\sum_{n=1}^{N} g_n$ is the mean quaternions feature vector.

The dimension of the covariance matrix is only related to the dimension of the quaternions feature vectors. For instance, if g_n is m-dimensional, the dimension for the covariance matrix is $m \times m$. The symmetric and positive definite covariance matrix of a given dimension does not form a vector space since it is not closed under multiplication with negative scalar. In fact, it forms a closed convex cone. Most of the common machine learning algorithms work with the features that exist in Euclidean space, and therefore mapping the convex cone of covariance matrix to the vector space of symmetric matrix is required. The mapping algorithm is performed by using the matrix logarithm [10].

The matrix logarithm of a covariance matrix is computed as follows. The eigendecomposition of covariance matrix, $\mathbb{C}$ is presented in Eq. 28.2.

$$\mathbb{C} = VDV^T \tag{28.2}$$

where the columns of V are orthonormal eigenvectors and D is the diagonal matrix of eigenvalues. Then, log-covariance matrix, $\mathbb{C}_{\log}$ can be obtained from Eq. 28.3.

$$\mathbb{C}_{\log} = VD_{\log}V^T \tag{28.3}$$

where D_{log} is the diagonal matrix after replacing each element of D by its natural logarithm. Due to its symmetry, $\mathbb{C}_{\log}$ has only $(m^2 + m)/2$ independent real numbers.

Support vector machine (SVM) [11] has gained a research hotspot in machine learning due to the perfect learning performance. SVM classifier deals with two-class classification problems by solving the following optimization problem (Eq. 28.4):

$$\begin{aligned}
\min_{w,b,\xi} \quad & \frac{1}{2}w^T w + C \sum_{i=1}^{j} \xi_i \\
\text{subject to} \quad & y_i(w^T K(x_i) + b) \geq 1 - \xi_i, \\
& \xi_i \geq 0.
\end{aligned} \tag{28.4}$$

where C is the penalty parameter of the error term and K is the kernel function.

In this work, multi-class SVM is employed as a classifier for gesture recognition. During gesture recognition training, a training set $\tau = \{(E1, L2), (E2, L2), \ldots, (En, Ln)\}$, with several examples of each gesture was created. Each vector $Ei \in \mathbb{R}^6$ is the gesture quaternions descriptor, while $Li \in \{1, \ldots, |n|\}$ is the label for the human gesture accordingly. Besides, radial basis function kernel (Eq. 28.5) is used in this research.

Table 28.1 Upper-body gesture recognition confusion matrix of our method

	LHSR	RHSL	HC	THHW	DC	DT
LHSR	30	0	0	0	0	0
RHSL	0	30	0	0	0	0
HC	0	0	29	1	0	0
THHW	0	0	1	29	0	0
DC	0	0	0	0	30	0
DT	0	0	0	0	0	30

Table 28.2 Upper-body gesture recognition confusion matrix of Lai et al. [5] method

	LHSR	RHSL	HC	THHW	DC	DT
LHSR	30	0	0	0	0	0
RHSL	0	30	0	0	0	0
HC	0	0	28	2	0	0
THHW	0	0	6	24	0	0
DC	0	0	0	0	30	0
DT	0	0	0	0	1	29

$$K(x_i, x_j) = \exp\left(-\frac{\|x_i - x_j\|^2}{2\sigma^2}\right) \quad (28.5)$$

We chose $\sigma = 18$ since it lead to satisfactory results in our experiments. Finally, the gesture classification is determined through one-versus-one approach (max-wins voting strategy).

28.3 Results and Discussions

In order to validate and compare the proposed upper-body gesture recognition algorithm, an experimental dataset of skeleton map sequences for 5 subjects was collected. The subjects vary in terms of skin colors, clothes, heights, and weights. Each subject performed the following 6 upper-body gestures: left hand swipe right (LHSR), right hand swipe left (RHSL), hand clap (HC), two hands horizontal waving (THHW), draw circle (DC), and draw tick (DT). Each gesture was repeated for 6 times by each subject. Thereafter, a leave-one-out cross-validation (LOOCV) procedure was conducted. Average correct classification rate (ACCR) was computed where the total number of correctly recognised samples divided by the total number of tested samples. Table 28.1 demonstrates the upper-body gesture recognition confusion matrix of our method with ACCR of 98.89 % as compared to state of the art method by Lai et al. [5], with ACCR of 95 % exhibited in Table 28.2.

28.4 Conclusion

In this paper, we present a vision-based human gesture recognition system using Kinect sensor. The arms orientations details are extracted from the skeleton map sequences and multi class SVM classifier is employed to train and recognize upper-body gesture. The proposed system was validated with a dataset with total 180 samples from 5 subjects. Comparison was made between our system and state of the art method. Our system achieved over nearly 99 % average correct recognition rate (ACCR) compared to 95 % ACCR for state of the art method. The system also works well for new users without the need to carry out additional training.

References

1. Moeslund TB, Hilton A, Krueger V (2006) A survey of advances in vision-based human motion capture and analysis. Comput Vis Image Underst 104:90–126
2. Shotton J, Fitzgibbon AW, Cook M, Sharp T, Finocchio M, Moore R, Kipman A, Blake A (2011) Real-time human pose recognition in parts from single depth images. In: Computer vision and pattern recognition, pp 1297–1304
3. Reyes M, Dominguez G, Escalera S (2011) Feature weighting in dynamic time warping for gesture recognition in depth data. In: Computer vision workshop, pp 1182–1188
4. Raptis M, Kirovski D, Hoppe H (2011) Real-time classification of dance gestures from skeleton animation. In: Eurographics symposium on computer animation, pp 147–156
5. Lai K, Konrad J, Ishwar P (2012) A gesture-driven computer interface using kinect. In: Image analysis and interpretation, pp 185–188
6. Biswas KK, Basu SK (2011) Gesture recognition using microsoft kinect. In: Automation, robotics and applications, pp 100–103
7. Gu Y, Do H, Ou YS, Sheng, WH (2012) Human gesture recognition through a kinect sensor. In: Robotics and biomimetics, pp 1379–1384
8. Saxena A, Driemeyer J, Ng A (2009) Learning 3-D object orientation from images. In: IEEE international conference on robotics and automation, pp 4266–4272
9. Tuzel O, Porikli F, Meer P (2008) Pedestrian detection via classification on riemannian manifolds. IEEE Trans Pattern Anal Mach Intell 30(10):1713–1727
10. Arsigny V, Pennec P, Ayache X (2006) Log-Euclidean metrics for fast and simple calculus on diffusion tensors. Magn Reson Med 56(2):411–421
11. Vapnik V (1998) Statistical learning theory. Wiley, New York

Chapter 29
3D Facial Expression Classification Using 3D Facial Surface Normals

Hamimah Ujir, Michael Spann and Irwandi Hipni Mohamad Hipiny

Abstract With current advanced 3D scanners technology, direct anthropometric measurements are easily obtainable and it offers 3D geometrical data suitable for 3D face processing studies. Instead of using the raw 3D facial points, we extracted one of its derivatives, 3D facial surface normals. We constructed a statistical model for variations in facial shape due to changes in six basic expressions using 3D facial surface normals as the feature vectors. In particular, we are interested in how such facial expression variations manifest themselves in terms of changes in the field of 3D facial surface normals. Using our approach, using 3D facial surface normal yields a better performance than 3D facial points and 3D distance measurements in facial expression classification. The attained results suggest surface normals do indeed produce a comparable result particularly for six basic facial expressions with no intensity information.

Keywords Facial expression classification · Principal component analysis · 3D facial features

---

H. Ujir (✉)
Department of Computational Science and Mathematics, Faculty of Computer Science and Information Technology, Universiti Malaysia Sarawak, Kota Samarahan, Sarawak, Malaysia
e-mail: uhamimah@fit.unimas.my

M. Spann
School of Electronic, Electrical and Computer Engineering, University of Birmingham, Birmingham, UK
e-mail: M.Spann@bham.ac.uk

I. H. M. Hipiny
Department of Computing and Software Engineering, Faculty of Computer Science and Information Technology, Universiti Malaysia Sarawak, Kota Samarahan, Sarawak, Malaysia
e-mail: mhihipni@fit.unimas.my

29.1 Introduction

Face processing studies have been carried out over the past few decades for different purposes which began with face recognition and now facial expression classification studies are also emerging. Classification of facial expressions is a challenging problem as the face is capable of complex motions and the range of possible expressions is wide [1]. Facial expression recognition is an emerging research area spanning several disciplines such as pattern recognition, computer vision and image processing. It also brings benefits in human centred multimodal human–computer interaction (HCI) whereas the user's affective states motivate human action and enrich the meaning of human communication.

The earlier study on the subject was dominated by 2D based approach. However, illumination changes and pose variation are the common challenges that remained unsolved by 2D modalities. 3D geometry contains ample information about human facial expression [2]. With the advantage of 3D domain, the level of facial expression subtlety explored between human to human interactions is able to be scrutinized.

With 3D scanners technology, direct facial anthropometric measurements can be carried out and it produces 3D facial landmarks as the output. Facial anthropometric refers to the comparative study of sizes and proportions of the human face which include the discriminatory structural characteristics of the human face [3]. These facial landmarks are the soft-tissue landmarks which lie on the skin and can be identified on the 3D point clouds. Moreover, these facial landmarks are the points where all faces join and that have a particular biological meaning [4]. For instance, the facial landmarks with a particular biological meaning such as the nose tip, inner-corner eyes and etc.

In this paper, six basic facial expressions (angry, disgust, fear, happy, sad, surprise) with no intensity information are classified. We presented 3D Facial Surface Normals (3DFSN) as the feature descriptor and these surface normals are then used in a statistical model to capture the variation of the shape due to facial expression changes. The statistical model generates the shape parameters which then are used as the feature vector to classify the facial expression using Support Vector Machines (SVM) as the classifier. Section 29.2 describes the related works in this field, followed by a discussion on surface normals in Sect. 29.3. Then, the experimental setting and results are discussed in Sect. 29.5 and finally, the conclusions are drawn.

29.2 Related Works

Combinations of facial features form a human facial expression. Therefore, the deformation of facial features should be a suitable approach in order to determine the facial expressions shown by the subjects. The question is which 3D properties

Table 29.1 Comparison of 3D facial features in facial expression classification rates

Author	Database	3D facial features	Classifier	Average (%)
[7]	BU-3DFE	Surface curvatures	Linear discriminant analysis	83.1
[6]	BU-3DFE	3D distance vector	FF neural network	92.07
[2]	BU-3DFE	Ratio of distances	Multiclass SVM (OVO)	87.1
[16]	BU-3DFE	3D deformable model	Particle swarm optimization	92.3
[13]	BU-3DFE	Surface depth changes	Multiclass SVM	76.2
[8]	BU-3DFE	Curve-based	Multiclass SVM (OVO) and AdaBoost	96.1
[17]	In-house dataset	2D and 3D wavelet	AdaBoost	94.8
[18]	BU-3DFE	3D SIFT	Multiclass SVM	78.4

best describe the deformation of facial features so that a higher success rate of facial expression classification can be achieved. These 3D properties should be the significant properties that involve in at least six basic facial expressions and therefore the facial deformation can be easily observe.

The use of 3D facial geometric data and extracted 3D features for facial expression classification has not been widely studied. According to [5], the most frequently used 3D facial features in 3D face classification are 3D point (also called point cloud feature), 3D feature distance [6], curvature-based descriptors [5] and facial profile curves and 3D shape analysis [6–8]. Table 29.1 shows a comparison of 3D facial features which denotes the difference in the aspect of database, classifiers and average classification rates. Generally, the 3D features are extracted and fed into the facial expression/face classification classifiers. In [7], distances between 3D facial landmarks were used directly as the input to classifiers in order to classify facial expression. However, all of these existing works used the highest intensity expression provided by BU-3DFE database [9]. With highest intensity expression, the deformations of facial features are more obvious and therefore the percentage of getting correct classification is higher.

29.3 3D Facial Surface Normals

Our focus is on the advantage of having 3D facial points which are easily provided by the technology of 3D scanners on the market. With the availability of raw 3D facial data, extraction of 3DFSN is a straightforward task. A surface normal is a vector that is perpendicular to the tangent plane to a surface at a point. In addition, surface normals are also the features that encode the local directional gradient. 3DFSN is suggested as another 3D geometric measurement feature that could improve facial expression classification rate.

We are motivated to observe the face deformations closely which happen when facial expression occurs. When the facial expression changes, the facial points positions also change which will cause its surface normal to change since surface normals is a derivative of facial point. We believe that each expression has a consistent distribution of surface normals which distinguish it from other expressions. Compared with facial points which solely depends on a facial point itself and also distance measurements which only compute the distance between each facial point and a pre-selected facial point (i.e. nose tip), surface normals are considered to be more accurate in describing facial surface changes. It is due to the fact that a surface normal is built from a facial point as well as its neighbouring facial points. This has to do with the computation of surface normals which includes every neighbouring facial points of those particular facial features.

The similar idea of taking into account the neighbouring facial points into the computation of a facial feature can be found in curvature-based descriptors and surface profiles. However, according to [10], the distribution of surface normals is more robust than the curvature which is caused by the computation of surface normals and curvature. In this work, we are interested in how such facial expression variations manifest themselves in terms of changes in the field of 3D facial surface normals.

Ceolin [11] in her work employed a 2.5D representation based on facial surface normals (also known as facial needle map) for gender and facial expression classification. The needle map used is a shape representation that is acquired from 2D intensity images using Shape-from-Shading. In our work, the surface normals are computed using the raw 3D facial points extracted from the Bosphorus database which is different from Ceolin's method. The surface normal of each triangular polygon is calculated using its corner points (i.e. three 3D facial points).

Let $\mathbf{F_i}$ be a 3D face of the ith subject. $\mathbf{F_i}$ is represented by the set of 3D facial points

$$\mathbf{F_i} = \{p_1^i, p_2^i, \ldots, p_N^i\} \quad (29.1)$$

where the p^is are the (x, y, z) coordinate of each 3D facial point and N is the number of 3D facial points in the face. At each of the 3D facial points on the facial surface, we encode the facial points using their unit surface normal vectors,

$$\mathbf{\Omega_i} = \{s_1^i, s_2^i, \ldots, s_N^i\} \quad (29.2)$$

where the s_k^is are the 3D unit normals $s_k^i = \{s_x, s_y, s_z\}$. Once all of the triangular polygon normals are calculated, the normal for each vertex in the triangulated face data is computed by averaging the normals of the neighbouring polygon surface normals. Figure 29.1 shows an example of the triangular polygon with its vertex normals and Fig. 29.2 shows an example of surface normals of a 3D facial surface. The red lines denote the surface normals on a 3D facial surface.

Fig. 29.1 Example of triangular polygons with its vertex normals

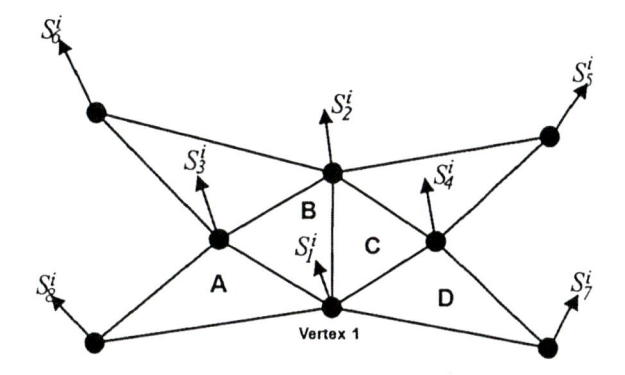

Fig. 29.2 Surface normals on a 3D facial surface

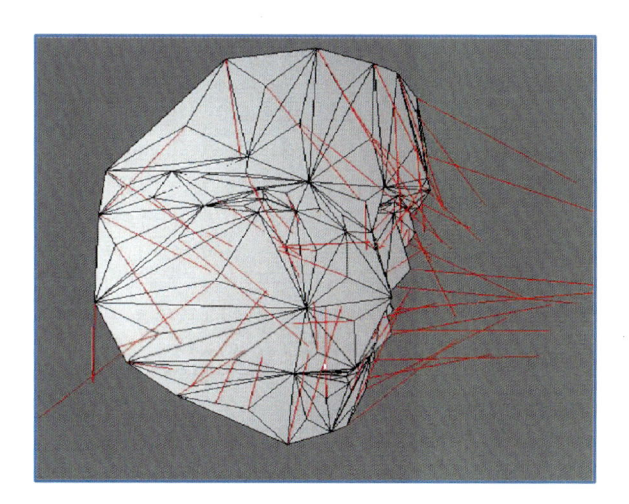

29.4 Methodology

This work begins with a pre-processing phase which involves 3D data extraction and 3D affine alignment. In this work, we used a statistical model to capture the variation of the shape due to facial expression changes. Statistical models attempt to model the shapes of objects found in images and the model characterizes the variation of shapes within a training set. In this work, we are using one of the famous unsupervised type of statistical model which is Principal Component Analysis (PCA). Unsupervised learning group is designed to extract common sets of features present in the input data and the examples given to the learner are unlabelled. In a sense, unsupervised learning can be thought of as finding patterns in the data above and beyond what would be considered pure unstructured noise [12].

After 3D facial surface normals of all training faces are extracted, they are used as the input to PCA and the probe 3D face is then projected to the face space defined by the eigenvectors of the PCA model. We apply SVM as the classifier and the shape weights from the PCA are used as the input to the classifiers.

29.5 Experimental Results

The aim of this experiment is to explore the potential of 3DFSN in classifying six basic facial expressions on the Bosphorus Database [9]. There are 64 subjects with 384 faces in total. 3-fold cross validation is chosen to maximize the number of subjects in one fold. Each fold have at least 21 subjects with six different facial expressions. All facial expressions of one subject belong to the same fold. This is to make sure when classification takes place, a facial expression of one particular subject will not be classified as another facial expression for the same subject. All facial expressions are used for both training and validation and each facial expression is used for validation exactly once.

For the purpose of evaluation, there are two feature vectors are used together with 3DFSN: (1) 3D facial points (3DFP) and (2) 3D distance measurements (3DDM). The reason we chose 3DFP point is simply because 3DFP is the raw information obtained from 3D space. 3DDM feature is used as to duplicate Soyel's [6] and Tang's [2] idea. Soyel and Demirel [6] used 83 facial features available in BU-3DFE database to find distances between points. Tang and Huang [2] extracted a set of 96 features which consist of the normalized distances and slopes of the line segments connecting a subset of the 83 facial feature points. However, there are few facial landmarks that were used in [6, 2] which are not provided in the database used in this work (i.e. facial points under the nose are not available for all 3D faces frontal profile) and therefore cannot be used as the feature vector. In our work, 3D distance measurements feature is the distance of each facial point to the nose tip.

Table 29.2 shows the results of using SVM with three different types of feature vectors as the baseline feature. For 3DFP, no misclassification error for Happy. Fear has the lowest score in which most of it has been misclassified as Surprise. Most Anger expressions are misclassified as Sad and vice versa. In contrast with other feature vectors in both classifiers, Disgust has the lowest rate of correctly classified expressions for 3DDM. Moreover, Fear and Sad expressions also have low rates of correct classification which is 15.38 % and 27.69 % respectively. Again, as opposed to other feature vectors, Happy is largely misclassified as Surprise when 3DDM is used. Similar to 3DFP, 3DFSN records a 100 % correct classification for the Happy expression. Disgust expression has a slightly lower rate of correctly classified than using 3DFP but it is still better than using 3DDM. 3DFSN the lowest rate of correctly classified for the Surprise expression, however the classification rate is still more than 50 %. The false positive rate for Surprise is 46.15 % and half of it comes from the Happy expression. Across all expressions, 3DFSN perform quite well compared to the two feature vectors.

References [2, 6, 7, 13] used BU-3DFE database [14] in their study and they only include facial expressions with the two highest levels of intensity. On the other hand, the Bosphorus database [9] does not provide facial expression with intensity information. We believe this is the reason for the significant difference in

Table 29.2 Confusion matrices of 3D facial expression classification using SVM

		Anger (%)	Disgust (%)	Fear (%)	Happy (%)	Sad (%)	Surprise (%)
3DFP	Anger	**56.92**	27.69	10.77	0.00	26.15	3.08
	Disgust	6.15	**27.69**	9.23	0.00	4.62	1.54
	Fear	3.08	3.08	**6.15**	0.00	6.15	1.54
	Happy	0.00	29.23	3.08	**100**	9.23	1.54
	Sad	30.77	6.15	7.69	0.00	**43.08**	1.54
	Surprise	3.08	6.15	63.08	0.00	10.77	**90.77**
3DDM	Anger	**52.31**	40.00	3.08	6.15	40.00	4.62
	Disgust	3.08	**1.54**	0.00	4.62	3.08	0.00
	Fear	3.08	7.69	**15.38**	9.23	3.08	4.62
	Happy	7.69	27.69	9.23	**56.92**	13.85	7.69
	Sad	32.31	15.38	12.31	1.54	**27.69**	4.62
	Surprise	1.54	7.69	60.00	21.54	12.31	**78.46**
3DFSN	Anger	**64.62**	27.69	7.69	0.00	23.08	6.15
	Disgust	7.69	**26.15**	9.23	0.00	9.23	3.08
	Fear	6.15	4.62	**29.23**	0.00	1.54	9.23
	Happy	3.08	26.15	10.77	**100**	13.85	20.00
	Sad	15.38	12.31	6.15	0.00	**47.69**	7.69
	Surprise	3.08	3.08	36.92	0.00	4.62	**53.85**

Recall rates for each expression are show in bold

the output as the facial expression with higher intensity means the deformation of each facial feature is obvious and easy to classify. Due to this, a comparison between the existing works and our work is indicated as unfair.

A similar concept of surface normals is also used in [11, 15]. Sandbach et al. [15] introduced LBNP which uses the same concept of surface normals for only AUs classification. Ceolin [11] used a 2.5D facial surface normals (or known as facial needle maps) which is acquired from 2D intensity images using Shape from Shading (SFS), referred to as Principal Geodesic Shape-From-Shading (PGSFS). In their work, the confusion matric of facial expression classification results are not provided therefore any comparison with their works cannot be carried out.

29.6 Conclusions

The key results of our work is we have shown that 3D facial surface normals outperformed 3DFP and 3DDM as the feature vectors in 3D facial expression classification without intensity information. In particular, we proved the feasibility of using 3DFSN to capture face deformation produced by six basic facial expressions compared to the two other 3D facial features. In addition, we proved that each expression has a consistent distribution of surface normals which distinguish it from other expressions and therefore the facial deformation of each facial expression is easily monitored.

For our future work, we would like to carry out facial expression classification experiments using 3D facial data with intensity information, specifically with the six basic facial expressions with the highest intensity level. We believe that our proposed approach will achieve a good result using this kind of data. Furthermore, we would like to train our system using facial expressions with different intensity levels and as a result, we will be able to classify the intensity level of facial expression.

Acknowledgments This preliminary work was supported by Malaysian Ministry of Higher Education through Exploratory Research Grant Scheme 1020/2013(17) to Dr Hamimah Ujir. Dr Hamimah Ujir is a scholar at the Faculty of Computer Science and Information Technology, UNIMAS where the work was carried out.

References

1. Sandbach G, Zafeiriou S, Pantic M, Rueckert D (2012) Recognition of 3D facial expression dynamics. J Image Vis Comput (in press)
2. Tang H, Huang TS (2008) 3D facial expression recognition based on properties of line segments connecting facial feature points. In: 8th IEEE international conference on automatic face & gesture recognition, pp 1–6
3. Gupta S, Markey MK, Bovik AC (2010) Anthropometric 3D face recognition. Int J Comput Vis 90(3):341–349 (Springer)
4. Vezzetti E, Marcolin F (2012) 3D human face description: landmarks measures and geometrical features. J Image Vis Comput 30(10):698–712
5. Gökberk B, İrfanoğlu MO, Akarun L (2006) 3D shape-based face representation and feature extraction for face recognition. J Image Vis Comput 24(8):857–869
6. Soyel H, Demirel H (2007) Facial expression recognition using 3D facial feature distances. LNCS book series. Image and analysis recognition. Springer, Berlin, pp 831–838
7. Wang J, Yin L, Wei X, Sun Y (2006) 3D facial expression based on primitive surface feature distribution. In: IEEE conference on computer vision and pattern recognition (CVPR), pp 1399–1406
8. Maalej A, Ben Amor B, Daoudi M, Srivastava A, Berreti S (2010) Local 3D shape analysis for 3D facial expression recognition. In: International conference on pattern recognition, pp 4129–4132
9. Savran A, Alyüz N, Dibeklioğlu H, Çeliktutan O, Gökberk B, Sankur B, Lale A (2008) Bosphorus database for 3D face analysis. In: Schouten B, Juul NC, Drygajlo A, Tistarelli M (eds) Biometrics and identity management. Springer, Berlin, Heidelberg, pp 47–56
10. Carmo MD (1976) Differential geometry of curves and surfaces. Prentice Hall, Englewood Cliffs
11. Ceolin SR (2012) Facial shape space using statistical models from surface normal. Ph.D., University of York
12. Ghahramani Z (2004) Unsupervised learning. In: Bousquet O, Raetsch G, von Luxburg U (eds) Advanced lectures on machine learning. Springer, Berlin
13. Gong B, Wang Y, Liu J, Tang X (2009) Automatic facial expression recognition on a single 3D face by exploring shape deformation. ACM Multimedia, pp 569–572
14. Yin L, Wei X, Sun Y, Wang J, Rosato MJ (2006) A 3D facial expression database for facial behavior research. In: 7th international conference on automatic face and gesture recognition (FGR06), pp 211–216

15. Sandbach G, Zafeiriou S, Pantic M (2012) Local normal binary patterns for 3D facial action unit detection. In: Proceedings of the IEEE international conference on image processing (ICIP 2012), Orlando, FL, USA, Oct 2012, pp 1813–1816
16. Mpiperis I, Malassiotis S, Strintzis MG (2008) Bilinear models for 3-D face and facial expression recognition. IEEE Trans Inf Forensics Secur 3(3):498–511
17. Pinto SCD, Mena-Chalco JP, Lopes FM, Velho L, Cesar RM (2011) 3D facial expression analysis by using 2D and 3D wavelet transforms. In: 18th IEEE international conference on image processing (ICIP), pp 1281–1284
18. Berretti S, Ben Amor B, Daoudi M, Del Bimbo A (2011) 3D facial expression recognition using SIFT descriptors of automatically detected keypoints. Vis Comput 27(11):1021–1036

Chapter 30
An Orchestrated Survey on T-Way Test Case Generation Strategies Based on Optimization Algorithms

AbdulRahman A. Al-Sewari and Kamal Z. Zamli

Abstract The test case construction is amongst the most labor-intensive tasks and has significant influence on the effectiveness and efficiency in software testing. Due to the market needed for diverse types of tests, recently, several number of t-way testing strategies (where t indicates the interaction strengths) have been developed adopting different approaches Algebraic, Pure computational, and Optimization Algorithms (OpA). This paper presents an orchestrated survey of the existing OpA t-way strategies as Simulated Annealing (SA), Genetic Algorithm (GA), Ant Colony Algorithm (ACA), Particle Swarm Optimization based strategy (PSTG), and Harmony Search Strategy (HSS). The results demonstrate the strength and the limitations of each strategy, thereby highlighting possible research for future work in this area.

Keywords T-way testing · Test case generation · Software and hardware testing · Optimization algorithms

30.1 Introduction

In the last 50 years, many new and useful techniques have been developed in the field of software testing for preventing bugs and for facilitating bug detection. Even with all these useful techniques and good practices are in place, there is no guarantee that the developed software is bug free. Here, only testing can demonstrate that quality has been achieved and identify the problems and the risks that remain.

A. A. Al-Sewari (✉) · K. Z. Zamli
Software Engineering Department, Faculty of Computer Systems and Software Engineering, Universiti Malaysia Pahang, 26300 Gambang, Kuantan, Pahang, Malaysia
e-mail: alsewari@ump.edu.my

Fig. 30.1 Simple software system

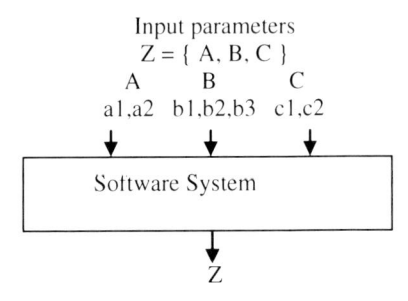

```
              Input parameters
              Z = { A, B, C }
              A      B      C
              a1,a2  b1,b2,b3  c1,c2
              ↓      ↓      ↓
         ┌─────────────────────────┐
         │     Software System     │
         └─────────────────────────┘
                    ↓
                    Z
```

Although desirable, exhaustive testing is often infeasible due to resource and timing constraints. To systematically minimize the test cases into manageable one, a new sampling technique, called t-way testing construction (where t indicates the interactions strength) has started to appear (e.g. AETG [1, 2], GTWay [3], IPOG families [4], PSTG [5–7], ITTDG [8], Aura [9], and Density [10–12]). In an effort to find the most efficient strategies capable of generating the most optimal test cases for every configuration (i.e., each combination is covered at most once), many new t-way testing construction strategies have been developed in the last 20 years.

T-way testing construction strategies based on OpA research have attracted much attention in recent years as part of a general interest in software testing approaches. Although useful, most existing OpA-based t-way testing strategies have not sufficiently supporting the (high interaction strength $t > 3$, variable interaction strength, constraint, sequence, Input output relations, and seeding). Here, this paper will survey the existing OpA strategies on t-way testing construction and elaborate the strength and the limitations of each of OpA-based strategy.

The rest of the paper is organized as follows. Section 30.2 illustrates the overview of t-way testing. Section 30.3 elaborates the t-way testing OpA-based strategies. Finally, Sect. 30.4 provides the conclusion.

30.2 Overview of *T*-Way Testing

T-way testing deals with interaction of parameters and their values. Consider a simple software system expressed as covering array, CA (N; 2, $2^2\ 3^1$) with strength $t = 2$ (i.e., 2 2-valued parameters and 1 3-valued parameter) as in Fig. 30.1 (Table 30.1).

Here, at full strength $t = 3$ (i.e., exhaustive combinations), the test suite will contain 12 test cases (see Fig. 30.2). Considering $t = 2$, there are three possible 2-way interactions between AB, AC, and BC respectively. Referring to Fig. 30.2, the most optimal 2-way test suite consists of only 7 test cases which cover all the

Table 30.1 Software system configuration

	Input parameters		
	A	B	C
System configurations	a1	b1	c1
	a2	b2	c2
		b3	

Combinatorial values for AB		
Input Parameters		
A	B	
a1	b1	
a1	b2	
a1	b3	
a2	b1	
a2	b2	
a2	b3	

Combinatorial values for AC		
Input Parameters		
A	C	
a1	c1	
a1	c2	
a2	c1	
a2	c2	

Combinatorial values for BC		
Input Parameters		
B	C	
b1	c1	
b1	c2	
b2	c1	
b2	c2	
b3	c1	
b3	c2	

Total test data = 12 for exhaustive test suite

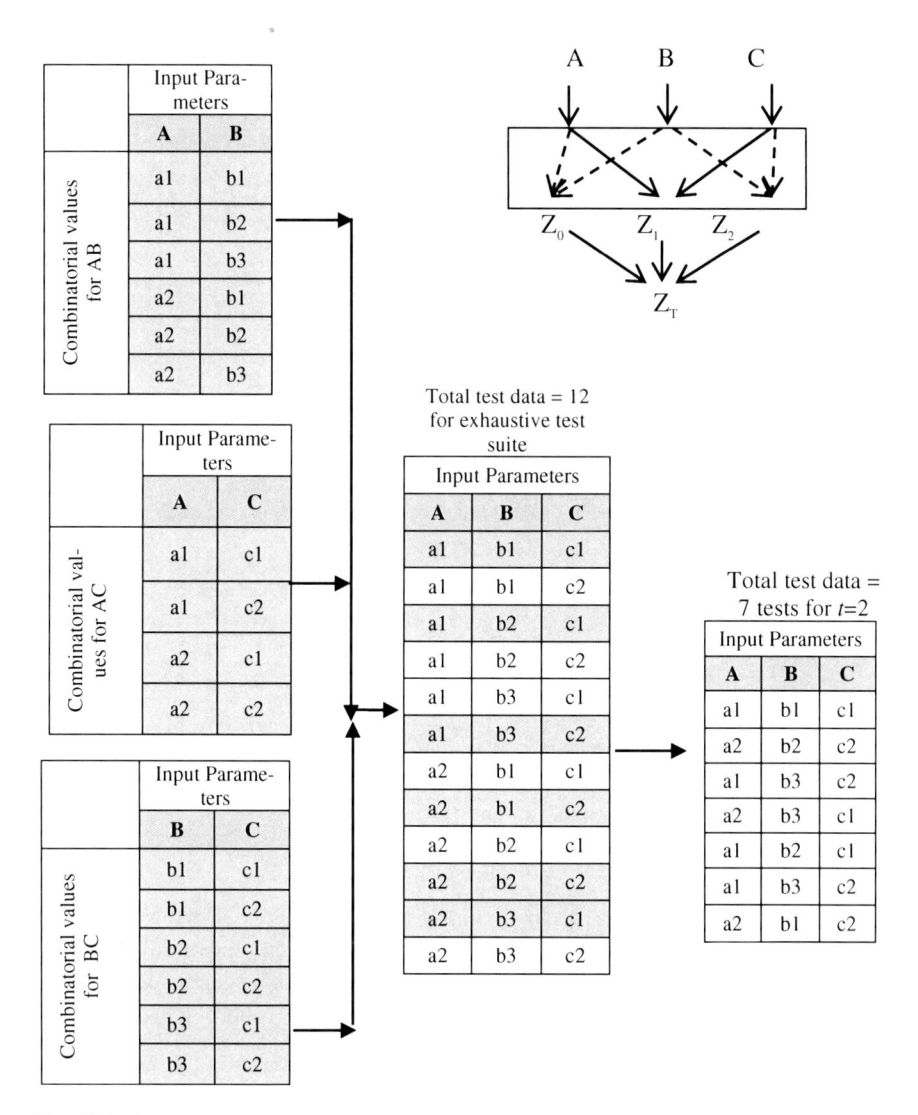

Input Parameters		
A	B	C
a1	b1	c1
a1	b1	c2
a1	b2	c1
a1	b2	c2
a1	b3	c1
a1	b3	c2
a2	b1	c1
a2	b1	c2
a2	b2	c1
a2	b2	c2
a2	b3	c1
a2	b3	c2

Total test data = 7 tests for $t=2$

Input Parameters		
A	B	C
a1	b1	c1
a2	b2	c2
a1	b3	c2
a2	b3	c1
a1	b2	c1
a1	b3	c2
a2	b1	c2

Fig. 30.2 Combining all the 2-way interaction tuples

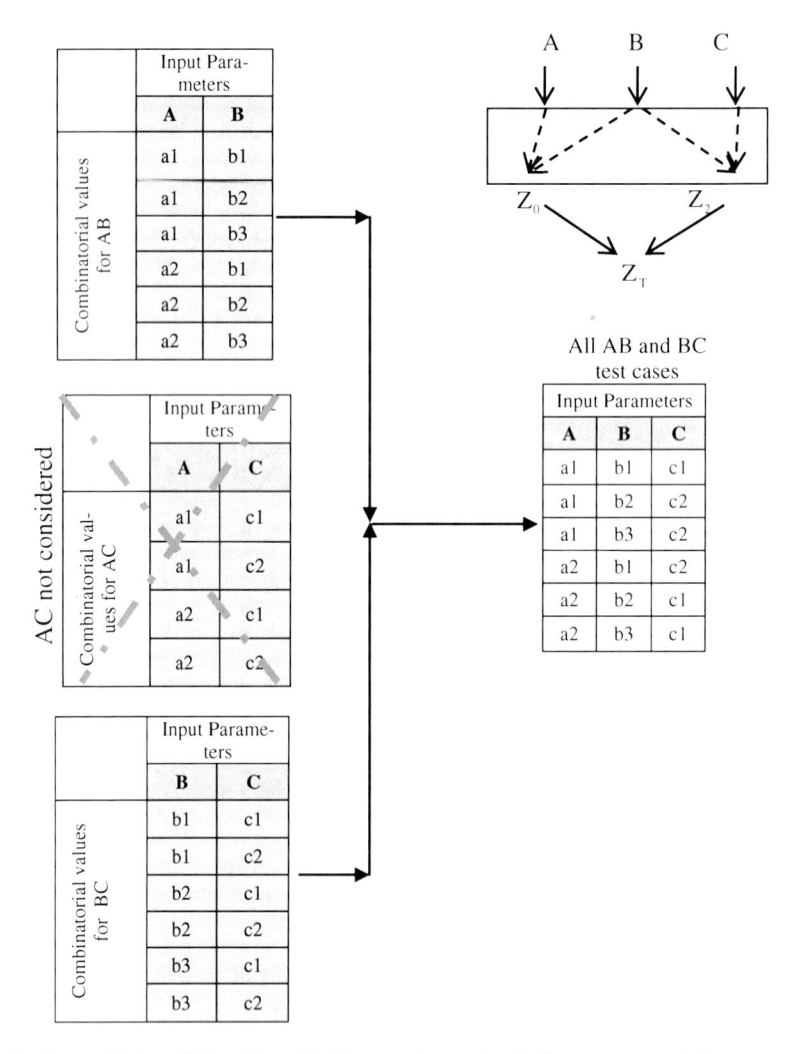

Fig. 30.3 Combining all the AB and BC interaction tuples for input output relation

required interactions [i.e., CA (6; 2, 2^2 3)]. In this example, there is a reduction of 58.3 % from exhaustive test suite.

To illustrate the concept of input output relation using the same example, assume that the combinations between AB and BC are of interests whilst the combination of AC is not important. Then, the test suite can be generated based on the combinations of only AB and BC as in Fig. 30.3. In this case, there is further reduction to 50 % (i.e. with 6 test cases).

30.3 Current State-of-the-Arts on T-Way OpA-Based Strategies

The generation of interaction test suite with optimal test size can be regarded as combinatorial optimization problem [1]. Naturally, OpA-based strategies excel in this respect. Many useful studies that adopt OPA-based strategies exclusively exploit the one-test-at-a-time approach. All the OPA-based strategies produce non-deterministic test suite since they count on some degree of randomization. In contrast, for pure computational-based counterparts, most studies are limited to small values of interaction strength (i.e., $2 \leq t \leq 6$). Furthermore, the support for variable strength constraints has also not been sufficiently investigated.

GA [2–5] and ACA [3, 5–7] represent early works in adopting OPA-based strategies for t-way test generation. The GA strategy mimics the natural selection process. GA begins with randomly created test cases, which are referred to as chromosomes. These chromosomes undergo a cycle of crossover and mutation processes until the predefined fitness function is met. In each cycle, the best chromosomes are selected and added to the final test suite. Up till now, existing strategies based on GA address only small values of uniform interaction strength t (i.e., $t \leq 3$). In fact, GA generally does not address variable strength interactions. Furthermore, no support is provided for constraints.

Unlike GA, ACA [3, 5–7] mimics the behavior of colonies of ants in search for food. Because colonies of ants travel from place to place (representing the parameter) to find food (representing the value selection of each parameter), the quality of the paths taken (representing the test case) is evaluated based on the amount of pheromones left behind (representing interaction coverage). The best path represents the best value of a test case to be added to the final test suite. Thus far, existing strategy based on ACA address only small values of interaction strength t (i.e., $t \leq 3$). Although, a variant of ACA, called ACS ACA addresses variable strength interactions, but with small value of interaction strength (i.e., $t \leq 3$) [5]. In addition, no support is provided for constraints.

SA [8, 9] relies on a large random search space for generating a t-way test suite. Using probability-based transformation equations, SA adopts binary search algorithm to find the best test case per iteration to be added to the final test suite. Apart from supporting variable strength interaction, SA also addresses constraints support through its variant, called SA_SAT [10]. Similar to other OPA-based counterparts, SA addresses small values of interaction strength (i.e., $t \leq 3$).

PSTG [11] is OpA-based t-way strategy for generating t-way test suite. PSTG is based on the Particle Swarm Optimization (PSO) algorithm, which mimics the swarm behavior of birds. Internally, PSTG iteratively performs local and global searches to find the candidate solution to be added to the final suite until all the interaction tuples are covered. Unlike other OPA-based strategies that address small values of t (i.e., $2 \leq t \leq 3$), PSTG can support up to $t = 6$. Although useful for some class systems, recent evidence has demonstrated the need to address $t > 6$ [12, 13]. As such, the interaction support provided by PSTG is still deemed limited. Although addressing variable strength interaction, in its new implementation called VS-PSTG [14], PSTG does not cater for the constraints support.

HSS [15–18] is the most recent OpA t-way testing strategies. HSS is based on the Harmony Search algorithm (HS), which mimics the harmonic between the missions team. Proposed by Geem and Kim [19], HS algorithm is derived from the natural process of searching and extemporizing for good harmony in music. Analogously, the search for best harmony (i.e., based on aesthetic estimation) mimics the optimization algorithms searching for a stable state (i.e., global optimum with minimum cost or maximum benefit) evaluated by some objective function. Often, aesthetic quality is judged by the sounds of various instruments, as objective functions evaluation is assessed based on the values of composed variables. The sounds for better aesthetic estimation can be enhanced practice after practice, as the values for better objective function's evaluation can be improved iteration-by-iteration. As will be elaborated in later, HS parameters include the number of improvisation, Harmony Memory, Harmony Memory Size (HMS), Harmony Memory Considering Rate (HMCR), and Pitch Adjustment Rate (PAR). Unlike other OPA-based strategies that address small values of t (i.e., $2 \leq t \leq 3$), HSS can support up to $t = 6$.

Summing up, Table 30.2 summarizes the description of existing t-way strategies.

Based on Table 30.2, all t-way testing OpA-based strategies are non-deterministic (every run generates different test suite). The existing t-way testing OpA-based strategies demonstrates the lack of support for input output relations. Only HSS addresses the supporting of high interaction strength ($t > 6$). Unlike HSS, GA, ACA, and SA support $t < 4$ due to the have computations in these algorithms, while PSTG has the ability to support interaction strength until 6. All the strategies in Table 30.2 are uniform interaction strategies. ACA, PSTG address the features of variable interaction strength support but with small value of interaction strength ACA with ($t < 4$) and PSTG with ($t < 7$). In contrast, HSS addresses the variable interaction strength and can go far for high interaction strength until ($t < 13$). SA and HSS are the only strategies addresses the constraints support while the strategy generates the test suite.

Table 30.2 Analysis of existing t-way strategies

		Randomness		Combination strength			Interaction support			
T-Way testing strategies		Non-deterministic	Deterministic	$t \leq 3$	$2 \leq t \leq 6$	$6 > t$	Uniform interaction strength	Variable interaction strength	Constraints support	Input output relations
OpA-based strategies	SA	√	X	√	X	X	√	X	X	X
	GA	√	X	√	X	X	√	X	X	X
	ACA	√	X	√	X	X	√	√	X	X
	PSTG	√	X	√	√	X	√	√	X	X
	SA_SAT	√	X	√	X	X	√	X	√	X
	HSS	√	√	√	√	√	√	√	√	X

√ Supported
X Not supported
N No information

30.4 Conclusion

Summing up, this paper presents the state-of-the-art of existing OpA-based *t*-way strategies, and highlighting the strengths and the limitations for each strategy. The market for software is vast, and hence, their required tests for ensuring conformance are diverse, thus, there is still need for new test ideas and test suite constructions. Also, there are many other OpA algorithms that have good potentials yet have not been adopted in this area.

Acknowledgments This research is partially funded by myGrants: A New Design of An Arti-fact-Attribute Social Research Networking Eco-System for Malaysian Greater Research Network, UMP RDU Short Term Grant: Development of a Pairwise Interaction Testing Strategy with Check-Pointing Recovery Support, and ERGS Grant: CSTWay: A Computational Strategy for Sequence Based *T*-Way Testing.

References

1. Floudas CA et al (1999) Handbook of test problems in local and global optimization, vol 33. Kluwer Academic Publishers, Dordrecht
2. Bryce R, Colbourn C (2007) One-test-at-a-time heuristic search for interaction test suites. In: Proceedings of the 9th annual conference on genetic and evolutionary computation. ACM, London, England
3. Shiba T, Tsuchiya T, Kikuno T (2004) Using artificial life techniques to generate test cases for combinatorial testing. In: Proceedings of the 28th annual international computer software and applications conference. IEEE Computer Society
4. McCaffrey J (2010) An empirical study of pairwise test set generation using a genetic algorithm. In: Proceedings of the 7th international conference on information technology. IEEE Computer Society
5. Chen X et al (2009) Variable strength interaction testing with an ant colony system approach. In: Proceedings of the 16th Asia-Pacific software engineering conference. IEEE Computer Society
6. Wang ZY, Xu BW, Nie CH (2008) Greedy Heuristic Algorithms to generate variable strength combinatorial test suite. In: Proceedings of the 8th international conference on quality software. IEEE Computer Society
7. Harman M, Jones BF (2001) Search-based software engineering. Inf Softw Technol 43(14):833–839
8. Stardom J (2001) Metaheuristics and the search for covering and packing array in department of mathematics. Simon Fraser University, Canada, p 89
9. Cohen MB, Colbourn CJ, Ling ACH (2008) Constructing strength three covering arrays with augmented annealing. Discrete Mathematics 308(13):2709–2722
10. Cohen MB, Dwyer MB, Shi J (2007) Interaction testing of highly-configurable systems in the presence of constraints. In: Proceeding of international symposium on software testing and analysis. ACM, London, UK
11. Ahmed BS, Zamli KZ, Lim CP (2012) Constructing a t-way interaction test suite using the particle swarm optimization approach. Int J Innovative Comput Inf Control 8(1):431–452
12. Younis MI, Zamli KZ (2009) ITTW: t-way minimization strategy based on intersection of tuples. In: Proceeding of IEEE symposium on industrial electronics and applications. IEEE Computer Society

13. Zamli KZ et al (2011) Design and implementation of a t-way test data generation strategy with automated execution tool support. Inf Sci 181(9):1741–1758
14. Ahmed BS, Zamli KZ (2011) A variable-strength interaction test suites generation strategy using particle swarm optimization. J Syst Softw 84(12):2171–2185
15. Alsewari ARA, Younis MI, Zamli KZ (2011) Generation of pairwise test sets using a harmony search algorithm. Comput Sci Lett 3(1)
16. Alsewari ARA, Zamli KZ (2011) Interaction test data generation using harmony search algorithm. In: Proceeding of IEEE symposium on industrial electronics and applications. IEEE Computer Society, Langkawi, Malaysia
17. Alsewari ARA, Zamli KZ (2012) Design and implementation of a harmony-search-based variable-strength t-way testing strategy with constraints support. Inf Softw Technol 54(6):553–568
18. Alsewari ARA, Zamli KZ (2012) A Harmony search based pairwise sampling strategy for combinatorial testing. Int J Phys Sci 7(7):1062–1072
19. Geem ZW, Kim JH (2001) A new heuristic optimization algorithm: harmony search. Simulation 76(2):60–68

Chapter 31
Adaptive Rate Mechanism for WLAN IEEE 802.11 Based on BPA-Artificial Neural Network

Jiwa Abdullah and A. M. I. Okaf

Abstract IEEE 802.11 WLANs provide multiple transmission rates to improve the system throughput by adapting the transmission rate to the current wireless channel conditions. The AutoRate Fallback (ARF) scheme is a simple and heuristic link adaptation approach and compliant with IEEE 802.11 standard, also most of commercial devices implement it but it's suffer from random packet collisions especially when the number of nodes increases and consequently cause a decline of the over all throughput. In this paper we propose rate adaptation in WLAN 802.11 based in neural network. The proposed rate adaptation scheme, appropriately adjust the data transmission rate based on the estimated wireless channel condition, specifically by dynamically adjusting the system parameters that determine the transmission rates according to the contention situations including the amount of contending nodes and traffic intensity. Through extensive simulation runs by using the Qualnet simulator, we evaluate our proposed scheme to show that our scheme yields higher throughput performance than the ARF scheme.

Keywords WLAN 802.11 · AutoRate fallback · Adaptive rate mechanism · Backpropagation · Neural network · Multilayer perceptron · Performance study

31.1 Introduction

The IEEE 802.11 WLAN has become an accepted standard for broadband wireless access. Its acceptance is attributed to several characteristics such as low cost, ease of deployment and the availability of multiple options of high data rates. Most

J. Abdullah (✉) · A. M. I. Okaf
Department of Communication Engineering, Faculty of Electrical and Electronic Engineering, Universiti Tun Hussein Onn Malaysia, Batu Pahat, Johor, Malaysia
e-mail: jiwa@uthm.edu.my

A. M. I. Okaf
e-mail: nasserokafe@yahoo.com

IEEE 802.11 WLAN devices use the Distributed Coordination Function (DCF) specified in the standard to coordinate channel access by means of carrier-sense multiple access with collision avoidance (CSMA/CA) [1]. In DCF, simultaneous transmission of multiple frames by different nodes resulted in collisions. In order to mitigate the effect of collisions, the standard employed retransmission scheme based on the binary slotted exponential backoff algorithm. Two access procedures are defined in the IEEE 802.11 DCF; basic access and RTS/CTS access. The specification provides multiple data rates by employing different modulation and channel coding schemes [2]. For example the standard specifies data rates of 11 and 54 Mbps. The rate adaptation scheme is a process of automatically switching the data rate to match the channel condition, selecting the data rate for that particular channel conditions, producing optimal quality performance. Hence, it maximizes the wireless link utilization as high as possible. Automatic Rate Fallback (ARF) is the original rate adaptation algorithm for WLAN implemented in the WaveLAN-II products [3]. It was designed to optimize the application throughput. In ARF, the sender decides on the channel quality by measuring the number of consecutive successed and failed transmissions. The sender then adjusts its modulation mode and data rate in accordance with these measurements. It counts the number of accumulated received and lost acknowledge frames (ACK frame) at the MAC layer to assess the current channel conditions. Then the new data rate can be adjusted. If *m* consecutive ACKs cannot be received correctly, the ARF algorithm decreases the current rate and starts a timer. When either the timer expires or the number of successfully received ACK reaches *n*, it will raise the current rate to a higher data rate and reset the timer. The value of *m* and *n* is 2 and 10 respectively adopted by the existing WLAN product. This scheme suffers from some problems; (1) If the channel condition varies rapidly, adaptation could not be effective; (2) If the channel condition changes very minimal, it tries using higher rate every ten successfully transmitted packets; resulted in an increased of retransmission attempts, thus decreases the application throughput; (3) Unable to distinguish between the lost ACKs caused by error-prone channel conditions and the one that caused by the packet collisions [4, 5], thus result in reduction of overall throughput.

An adaptive ARF algorithm (AARF) was developed [6] to alleviate the inefficiency of ARF as a result of an automatic attempt of new rates every ten successfully transmissions. This algorithm operates similar to ARF, but instead of trying the next higher rate every ten successfully transmissions, it doubles this number whenever the first transmission attempts with the higher rate failed. The number of successfully transmission attempts required to increase the rate is reset to ten every time the rate is decreased and it does not exceed a given threshold (i.e. 50). Nonetheless, it may also cause throughput degradation for very high contentions scenario where collision probabilities are peaking. Other schemes are the ARF-CD and AARF-CD [5]. These schemes apply the Request To Send/Clear To Send (RTS/CTS) mechanism to avoid packet collisions, hence an improve throughput performances, but increased the transmission overheads.

31.2 Artificial Neural Network-Based Rate Adaptation

31.2.1 Concepts

In this paper we propose rate adaptation scheme based on Backpropagation Algorithm (BPA) ANN. The ARF scheme typically depends on the successful or failure of ACK frame reception at the MAC layer, to access the current channel condition, which forms a basis of adjusting the transmission rates. Actually, when the number of contending nodes increased, the overall throughput degraded. The throughput of IEEE 802.11 WLAN could be influenced by not only the increase in number of contending nodes but also on factors such as channel conditions and traffic load. The correlation between them is nonlinear and quite complex. Thus, it is rather difficult to thoroughly consider the correlation function with analytical formulation in order to provide the optimum throughput with arbitrary number of nodes, channel condition and traffic load. We proposed the use of Back Propagation Algorithm Neural Network (BPA-NN), to model the function between the optimum values of consecutive acknowledges (ACKs) that could not be received correctly, m, the number of successfully received acknowledges (ACKs), n and other network parameters such as throughput, number of contending nodes, channel conditions and traffic intensity. It is expected that there will be an improved performance of aggregate throughput in 802.11 WLAN with multiple nodes.

Qualnet 5.0 simulator is used to run the simulation experiments by varying the transmission parameters. The values of throughput are obtained with different values of thresholds of m and n. In each transmission scenario, the threshold of m varies from 1 to 4 while the threshold of n varies from 8 to 12. Therefore, there will be 20 output values in each, for a given number of nodes. By examining all 20 values it is expected that the optimum throughput can be obtained which corresponds to each of the sets m and n. The number of nodes k will be varied from 2 to 10 in each transmission scenario. Thus, we will get the optimum throughput corresponding to a duple set of (m, n) in each number of nodes, k instead of the default value of m and n that equals to 2 and 10 respectively. By using the techniques offered by BPA-NN to model the function between throughput metrics, and the optimum values of m and n can then be determined, resulted in an optimum throughput. The procedure for this is shown below.

Step 1: Examine the optimum values of (m, n) with various k and different channel conditions.
Step 2: Train the BPA-NN to model the function between the context metrics of throughput and optimal values of (m, n).
Step 3: Generalize the output.
Step 4: Examine the throughput by using the optimal values of (m, n) that were obtained after generalization.

Table 31.1 IEEE 802.11b parameters

Parameter	Value
Auto rate fallback	Yes
Simulation time	25 Min
Transmission rates	1, 2, 5.5, 11 Mbps
Slot time	20 μs
SIFS	10 μs
DIFS	50 μs
Payload	1,972 bytes
802_11b_CW_MIN	31
802_11b_CW_MAX	1,023
SHORT_RETRY_LIMIT	7
LONG_RETRY_LIMIT	4

31.2.2 Simulation Setup

The location of each node is 25 m apart from the access point (AP), while all other stations are within the transmission and collision range of each other. It is assumed that all nodes operated with ideal channel condition, with BER is zero. Each node transmits CBR flow to the AP with the rate of 1,000 packets per second with fixed packet size of 1,972 bytes of payload. The system parameters adopted in the simulation scenarios are shown in Table 31.1.

31.2.3 Simulation Results with Number of Nodes as 2, 5, 8 and 10 Respectively and Various Number of (m, n)

Table 31.2 shows the throughput corresponding to the default value and optimum value of (m, n) for the given number of nodes with BER = 0 and traffic intensity, 1,000 packet per second. It can be seen that for default value (2, 10) and the number of nodes as 10, the throughput is 1.4616 Mbps while the optimum throughput, 1.8754 Mbps corresponding to (m, n) = (4, 8), Thus, choosing the optimum value of (m, n), the throughput is improved by 28.31 % in comparison with using default value of (m, n). When number of nodes is 8, the throughput with default values of (2, 10) is 2.3145 Mbps while the optimum value of throughput is 2.9641 Mbps corresponding to (m, n) as (4, 8). The improvement of throughput is 28.06 % in comparison to the default value of (2, 10). For 5 nodes, the throughput is improved by 13.64 % and for 2 nodes, the throughput is improved by 3.303 %. Figure 31.1 shows the formal comparisons of throughput corresponding to the default and optimum values of (m, n) when Channel condition BER = 0 and traffic intensity 1,000 packet per second.

Table 31.3 shows the throughput corresponding to default value and optimum value of (m, n) with 2, 5, 8 and 10 nodes when channel condition BER = 0 and

Table 31.2 Throughput corresponding to default value and optimum value of (m, n) with 2, 5, 8 and 10 nodes with BER = 0 and traffic intensity 1,000 packet per second

Number of nodes	Throughput corresponding to default value of $(m, n) = (2, 10)$ (Mbps)	The optimum value of throughput in the varied number of (m, n) (Mbps)	The optimum value of (m, n) corresponding to the optimum value of throughput
2	6.4120	6.6238	(2, 11)
5	3.2961	3.7458	(4, 9)
8	2.3145	2.9641	(4, 8)
10	1.4616	1.8754	(4, 8)

Fig. 31.1 Throughput corresponding to default and optimal values of (m, n)

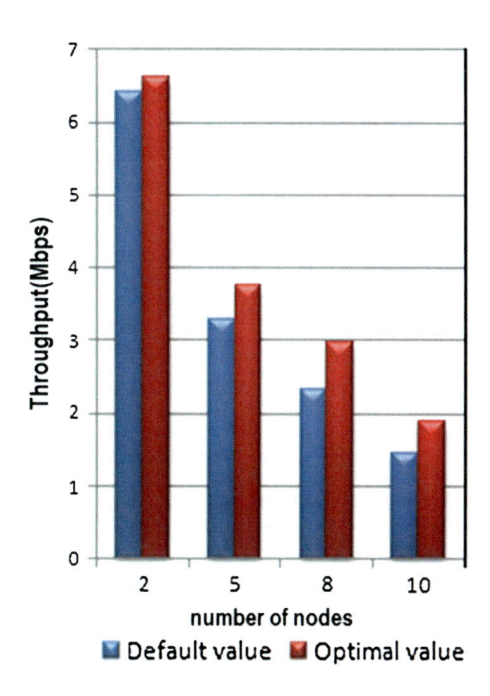

traffic intensity 10 packet per second. Figure 31.2 shows the throughput corresponding to default and optimal values of (m, n) when channel condition BER = 0 and traffic intensity 10 packet per second. The throughput improved by 11.17, 11.43, and 6.46 % when number of nodes = 10, 8, and 5 respectively. From these observations it is demonstrated that the ARF algorithm can't achieve the optimum throughput by the approach of fixed parameters in a contention-varying wireless network. Instead, the choices of (m, n) should be adaptive to the contention situations such as the number of nodes and traffic intensity in order to provide the best performance. Table 31.4 shows conclusively the optimum values of (m, n) with respect to the number of contending nodes, channel conditions and traffic intensity.

Table 31.3 Throughput corresponding to default value and optimum value of (m, n) with 2, 5, 8 and 10 nodes when channel condition BER = 0 and Traffic intensity 10 packet per second

Number of nodes	Throughput corresponding to default value of (m, n) = (2, 10) (Mbps)	The optimum throughput (Mbps)	The optimum value of (m, n)
2	0.2816	0.2816	(2, 10)
5	0.7412	0.7891	(4, 8)
8	1.2241	1.3641	(4, 8)
10	1.3426	1.4926	(4, 8)

Fig. 31.2 Throughput corresponding to default and optimal values of (m, n)

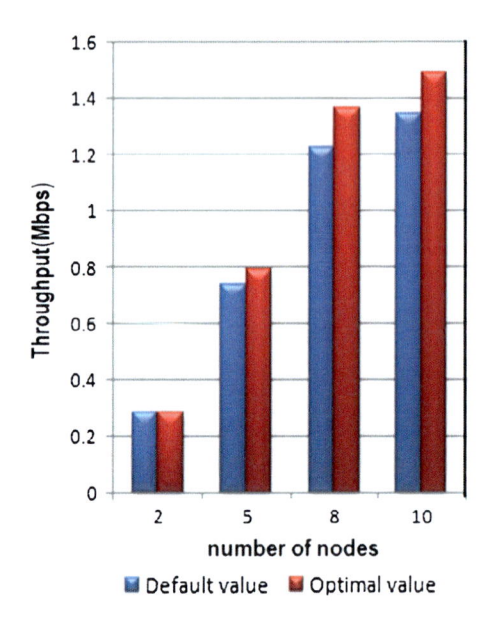

The throughput of IEEE 802.11 WLAN could be affected by the number of contending nodes, channel conditions and traffic loads. The correlation function between these parameters is nonlinear and complex. Hence, it is rather difficult to thoroughly depict the correlation function with analytical formulas in order to provide the optimum values of (m, n) with an arbitrary number of nodes, channel conditions and traffic intensities. Hence, the Artificial Neural Network (ANN), Multilayer Perceptron (MLP) type, is exploited to model the operation on the input variables and then be able to predict the outcome of m and n. The input includes the number of contending nodes, channel condition and traffic intensity [7, 8]. After obtaining the optimum values of (m, n) Matlab simulation using Qualnet simulator with different number of nodes and channel conditions in off-line operation, we will exploit multilayer perceptron (MLP) to learn the correlation function between them and to generalize that to an arbitrary context elsewhere. At runtime, the optimal values of (m, n) can be chosen according to the current metrics by using a simple table lookup which can be process rapidly to achieve the

Table 31.4 The optimal values of (m, n) with respect to the number of contending nodes, channel conditions and traffic intensity

Channel condition and traffic intensity	Number of nodes			
	2	5	8	10
BER = 0 and 1,000 packet sent per second	(2, 11)	(4, 9)	(4, 8)	(4, 8)
BER = 0 and 10 packet sent per second	(2, 10)	(4, 8)	(4, 8)	(4, 8)

best system throughput without much time spent on learning about the nonlinear and complicated correlation functions. A multilayer perceptron (MLP) is a feedforward ANN model that operate on sets of input data onto a set of appropriate output. An MLP consists of multiple layers of nodes in a directed graph, with each layer fully connected to the next one. Except for the input nodes, each node consists of a neuron with a nonlinear activation function. In this paper, MLP is utilized as a BPA to the network. MLP is an improved version of the standard linear perceptron, which can distinguish data that is not linearly separable [7]. The network is supplied with a sequence of both input data and target output data. The network is being told precisely by a trainer what should be emitted as output. During the learning phase, the trainer can tell the network how well it performs (reinforcement learning) or what the correct behaviour would have been (fully supervised learning). Feedforward networks often have one or more hidden layers of sigmoid neurons followed by an output layer of linear neurons. Multiple layers of neurons with nonlinear transfer functions allow the network to learn nonlinear and linear relationships between input and output vectors.

Figure 31.3 shows the architecture of the BPA-NN. The input signals is the number of contending nodes, channel conditions and traffic intensity. They enter the network by the the input layer and propagate through the network layer by layer to the output. The ith neuron at the lth layer can be described as,

$$u_i(l) = \sum_{j=1}^{N_{l-1}} w_{ij}(l) a_j(l-1) + \theta_i(l) \tag{31.1}$$

$$a_i(l) = h_l(u_i(l)) \quad 1 \leq i \leq N_l; \; l = 1, 2 \tag{32.2}$$

where N_l is the number of neurons at the lth layer; $u_i(l)$ and $a_i(l)$ are correspondingly the activation and output values of the ith neuron at the lth layer. The input units are represented by $a_i(0)$ and the output units by $a_i(2)$. The weight, $w_{ij}(l)$ refers to the weight connecting the output of the jth neuron at the $(l-1)$th layer to the activation function of the ith neuron at the lth layer. The bias associated with the ith neuron at the lth layer is given by $\theta_i(l)$. The transfer function, $h_l(\cdot)$ is activated to start the operation of the BPA to produce the predicted output of the neuron. The nonlinear function can be modeled with MLP by recursively adjusting $w_{ij}(l)$ and $u_i(l)$ to minimize the Mean Squares Error (MSE) between the

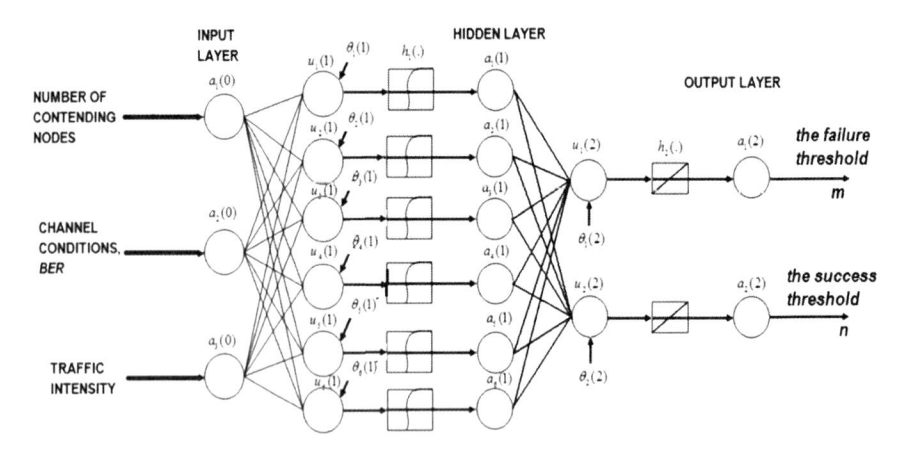

Fig. 31.3 The BPA-NN architecture

targets, tar_i (the optimum values of m and n in this case) and actual outputs, $a_j(2)$. That is,

$$E = \frac{1}{2}\sum_{j=1}^{M}\sum_{i=1}^{N_3}\left(tar_i^{(j)} - a_i^{(j)}(2)\right)^2 \quad (31.3)$$

where M is the number of training patterns. Matlab's neural network toolbox is used to implement the simulations [9] of the BPA. The input signals which represent the number of contending nodes, channel conditions, and traffic intensity and output signals represent the optimum values of (m, n). Table 31.4 shows the quantitative parameters. Eight trainer patterns are provided in total, six for training presented as a regular type and two for testing, representing staff. Initially, the epoch was set with value of 400 with the learning rate of 0.05. Finally, the MSE comes to 1.0658×10^{-14} which can provide great precision of the nonlinear function approximation. By testing the trained MLP, the network is sufficiently accurate to predict the output and the input–output correlations can be generalized for other input signals.

31.3 Performance Evaluations

31.3.1 BER = 0 and 1,000 Packet Sent Per Second

In this scenario the simulation setup assumes an ideal channel condition (BER = 0) and traffic intensity is heavy (1,000 packets per second) for data transmissions with varied number of nodes (from 2 to 10 nodes). Table 31.5 shows the throughput corresponding to the default and optimum values of (m, n) for different number of nodes (from 2 to 10) when the BER = 0 and traffic intensity is

Table 31.5 The throughput corresponding to the default and optimal values of (m, n) with varied number of nodes in case BER = 0 and 1,000

Number of nodes	Throughput, for default value of (m, n) = (2, 10) (Mbps)	The optimal value of (m, n)	Throughput corresponding to optimal value of (m, n) (Mbps)
2	6.4120	(2, 11)	6.6238
3	5.8231	(2, 11)	5.9981
4	4.6152	(4, 10)	4.8241
5	3.2961	(4, 9)	3.7458
6	3.021	(4, 8)	3.5891
7	2.8632	(4, 8)	3.3415
8	2.3145	(4, 8)	2.9641
9	1.8642	(4, 8)	2.5418
10	1.4616	(4, 8)	1.8754

heavy (1,000 packet per second). It is shown that the values of (m, n) have changed according to new channel conditions. It is shown that the throughput reduced as the number of nodes increase due to packet collision among the contention nodes. It is shown also that the throughput has increased by 3, 4.5, 18.8, 16.7, 36.3 % when the number of nodes are set as 3, 4, 6, 7, and 9 respectively.

Hence, the new auto rate fallback based on BPA-NN scheme gives throughput better than the normal auto rate fallback. In the normal operation, when the number of nodes is increased, the data traffic will be congested and resulted in an increased transmission failures due to increase in packet collisions. Thus, adapting the number of (m, n) according to contention situation such as the number of contending nodes, channel conditions and traffic intensity dynamically using BPA-NN resulted in better system throughput in comparison when default values of (m, n) are used.

31.3.2 BER = 0 and 10 Packet Sent Per Second

The simulation scenario considers an ideal channel condition BER = 0 and light traffic intensity (10 packets sent per second) for data transmissions with varied number of nodes from 2 to 10 nodes. Table 31.6 shows the throughput corresponding to the default and optimum values of (m, n) for different number of nodes (from 2 to 10) when BER = 0 and traffic intensity is light (10 packet per second). It is also shown that the throughput changed with changing value of (m, n) although the traffic intensity is low. This changes is may influence the performance especially when the number of nodes is increased. Figure 31.5 shows the throughput corresponding to the default and optimal values of (m, n) with varying number of nodes in the case of BER = 0 and 10 packet per second. It is shown that the throughput with both the default values of (m, n) and the optimum values of (m, n) increases as the number of nodes is increased, while the total

Table 31.6 The throughput corresponding to the default and optimal values of (m, n) with varied number of nodes in case of BER = 0 and 10 packet sent per second

Number of nodes	Throughput corresponded to default value of (m, n) = (2, 10) (Mbps)	The optimal value of (m, n)	Throughput corresponded to optimal value of (m, n) (Mbps)
2	0.2816	(2, 10)	0.2816
3	0.4167	(2, 9)	0.4295
4	0.5816	(4, 8)	0.5928
5	0.7412	(4, 8)	0.7891
6	0.8928	(4, 8)	0.9216
7	1.1261	(4, 8)	1.1886
8	1.2241	(4, 8)	1.3641
9	1.2541	(4, 8)	1.4211
10	1.3426	(4, 8)	1.4926

Fig. 31.4 The throughput corresponding to the default and optimal values of (m, n) with varying number of nodes in case of BER = 0 and 1,000 packet sent per second

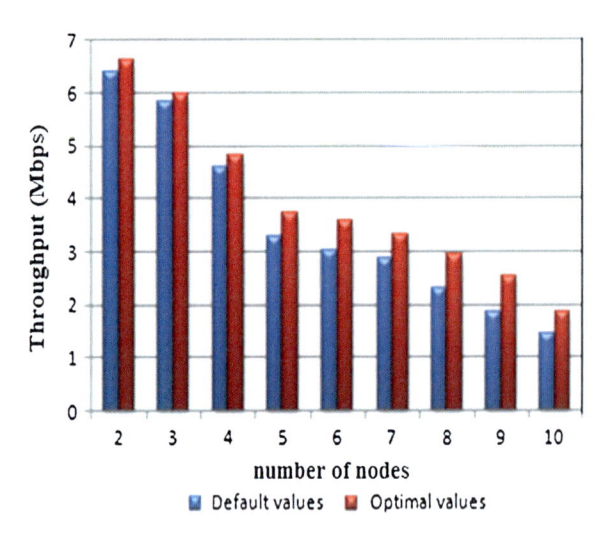

channel utilization is reduced from the network capacity even though number of nodes reach 10. It is shown that when the number of nodes gets large the occurrences of packet collisions become more frequently, the auto rate fall back scheme based on neural network provides slightly higher throughput than the original one by means of an adaptive manner to reflect the change of contention situation. It is shown that the throughput increased by 11.4, 13.3, 11.1 % when the number of nodes is 8, 9, and 10 respectively. The results shown in Figs. 31.4 and 31.5 demonstrated that the auto rate fallback scheme based on BPA-NN can adapt to different contention situations (i.e. the number of nodes and traffic intensity) and improve the system throughput as fast as possible in accordance with the current contention situations.

Fig. 31.5 The throughput corresponding to the default and optimal values of (m, n) with varying number of nodes in case of BER = 0 and 10 packet sent per second

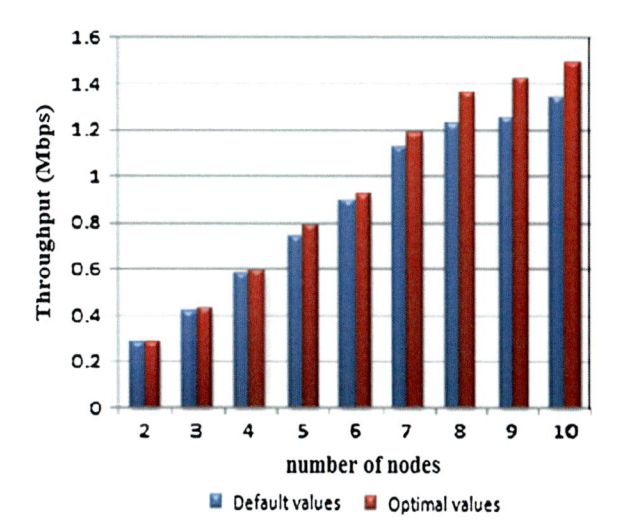

31.4 Conclusions

This paper presents rate adaptation scheme based on neural network to improve the system throughput in IEEE 802.11 Wireless Local Area Network by adapting the transmission rates to the current channel conditions and contention situation. Neural networks are exploited to learn the correlation function of the optimal success and failure thresholds with respect to the corresponding contention situations including the amount of contending nodes and traffic intensity in the off-line stage. At runtime, the optimal threshold values can be chosen according to the current contention situations rapidly by using a simple table lookup to achieve the best system throughput. Qualnet 5.0 simulator is used to evaluate and compare the system performance of new auto rate fallback scheme based on neural network and the standard auto rate fallback scheme. Simulation results illustrate that the auto rate fall back scheme based on neural network improve the achievable performance of the aggregate throughput in a variety of IEEE 802.11 Wireless Local Area Network environments.

References

1. IEEE Std 802.11-1999 (2003) Wireless LAN medium access control (MAC) and physical layer (PHY) specifications, Jun 2003
2. IEEE 802.11 a/b (1999) Wireless LAN medium access control (MAC) and physical layer (PHY) specifications. IEEE Standard, Aug 1999
3. Kamerman A, Monteban L (1997) WaveLAN-II: a high-performance wireless LAN for the unlicensed band. Bell Labs Tech J 2:118–133

4. Kim J, Kim S, Choi S, Qiao D (2006) CARA: collision aware rate adaptation for IEEE 802.11 WLANs. IEEE INFOCOM
5. Maguolo F, Lacage M, Turletti T (2008) Efficient collision detection for auto rate fallback Algorithm. IEEE INFOCOM
6. Lacage M, Manshaei MH, Turletti T (2004) IEEE 802.11 rate adaptation: a practical approach. In: Proceedings of the 7th ACM international symposium on modeling, analysis and simulation of wireless and mobile systems, pp 126–134
7. Haykin S (1999) Neural networks: a comprehensive foundation, 2nd edn. Prentice Hall, Englewood Cliffs
8. Faucett L (1994) Fundamentals of neural networks architecture, algorithms, and applications. Prentice-Hall, Englewood Cliffs
9. Matlab (2012) MATLAB and Statistics Toolbox Release 2012b, The Mathworks, Inc., Natick, Massachusetts, US

Chapter 32
Traffic Sign Detection and Classification for Driver Assistant System

Nursabillilah Mohd Ali, Nur Maisarah Mohd Sobran,
M. M. Ghazaly, S. A. Shukor and A. F. Tuani Ibrahim

Abstract In this paper we explain the proposed method of traffic sign detection and classification for driver assistant system (DAS). Color detection framework using RGB method is utilized in this study, whereas an artificial neural network (ANN) has been used as classifiers for classification. There are at least 100 types of Malaysian Traffic Signs have been employed in this research. Most of the images are taken at various places throughout the urban and suburban areas involved with scale, illumination and rotational changes as well as occlusion images. The experimental results are shown that the proposed framework achieved at least 80 % successful detection with 21 false positive images. On the other hand, the ANN gives strong rates where at least most of the signs can be classify with more than 85 % success.

Keywords Color detection · Illumination-invariant · Classification of occlusion images

32.1 Introduction

Of late, Driver Assistance System (DAS) was created and become the most popular topic in computer vision application, i.e. traffic sign detection and recognition. It has contributes to road safety and awareness among drivers for the last 15 years. Since then, many techniques have been created and implemented to achieve accurate and robust performance towards traffic sign identification system [1].

N. M. Ali (✉) · N. M. M. Sobran · M. M. Ghazaly · S. A. Shukor · A. F. T. Ibrahim
Department of Mechatronics, Faculty of Electrical Engineering,
Universiti Teknikal Malaysia Melaka, Durian Tunggal, Melaka, Malaysia
e-mail: nursabillilah@utem.edu.my; ilahnur203@gmail.com

N. M. M. Sobran
e-mail: nurmaisarah@utem.edu.my

Even though many techniques have made various algorithms and equipped with good system performance, the disorganized comparisons are still missing and datasets are not easily available. Detection and recognition stages are a multitask-based category as many datasets are used as a training set to achieve the result. Driver's vision is easily distracted by the headlight of the oncoming vehicles at night, which make driving more difficult and hence, lead to more traffic accidents. Traffic signs are not only used to regulate congested traffic but also give useful information in order to avoid accidents that might happen. Traffic signs can be used to differentiate their information based on colours and shapes. They are designed so that drivers can easily detect and recognize them. In Malaysia for instance, more than 100 road signs classes can be investigated [2]. The paper is organized as follows: Sect. 32.2 entails the survey on traffic signs. Section 32.3 discusses the system overview about the proposed system and addressed the result. Finally the analysis made before conclusions in Sect. 32.4.

32.2 Literature Survey

There are many existing techniques for traffic sign identification that has been used. Traffic sign detection has been treated with simple background images by many existing research. However, complex background is an important aspect as we need to observe the performance result. Color is an important part in detection phase. Several researchers make used of the advantages of color model to build algorithm for traffic sign detection. It is due to several variances that might affect the appearance of the signs. There are many color model that have been used has compromised HSV [3] or HSI [4], YUV [5] and YCbCr. Besides that, several researchers have developed color of databases, look-up tables and hierarchy region growing technique so that the detection technique would be much easier.

On the other hand, in traffic sign classification and recognition, many researchers used template matching, Latent Dirichlet Allocation (LDA), Support Vector Machine (SVM), and other learning-based method. Pictogram have been classified using template matching and cross correlation using OCR (optical character recognition) systems. Moutarde et al., employed traffic system for speed limit signs based on single digit recognition using neural network technique. However, the recognition result was not represented by the developed system. The system achieved almost 89 and 90 % for U.S. and European, respectively using 281 traffic signs. Yet, this method still using certain signs such as speed limit and not focusing on the all classes.

32.3 Proposed Framework

The paper focused on development of detection using color segmentation and classification using neural network. Figure 32.1 shows the proposed technique. It consists of 3 main steps. Twelve types of traffic signs were used: *Uneven Road, No*

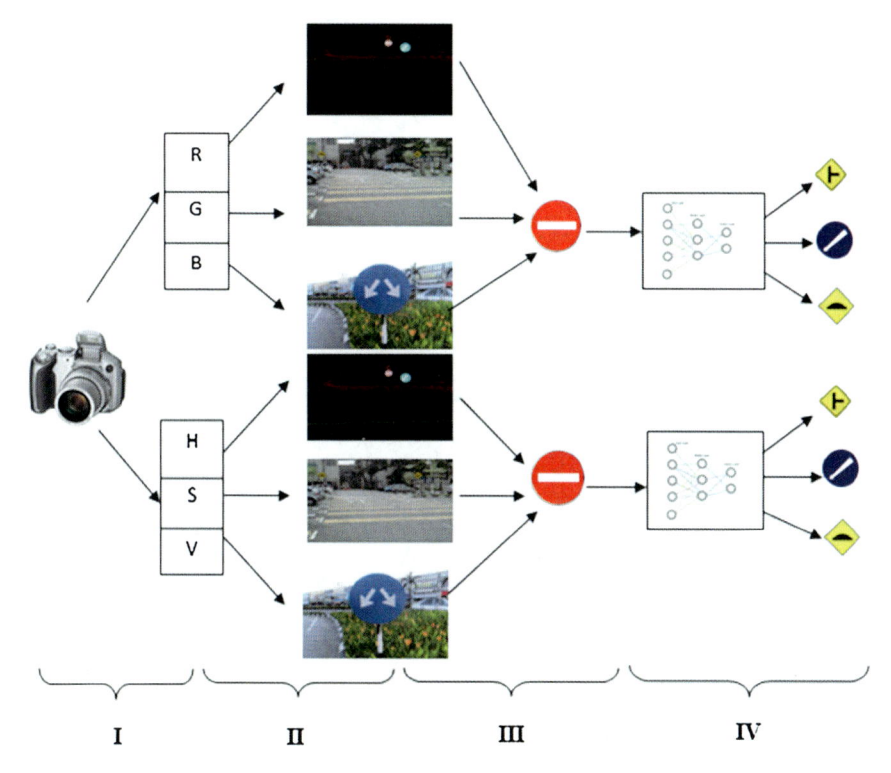

Fig. 32.1 Algorithm development process. (*I*) Image acquisition; (*II*) color segmentation; (*III*) detection process; (*IV*) classification/recognition process

Entry, No-U-Turn, Right Junction, Yield/Give Way, Keep Both Side, Stop Sign, Keep Left, Left Junction, Speed Limit 30 km/h, *Speed Limit* 50 km/h *and Speed Limit* 60 km/h *signs*. These signs were chosen because they were the most common traffic signs. For recognition stage, experiments conducted included partial occluded signs, illumination changes and rotational changes.

32.3.1 RGB Color Segmentation Utilized by Aspect Ratio Determination

This research utilized algorithms between using RGB color space. The algorithms were able to detect red, blue and yellow traffic signs, respectively. Algorithms were tested on static images separately. Images were mostly taken during sunny day and at night. Tests were carried out over 403 images which represented 101 partial occluded and 253 non-partial occluded signs with 49 rotational changes signs.

Table 32.1 RGB color ratio

No.	Traffic signs	Color channel
1	Red	Red/Green
2	Blue	Blue/Green
3	Yellow	Green/Blue

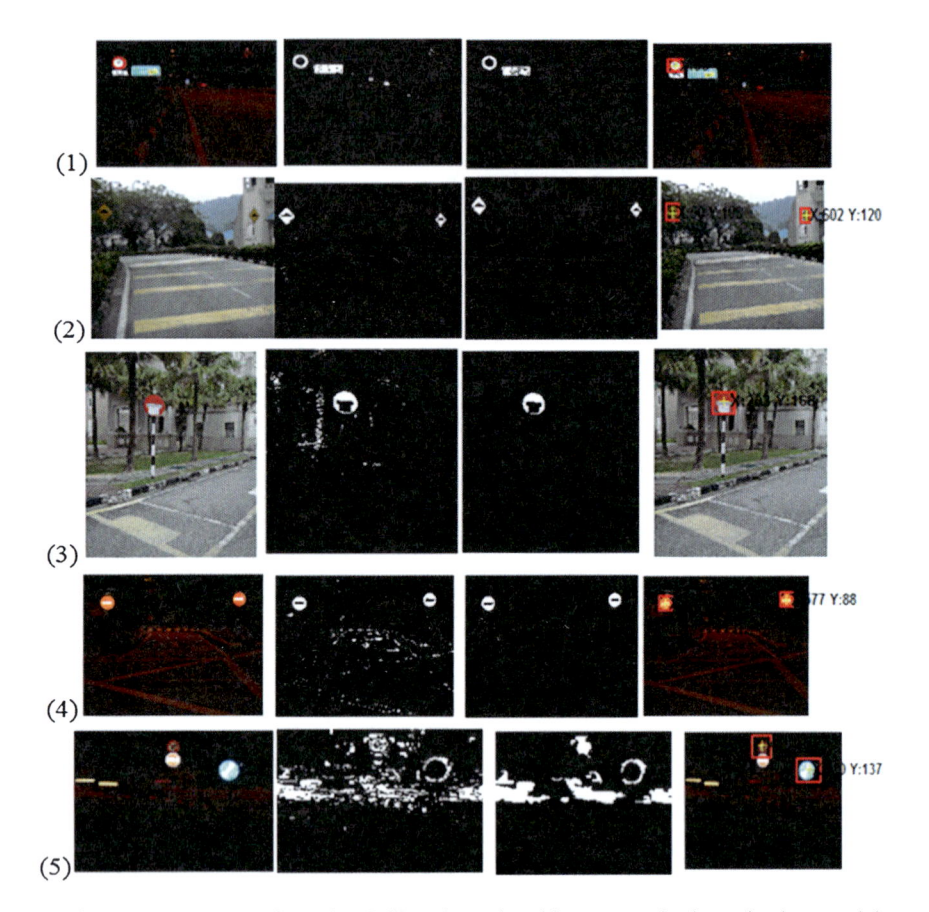

Fig. 32.2 Color segmentation using RGB color ratio with aspect ratio determination, at night (*1, 4 and 5*) and daytime (*2 and 3*)

The traffic signs can be affected by the illumination conditions such as direct sunlight, shadows and reflected sunlight which make detection more difficult. In this research, the images used were in resolution of 500 × 667 pixels using digital camera. The objective of the research is to detect any sign of given colour regardless of the illumination changes with partial occlusion.

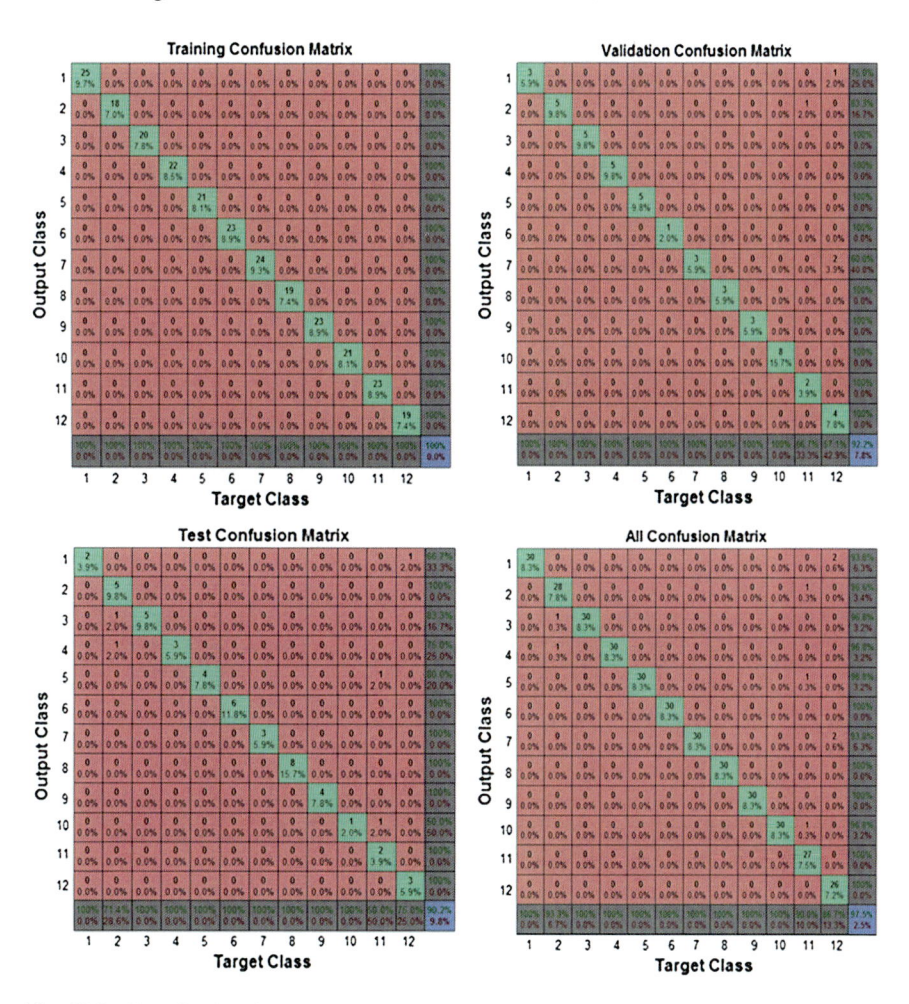

Fig. 32.3 Classification for each signs classes for test and all confusion matrixes with their respective percentages

Color segmentation using RGB algorithm was implemented based on Table 32.1. The color ratio in Fig. 32.2 was able to detect traffic signs with red, blue, yellow, white and black colours, respectively. However, achromatic color which is represented by dominant colour of hue is used to search the pixel of interest (POI) within traffic signs. The equation is as the following

$$f(R,G,B) = \frac{(|R-G|+|G-B|+|B-R|)}{3D} \qquad (32.1)$$

The aspect ratio utilized in the color detection if the minimum bounding box size is less than 300 pixels and ratio is less than 1.2. The determination was able to detect at specific traffic sign using RGB color segmentation.

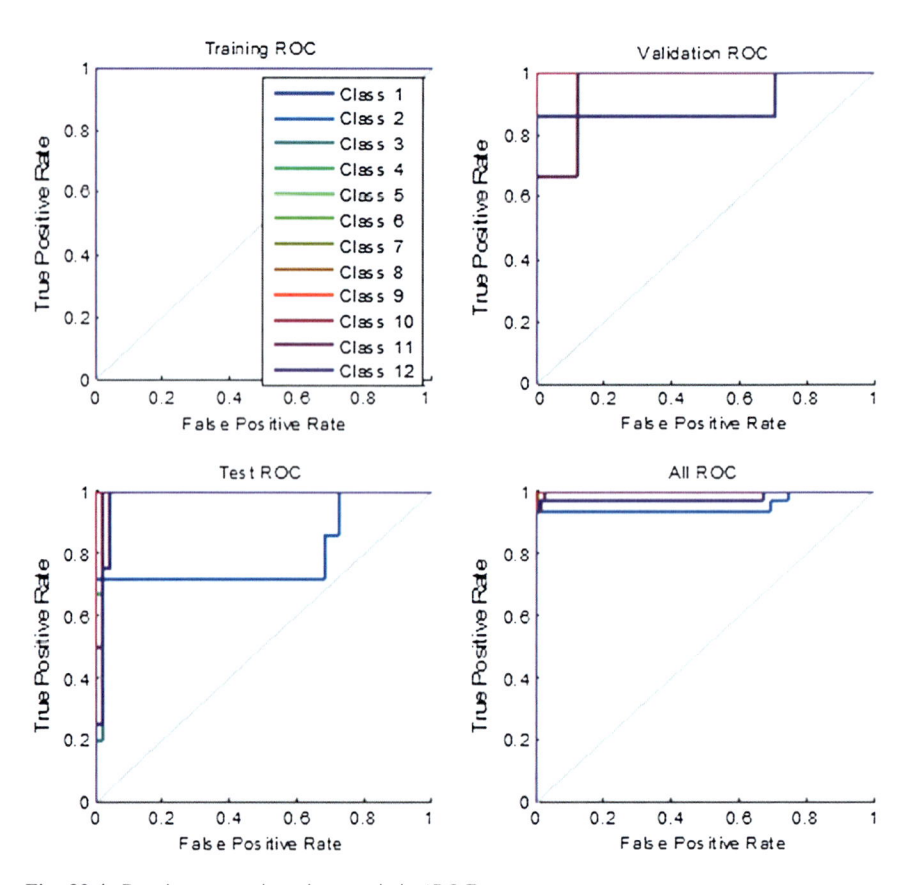

Fig. 32.4 Receiver operating characteristic (*ROC*) rate

32.3.2 Classification Stage: An Neural Network Approach

A supervised feed forward neural network was utilized in this study. It is a recursive tool that really helps in object classification. It can also be regarded as a pattern recognition tool. It extracts patterns when given input based on specific conditions of target. The flow of neural network will basically cover generation of weight and bias, trained, validated and tested. The neural network was trained using scale conjugate gradient (SCG) back propagation algorithm using log sigmoid function with 100 numbers of epochs.

Figure 32.3 illustrates the 12 sign classes that are classified according to its class by dividing the network for training, testing and validation. Based on the training data, the network divides 70 % for training, 15 % for testing and 15 % for validation's confusion matrix. After running the ANN algorithm, for instance, the percentage of confusion matrix of training uneven signs target which is represented at first row at Fig. 32.3 (*training confusion matrix*) obtained 100 %

accuracy. However, when the network tried to validate the data, it gives 75 % correctness whereas 66.7 % correctness for testing part. Based on the all confusion matrix data as shown at Fig. 32.3, 93.8 % classifies the first class as uneven signs target whereas 6.3 % produces wrong classification.

Based on this classification result, Fig. 32.4 shows the 12 sign classes of ROC value that are respectively reserved for training, testing and validation. ROC is widely used in predicting the percentage of accuracy in pattern recognition technique. As can be seen from the all ROC graph, the all classes approach true positive (TP) rate which is approximately one. TP is number of correctly predicted meaning that less false positive rate occurred in classification the signs classes.

32.4 Conclusion

In a nutshell, we have explained an algorithm using RGB color space in traffic sign detection. It was shown that the algorithm is able to detect red, yellow and blue traffic signs respectively using RGB color space. This frame able to detect more than one sign and reduced the false positive rates. Based on the result, the algorithm not only can detect partially occluded sign, but also all colors representing traffic signs involved with illumination and rotational changes. As for classification stage, a supervised neural network is used to classify the sign according to its class. This classifier can be extending for future research such as identification of traffic sign. In future research, we are planning to optimize the detection and classification algorithm to be applied in real time traffic sign classification and recognition application.

References

1. Nguwi YY, Kouzani AZ (2006) A study on automatic recognition of road signs. In: 2006 IEEE conference on cybernetics and intelligent systems, pp 1–6
2. Lim KH, Seng KP, Ang LM (2010) Intra color-shape classification for traffic sign recognition. In: International computer symposium (ICS) 2010, pp 642–647
3. Hua H, Chao C, Yulan J, Shinning T (2008) Automatic detection and recognition of circular road sign. In: IEEE/ASME international conference on mechatronics and embedded systems and applications (MESA) 2008, pp 626–630
4. Pazhoumand-Dar H, Yaghobi M (2010) DTBSVMs: a new approach for road sign recognition. In: Proceedings of the 2nd international conference on computational intelligence, communication systems and networks (CICSyN) 2010, pp 314–319
5. Vitabile S, Pollaccia G, Pilato G, Sorbello F (2001) Road signs recognition using a dynamic pixel aggregation technique in the HSV color space. In: Proceedings of the 11th international conference on image analysis and processing, pp 572–577

Chapter 33
Comparison of Multilayer Perceptron and Radial Basis Function Neural Networks for EMG-Based Facial Gesture Recognition

Mahyar Hamedi, Sh-Hussain Salleh, Mehdi Astaraki, Alias Mohd Noor and Arief Ruhullah A. Harris

Abstract This paper compared the application of multilayer perceptron (MLP) and radial basis function (RBF) neural networks on a facial gesture recognition system. Electromyogram (EMG) signals generated by ten different facial gestures were recorded through three pairs of electrodes. EMGs were filtered and segmented into non-overlapped portions. The time-domain feature mean absolute value (MAV) and its two modified derivatives MMAV1 and MMAV2 were extracted. MLP and RBF were used to classify the EMG features while six types of activation functions were evaluated for MLP architecture. The discriminating power of single/multi features was also investigated. The results of this study showed that symmetric saturating linear was the most effective activation function for MLP; the feature set MAV + MMAV1 provided the highest accuracy by both classifiers; MLP reached higher recognition ratio for most of features; RBF was the faster algorithm which also offered a reliable trade-off between the two key metrics, accuracy and time.

M. Hamedi (✉) · A. R. A. Harris
Centre for Biomedical Engineering, Faculty of Bioscience and Medical Engineering,
Universiti Teknologi Malaysia, Johor Bahru, Malaysia
e-mail: hamedi.mahyar@ieee.org; hamedi.mahyar@gmail.com

A. R. A. Harris
e-mail: ruhullah@utm.my

S.-H. Salleh · A. M. Noor
Centre for Biomedical Engineering, Transportation Research Alliance,
Universiti Teknologi Malaysia, Johor Bahru, Malaysia
e-mail: hussain@fke.utm.my

A. M. Noor
e-mail: alias@mail.fkm.utm.my

M. Astaraki
Department of Biomedical Engineering, Science and Research Branch,
Islamic Azad University, Tehran, Iran
e-mail: astarakee@yahoo.ca

Keywords Facial gesture recognition · EMG classification · Feature extraction · Multilayer perceptron neural network · Radial basis function neural network

33.1 Introduction

The human face is one of the most informative sources for conveying people's internal states. Facial expressions/gestures are counted as effective tools to communicate non-verbally. They have recently taken the researchers' attention to be applied in various fields such as system verification, human identification, security systems, and human machine interaction. Recent facial gesture recognition systems rely on image- [1], video- [2], and myoelectic-based [3] techniques. It has been reported that the image/video-based methods resulted in various drawbacks such as drifts, loss of communication, high sensitivity, high implementation cost, slow communication rates, and high complexity for analysis. Therefore, myoelectric-based technique in the form of EMG has been suggested and focused [3–11]. Designing such systems require considering a correct signal recording protocol, and using robust methods for preprocessing, processing and classification. Ang et al. [4] presented a recognition system based on the facial expressions, happiness, anger and sadness. The signals were recorded by 3 pairs of electrodes, and they were rectified. Then, mean value, standard deviation, RMS, and power spectrum density (PSD) of EMGs were computed as the features. By using minimum distance classifier, the facial gestures were recognized with 94.44 % accuracy. Arjunan and Kumar [5] studied the recognition of facial muscle movements during unvoiced speech by acquiring the EMG signals through 4 channels and characterized the muscle activation by extracting normalized integral moving RMS (MRMS) features. In the field of human machine interaction (HMI), Firoozabadi et al. [6] proposed an interface to be used for controlling a virtual robotic wheelchair. Three pairs of bipolar electrodes were employed to capture the signals from five facial expressions. They reported 89.75–100 % classification accuracy where MAV and SVM were utilized for feature extraction and classification respectively. Rezazadeh et al. [7] designed an interface to control a virtual interactive tower crane through five different facial EMGs. They used RMS features and subtractive fuzzy c-means (SFCM) clustering algorithm as a classifier and obtained 92.6 % recognition ratio. They expanded their study in [8] by considering eight facial gestures and employing SFCM plus adaptive neuro-fuzzy inference system (ANFIS) for classification. In our previous studies, a two channel-based facial gesture recognition system was proposed where 90.8 % accuracy was achieved using FCM for recognizing five gestures [9]. A comparative study was carried out where FCM outperformed SVM in the classification of eight facial expressions [10]. A multipurpose interface was designed through ten facial gestures for HMI applications and 90.41 % discrimination rate was reached by FCM [3]. The effectiveness of six different facial EMG features, integrated EMG (IEMG), variance, wave length (WL), mean absolute value slope (MAVS), MAV, and RMS were compared and evaluated. It was concluded that, RMS was the most discriminative one during classification of ten facial gestures [11]. Although most of mentioned studies applied FCM as a robust classifier, it was reported that its computational load during the training stage is high which is not proper for real-time applications. To address this obstacle, we suggested a very fast elliptic basis function neural network (VEBFNN) to provide a reliable trade-off between accuracy and speed [12]. The results indicated a remarkable decrease in training time while the average recognition accuracy was reduced about 3.3 %.

According to the literature, there were a few methods used and examined for facial EMG signal feature extraction and classification. And since the final performance of the systems is highly dependent on these two factors, more investigation is required in this area to find more robust methods. In this paper, ten different facial EMG signals were recorded by considering a similar setup as in [3]. Theoretically, when a signal is modeled as a Laplacian, a maximum likelihood of amplitude can be estimated from MAV [13]. Therefore, MAV feature and its two recent modified derivatives MMAV1 and MMAV2 were extracted from the EMGs and multifeature sets including two and three features were formed. For the purpose of classification, Multilayer Perceptron (MLP) and Radial Basis Function (RBF) Neural Networks (NN) were applied while six different types of activation functions were considered in MLP structure. In a comparative study, the impact of the considered features as well as the classifiers were investigated on the system performance in order to find the most discriminating feature and the more efficient classifier.

33.2 Methods and Materials

33.2.1 System Setup and Data Acquisition

The same setup as presented in [3] was considered in this study. Ten healthy volunteers in the age range of 26–41 participated in this experiment. The subjects' face skin was cleaned by alcohol pad; then, three pairs of pre-gelled Ag/AgCl electrodes were placed in bipolar configuration, one pair on the Frontalis muscle within a 2 cm distance and the other two pairs on the left and right Temporalis muscles. Moreover, an electrode was placed on the boney part of the left wrist as a ground. The facial EMGs were captured via BioRadio 150 (Clevemed) and the signals were recorded at the rate of $\sim 1,000$ Hz sampling frequency. Through the activation of filters with a low cut-off frequency 0.1 Hz and a notch filter of 50 Hz, unwanted artifacts from user movements and inference noises were removed by the device software itself. The ten facial gestures considered in this research were smiling with both sides of the mouth, smiling with left side of the mouth, smiling with right side of the mouth, opening the mouth (saying 'a' in the word apple), clenching the molars, gesturing 'notch' by raising the eyebrows, frowning, closing both eyes, closing the right eye and closing the left eye. All subjects were asked to perform each facial gesture five times for two seconds (active signal), and with 5 s rest between to reduce the effect of muscle exhaustion. Totally, ten (number of gestures) series of three (number of channels) dimensional data sets with 10,000 ms length were captured for each subject.

33.2.2 Preprocessing

Recorded EMGs were passed through a band-pass filter in the range of 30–450 Hz to envelope the most significant spectrum of signals. Then, filtered EMGs were segmented into non-overlapped windows with 256 ms length which resulted in 39 portions (10,000 ÷ 256 ≈ 39) for each gesture in each channel.

33.2.3 Feature Extraction

This stage aimed to extract the significant properties of facial EMGs in order to characterize the facial gestures into distinguishing categories. The features should have enough information for classification and be easy to compute so as to provide better performance and faster processing. In this study, the time-domain feature MAV and its two modified derivatives MMAV1 and MMAV2 [14] were computed and extracted from the segmented EMGs through the following equations:

$$MAV_k = \frac{1}{N} \sum_{i=1}^{N} |x_i| \tag{33.1}$$

$$MMAV1_k = \frac{1}{N} \sum_{i=1}^{N} \omega_i |x_i|, \quad \omega(i) = \begin{cases} 1, & 0.25N \leq i \leq 0.75N \\ 0.5, & otherwise \end{cases} \tag{33.2}$$

$$MMAV2_k = \frac{1}{N} \sum_{i=1}^{N} \omega_i |x_i|, \quad \omega(i) = \begin{cases} 1, & 0.25N \leq i \leq 0.75N \\ \frac{4i}{N}, & 0.25N > i \\ \frac{4(i-N)}{N}, & 0.75N < i \end{cases} \tag{33.3}$$

where x_i is the value of each part of the segment k, N is the segment length (256), and ω_i is the weighting window function which is continuous in MMAV2.

By considering three channels, a single feature vector composed of 3 × 390 features (for 10 gestures) was achieved for each subject. In addition to the single features, multifeature sets including two and three features were considered for characterizing the facial gestures. To do so, the single features were concatenated as MAV + MMAV1, MAV + MMAV2, MMAV1 + MMAV2, and MAV + MMAV1 + MMAV2. The dimensions of constructed multifeature sets including 2 and 3 features were increased to 6, and 9 respectively.

33.2.4 Classification

In order to recognize the facial gestures accurately, extracted features must be classified into distinct groups through a robust classifier. In this study, MLP and RBF neural networks were applied to classify the facial EMG features. Moreover, these classifiers were used to evaluate the impact of single/multi feature sets on the system performance.

MLP is a feed-forward artificial NN extensively employed for pattern recognition-based problems. Recently, it was adopted for classifying and detecting different states of human emotions through biomedical signals [15]. This NN applies a unit (neuron) whose output is a nonlinear differentiable function of its input. This method is highly flexible due to its nonlinear structure and since various types of activation functions can be utilized. MLP employs the supervised learning Backpropagation (BP) algorithm for network training and the gradient descent weight update rule for computing new weights during the learning process [16]. Training errors over all network output units are calculated by the following equation:

$$E(\vec{\omega}) \equiv \frac{1}{2} \sum_{s \in T} \sum_{n \in R} (t_{ns} - o_{ns})^2 \qquad (33.4)$$

where $\vec{\omega}$ is the weight vector, T indicates the training samples set, R represents the output units set, t_{ns} and o_{ns} specify the target and output values related to the nth output unit and training sample s.

In this work, the MLP comprised of three layers where the number of neurons in the input layer was equal to the feature vector dimensions (3, 6, and 9). In the hidden layer, this number was selected manually in each run so as to reach the best performance. And the number of neurons in the output layer was the same as the number of classes in the training data set (10). Furthermore, the capability of several activation functions, linear (*purelin*), saturating linear (*satlin*), symmetric saturating linear (*satlins*), positive linear (*poslin*), log-sigmoid (*logsig*), and hyperbolic tangent sigmoid (*tansig*) were examined in order to find the most efficient one.

RBF is a mixture of instance-based and probabilistic artificial NN approaches. It is a three layer NN with spatially localized kernel functions (radial basis functions) as activation functions in the hidden layer. The output of the network is a linear combination of RBFs of the inputs and neuron parameters. In our study, the input layer of RBF network included the same number of neurons as in MLP and the neuron parameters known as spread of radial basis functions were adjusted manually for each run to gain the highest performance. The output of the network which is then a scalar function of the input vector x is computed by

$$f(x) = \sum_{i=1}^{k} \alpha_i \rho(\|x - c_i\|) \qquad (33.5)$$

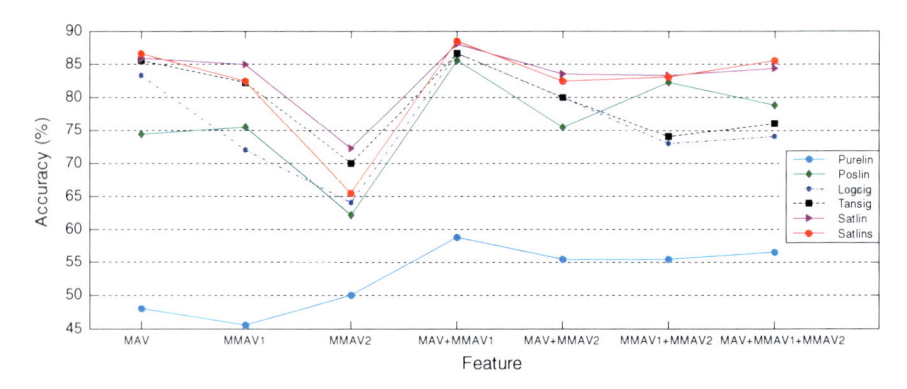

Fig. 33.1 Effectiveness of different single/multi features and various activation functions of MLP on the recognition accuracy

where α_i is the weight of neuron i in the linear output neuron, and k is a constant that specifies the number of kernel functions (number of neurons in the hidden layer) and it is adjusted differently for each subject. Here $\rho(\|x - c_i\|)$ is the Gaussian kernel function centered at the point c_i with variance σ_i^2. The details of RBFNN training procedure can be found in [17].

The proposed methods of classification were implemented by means of the MATLAB software package version 7.12.0 with NN toolbox. In order to construct the NN, each feature set was shuffled and divided into training and testing data sets which were 300×3, 90×3 for single features; 300×6, 90×6 for the sets containing two features, and 300×9, 90×9 for the set with three features.

33.3 Results and Discussion

This section discusses the results of several experiments carried out throughout the study. Figure 33.1 investigates the impact of single/multi features on the recognition performance achieved by MLP when considering different activation functions. As can be seen, among the single features, generally MAV provided better discriminative information comparing with its two modified versions.

The results from multifeature sets in Fig. 33.1 showed that MAV + MMAV1 outperformed other sets which indicated that MMAV1 could provide complementary information when concatenated with MAV. On the other hand, since MMAV2 achieved low accuracy as a single feature which proved its weakness in discriminating the facial gestures, it undesirably affected the other three sets. Figure 33.1 also demonstrates the effectiveness of six different activation functions utilized in MLP architecture on the system accuracy over the considered features. Obviously, MLP behaved differently by applying different activation function over the features. Amongst all, *purelin* resulted in significant misclassification over all

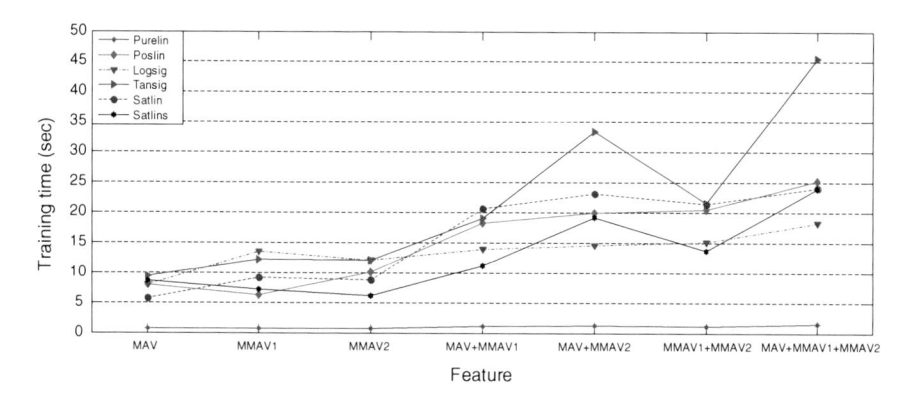

Fig. 33.2 The training time consumed during the classification of different single/multi features by various activation functions of MLP

features whereas *satlin* and *satlins* led to higher recognition accuracy in comparison with other activation functions. According to the results, when MLP classified the facial gesture feature set MAV + MMAV1 by means of *satlins* activation function, the highest recognition accuracy of 88.6 % was obtained.

Figure 33.2 depicts the influence of different features and various activation functions used in MLP classifier on the computational load consumed during the training stage. It can be observed that there was an increase in the time when applying more features and the maximum time was spent during the training of feature set MAV + MMAV1 + MMAV2. Considering the set MAV + MMAV1 which achieved the highest recognition accuracy in the previous section, it is obvious that the training time for this multifeature set was lower than the other sets when being classified by MLP with all activation functions. According to this figure, *purelin* and *tansing* activation functions resulted in the minimum and the maximum computational cost respectively which implied the rate of simplicity and complexity of these functions during the processing. The remarkable point was where *satlins* performed very fast in comparison with other activation functions.

The statistical analysis applied to the results in Figs. 33.1 and 33.2 showed that MLP achieved the best performance, 88.6 % recognition accuracy and 11.2 s training time, when utilizing the feature set MAV + MMAV1 and the activation function *satlins*.

The results provided by RBF over all features are illustrated in Fig. 33.3. As can be seen, facial gestures were recognized within the accuracy range of 81.1 % (with MAV + MMAV1) to 85.5 % (with MMAV1 + MMAV2). It was also demonstrated that the training time did not noticeably vary and the maximum time (0.53 s) was consumed for the classification of MAV + MMAV1 + MMAV2. Figure 33.3 also compares the performance of the best MLP architecture (with *satlins* activation function) with RBF algorithm. Overall, MLP outperformed RBF in recognizing the facial gestures since it obtained higher accuracy using most of single/multi feature sets. However, less variation in recognition ratio through RBF

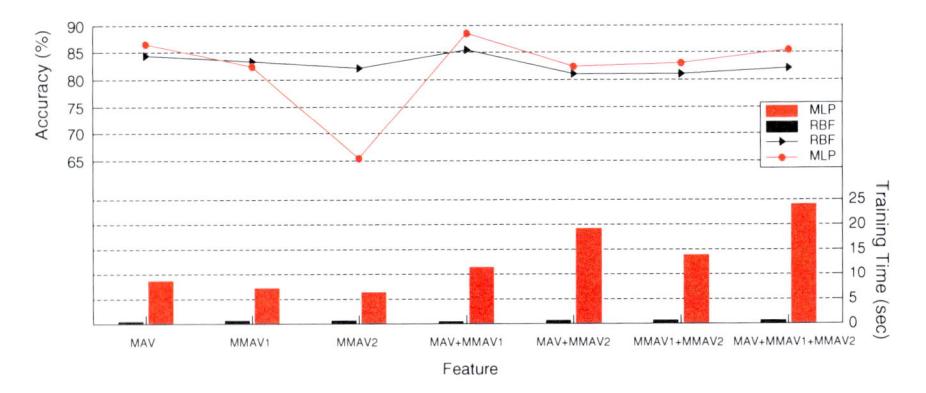

Fig. 33.3 Comparison of MLP and RBF in terms of recognition accuracy and computational cost over different single/multi features

indicates more stability of this classifier over different features. As a significant point, the maximum accuracy was achieved by both algorithms when MAV + MMAV1 was applied which again designates the high discriminating power of this feature set. In terms of training time, it is clear that RBF was much more stable and faster than MLP due to the fact that the hidden layer in RBF is computed only through a function of the distance, not a series of weights.

Since the appropriate classifier must be selected based on the application, MLP can be used when higher accuracy is required and RBF can be adopted in real-time processing where high speed is necessary. Moreover, RBF is preferred in the systems which desire a reasonable trade-off between the two key factors.

33.4 Conclusion and Future Work

This paper investigated and compared the usage of two well-known neural networks MLP and RBF for recognizing ten different facial gestures through facial myoelectric signals. Accordingly, the time-domain feature MAV and its two modified derivatives MMAV1 and MMAV2 were extracted and applied in the forms of single and multifeature sets for characterizing the facial EMGs. The efficiency of the features was evaluated and the set MAV + MMAV1 was reported as the most efficient one. In this study, six different types of activation functions were tested in MLP architecture and the best performance was provided by using *satlins*. Finally, it was clarified that MLP generally achieved better recognition accuracy while RBF delivered a much faster training procedure. Therefore, it was recommended to choose the classifier based on the application's target that is accuracy, speed or a trade-off between them.

This research compared the capability of two neural networks for classification of facial myoelectric signals by just considering a few types of EMG features and without concentrating on other steps of signal processing which can affect the system performance. So, extracting more features and utilizing other methods for preprocessing such as different segmentation and filtering techniques will be studied in future.

Acknowledgments This research project is supported by Center of Biomedical Engineering (CBE), Transport Research Alliance, Universiti Teknologi Malaysia research university grant (Q.J130000.2436.00G32) and funded by Ministry of Higher Education (MOHE).

References

1. Kotsia I, Pitas I (2007) Facial expression recognition in image sequences using geometric deformation features and support vector machines. IEEE Trans Image Process 16(1):172–187
2. Cohen I, Sebe N, Garg A, Chen LS, Huang TS (2003) Facial expression recognition from video sequences: temporal and static modeling. Comput Vis Image Underst 91(1):160–187
3. Hamedi M, Salleh SH, Tan TS, Ismail K, Ali J, Dee-Uam C, Pavaganun C, Yupapin PP (2011) Human facial neural activities and gesture recognition for machine-interfacing applications. Int J Nanomed 6:3461–3472
4. Ang LBP, Belen EF, Bernardo RA Jr, Boongaling ER, Briones GH, Coronel JB (2004) Facial expression recognition through pattern analysis of facial muscle movements utilizing electromyogram sensors. In: TENCON 2004, IEEE region 10 conference, vol 100. IEEE Press, pp 600–603
5. Arjunan S, Kumar DK (2007) Recognition of facial movements and hand gestures using surface Electromyogram (sEMG) for HCI based applications. In: 9th Biennial conference of the Australian Pattern Recognition Society on digital image computing techniques and applications. IEEE Press, pp 1–6
6. Firoozabadi SMP, Oskoei MA, Hu H (2008) A human-computer interface based on forehead multi-channel bio-signals to control a virtual wheelchair. In: Proceedings of the 14th Iranian conference on biomedical engineering (ICBME), pp 272–277
7. Rezazadeh IM, Wang X, Firoozabadi SMP, Hashemi Golpayegani MR (2010) Using affective human machine interface to increase the operation performance in virtual construction crane training system: a novel approach. Autom Construct J 20:289–298
8. Rezazadeh IM, Firoozabadi SMP, Hu H, Hashemi Golpayegani SMR (2011) A novel human-machine interface based on recognition of multi-channel facial bioelectric signals. Australas J Phys Eng Sci Med 34(4):497–513
9. Hamedi M, Rezazadeh IM, Firoozabadi SMP (2011) Facial gesture recognition using two-channel biosensors configuration and fuzzy classifier: a pilot study. In: International conference on electrical, control and computer engineering (INECCE). IEEE Press, pp 338–343
10. Hamedi M, Salleh SH, Tan TS, Ismail K (2011) Surface electromyography-based facial expression recognition in bipolar configuration. J Comput Sci 7(9):1407–1415
11. Hamedi M, Salleh SH, Noor AM, Tan TS, Ismail K (2012) Comparison of different time-domain feature extraction methods on facial gestures' EMGs. In: Electromagnetics research symposium proceedings, KL. PIER, pp 1897–1900
12. Hamedi M, Salleh SH, Astaraki M, Noor AM (2013) EMG-based facial gesture recognition through versatile elliptic basis function neural network. Biomed Eng Online 12(1):73
13. Asghari Oskoei M, Hu M (2007) Myoelectric control systems—A survey. Biomed Signal Process Control 2(4):275–294
14. Phinyomark A, Limsakul C, Phukpattaranont P (2009) A novel feature extraction for robust EMG pattern recognition. J Comput 1(1):71–80

15. Van den Broek EL, Lisy V, Janssen JH, Westerink JHDM, Schut MH, Tuinenbreijer K (2010) Affective man-machine interface: unveiling human emotions through biosignals. In: Biomedical engineering systems and technologies. Communications in computer and information science, vol 52. Springer, New York, pp 21–47
16. Mitchell TM (1997) Machine learning. McGraw Hill, New York
17. Schwenker F, Kestler HA, Palm G (2001) Three learning phases for radial-basis-function networks. Neural Netw 14(4):439–458

Chapter 34
The Effect of Bat Population in Bat-BP Algorithm

Nazri Mohd. Nawi, Muhammad Zubair Rehman and Abdullah Khan

Abstract A new metaheuristic based back-propagation algorithm known as Bat-BP is presented in this paper. The proposed Bat-BP algorithm successfully solves the problems like slow convergence to global minima and network stagnancy in back-propagation neural network (BPNN) algorithm. In this paper, the bat population is increased from 10 to 500 bats to detect the performance decline or incline in the Bat-BP algorithm by performing simulations on XOR and OR datasets. The simulation results show that the convergence rate to global minimum in Bat-BP is directly proportional with an increase in bats on 2-bit XOR dataset. In case of 3-bit XOR and 4-bit OR datasets, the results deteriorated with an increase in the bat population.

Keywords Bat population · Metaheuristics · Slow convergence · Global minima · Bat algorithm · Back-propagation neural network algorithm · Bat-BP algorithm · Momentum

34.1 Introduction

Back-propagation Neural Network (BPNN) is a very old optimization technique applied on the Artificial Neural Networks (ANN) to speed up the network convergence to global minima during training process [1–3]. BPNN follows the basic

---

N. M. Nawi · M. Z. Rehman (✉) · A. Khan
Software and Multimedia Centre, Faculty of Computer Science
and Information Technology, Universiti Tun Hussein Onn Malaysia (UTHM),
P.O. Box 101, 86400 Parit Raja, Batu Pahat, Johor, Malaysia
e-mail: zrehman862060@gmail.com

N. M Nawi
e-mail: nazri@uthm.edu.my

A. Khan
e-mail: hi100010@siswa.uthm.edu.my

principles of ANN which mimics the learning ability of a human brain. Similar to ANN architecture, BPNN consists of an input layer, one or more hidden layers and an output layer of neurons. In BPNN, every node in a layer is connected to every other node in the adjacent layer. Unlike normal ANN architecture, BPNN learns by calculating the errors of the output layer to find the errors in the hidden layers [4]. This qualitative ability makes it highly suitable to be applied on problems in which no relationship is found between the output and the inputs. Due to its high rate of plasticity and learning capabilities, it has been successfully implemented in wide range of applications [5].

Despite providing successful solutions BPNN has some limitations. Since, it uses gradient descent learning which requires careful selection of parameters such as network topology, initial weights and biases, learning rate, activation function, and value for the gain in the activation function. An improper use of these parameters can lead to slow network convergence or even network stagnancy [5, 6]. Previous researchers have suggested some modifications to improve the training time of the network. Some of the variations suggested are the use of learning rate and momentum to stop network stagnancy and to speed-up the network convergence to global minima. These two parameters are frequently used in the control of weight adjustments along the steepest descent and for controlling oscillations [7].

Besides setting network parameters in BPNN, evolutionary computation is also used to train the weights to avoid local minima. To overcome the weaknesses of gradient-based techniques, many new algorithms have been proposed recently. These algorithms include global search techniques such as hybrid PSO-BP [8], artificial bee colony back-propagation (ABC-BP) algorithm [9, 10], evolutionary artificial neural networks algorithm (EA) [11], and genetic algorithms (GA) [12].

In this paper, Bat-BP algorithm is proposed. The Bat-BP with varying population of bats is simulated on XOR and OR datasets and shows high accuracy and avoids local minima. We find that by using an appropriate metaheuristic technique such as BAT with an equally intelligent BPNN can answer many limitations of gradient descent efficiently. The next two sections provide a brief discussion of BAT algorithms, followed by the proposed BAT-BP algorithm, simulation results, and finally the conclusion.

34.2 The Bat Algorithm

Bat is a metaheuristic optimization algorithm developed by Yang in 2010 [13]. Bat algorithm is based on the echolocation behavior of microbats with varying pulse rates of emission and loudness. Yang [13] has idealized the following rules to model Bat algorithm;

1. All bats use echolocation to sense distance, and they also "know" the difference between food/prey and back-ground barriers in some magical way.

> **Step 1:** BAT initializes and passes the best weights to BPNN
> **Step 2:** Load the training data
> **Step 3: While** MSE < Stopping Criteria
> **Step 4:** Initialize all BAT Population
> **Step 5:** Bat Population finds the best weight in Equation 4 and pass it on to the network, the weights, w_{ij} **with momentum, α** and biases, b_i in BPNN are then adjusted using the following formulae;
> $$w_{ij}(k+1) = (w_{ij}k + \mu\partial_j y_i)$$
> $$b_i(k+1) = b_i k + \mu\partial_j$$
> **Step 6:** Feed forward neural network runs using the weights initialized with BAT
> **Step 7:** Calculate the backward error
> **Step 8:** Bat keeps on calculating the best possible weight at each epoch until the Network is converged.
> **End While**

Fig. 34.1 The proposed Bat-BP algorithm

2. A bat fly randomly with velocity (v_i) at position (x_i) with a fixed frequency (f_{min}), varying wavelength λ and loudness A_0 to search for prey. They can automatically adjust the wavelength (or frequency) of their emitted pulses and adjust the rate of pulse emission $r \in [0,1]$, depending on the proximity of their target.
3. Although the loudness can vary in many ways, Yang [13] assume that the loudness varies from a large (positive) A_0 to a minimum constant value A_{min}. The pseudo code for the Bat algorithm can be seen in Fig. 34.1.

Firstly, the initial position x_i, velocity v_i and frequency f_i are initialized for each bat b_i. For each time step t, the movement of the virtual bats is given by updating their velocity and position using Eqs. (34.2), and (34.3) as follows:

$$f_i = f_{min} + (f_{max} + f_{min})\beta \qquad (34.1)$$

$$v_i^t = v_i^{t-1} + (x_i^t + x_*)f_i \qquad (34.2)$$

$$x_i^t = x_i^{t-1} + v_i^t \qquad (34.3)$$

where β denotes a randomly generated number within the interval [0,1]. Recall that x_i^t denotes the value of decision variable j for bat i at time step t. The result of f_i in Eq. (34.1) is used to control the pace and range of the movement of the bats. The variable $x*$ represents the current global best location (solution) which is located after comparing all the solutions among all the n bats. In order to improve the variability of the possible solutions, Yang [13] has employed random walks. Primarily, one solution is selected among the current best solutions for local search and then the random walk is applied in order to generate a new solution for each bat;

$$x_{new} = x_{old} + \in A^t \qquad (34.4)$$

where, A^t stands for the average loudness of all the bats at time t, and $\varepsilon \in [-1,1]$ is a random number. For each iteration of the algorithm, the loudness A_i and the emission pulse rate r_i are updated, as follows:

$$A_i^{t+1} = \alpha A_i^t \qquad (34.5)$$

$$r_i^{t+1} = r_i^0[1 - exp(-\gamma t)] \qquad (34.6)$$

where α and γ are constants. At the first step of the algorithm, the emission rate, r_i^0 and the loudness, A_i^0 are often randomly chosen. Generally, $A_i^0 \in [1,2]$ and $r_i^0 \in [0,1]$ [12].

34.3 The Proposed BAT-BP Algorithm

In the proposed BAT-BP algorithm, each position represents a possible solution (i.e., the weight space and the corresponding biases for BPNN optimization in this paper). The weight optimization problem and the position of a food source represent the quality of the solution. In the first epoch, the best weights and biases are initialized with BAT and are passed on to the BPNN. The weights in BPNN are calculated and compared in the reverse cycle. In the next cycle BAT again update the weights with the best possible solution and BAT will continue to search the best weights until the last epoch is reached or the MSE is achieved. The proposed Bat-BP algorithm is shown in Fig. 34.1.

34.4 Results and Discussions

The Workstation used for carrying out experimentation comes equipped with a 2.33 GHz Core-i5 processor, 4-GB of RAM and Microsoft Windows 7. The simulations are carried-out using MATLAB 2010 software on three datasets such as 2-bit XOR, 3-bit XOR and 4-bit OR. Three layer back-propagation neural networks is used for training, the hidden layer is kept fixed to 10-nodes while output and input layers nodes vary according to the datasets given. Pulse rate of 0.9 and loudness of 0.7 is found to be optimal in Bat-BP algorithm and Log-sigmoid activation function is used. The bat population is varied from 10 to 500 bats and a total of 20 trials run on each dataset. Each trial is limited to 1,000 epoch CPU time, average accuracy, and Mean Square Error (MSE) are recorded for each independent trials on XOR and OR datasets and stored in a separate file.

Table 34.1 CPU time, epochs and MSE for 2-bit XOR dataset with varying population of BATS and **2-10-1** ANN architecture

Algorithms	10	30	50	70	100	300	500
CPU time	2.39	0.56	0.71	0.52	0.8	1.23	2.05
Epochs	23.25	8.1	5	3	3	2	1
MSE	0	0	0	0	0	0	0
Accuracy (%)	100	100	100	100	100	100	100

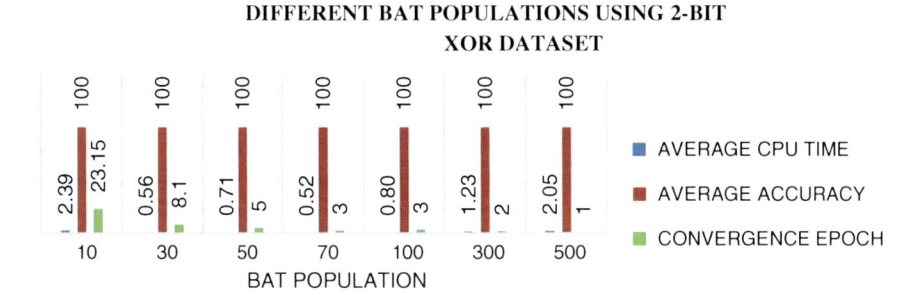

DIFFERENT BAT POPULATIONS USING 2-BIT XOR DATASET

Fig. 34.2 Bat-BP convergence performance on 2-bit XOR with 2-10-1 ANN architecture and varying population of bats

34.4.1 2-Bit XOR Dataset

The first test problem is the 2 bit XOR Boolean function consisting of two binary inputs and a single binary output. In simulations, we used 2-10-1 network architecture for two bit XOR. For the simple Bat-BP and modified Bat-BP, Table 34.1, shows the CPU time, number of epochs and the MSE for the 2 bit XOR test problem. Figure 34.2 shows, Bat-BP convergence performance on 2-bit XOR with 2-10-1 ANN Architecture.

The BAT-BP algorithm successfully avoids the local minima and converges on the provided network successfully within 100 epochs, as shown in Table 34.1. The convergence rate in Bat-BP increases with an increase in Bat population. The best results are obtained when the bats are increased to 500 as shown in Fig. 34.3. With 500 bats, Bat-BP was more stable and converged to global minima in 2.05 CPU cycles with 100 % accuracy, 0 MSE and a mere 1 epoch.

34.4.2 3-Bit XOR Dataset

In the second phase, we used 3 bit XOR dataset consisting of three inputs and a single binary output. For the three bit input we apply 3-10-1, network architecture. For the 3-10-1 the network it has forty connection weights and eleven biases.

DIFFERENT BAT POPULATIONS USING 3-BIT XOR DATASET

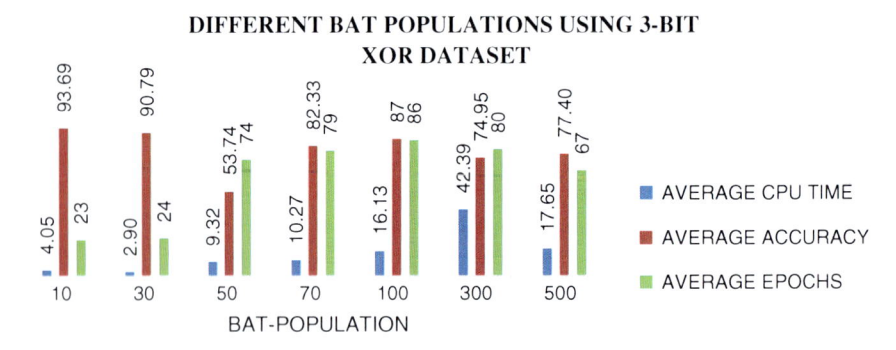

Fig. 34.3 Bat-BP convergence performance on 3-bit XOR with 3-10-1 ANN architecture and varying population of bats

Table 34.2 CPU time, epochs and MSE for 3-bit XOR dataset with varying population of BATS and **3-10-1** ANN architecture

Algorithms	10	30	50	70	100	300	500
CPU time	4.05	2.90	9.32	10.27	16.13	42.39	17.65
Epochs	23	24	74	79	86	80	65
MSE	0.0625	0.1240	0.4550	0.1750	0.1006	0.2366	0.2250
Accuracy (%)	93.69	90.79	53.74	82.33	87.00	42.39	17.65

For the Bat-BP, Table 34.2 shows the CPU time, number of epochs and the MSE for the 3 bit XOR test problem.

As can be seen in Fig. 34.3, 3-bit XOR dataset performs better with 10 bats and converged to global minima within 23 epochs, 93.69 % accuracy and 4.05 CPU cycles. During the simulations, it is also observed that 3-bit XOR's performance becomes worse with an increase in the bat population. With an increase in the bats, a sudden downward spike in the MSE with respect to accuracy can be observed in Table 34.2.

34.4.3 4-Bit OR Dataset

The third dataset is based on the logical operator OR which indicates whether either operand is true. If one of the operand has a nonzero value, the result has the value 1. Otherwise, the result has the value 0. The network architecture used here is 4-10-1 in which the network has fifty connection weights and eleven biases. Table 34.3, illustrates the CPU time, epochs, and MSE performance of the Bat-BP algorithm. Figure 34.4, shows the Bat-BP algorithm convergence performance on 4-bit OR.

Table 34.3 CPU time, epochs and MSE for 4-bit OR dataset with varying population of BATS and **4-10-1** ANN architecture

Algorithms	10	30	50	70	100	300	500
CPU time	2.88	0.20	0.30	0.75	0.41	1.12	1.94
Epochs	46.8	2.6	3	5	7	2	1
MSE	0	0.0063	0	0.1250	0.0125	0.0001	0
Accuracy (%)	100	99.38	100	87.50	98.75	99.92	100

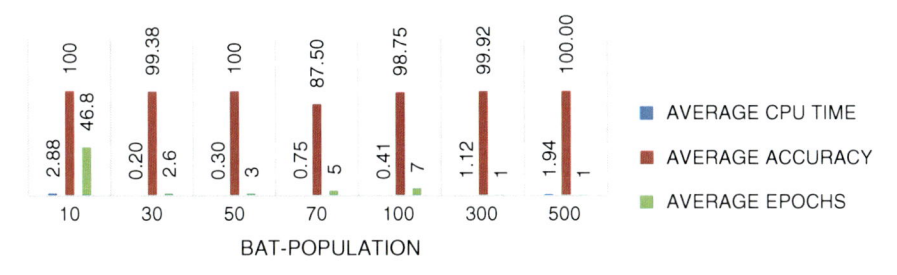

Fig. 34.4 Bat-BP convergence performance on 4-bit OR with 4-10-1 ANN architecture and varying population of bats

Unlike 3-bit XOR dataset, 4-bit OR dataset performed well with 10 and 50 and 500 bat population and showed a continuous decrease in no of epochs with respect to an increase in bats, as seen in Table 34.3. Overall, Bat-BP performed well in terms of accuracy and stayed with an acceptable range of 87–100 % as can be seen in Fig. 34.4.

34.5 Conclusions

BPNN algorithm is one of the most widely used procedure to train Artificial Neural Networks (ANN). But BPNN algorithm has some drawbacks, such as getting stuck in local minima and slow speed of convergence. Nature inspired meta-heuristic algorithms provide derivative-free solution to optimize complex problems. In this paper, a meta-heuristic search algorithm, called Bat was used to train BPNN to achieve fast convergence rate in the neural network. The proposed Bat-BP was trained on XOR and OR datasets with varying population of Bats. In Bat-BP, each bat represents the best solution or connection weights to the network and an increase in the bats increases the possibility of finding the optimal weight/solution. So, the idea was to increase the bat population and to check its effect on the overall performance of the BAT-BP algorithm. During the simulation results, it was found that the convergence rate to global minimum in Bat-BP is directly proportional with an increase in bats on 2-bit XOR dataset. In case of 3-bit XOR and 4-bit OR datasets, the results deteriorated with an increase in the bat population.

Acknowledgments The Authors would like to thank Office of Research, Innovation, Commercialization and Consultancy Office (ORICC), Universiti Tun Hussein Onn Malaysia (UTHM) and Ministry of Higher Education (MOHE) Malaysia for financially supporting this Research under Fundamental Research Grant Scheme (FRGS) vote no. 1236.

References

1. Deng WJ, Chen WC, Pei W (2008) Back-propagation neural network based importance-performance analysis for determining critical service attributes. J Expert Syst Appl 34(2)
2. Kosko B (1992) Neural network and fuzzy systems, 1st edn. Prentice Hall, Englewood Cliffs
3. Rumelhart DE, Hinton GE, Williams RJ (1986) Learning internal representations by error propagation. Parallel Distrib Process Explor Microstruct Cogn
4. Rehman MZ, Nawi NM, Ghazali R (2012) Studying the effect of adaptive momentum in improving the accuracy of gradient descent back propagation algorithm on classification problems. Int J Mod Phys (IJMPCS), 1(1)
5. Nawi NM, Ransing MR, Ransing RS (2007) An improved conjugate gradient based learning algorithm for back propagation neural networks. Int J Comput Intell 4(1):46–55
6. Nawi NM, Rehman MZ, Ghazali MI (2011) Noise-induced hearing loss prediction in Malaysian industrial workers using gradient descent with adaptive momentum algorithm. Int Rev Comput Softw (IRECOS), 6(5):740–749
7. Lee K, Booth D, Alam PA (2005) Comparison of supervised and unsupervised neural networks in predicting bankruptcy of Korean firms. Expert Syst Appl 29(1):1–16
8. Mendes R, Cortez P, Rocha M, Neves J (2002) Particle swarm for feed forward neural network training. In: Proceedings of the international joint conference on neural networks, vol 2, pp 1895–1899
9. Nandy S, Sarkar PP, Das A (2012) Training a feed-forward neural network with artificial bee colony based backpropagation method. Int J Comput Sci Inf Technol (IJCSIT),4(4):33–46
10. Karaboga D, Akay B, Ozturk C (2007) Artificial bee colony (ABC) optimization algorithm for training feed-forward neural networks. In: 4th international conference on modeling decisions for artificial intelligence (MDAI 2007), Kitakyushu, Japan, 16–18 Aug 2007
11. Yao X (1993) Evolutionary artificial neural networks. Int J Neural Syst 4(3):203–222
12. Montana DJ, Davis L (1989) Training feedforward neural networks using genetic algorithms. In: Proceedings of the eleventh international joint conference on artificial Intelligence, vol 1, pp 762–767
13. Yang XS (2010) A new metaheuristic bat-inspired algorithm. In: Nature inspired cooperative strategies for optimization (NICSO 2010), pp 65–74

Chapter 35
Synthesizing *Asli* Malay Song: Transforming Spoken Voices into Singing Voices

Nurmaisara Za'ba, Nursuriati Jamil, Siti Salwa Salleh and Nurulhamimi Abdul Rahman

Abstract Singing voices are consist of frequency related to musical pitch and fundamental frequency (F0) fluctuation such as vibrato, overshoot, preparation and fine fluctuation. Singing Malay *asli* music requires the singer to perform another type of fluctuation in singing called *patah lagu*, at every central note or any suitable longer notes. This paper discusses the construction of speech and singing corpus, and the F0 analysis of *patah lagu* for speech to singing synthesis application.

Keywords Fundamental frequency (F0) · *Patah lagu* · Singing synthesis

35.1 Introduction

Speech and singing have many differences in terms of production, characteristic and perception. The range of frequencies and amplitudes in singing voice is much larger than in speech. When singing a song, a singer expresses lyrics by changing notes according to the melody of the musical score. According to Saitou et al. [7],

N. Za'ba · N. Jamil (✉) · S. S. Salleh
Faculty of Computer and Mathematical Sciences,
MARA University of Technology, Shah Alam, Malaysia
e-mail: liza@tmsk.uitm.edu.my

N. Za'ba
e-mail: maisara@tmsk.uitm.edu.my

S. S. Salleh
e-mail: ssalwa@tmsk.uitm.edu.my

N. A. Rahman
Faculty of Music, MARA University of Technology, Shah Alam, Malaysia
e-mail: nurulhamimi@salam.uitm.edu.my

there are several types of F0 fluctuations such as overshoot, preparation, vibrato and fine fluctuations in a singing voice. These fluctuations are important to singing voice characteristics [11]. However performing traditional *asli* music requires traditional melodic ornamentation called *patah lagu,* which is also another type of fluctuation needs to be considered.

Patah lagu (also known as *lenggok, bunga melodi, penyedap lagu* in Malay) is a form of ornamentation and widely used in most of *asli* music. The presence of *patah lagu* in *asli* music is very important; without it, a song will sound dreary and dull [6]. Za'ba et al. [12] discussed Traditional Malay *asli* music and *patah lagu* in detail, and reveals the variation of F0 contour in *patah lagu* performed by singers. *Patah lagu* in Malay *asli* music is a unique type of ornamentation, that is turns, mellismetic rendering, embellishing the longer notes with new ideas and accent at the end of a phrase or cadence [5, 6]. All *asli* music should be sung with *patah lagu,* even though the details of the *patah lagu* is not properly written or described in the score.

Four types of F0 variations in a Japanese song, namely: overshoot, vibrato, preparation and fine fluctuation were identified and the importance of these F0 fluctuations were demonstrated through a psychoacoustic experiment [8]. The study indicated that overshoot has the largest effect on singing voice perception and preparation as a new important fluctuation.

The presence of ornamentation is not only important to western music but to all Indian music. Hindustani and Carnatic music in India are considered as two distinct musical areas with differences in nomenclature, style and grammar. Arora et al. [2] investigated vocal performance in Indian Classical Raga using PRAAT speech processing system, by introducing Indian ornamentation such as *gamak* and *meend* in the system.

The goal of this paper is to discuss the construction of speech and singing corpus as the primary subject for developing a speech to singing synthesis system for Malay *asli* music. The discussion includes the methods for constructing databank of speech and singing voices, analysis and synchronization of lyrics and musical information. Experiment to reveal F0 fluctuations in *patah lagu* is also presented and discussed in this paper.

The paper is organized as follows. Section 35.2 introduces speech and singing corpus. Section 35.3 presents the observation of the fluctuation phenomenon in corpus. Section 35.3.1 shows the experimental results. Conclusions are given in Sect. 35.4.

35.2 Speech and Singing Corpus

The song of the singing corpus used in this study is called *Seri Mersing,* a popular traditional Malay *asli* music in Malaysia. The recorded voice data were separated into two; speech and singing voices. The musical composition and information is also recorded and discussed in the following section.

35 Synthesizing *Asli* Malay Song

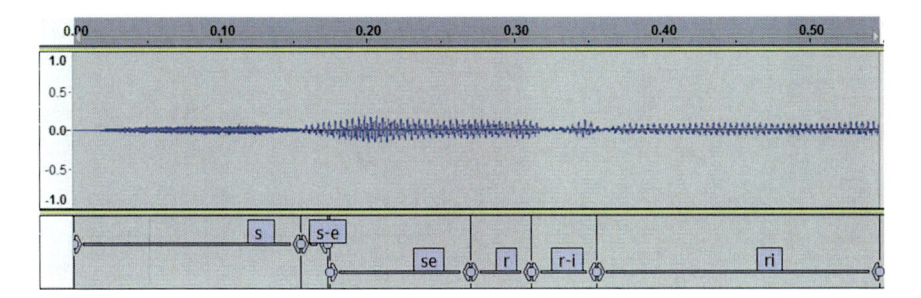

Fig. 35.1 Annotated speech waveform

35.2.1 Speech Corpus

Speaking voices were recorded as the input for the speech to singing synthesis system. Four speakers (two males and two females) read the given song lyrics of *Seri Mersing* in a neutral reading style. The speaking voices were digitized at 16 bit/48 kHz. The speaking voices were manually annotated into five kinds of labels; vowel, consonant, boundary (transition region from vowel to consonant or vice versa), syllable, and silence. This is illustrated in Fig. 35.1.

The syllables (e.g. 'se', 'ri') were then manually segmented and stored in a data bank as an input data for the speech to singing synthesis system.

35.2.2 Singing Corpus

Four singers (two males and two females) were asked to sing *Seri Mersing* using the same lyrics with 48-kz sampling and 16-bit accuracy. All four singers are well-trained and have experience in performing Malay *asli* music. The singers were required to perform the song with their own individual *patah lagu* improvising style. The singers listen to the melody of *Seri Mersing* from a headphone. The lyrics or *pantun* for *Seri Mersing* are as follows:

Seri Meresing,
Lagu lah Melayu,
Digubahlah oleh,
Biduan dahulu.

Table 35.1 Synchronization information

Syllable	Se	ri	Me	re	sing
Music note	D^4	D^4	D^4	D^4	D^4
Note value	Quaver	Dotted quaver	Semi quaver	Crotchet	Minim (*patah lagu*)

Table 35.2 Musical note values in time (s)

Musical note values	Note values in time (s)
Semi quaver	0.25
Quaver	0.5
Dotted quaver	0.75
Crochet	1
Dotted crochet	1.5
Minim	2

35.2.3 Musical Composition Information

Musical information of the song is synchronized with the syllables in the lyrics of *Seri Mersing*. The written musical note for each syllable and its synchronized musical note value/beat in second(s)/ms are shown in Table 35.1.

Music tempo for *Seri Mersing* is 60 beat per minute (BPM), with the following note values information converted into time in seconds (Table 35.2).

35.3 F0 Fluctuations

F0 extraction is important to examine how the feature values change over time, as well as how much of the total signal is taken up with periodic of voiced sound. F0 is perceived as pitch by human [9]. Variation of F0, if any, dictates the existence of *patah lagu*. Here, the F0 contours of the *patah lagu* were estimated using F0 extraction method, TEMPO in STRAIGHT [4]. F0 of a signal is the lowest frequency at which the signal repeats and is only relevant for periodic signals. The extracted F0 in log of frequency represent the F0 contour in Hz. STRAIGHT uses algorithm based on instantaneous frequency to provide source information. The method extracts F0 as the instantaneous frequency of fundamental component of a complex sound.

The extracted pitch from STRAIGHT were then converted into musical interval into 12 semitones by 100 cent each. The conversion is done here to detect the connection between the frequencies of two notes, or the perceived pitches of the notes, at the beginning and at the end of the interval. The fundamental frequency (F0) in Hz is converted into cent scale according to Kako et al. [3] using;

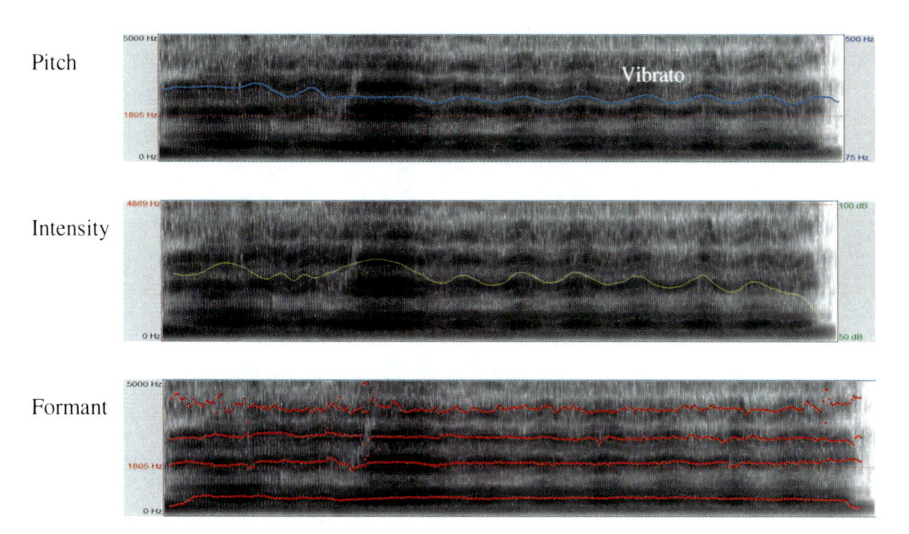

Fig. 35.2 Spectrograms analysis on pitch, formant, and intensity of *patah lagu* "Sing" performed by a male singer

$$\text{Pitch (cents)} = 1{,}200 \times \log_2 \frac{F0}{440 \times 2^{3/12-5}} \qquad (35.1)$$

STRAIGHT is confirmed beforehand can extract fine fluctuations in the F0 contours accurately than other methods can [7]. It was reported that TEMPO in STRAIGHT is also one of the most useful methods for estimating F0 contours [4].

35.3.1 Analysis of F0 Contour of Patah Lagu

From the F0 extraction experimentation using STRAIGHT, about 55–60 *patah lagu* were analyzed and observed to appear on almost all sustain crochet, minim and semibreve notes by each of the four singers, throughout the song. It shows an average of 79 % use of *patah lagu* in singing *Seri Mersing* on every central or long note. Some of the *patah lagu* were accompanied with natural vibrato, and the power changes are synchronized with pitch. The F0 contour of the *patah lagu* appeared to be rendered dissimilarly in term of pitch and time/duration of rendition. It shows that singers may perform *patah lagu* in their own individual style while singing.

Figure 35.2 is an example of a spectrogram for *patah lagu* "Sing" rendered continuously on a steady pitch by a male singer. As shown in Fig. 35.2a, the *patah lagu* is not heavily performed in "Sing", but steady vibrato can be seen. Figure 35.2b shows the energy (or perceived as loudness) increases as the vowel is sustained, and continues to synchronize with pitch as the pitch changes in time. The spectrogram also shows the formants (darker bands running horizontally across the spectrogram) changing rapidly as the vocal tract vibrates to sustain the vowel.

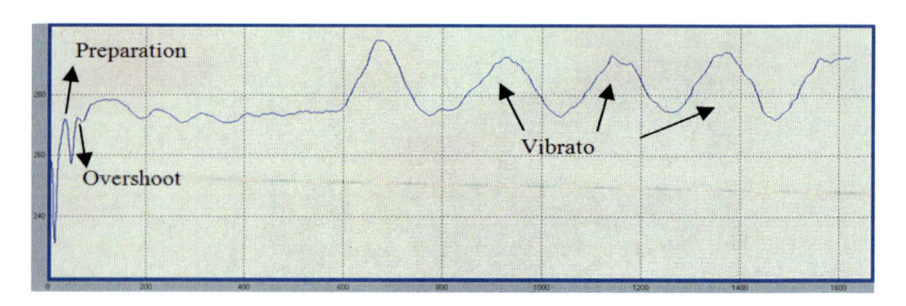

Fig. 35.3 F0 contour of a singing voice

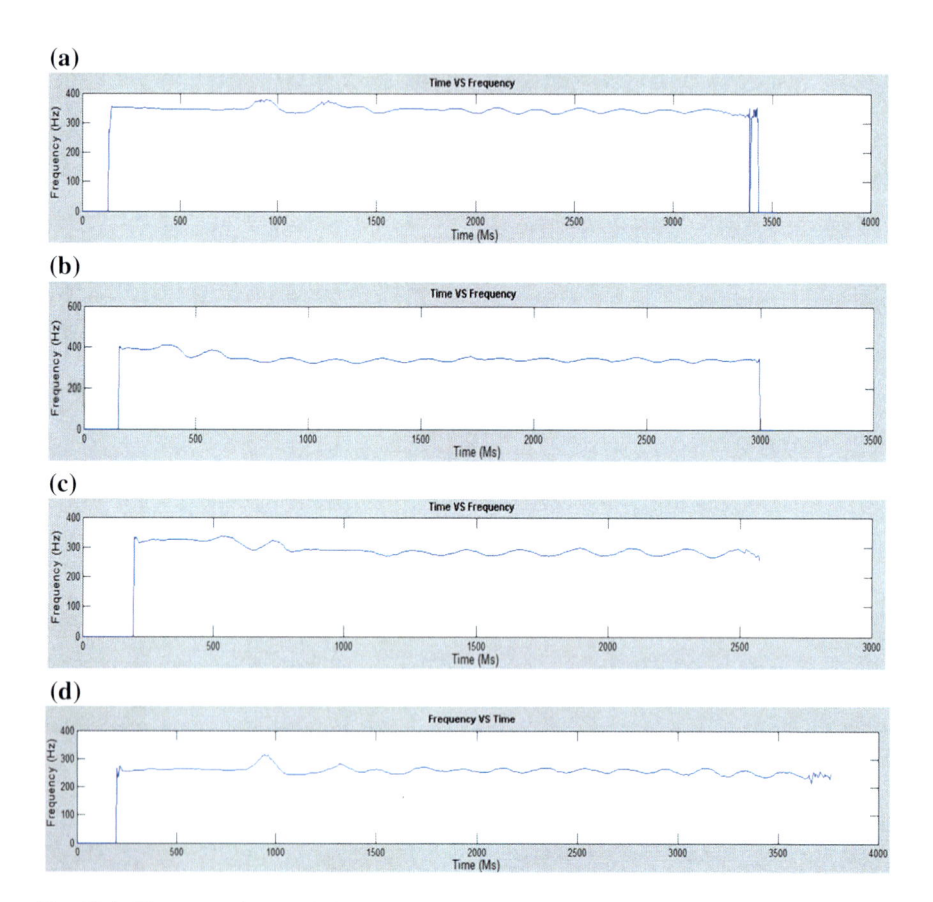

Fig. 35.4 F0 contour for *patah lagu* "Sing" by four different singers

Figure 35.3 shows an estimated F0 contour along the logarithmic axis using STRAIGHT based on an original singing voice. As shown, dynamic acoustic features such a preparation, overshoot and vibrato is performed by the singer and

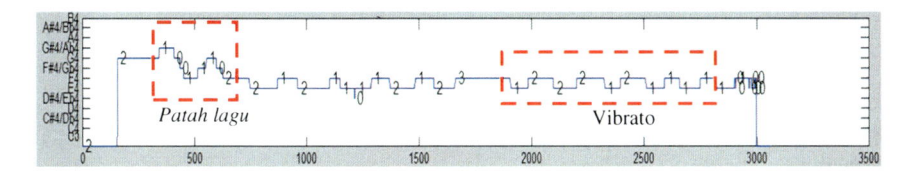

Fig. 35.5 Note transition contour for *patah lagu* "Sing" by a female singer. *The vertical axis* indicates musical notes and *the horizontal axis* indicates time in second. Note value 0 = Undefined (less than 0.0625 s), 1 = Hemidemisemiquaver (0.0625 s), 2 = Demisemiquaver (0.125 s)

also observed on other singing voices. Overshoot is deflection exceeding the target note after note changes, vibrato is periodic frequency modulation around 4–8 Hz, and preparation is deflection of the opposite direction of note change just before note changes. Fine-fluctuation is irregular fine fluctuation higher than 10 Hz observed throughout the contour.

Figure 35.4 presents the F0 contours for *patah lagu* "Sing" performed by four different singers. These contours show variation of individual style in improvising *patah lagu* "Sing" by all the singers. Female singers sing at a higher pitch range and especially the younger ones, they tend to improvise *patah lagu* more creatively and longer in duration compared to the males.

Figure 35.5 shows a stepwise note transition contour to represent melody component note change of the extracted F0 after conversion from Hz into musical notation and values. The vertical axis indicates mucial note and the horizontal axis indicates time in second. The pitch shows changes around 3–5 musical notes within a *patah lagu*. For example in the *patah lagu*, the voice glides from note G#4 down to F4, and then glides back up to G4 before going down again to F4. The variation of pitch significantly relates to the use of natural vibrato (100 cent) and *patah lagu* or ornamentation in singing. A natural vibrato is usually small, it does not alter the pitch of a note to a discernible degree. A value 0–5 was given to indicate the duration for a note in the singing. Proper discussion on its method and procedure was discussed in [12].

35.4 Conclusion

The existence of *patah lagu* is proven through F0 extraction experimentation where results reveal variation of pitch changes in singing for a crochet, minim or semibreve notes. The variation of pitch significantly relates to the use of natural vibrato and *patah lagu* in singing.

The analysis results indicate that the F0 fluctuations such as the vibrato, overshoot, and preparation are important in singing voice. It is also important that *patah lagu* fluctuation to be embedded in the traditional Malay *asli* speech to singing synthesis system. Hence, we have to consider these F0 fluctuations to be applied in the contour to construct an F0 control model for natural Malay singing voices in our future work.

Acknowledgments The authors would like to thank Fundamental Research Grant Scheme (FRGS)—MOHE and University Technology of MARA (UiTM) for research funding and support.

References

1. Akagi M et al (2000) Perception of synthesized singing voices with fine fluctuations in their fundamental frequency fluctuations. In: Proceedings of the ICSLP2000, vol 3, pp 458–461
2. Arora V, Behera L, Sircar P (2009) Singing voice synthesis for Indian Classical Raga System. In: ISSC 2009, UCD, 10th–11th June 2009
3. Kako T, Ohishi Y, Kameoka H, Kashino K, Takeda K (2009) Automatic identification for singing style based on sung melodic contour characterized in phase plane. In: 10th International Society for Music information retrieval conference (ISMIR 2009), pp 393–397
4. Kawahara H, Masuda-Katsuse I, de Cheveigne A (1999) Restructuring speech representations using a pitch adaptive time-frequency smoothing and an instantaneous-frequency based on F0 extraction: possible role of a repetitive structure in sounds. Speech Commun 27:187–207
5. Matusky P, Beng TS (1997) Muzik Malaysia: Tradisi Klasik, Rakyat dan Sinkretik. The Asian Centre
6. Nasuruddin MG (2007) Traditional Malaysian music. Dewan Bahasa dan Pustaka, Kuala Lumpur
7. Saitou T, Unoki M, Akagi M (2005) Development of an F0 control model based on F0 dynamic characteristics for singing voice synthesis. Speech Commun 405–417
8. Saitou T, Goto M, Unoki M, Akagi M (2007) Vocal conversion from speaking voice to singing voice using STRAIGHT. In: Proceedings of INTERSPEECH, Antwerp, Belgium
9. Seashore C (1967) Psychology of music. Dover Publications Inc, New York
10. Soliano ABA (1988) Seri mersing. Dewan Bahasa dan Pustaka
11. Yatabe M, Kasuya H (1998) Dynamic characteristics of fundamental frequency in singing. In: Proceedings of the autumn meeting of the acoustical society of Japan, 3-8-6
12. Za'ba N, Jamil N, Salleh SS (2011) Investigating ornamentation in Malay traditional, Asli Music. In: WSEAS international conference/EURO-SIAM/EUROPMENT

Chapter 36
Filled Pause Classification Using Energy-Boosted Mel-Frequency Cepstrum Coefficients

Raseeda Hamzah, Nursuriati Jamil and Noraini Seman

Abstract Filled pause is one type of disfluency, identified as the often occurred disfluency in spontaneous speech and known to affect Automatic Speech Recognition accuracy. The purpose of this study is to analyze the impact of boosting Mel-Frequency Cepstral Coefficients with energy feature in classifying filled pause. A total of 828 filled pauses comprising a mixture of 62 male and female speakers are classified into /mhm/, /aaa/ and /eer/. A back-propagation neural network using fusion of gradient descent with momentum and adaptive learning rate is used as the classifier. The results revealed that energy-boosted Mel-Frequency Cepstral Coefficients produced a higher accuracy rate of 77 % in classifying filled pauses.

Keywords Malay filled pause · Energy · Mel-frequency cepstral coefficients · Energy-boosted MFCC · Artificial neural network · Gradient descent momentum

36.1 Introduction

An Automatic Speech Recognition (ASR) system mainly processes read speech or spontaneous speech and converts them into text. ASR for read speech is considerably less complicated than spontaneous speech. One of the many hurdles of

---

R. Hamzah (✉) · N. Jamil · N. Seman
Faculty of Computer and Mathematical Sciences, MARA University
of Technology, Shah Alam, 40450 Selangor, Malaysia
e-mail: rashamzah82@gmail.com

N. Jamil
e-mail: liza@tmsk.uitm.edu.my

N. Seman
e-mail: aini@tmsk.uitm.edu.my

processing spontaneous speech is the occurrence of disfluencies such as filled pause, repetition and sentence restart; filled pause being the highest occurred disfluency. In this paper, spontaneous speech is defined as unprepared speech, in opposition to read speech where utterances contain formal, well-formed sentences close to those that can be found in written documents. Many studies have focused on the detection and the correction of these disfluencies. Our research focuses on classification of filled pauses as it is the most common type of disfluencies occurring in spontaneous speeches.

Mel-Frequency Cepstral Coefficients (MFCC) feature is one of the best known and most commonly used features for speech recognition [1]. It is a type of speech feature involving coefficients that represent audio which are derived from a type of cepstral representation of the audio clip [2]. MFCC has been successfully used in recent speech processing related work such as in non-speech detection of dysarthric speech [3]; isolated spoken speech recognition [4] and continuous speech recognition [5]. Prosodic features are also widely used in acoustical-based filled pause detection. Energy, pitch and formant frequency are three prosodic features used in filled pause detection by [6–8].

In this research, energy is chosen as the acoustical feature to be integrated with Mel-Frequency Cepstral Coefficient (MFCC). Energy is chosen rather than pitch and formant frequency because pitch shows no greater result of performance [9]. Formant frequency is said to be better than pitch [10]. However, formant frequency is highly computational as it involves complicated calculations. Energy is a very good acoustical representation of noisy data as the energy of the speech is higher compared to energy of speech with background noise [11].

Several well-known modeling methods such as Hidden Markov Model (HMM), Gaussian Mixture Model (GMM), Support Vector Machine (SVM) and Artificial Neural Network (ANN) have been used as classifiers in ASR. Among these classifiers, ANN is the most efficient in speech recognition as stated in previous work by [12, 13]. ANN also has the ability to generalize after learning from the sample given to its network. The generalization ability makes ANN to properly understand the hidden part of the population even if the sample data contain noisy information [14]. The raw data used in this research is gathered from Malaysia Parliamentary Hansard Document (MPHD) debate session of the year 2008. Since the spontaneous speech data is recorded live, it is surrounded with background noise, interruption, and various speaking style (low, medium and high intonation). Therefore, ANN is preferred as the classifier model in this research.

36.1.1 Data Preparation and Methodology

The Malaysia Parliamentary Hansard Document (MPHD) spoken data used in this research consists of 51 video files (.avi) and 42 text files (.pdf). Each session of the debate contains 6–8 hours of spoken speeches [4]. It comprises 253 topics with 222 selected male and female speakers. However, for the purpose of this research the selection of 10 topics with 35 speakers are used. Thus, the data collection contains 828 filled pauses. The Malay filled pauses that are analyzed in this research are defined as /MHM/, /AAA/ and /EER/. Based on the filled pause collection, there are 82 /MHM/, 484 /EER/ and 272 /AAA/.

For the initial stage of data collection, 562 sentences have been extracted from the audio MPHD. Filled pause from each sentence is manually segmented and is grouped based on its type (/EER/, /AAA/, and /MHM/) and labeled as Manual Datasets (MDs).

36.1.2 Pre-processing

Pre-processing is a crucial step in signal processing especially in automatic speech recognition. Without a proper pre-processing on the recorded speech input, the classification performance or recognition rate will be decreased. In general, the pre-processing operations involved in this research are filtering, pre-emphasis, frame blocking, and windowing and feature extraction. The input of the speech waveform is initially sampled at a sampling rate of 16 kHz. Then, it is filtered using high pass filter to suppress the noises and accentuate only the high frequency speech waveform.

A pre-emphasis process is then applied to the input speech waveform. The pre-emphasis is defined by:

$$y[n] = s[n] - P * s[n-1] \tag{36.1}$$

where, s[n] is the nth speech sample, y[n] is the corresponding pre-emphasized sample and P is the pre-emphasis factor typically having a value between 0:9 and 1. The input speech waveform then will pass through frame blocking and windowing processes. The common Hamming Window is applied by applying the formula:-

$$\left\{ \begin{array}{ll} 0.54 - 0.46\cos(2\pi n|(N+1)), & 0 \leq n \leq N \\ 0 & otherwise \end{array} \right\} \tag{36.2}$$

36.1.3 Features Extraction

The front-end signal processing used in ASR is feature extraction. Feature extraction involved a process of converting the speech waveform to some type of parametric representation.

36.1.3.1 Energy

Let $x(i)$ be the ith sample of a speech signal. If the length of the frame is N samples, then the jth frame can be represented as Eq. (36.3).

$$fj = \{x(i)\}_{i=(j-1).(N+1)}^{j.N} \tag{36.3}$$

The most common way of calculating energy of a speech signal is by calculating its frame by frame energy [11] as shown Eq. (36.4).

$$Ej = \frac{1}{N} \cdot \sum_{i=(j-1).N+1}^{j.N} x^2(i) \tag{36.4}$$

where Ej = Energy of the jth frame and fj = jth frame is the considered frame where the energy is calculated. The energy for 272 /AAA/, 484 /EER/ and 82 /MHM/ and are calculated and the average energy of each filled pause is 0.69, 10.15 and 11.66 respectively. It shows that the /AAA/ type of filled pause has the lowest energy while /EER/and /MHM/ produce almost the same energy value.

36.1.3.2 Mel-Frequency Cepstral Coefficient Extraction

In order to compute the MFCC, a Discrete Fourier Transform (DFT) is performed on each of the windowed speech waveform with 512 DFT. DFT is used to calculate short-term power spectrum of speech signal. The Mels-scale is then calculated by using Eq. (36.5).

$$mel(f) = 2595 \cdot ^{10}\log\left(1 + \frac{f}{700}\right) \tag{36.5}$$

The derived log amplitudes of the spectrum are mapped onto the Mel scale using triangular overlapping windows. After doing some analysis, a 12-cepstral coefficient of MFCC is chosen. The Discrete Cosine Transform (DCT) of the list of Mel log amplitudes is further calculated. The MFCCs are the amplitudes of the resulting spectrum. The flow of MFCC feature extraction is illustrated as in Fig. 36.1.

36.1.3.3 Data Normalization

Data normalization needs to be performed prior to feeding into Artificial Neural Network model [4]. Raw recorded input speech is in the form of unequal waveform, length and amplitude. In this research, data normalization is done by taking the mean of each MFCC and energy features extracted from the Manual Datasets filled pause.

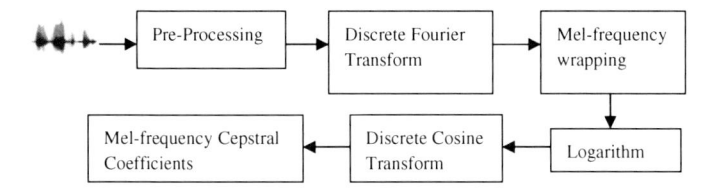

Fig. 36.1 MFCC extraction

Table 36.1 The MLP-ANN structure of the research

Hidden neuron	Features		Training function
	MFCC	Energy-boosted MFCC	
7	12	13	GDM
12	12	13	GDM
24	12	13	GDM

36.1.4 Classification and Evaluation

Multilayer Perceptron (MLP) is one of ANN's architecture. In this research, MLP is chosen because its ability to function well for non-linear phenomena [15]. Multilayer neural networks with the backpropagation algorithm are used in this research to classify the three types of filled pause /AAA/, /EER/ and /MHM/. A three-layer feed-forward with three different structure neural network with one hidden and one output layer is applied. The three structures that is implemented in this research use same inputs, target assignments, activation functions, output layer structure, network parameters, and differ only in the number of hidden neurons (HN) in the hidden layers. The output layer of three neurons is corresponds to one filled pause of each in the three-layer feed-forward network structure. Each filled pause is assumed to correspond to a class, and each filled pause belonging to its respective class is labeled with an integer number from one to three for the classification purpose. The number of input layer is chosen by calculating by multiplying the cepstral order with the total frame number as below:

$$Total\_Frame\_Number = Signal\_length/Shift - (Frame\_length/Shift) \quad (36.6)$$

$$Input\_Neuron\_Number = Cepstral\_order * Total\_Frame\_Number \quad (36.7)$$

The summarized structure of MLP used in this research is shown as in Table 36.1.

In order to evaluate the performance each MLP structure used in this research, different numbers of hidden neurons in the hidden layers were going through a trial an error process. The number of hidden neuron cannot be too many, otherwise, it cannot obtain good convergence rate [4]. A too small number of hidden neuron will causes large classification error. One of the guideline for choosing the number of hidden neuron is by implementing Geometric Pyramid Rule (GPR) that is calculated by the following equation

$$HNN = (Input\_number * Output\_number)^{1/2} \qquad (36.8)$$

The training algorithm that has been chosen is Gradient Descent with Momentum (GDM) which is discussed further in the following subsection.

36.1.4.1 Training the Artificial Neural Network

Training of MLP-ANN is done using the standard Gradient Descent (GD) algorithm. The training goal is set to 0.001 and to achieve this goal, 1,800 iterations of gain 300 are done. One of the disadvantage of GD is it uses a fix step-size [16]. This procedure makes the convergence time longer causing insufficient training on MLP-ANN. To overcome this limitation, GDM is chosen as the training algorithm in this research due to its ability to provide variable step size thus increases the convergence time [16].

36.1.4.2 Testing and Evaluation

In this research, a ratio of 70 % for training data and 30 % of testing data is used in the experiment. Accuracy rate is calculated as in Eq. (36.9) to measure the closeness of the tested data to the actual data.

$$Accuracy = \frac{Actual\_data - tested\_data}{Actual\_data} \qquad (36.9)$$

Another evaluation is done to assess the maximum acceptance of the filled pauses in ANN model. This is done by calculating the overall performance scores of true positive and false positive rates of the filled pauses. True positive is defined as the rate of correctly accepted filled pause while false positive is defined as incorrectly rejected filled pause in ANN. Equation (36.10) shows the equation of score measurements.

$$Score = True\_positive\_rate * 0.7 + (1 - False\_positive\_rate) * 0.3 \qquad (36.10)$$

36.2 Results

The results of filled pause classification using MFCC and energy-boosted MFCC are presented in Tables 36.2 and 36.3. It can be observed that the energy-boosted MFCC outperformed the classification done by MFCC only. Among the three

Table 36.2 Performance result of MFCC-based feature

HN	Epoch	Time	Performances						
			MSE	Accuracy (%)			Score (%)		
				AAA	EER	MHM	TP	FP	Score
7	300	0:00:02	0.029500	60	60	20	53	47	53
	600	0:06:36	0.009920	60	40	20	40	60	40
	900	0:12:21	0.008910	60	40	40	47	53	47
	1,200	0:10:03	0.000996	20	40	20	47	53	47
	1,500	0:13:11	0.000997	80	40	40	53	47	53
12	300	0:03:03	0.041100	40	60	20	40	60	40
	600	0:07:27	0.001750	80	40	60	60	40	60
	900	0:10:20	0.006560	60	40	40	47	53	47
	1,200	0:16:48	0.000999	80	40	20	47	53	47
24	300	0:03:11	0.029700	40	60	20	40	60	40
	600	0:08:02	0.018100	60	60	20	47	53	47
	900	0:12:14	0.003690	40	60	60	53	47	53
	1,200	0:15:13	0.001770	80	40	20	47	53	47
	1,500	0:17:06	0.000999	40	60	40	47	53	47

Table 36.3 Performance result of Energy-boosted MFCC

HN	Epoch	Time	Performances						
			MSE	Accuracy (%)			Score (%)		
				AAA	EER	MHM	TP	FP	Score
7	300	0:03:36	0.022000	80	60	40	60	40	60
	600	0:09:39	0.004800	40	60	40	53	47	53
	900	0:12:54	0.002190	60	80	40	60	40	60
	1,200	0:11:32	0.000998	80	80	40	67	33	67
12	300	0:03:18	0.030300	60	80	40	60	40	60
	600	0:06:25	0.001470	100	60	40	67	33	67
	900	0:09:42	0.003320	100	80	40	73	27	73
	1,200	0:12:37	0.001880	100	60	40	67	33	67
	1,500	0:15:49	0.001100	100	80	40	73	27	73
	1,800	0:12:28	0.001000	80	60	40	60	40	60
24	300	0:03:21	0.015900	80	60	40	67	33	67
	600	0:06:48	0.001300	60	60	40	53	47	53
	900	0:10:29	0.000997	60	20	80	53	47	53

types of filled pauses, the highest accuracy rate is achieved by energy-boosted MFCC for /AAA/ filled pause with average of 77 % accuracy rate and 63 % score. The average energy of /AAA/ filled pause is much lower than /EER/ and /MHM/ as discussed earlier in Sect. 36.1.3.1. The integration of MFCC and energy on each filled pause boosted the pattern of the feature vector, thus increasing the discriminations of patterns among the filled pauses. The ratio of average accuracy for each type of filled pause is calculated by finding its average accuracy differences

Table 36.4 The ratio of average accuracy for MFCC and energy-boosted MFCC feature

Filled pause	MFCC	MFCC + ENERGY
/AAA/ : /EER/	8	12
/AAA/ : /MHM/	26	34
/EER/ : /MHM/	18	22

Table 36.5 Performance comparison between MFCC-based and energy-boosted MFCC

Features	Average accuracy	Average MSE	Average score (%)
MFCC	57 % (AAA), 49 % (EER) and 31 % (MHM)	0.011071	48
MFCC + energy	77 % (AAA), 65 % (EER) and 43 % (MHM)	0.006712	63

among filled pause types. The purpose of calculating the ratio is to show the impact of energy-boosted MFCC towards the classification of filled pause. The ratio of average accuracy between all types of filled pauses is summarized in Table 36.4. From the table, it shows that the ratio becomes larger as MFCC features of the filled pauses are integrated with energy. Therefore, improving the classification accuracy of each filled pause types.

The summarized comparison performance rate is described as Table 36.5. It shows the average performance comparisons of MFCC and energy-boosted MFCC in terms of score and Mean Square Error (MSE). MSE is the standard criterion for the assessment of signal quality [17]. The MSE of energy-boosted MFCC is lower than MFCC-based features. The score of the ANN performance that is used in this research is also calculated to show the acceptance of filled pause throughout the ANN. It is proven that the energy-boosted MFCC achieved higher score rate and accuracy.

Acknowledgments The authors thankfully acknowledge Ministry of Higher Education Malaysia for Fundamental Research Grant Scheme (FRGS, Grant No: 600-RMI/FRGS 5/3(48/2013) and MARA University of Technology for providing research facilities throughout this research.

References

1. Mahesha P, Vinod DS (2012) Feature based classification of dysfluent and normal speech. In: Proceedings of the second international conference on computational science, engineering and information technology (CCSEIT' 12). ACM, New York, pp 594–597
2. Rosdi F, Ainon RN (2008) Isolated Malay speech recognition using Hidden Markov Models. In: Proceedings of the international conference on computer and communications engineering (ICCCE08), Malaysia, pp 721–725
3. Hu Y (2009) Detecting non-speech in dysarthric speech. Master thesis. University of Sheffield
4. Seman N (2012) Coalition of artificial intelligent (AI) algorithms for isolated spoken Malay speech recognition. PhD thesis, UniversitiTeknologi Mara, Shah Alam

5. Horia IC (2011) Towards a speaker-independent, large-vocabulary continuous speech recognition system for Romanian. PhD thesis, University of Politehnica Din Bucureşti
6. Garg G, Ward N (2006) Detecting filled pauses in tutorial dialogs. In: Report of University of Texas at El Paso, El Paso
7. Kaushik M (2010) Automatic detection and removal of disfluencies from spontaneous speech. In: Proceedings 13th Australasian international conference on speech science and technology Melbourne, pp 98–101
8. Veiga A, Candeias S, Lopes C, Perdigão F (2011) Characterization of hesitations using acoustic models. In: ICPhS XVII, Hong Kong, pp 17–21
9. Stouten F (2006) Coping with disfluencies in spontaneous speech recognition: acoustic detection and linguistic context manipulation. Speech Commun 48(11):1590–1606
10. Audhkhasi K (2009) Formant-based technique for automatic filled-pause detection in spontaneous spoken English. In: Acoustics, speech and signal processing, IEEE international conference ICASSP, IEEE, pp 4857–4860
11. Sakhnov KE, Verteletskaya E, Simak B (2009) Approach for energy-based voice detector with adaptive scaling factor. IAENG Int J Comput Sci 36(4):394–399, IJCS_36_4_16
12. Majeed SA, Husain HS, Samad A, Hussain A (2012) Hierarchical K-Means algorithm applied on isolated Malay digit speech recognition. In: International conference on system engineering and modeling (ICSEM 2012) IPCSIT, vol 34. IACSIT Press, Singapore
13. Dede G, Sazli MH (2010) Speech recognition with artificial neural network. Digit Signal Proc 20(3):763–768
14. Zhang GP (2000) Neural networks for classification: a survey. IEEE Trans Syst Man Cybern Part C Appl Rev 30:451–462
15. Ayoubi S, Shahri AP, Karchegani PM, Sahrawat A (2011) Application of artificial neural network (ANN) to predict soil organic matter using remote sensing data in two ecosystems. In: Atazadeh I (ed) Biomass and remote sensing of biomass, ISBN: 978-953-307-490-0
16. Gong L, Liu C, Li Y, Yuan F (2012) Training feed-forward neural networks using the gradient descent method with the optimal stepsize. J Comput Inf Syst 8(4):1359–1371
17. Wang Z, Bovik AC (2009) Mean square error: love it or leave it? IEEE Sig Proc Mag 26(1):98–117

Chapter 37
Developing a Rule Base for Recommending Insurance Products

Siti Fatimah Abdul Razak, Shing Chiang Tan and Way-Soong Lim

Abstract This paper presents a fuzzy rule base for insurance prospects profiling based on two main sources, i.e. from dataset and experts. A historical dataset of Islamic insurance policyholders was used as a case study for this purpose. The Wang and Mendel algorithm (WMA) (Wang and Mendel, IEEE Trans Syst Man Cybern 22:1414–1472, [1]) was applied on the dataset to generate fuzzy rules which includes five types of policies with five pre-determined most influential variables, namely *age*, *gender*, *marital status*, *monthly salary* and *job class*. Fuzzy rules from the dataset and expert were compared based on similarity and accuracy measures. The rule base is proposed to be used by insurance advisors to make decisions and provide appropriate proposals according to the prospect's profile.

Keywords Rule base · Insurance · Decision support

37.1 Introduction

Insurance advisors are intermediaries between insurance companies and prospects [1, 2]. They often employ subjective criteria in profiling prospects and suggesting suitable policies based on the prospects' profile. Their judgment is mainly based on the knowledge they have about the insurance products as well as previous

---

S. F. A. Razak (✉) · S. C. Tan
Faculty of Information Science and Technology, Multimedia University, Melaka, Malaysia
e-mail: fatimah.razak@mmu.edu.my

S. C. Tan
e-mail: sctan@mmu.edu.my

W.-S. Lim
Faculty of Engineering and Technology, Multimedia University, Melaka, Malaysia
e-mail: wslim@mmu.edu.my

experience. Therefore, fuzzy *IF–THEN* rules which have been applied to represent uncertainties and ambiguity in various disciplines and industries including insurance [3] is proposed to represent the advisors knowledge. Fuzzy *IF–THEN* rules are amongst the most expedient form of knowledge representation [4].

The Wang and Mendel algorithm is the most cited algorithm for generating fuzzy rules from a dataset [1]. Since its inception, it has been cited by more than 250 publications in the Scopus repository and applied to various types of dataset in insurance including life insurance, health insurance, general motor insurance, investment-oriented insurance and etc. [5–7]. However, none has specifically attempted to study the applicability of this algorithm to Islamic insurance products.

Furthermore, to the best of our knowledge, no previous research has applied fuzzy logic approach to address Islamic insurance industry needs or represent the insurance advisors knowledge either in Malaysia or abroad. Previous research was conducted in the form of empirical studies related to suitable operating model [8], factors of Islamic insurance demand [9, 10], factors leading to consumer's acceptance of Islamic insurance [11], influence of agents positive attitudes, high commitment and motivation on the performance of the Islamic insurance industry in Malaysia [12] and etc. Therefore, this study aims to fill the gap by proposing a fuzzy rule base which brings together the knowledge from Islamic insurance dataset and represents insurance advisors knowledge.

This paper is organized as follows. Section 37.2 describes the proposed rule base. Section 37.3 discusses the experiments and results of the implementation based on a case study. Finally, Sect. 37.4 concludes the study and describes future work.

37.2 The Rule Base

The rule base is characterized by a set of IF–THEN linguistic rules derived from a dataset and defined by expert. A fuzzy rule, *Rule i*-th in a rule base of size N can be represented in the form of *IF x is A_i THEN y is B_i* where x is the antecedent part which provides the input fuzzy linguistic variables conditions and y is the consequent (output variable conditions). Both A_i and B_i are fuzzy sets representing linguistic values of x and y.

Wang and Mendel [1] proposed a five step procedure to generate fuzzy rules from dataset and group these fuzzy rules and linguistic fuzzy rules from human expert into one joint fuzzy rule base. In general, the number of rules is based on the number of training pairs. New rules are allowed to compete with existing rules, making the rule base adaptive. This algorithm generates complete rules while instantaneously considering all available variables. The steps are briefly described as follows:

- Step 1: Divide each variable of the input space into a user defined number of triangular membership fuzzy sets automatically.

37 Developing a Rule Base for Recommending Insurance Products

- Step 2: Generate a fuzzy rule for each i-th data pair in the form of *IF X_1 is A_1^i and X_2 is A_2^i and X_p is A_p^i THEN y is C_i*. The A_j^i are those for which the degree of match of x_j^i is maximum for each input variable j from pair i. The fuzzy set C_i is the one for which the degree of match of the observed output y_i, is maximum.
- Step 3: Assign a degree to each generated rule. For a given rule, it is equal to the rule weightage for the considered pair. If any a priori information is available, the degree will be the product of the weightage by the confidence level. If there is any identical premise for two rules, only one rule with a higher degree is maintained.
- Step 4: Form the joint fuzzy rule base from generated rules from dataset or linguistic rules from expert. If there is more than one rule in one box of the fuzzy rule base, use the rule that has a maximum degree.
- Step 5: Obtain a mapping based on the joint fuzzy rule base using a defuzzification strategy.

Based on Ketata et al. [13], two fuzzy rules are considered similar based on the degree to which the fuzzy sets are equal. The similarity relations, $S(A_{ik}, A_{ij})$ between rules induced from dataset and linguistic rules is defined by Eqs. (37.1) and (37.2) respectively [13]. $S(A_{ik}, A_{ij})$ is equal to zero when the fuzzy sets differ in two compared rules.

$$S(A,B) = \frac{|A \cap B|}{|A \cap B|} + \frac{|A \cap B|}{|A| + |B| - |A \cap B|} \quad (37.1)$$

$$d_s = \frac{\sum_{k \in input\ and\ output\ parameters} S(A_{ik}, A_{ij})}{Total\ number\ of\ input\ and\ output\ parameters} \quad (37.2)$$

$$\begin{aligned} D_e &= \{(X_{e1}^1, X_{e2}^1, X_{e3}^1, X_{e4}^1, X_{en}^1, Y_{e1}^1), (X_{e1}^2, X_{e2}^2, X_{e3}^2, X_{e4}^2, Y_{e1}^2), \\ &\quad (X_{e1}^3, X_{e2}^3, X_{e3}^3, X_{e4}^3, Y_{e1}^3)\} \\ D_a &= \{(X_{a1}^1, X_{a2}^1, X_{a3}^1, X_{a4}^1, Y_{a1}^1), (X_{a1}^2, X_{a2}^2, X_{a3}^2, X_{a4}^2, Y_{a1}^2)\} \end{aligned} \quad (37.3)$$

with

$$X_{ab}^c = \begin{cases} a\ or\ e & dataset\ or\ expert\text{-}induced\ rule \\ b & parameter\ number\ in\ the\ rule \\ c & data\ pair\ number \end{cases}$$

There are three main steps involved. One, generate fuzzy rule base. Two, calculate the similarity measure between two fuzzy rules using Eq. (37.2). Three, set the minimal value of the degree of similarity between two rules as the threshold. Rules with similarity measures above threshold value are merged. This is further shown in Eq. (37.3) where X refers to input parameters while Y refers to output parameters. By considering the degree of similarity between rules, redundant and inconsistent rule can be removed, thus simplifying the rule base. Rules which are too specific may also be removed. Even though expert rules are given a higher priority, rules induced from the dataset are necessary to represent hidden knowledge which is not acquired from the expert. Rules from both sources complement one another, making the rule base more complete.

37.3 Case Study

The fuzzy rule base is developed to bring together rules induced using the Wang and Mendel algorithm [1] and linguistic rules from expert based on a historical dataset of existing profiles from an Islamic insurance operator. Due to restricted access from the operator, this study only considers 747 records of five different policy (product) type—Policy A, B, C, D and E subscribers with 33 variables which represent prospects and payers personal details, i.e. *age, gender, occupation, address, marital status, monthly salary* and etc. as well as product details, i.e. *policy type, policy effective date, sum coverage, mode of payment, participants details* and etc. The records are captured with no particular order from the operator.

Knowledge from a product expert was acquired via interview sessions to determine the information or variables which matters most during a prospect's profiling process. The expert mentioned explicitly during the interview sessions that the profiling process is done subjectively and usually influenced by an advisor's experience, product information and perception of the prospect life style. However, these factors are subject to his opinion. The expert outlined five important variables which normally serve as a base guideline when considering which product to recommend a prospect. The variables are *prospect's age, gender, occupation, marital status* and *monthly salary*. Other information like desired monthly contribution, payment mode, total coverage distribution, year of product maturity, policy rider contribution and so on will only be considered according to the product type which appeals to the prospect.

Since the study considers a single product expert, additional measures seems necessary to justify and support his claims. The relationship of those variables with product type was computed based on correlation coefficients value. Result revealed that those variables have positive relationship among them with a positive high correlation coefficient value of 0.9728. Hence, the five variables i.e. *age, gender, occupation, marital status* and *monthly salary* is worth to be considered for prospects profiling process and recommending suitable product type. However, taking into account a suggestion made by the product expert, a new variable called *job class* was introduced to classify prospects occupation based on the occupational risk. For example, occupations with high risk are policeman, soldiers, enforcement officers; occupations with moderate risk are engineers, salesman, and factory workers; and lowest risk is for those like academics, clerks, managers and caregivers. Therefore, the five most influential variables to recommend a suitable product type are *age, gender, marital status, monthly salary* and *job class*.

Next, the expert listed 21 ground rules which he has practiced when approaching new prospects. Among those rules are listed as follows.

- IF a prospect is a married male adult with a high risk job and has a high monthly salary, THEN the prospect is most likely to subscribe Policy B.
- IF a prospect is single and have a moderate monthly salary, THEN the prospect is likely to subscribe either Policy C or Policy D or both.
- IF a prospect is a single parent with moderate monthly salary, THEN the prospect is most likely to subscribe Policy B and C.

Then, Wang and Mendel algorithm [13] is applied to the dataset, also restricting the input variables to *age, gender, marital status, monthly salary* and *job class*. Membership functions were assigned to the variables. The variable *policy type* was set as the output variable with less likely, likely and most likely membership functions. The output result was 54 fuzzy rules, a few listed below. These rules were then compared with 21 ground rules stipulated by the expert earlier. Moreover for the purpose of simplicity, the ground rules are transformed into similar form as the algorithm induced rules.

- IF (Age = adult) AND (Gender = Male) AND (Marital status = married) AND (Job Class = high) AND (Monthly Salary = low) THEN (Policy B = most likely) AND (Policy D = most likely).
- IF (Age = young adult) AND (Marital status = single) AND (Monthly Salary = high) THEN (Policy C = most likely) AND (Policy D = most likely).
- IF (Age = young adult) AND (Marital status = married) AND (Gender = Female) THEN (Policy E = most likely).

Rules from both sources are evaluated and compared based on their similarity relations described earlier. The threshold value is set as 50 % as proposed by Ketata et al. [13]. Rules with a degree of similarity, d_s higher than the threshold value are merged to reduce the number of rules in the rule base. The final rule base were reduced from 78 (57 algorithm induced rules and 21 expert ground rules) to 63 rules. Rules identified by expert are given priority [14]. Out of 63 rules, 20 rules are those originally identified by the expert. One rule is removed since it is a subset of a more general form of rule which is also identified by expert. The remaining rules are either originally induced by the algorithm or a result of merged rules due to high similarity.

The accuracy of the fuzzy rules in the rule base is later measured based on correctly classified profiles. The minimum threshold is set as 30 % accuracy [14]. The predictive verification technique was performed on fifty new profiles. The input variables of each profile were then entered. The fuzzy rules which are fired and the output variable (*policy type*) recommended were compared with existing profile in the dataset. The result was 74.5 % accurate which is above the threshold value.

37.4 Conclusions and Future Work

Prospects profiling in Islamic insurance area using fuzzy logic is considered very new. Initially, the well-known Wang and Mendel algorithm [1] is applied to this dataset to examine the possibility of this approach. The proposed rule base is hoped to help ease the process of profiling prospects for Islamic insurance products. Additionally, expert knowledge may also be preserved in the rule base for future advisors training, as a guideline before approaching a potential prospect, allowing insurance operators to understand their prospects better and etc. However, the fuzzy rule base requires additional experiments to evaluate the completeness of the rule base and increase the adaptability of the rules. More experts should be involved in validating and verifying the rule base as well. In future, we plan to compare the proposed rule base with other method from other researcher. Future work will also include not only the recommendation of products (*policy type*) but also determine suitable weightage of the policy details.

References

1. Wang L-X, Mendel JM (1992) Generating fuzzy rules by learning from examples. IEEE Trans Syst Man Cybern 22(6):1414–1472
2. Alias NZ, Hussein N, Mohamad AA (2012) Malaysian Insurance Industry—a macro perspective. Econ Res ER/008/2012:5–7
3. Shapiro A (2004) Fuzzy logic in insurance: the first 20 years. J Insur Math Econ 35:399–424
4. Zadeh L (1965) Fuzzy sets. Inf Contr Elsevier Sci 8:338–353
5. Huang C-S, Lin Y-J, Lin C-C (2007) An evaluation model for determining insurance policy using AHP and fuzzy logic: case studies of life and annuity insurances. Proc WSEAS Int Conf Fuzzy Syst 8:126–137
6. Huang C-S, Lin Y-J, Lin C-C (2008) Determination of insurance policy using a hybrid model of AHP, fuzzy logic and Delphi technique: a case study. WSEAS Trans Comput 7(6):660–669
7. Abdullah L, Rahman MNA (2012) Employee likelihood of purchasing health insurance using fuzzy inference system. Int J Comput Sci Issues 9(1):112–116
8. Htay SNN, Zaharin HR (2011) Critical analysis on the choice of Takaful (Islamic Insurance) operating models in Malaysia. IREP
9. Hamid MA, Osman J, Nordin BAA (2009) Determinants of corporate demand for Islamic Insurance in Malaysia. Int J Econ Manag 3(2):278–296
10. Arifin J, Yazid AS, Sulong Z (2013) A conceptual model of literature review for family Takaful (Islamic Life Insurance) demand in Malaysia. Int Bus Res 6(3):210–216
11. Rahim FA, Amin H (2012) Determinants of Islamic Insurance acceptance: an empirical analysis. Int J Bus Soc 12(2):37–54
12. Hamid MA, Ab Rahman NMN (2011) Commitment and performance: a case of Takaful (Islamic Insurance) representatives in Malaysia. Aust J Basic Appl Sci 5(10):777–785
13. Ketata R, Bellaaj H, Chtourou M, Amer MB (2007) Adjustment of membership functions, generation and reduction of fuzzy rule base from numerical data. Malays J Comput Sci 20(2):147–169
14. Guillaume S (2001) Designing fuzzy inference systems from data: an interpretability-oriented review. IEEE Trans Fuzzy Syst 9(3):426–443

Chapter 38
Modeling and Path Planning Simulation of a Robotic Arc Welding System

Muhammad Hafidz Fazli Bin Md Fauadi, Fairul Azni Jafar, Lau Ong Yee, Mohd Nazmin Bin Maslan and Saifudin Hafiz Yahya

Abstract The utilization rate of robots to execute manufacturing processes in Small and Medium Enterprises are steadily increasing not only due to the demand for higher productivity and higher quality but also for the sake of stable throughput and minimizing turnover risk. Prior to the implementation of robotic manufacturing system (RMS), it is very important to properly plan the specific jobs that need to be carried out. Additionally, in order to optimize the efficiency of an industrial robot, the entire operation parameters of the particular purpose should be carefully taken into consideration starting from the planning stage. One of an effective method of planning RMS is through the simulation of the operation, which could be done virtually. This paper proposes the welding work flow and workcell design of an arc welding robot system. Analysis that has been carried out to establish the workcell design includes forward kinematic analysis and robot path planning analysis.

Keywords Robotic manufacturing system · Industrial automation

38.1 Introduction

Robots have long been deployed in factory to execute numerous manufacturing processes. Recently, robots are no longer exclusive for large enterprises as a number of Small and Medium Enterprises (SME) have already starting to deploy

---

M. H. F. B. M. Fauadi (✉) · F. A. Jafar · M. N. B. Maslan · S. H. Yahya
Faculty of Manufacturing Engineering, Universiti Teknikal Malaysia Melaka (UTeM), 76100 Hang Tuah Jaya, Durian Tunggal, Malaysia
e-mail: hafidz@utem.edu.my

L. O. Yee
Department of Mechanical Engineering, Politeknik Kuching Sarawak (PKS), Kuching, Malaysia

automated system in their shop floor. Among the key advantages of having robot in SME include the ability to generate stable throughput and risk minimization of staff turnover especially with the skilled workers. Staff turnover is a critical factor particularly due to the reason that skilled and experienced workers naturally tend to choose bigger organization for better remuneration. Based on the motivation, this study focuses on the design of a robotic arc welding system.

Robot modeling and simulation is one of the critical stages in designing and developing any industrial robotic application. Among the elements that need to be considered when modeling a robotic application include kinematic, dynamic and the related control system [1, 2]. Understanding these fundamental issues will enable researcher to design and develop an efficient and reliable robot manufacturing system. In order to describe in detail, the first issue for industrial robot is on how to translate the end-effectors' position in Cartesian space from encoder reading in joint space coordinate. While the tasks are planned in Cartesian space the robot joints' should still being controlled at joint space. Inverse kinematic solutions are required to acquire joint space reference input from the tasks planned earlier in Cartesian Space. In order to generate smooth trajectory, cubic polynomial has been used to model the motor angular position from initial angular position to its final destination value. The first and second time derivative of angular position can be developed in order to derive the angular speed and acceleration input. These aspects appeared either entirely or partially in the previous research works that have been conducted [3, 4, 5].

As such, the objectives of this research include

- To propose a suitable RMS for an arc welding robot that includes utility robot for material handling and fixtures design and material transfer.
- To simulate and analyze the designed RMS.

38.2 Robot Arc Welding Operation

Modern welding operations need to be conducted in a precise manner to ensure the joining quality produced. As such, a fixture and positioning equipment is typically required to establish a complete robotic welding system [1, 6]. Without a fixture, position of the workpiece could be easily misaligned during the transferring or welding process. On the other hand, a utility robot may also be used to place and hold the workpiece so that the welding robot can weld it at required weld joints. Workpiece that was selected for this project is a muffler as shown in Fig. 38.1 where it consists of two weld points that are labeled as Weld Point B and Weld Point C. The operation starts with conveyor *A* and *B* move the workpiece to supply three muffler parts to be welded. Positioning robot will transfer the workpiece to the workstation. Welding robot will move the end-effector to the weld joint and execute the welding operations. After careful consideration, the research proposes a layout arrangement for the RMS as shown in Fig. 38.2 where it consists of two

Fig. 38.1 Components of a muffler

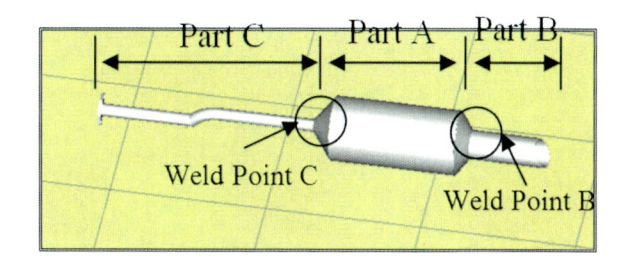

Fig. 38.2 Proposed work cell design of robotic arc welding operation

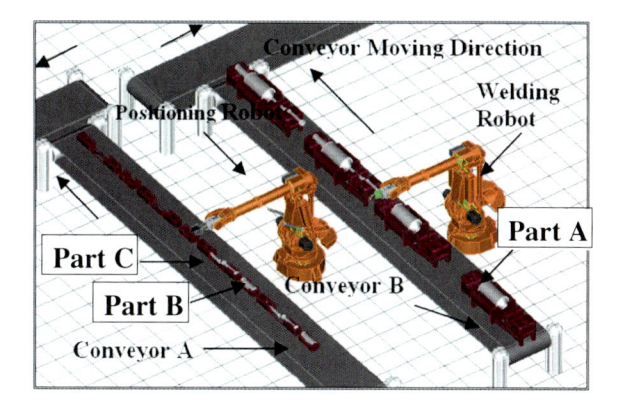

robots and conveyor system. Conveyors operate based on intermittent synchronous transport approach that stops at each station for specified time duration. Two robots are employed:

- Welding robot to carry out the welding operation.
- Utility robot to position the workpiece.

The sequence of operation goes as follow. First, positioning robot picks Part C from Conveyor A and places it to Conveyor B. Positioning robot places Part C to the weld joint C coordinate and welding robot starts to weld first half circle. Welding robot arm will be retracted when the first half circle has been welded. Consequently, Positioning robot un-grips Part C on Conveyor B and moves towards Part B on Conveyor A. When welding robot starts to weld second half circle, positioning robot moves and grasps Part B before place it at Weld Point B. Welding robot arm will be retracted to home position when Weld Point C has been welded. It will then move to Weld Point B and start welding the first half of the circle.

Welding robot finished welding for first half circle and positioning robot un-grasped Part B and retracted to home position. Welding robot starts to weld second half circle. Welding robot finished welding for second half circle and move back to home position. Conveyor A and B then would transfer next workpiece to both welding and positioning robot accordingly. When the welding robot has finished

Table 38.1 Arm parameters for (a) welding robot; and (b) positioning robot

(a) Arm parameter for welding robot

Welding robot

Joint	θ	d	a	α
1	0	900	188	−90
2	90	0	−950	0
3	0	0	−225	90
4	0	1,705	0	−90
5	0	0	0	90
6	0	200	0	0

(b) Arm parameter for positioning robot

Positioning robot

Joint	θ	d	A	α
1	0	900	188	−90
2	90	0	−950	0
3	0	0	−225	90
4	0	1,906	0	−90
5	0	0	0	90
6	0	200	0	0

welding both weld joint B and C, it will return to its home position. Same rule applies for the positioning robot. Conveyor B will start moving and transfers the completed work piece or muffler to the unloading area.

38.3 Modeling of a Robotic Arc Welding System

In order to analyze the forward kinematics of the welding robot, firstly arm parameter from the simulation software need to be obtained. The parameters are summarized in Table 38.1.

There are four variables that need to be determined in analyzing robot's forward kinematics, those are approach vector (a), sliding vector (s), normal vector (n) and position vector of the robot hand (p). By replacing all the information for the arm parameter into the D-H transformation matrix show in the Eqs. (38.1) and (38.2) [1], solution for the welding robot forward kinematics can be obtained. Meanwhile, the forward kinematics for this robot arm is given by Eq. (38.2).

$$H_{i-1}^{i} = \begin{pmatrix} C\theta_i & -C\alpha_i S\theta_i & S\alpha_i S\theta_i & \alpha_i C\theta_i \\ S\theta_i & C\alpha_i C\theta_i & -S\alpha_i C\theta_i & \alpha_i S\theta_i \\ 0 & S\alpha_i & C\alpha_i & d_i \\ 0 & 0 & 0 & 1 \end{pmatrix} \quad (38.1)$$

$$H := H_1 \cdot H_2 \cdot H_3 \cdot H_4 \cdot H_5 \cdot H_6$$

$$H = \begin{pmatrix} -0.448 & 0.401 & 0.799 & 2.397 \times 10^3 \\ -0.401 & 0.709 & -0.58 & -750.358 \\ -0.799 & -0.58 & -0.157 & 1.508 \times 10^3 \\ 0 & 0 & 0 & 1 \end{pmatrix} \quad (38.2)$$

So, transformation matrix can be expressed as in Eq. (38.3).

$$H := \begin{pmatrix} n & s & a & p \\ 0 & 0 & 0 & 1 \end{pmatrix} \quad (38.3)$$

With regards to the problem, the matrices can be established for the welding robot complies to Eq. (38.4).

$$\begin{pmatrix} n & s & a & p \\ 0 & 0 & 0 & 1 \end{pmatrix} := \begin{pmatrix} -0.448 & 0.401 & 0.799 & 2.397 \times 10^3 \\ -0.401 & 0.709 & -0.58 & -750.358 \\ -0.799 & -0.58 & -0.157 & 1.508 \times 10^3 \\ 0 & 0 & 0 & 1 \end{pmatrix} \quad (38.4)$$

Path planning is a method of deciding the route a robot should take and expressed in terms of the current internal representation of the job contour. And it's important for the industrial robot user to understanding it where it can reduce the time required to reprogram the path. Using Workspace 5 Simulation software:

1. Initial position, $p_1 = 3{,}700$ mm
2. Final position, $p_2 = 3{,}354$ mm
3. Initial speed, $s_1 = 0$ mm/s
4. Final speed, $s_2 = 100$ mm/s
5. Maximum acceleration, $a_{max} = 10$ mm/s^2
6. Time taken to ramp to:

- Maximum acceleration, $dt_{max} = 19$ s
- Maximum allowable jerk, $j_{max} = 20$ mm/s^3

With regards to the p_1 and p_2 data respectively, the initial and final positions, and s_1 and s_2 are the initial and final speeds. The speed that is reached at the end of the ramp from zero acceleration to the maximum acceleration, a_{max}, is s_a. The time taken to ramp up to a_{max} is dt_{max}. The speed that is reached at the end of the acceleration cruise at a_{max} is s_b. The maximum allowable jerk is j_{max}. The acceleration ramp up to a_{max} is followed by an acceleration cruise and then an acceleration ramp down to zero.

Based on the result obtained from Workspace simulation, it is feasible to calculate the optimal path planning as shown in following Eqs. (38.5)–(38.10):

(a) Total way-point to way-point distance,

$$D = p_2 - p_1 \quad (38.5)$$

(b) Traveling speed between two unknown control points,

$$s_a = s_1 + \frac{a_{max} \cdot dt_{max}}{2} \qquad (38.6)$$

$$s_b = s_2 + \frac{a_{max} \cdot dt_{max}}{2} \qquad (38.7)$$

(c) Distances the robot arm covered during:

1. The ramp up to a_{max},

$$D_1 = a_{max} dt_{max}^2 \left(\frac{1}{4} - \frac{1}{\pi^2} \right) + s_1 dt_{max} \qquad (38.8)$$

2. The acceleration cruise,

$$D_2 = \frac{s_b^2 - s_a^2}{2 a_{max}} \qquad (38.9)$$

3. The ramp down from a_{max} to zero acceleration

$$D_{max} = a_{max} dt_{max}^2 \left(\frac{1}{4} + \frac{1}{\pi^2} \right) + s_b dt_{max} \qquad (38.10)$$

38.4 Programming of End Effector Class Modules

Another important aspect of this research is to program the behavior of the end effectors involved. This is carried out by developing Visual Basic class modules for both the arc weld torch and gripper. Both class modules were then embedded in the Workspace Simulation software. Due to limited space, brief discussion is included.

- Arc Weld Torch module—Several commands have been developed. In order to execute the operation, *ArcOn* method is called. Then the *DrawToolSpark* command turns on the drawing of yellow sparks from the end of the tool to represent the weld. This only occurs if the object *rParentRobot* is set to true. The number and length of the sparks is determined by the *NumLines* and *MaxSparkLength* variables.
- Gripper module – *GraspedObject = GraspObj* which defines the target object to be attached to the tool. Meanwhile for un-grasp, it uses if statement by determining whether the tool has grasped any object or not before it executes the module. Once the required condition has been met, *Detach* method is used to detach the grasped object from the tool (gripper).

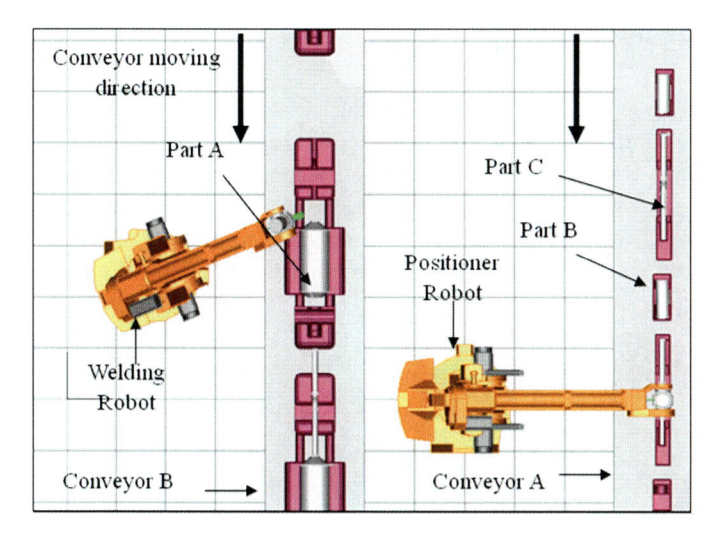

Fig. 38.3 *Welding robot* moves to the weld point b and positioning robot grasps *part c*

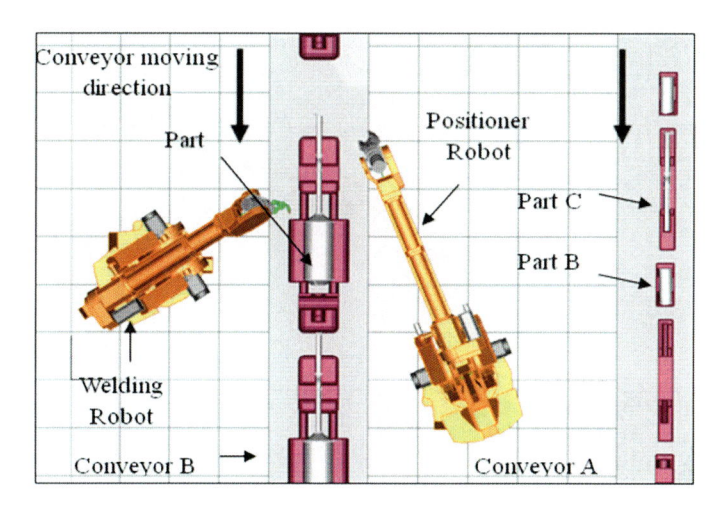

fig. 38.4 *Welding robot* welds weld point b and positioning robot un-grasps *part c*

The involved robot welding process consists of twenty Teaching Point Coordinates (TCP) and each step has been virtually simulated. Screen shots of the simulation results are shows in Figs. 38.3 and 38.4 for two specific TCP of both robots.

38.5 Conclusion

The research has successfully mapped arc welding requirement to establish a robotic arc welding system. Utilization of an automated welding station would also optimize the cycle time by using the time optimal method to generate the robot's trajectory plan. Ability to accurately translate the robot welding requirement into specific automated process is important in order to optimize facility utilization and consequently increase productivity for the industry that applies it.

References

1. Man Z (2005) Robotics. Prentice-Hall, Pearson Asia. ISBN 981-244-756-3
2. Niku S (2001) Introduction to robotics: analysis, systems, applications. Prentice Hall, Englewood Cliffs
3. Liu Z, Bu W, Tan J (2010) Motion navigation for arc welding robots based on feature mapping in a simulation environment. Robot Comput Integr Manuf 26(2):137–144
4. Jia J, Zhang H, Xiong Z (2006) A fuzzy tracking control system for arc welding robot based on rotating arc sensor. In: Proceedings of 2006 IEEE international conference on information acquisition, pp 967–971
5. Fauadi HF, Jumali MS (2008) Modeling and simulation of programmable universal machine for assembly (PUMA) industrial robot for automotive-related assembly process. In: International symposium on information technology 2008 (ITSim 2008), pp 1–5
6. Chen XZ, Chen SB (2010) The autonomous detection and guiding of start welding position for arc welding robot. Ind Robot: Int J 37(1):70–78

Chapter 39
A New Optimized Cuckoo Search Recurrent Neural Network (CSRNN) Algorithm

Nazri Mohd Nawi, Abdullah Khan and Muhammad Zubair Rehman

Abstract Selecting the optimal topology of neural network for a particular application is a difficult task. In case of recurrent neural networks (RNN), most methods only introduce topologies in which their neurons are fully connected. However, recurrent neural network training algorithm has some drawbacks such as getting stuck in local minima, slow speed of convergence and network stagnancy. This paper propose an improved recurrent neural network trained with Cuckoo Search (CS) algorithm to achieve fast convergence and high accuracy. The performance of the proposed Cuckoo Search Recurrent Neural Network (CSRNN) algorithm is compared with Artificial Bee Colony (ABC) and similar hybrid variants. The simulation results show that the proposed CSRNN algorithm performs better than other algorithms used in this study in terms of convergence rate and accuracy.

Keywords Recurrent neural network · Local minima · Artificial bee colony · Cuckoo search algorithm · Hybrid neural networks · Swarm optimization

---

N. M. Nawi · A. Khan · M. Z. Rehman (✉)
Software and Multimedia Centre, Faculty of Computer Science and Information Technology, Universiti Tun Hussein Onn Malaysia (UTHM), Parit Raja, P.O. Box 101 86400 Batu Pahat, Johor, Malaysia
e-mail: zrehman862060@gmail.com

N. M. Nawi
e-mail: nazri@uthm.edu.my

A. Khan
e-mail: hi100010@siswa.uthm.edu.my

39.1 Introduction

The current advances in the working principles of Artificial Neural Networks (ANN) have paved way to classify, predict and forecast on nonlinear systems and huge datasets with high accuracy [1–5]. This makes ANN very attractive and considered superior to the classical modeling and control techniques. Among several possible network architectures, the feed forward and recurrent neural networks (RNN) are most commonly used [6]. In a feed forward neural network the signals are transmitted only in one direction, starting from the input layer, consequently through the hidden layers to the output layer. A recurrent neural network (RNN) has local feedback connections to some of the previous layers. It is different from feed forward network architecture in the sense that there is at least one feedback loop.

RNN are topologically more compact than pure feed forward networks for applications such as time-series prediction and nonlinear system modeling. The modified Elman network implements full return in the hidden layer, enabling this network to realize complex temporal dynamics [7]. Elman networks are generally trained with back propagation algorithms, and more recently with quasi-Newton algorithms such as the self-scaling method of Oren and Spedicato [8] and the Levenberg–Marquradt algorithm [9]. Researchers have developed a variety of schemes by which gradient methods, and in particular back propagation learning, can be extended to recurrent neural networks. A popular expression of gradient descent is error-back propagation introduced by Rumelhart [10] and Werbos [11]. While back propagation is simple to implement but several problems can occur in its use in practical applications, such as local minima entrapment and too much time consumption. An important step of the RNN based impedance mining method is to train the neural network in such a way that it learns the dynamic performance of the tested system [9]. However, certain property of RNN makes many of algorithms less efficient, and it often takes an enormous amount of time to train a network of even a reasonable size. In addition, the complex error surface of the RNN network makes many training algorithms more flat to being intent in local minima. Thus the main disadvantage of the RNN is that they require substantially more connections, and more memory in simulation, then standard back propagation network, thus resulting in a substantial increase in computational times.

To overcome these drawback many evolutionary computing technique have been used. Evolutionary computation is often used to train the weights and parameters of the networks. In recent years, many improved learning algorithms have been proposed to overcome the weaknesses of RNN. These algorithms include global search technique such as hybrid PSO-BP [10], particle swarm optimization (PSO), genetic algorithms (GA) [11]. In this paper, we propose a new meta-heuristic search algorithm, called cuckoo search recurrent neural network (CSRNN). Cuckoo search (CS) is developed by Yang and Deb [12] which imitates animal behavior and is constructive for global optimization [13–15]. The CS algorithm has been applied independently to solve several engineering design optimization problems, such as the design of springs and welded beam structures [16], and forecasting [17]. For these problems, Yang and Deb showed that the optimal solutions obtained by CS are far better than the best solutions obtained by an efficient particle swarm optimizer or genetic algorithms. In particular, CS can be modified to provide a relatively high convergence rate to the true global minimum [16].

In this paper, the convergence performance of the proposed Cuckoo Search Recurrent neural network (CSRNN) algorithm is analyzed on XOR and OR datasets. The results are compared with artificial bee colony sing back-propagation (ABC-BP) algorithm, and similar hybrid variants. The main goals are to decrease the computational cost and to accelerate the learning process using a hybridization method.

The remaining paper is organized as follows: Sect. 39.2 gives literature review on RNN. Section 39.3, explains the proposed CSRNN algorithm and the simulation results are discussed in Sect. 39.4. Finally, the paper is concluded in the Sect. 39.5.

39.2 Recurrent Neural Network

A simple Recurrent Neural Network (RNN) [18] has activation feedback which has short-term memory. A state layer is updated not only with the external input of the network but also with activation from the previous forward propagation. The feedback is modified by a set of weights to enable regular adaptation through learning. In this papers we used three layer network with one input layer, one hidden or 'state' layer, and one 'output' layer. Each layer will have its own index variable: k for output nodes, j and l for hidden, and i for input nodes. In a feed forward network, the input vector, x is propagated through a weight layer, V;

$$net_j(t) = \sum_i^n x_i(t)v_{ji} + b_j \qquad (39.1)$$

where, **n** the number of inputs is, b_j is a bias, and **f** is an output function. In a simple recurrent network, the input vector is similarly propagated through a weight layer, but also combined with the previous state activation through an additional recurrent weight layer, U;

$$y_j(t) = f(net_j(t)) \qquad (39.2)$$

$$net_j(t-1) = \sum_i^n x_i(t)v_{ji} + \sum_l^m y_l(t-1)u_{jl} + b_j(t) \qquad (39.3)$$

where, m is the number of 'state' nodes. The output of the network is in both cases determined by the state and a set of output weights, W;

$$y_k(t-1) = g(net_j(t-1)) \tag{39.4}$$

$$net_k(t) = \sum_{j}^{m} y_k(t-1)w_{kj} + b_k \tag{39.5}$$

where, g is an output function and w_{kj} represents weights from hidden to output layer.

39.3 The Proposed CSRNN Algorithm

Step 1: CS is Initialized and passes the weights to RNN
Step 2: Load the training data
Step 3: **While MSE**<*stopping criteria*
Step 4: Initialize all cuckoo nests
Step 5: Pass the cuckoo nests as weights to network
Step 6: Feed forward network runs using the weights initialized with CS
Step 7: Calculate the error
Step 8: CS keeps on calculating the best possible weight at each epoch until the network is converged
End While

In the proposed CSRNN algorithm, each best nest represents a possible solution (i.e., the weight space and the corresponding biases for RNN optimization in this paper). The weight optimization problem and the size of a population represent the quality of the solution. In the first epoch, the best weights and biases are initialized with CS and then those weights are passed on to the RNN. The weights in RNN are calculated. In the next cycle CS will updated the weights with the best possible solution and CS will continue searching the best weights until the last cycle/epoch of the network is reached or either the MSE is achieved.

39.4 Results and Discussions

39.4.1 Preliminary Study

The simulation experiments are performed on an AMD E-450, 1.66 GHz CPU with 2-GB RAM and MATLAB 2009b software is used. For performing simulations, two dataset like 2-bit XOR and 4-bit dataset are selected. The following four algorithms are analyzed and simulated on the problems:

Table 39.1 CPU time, epochs MSE and SD for 2-5-1 ANN structure

Algorithm	ABC-BP	ABC-LM	BPNN	CSRNN
CPUTIME	172.3388	123.9488	42.643	0.7884
EPOCHS	1,000	1,000	1,000	7
MSE	2.39E−04	0.1250028	0.22066	0
MSE (SD)	6.70E−05	1.551E−06	0.01058	0
Accuracy (%)	96.47231	71.69041	54.6137	100

1. Artificial Bee Colony Back-Propagation (ABC-BP),
2. Artificial Bee Colony Levenberg–Marquardt (ABC-LM),
3. Back-Propagation Neural Network (BPNN), and
4. Proposed Cuckoo Search Recurrent Neural Network (CSRNN).

Three layer neural networks is used for testing of the models, the hidden layer is kept fixed to 5-nodes while output and input layers nodes vary according to the data set given. Log-sigmoid activation function is used. A total of 20 trials are run with each trial consisting of 1,000 epochs for each case.

39.4.2 The 2-bit XOR Problem

The first test problem is the 2-bit XOR Boolean function of two binary inputs and a single binary output. In the simulations, we used 2-5-1, RNN network for two bit XOR problem. The parameters range for the upper and lower band is used [5, −5], [5, −5], [5, −5], [1, −1] respectively. Table 39.1, shows the CPU time, number of epochs, MSE and the Standard Deviation (SD) for the 2 bit XOR with 5 hidden neurons when tested on CSRNN, ABC-LM, ABC-BP and BPNN algorithms. The CSRNN algorithm successfully avoided the local minima and trained the network. From this, we can easily see that the proposed CSRNN can converge successfully for almost every kind of network structure. Also CSRNN has less MSE with high accuracy when compared with other algorithms.

In Table 39.1, the CSRNN algorithm can be seen to converge within 7 epochs, which is quite superior convergence rate as compared to the other algorithms. The ABC-BP and ABC-LM algorithm are not converging within 1,000 epochs. BPNN also shows many failures in convergence to the global solution. CSRNN can be seen outperforming the ABC-BP, ABC-LM, and BPNN algorithms with 100 % accuracy.

39.4.3 The 4-bit OR Problem

The second problem selected for training the proposed CSRNN is 4-bit OR dataset. In 4-bit OR, if the number of inputs all is 0, the output is 0, otherwise the output is 1. Again for the four bit input we applied 4-5-1, Elman recurrent neural network

Table 39.2 CPU time, epochs and MSE error for 4-5-1 ANN structure

Algorithm	ABC-BP	ABC-LM	BPNN	CSRNN
CPU TIME	162.4945	118.7274	63.28089	1.822
EPOCHS	1000	1,000	1,000	13
MSE	1.91E−10	1.82E−10	0.05277	0
MSE (SD)	1.5E−10	2.1E−11	0.00843	0
Accuracy (%)	99.97	99.99	89.8349	100

structure. For the 4-5-1 network structure it has thirty connection weights and six biases. Table 39.2 illustrates the CPU time, epochs, and MSE performance accuracy and SD of the proposed CSRNN algorithm, ABC-BP, ABC-LM and BPNN algorithms respectively. From Table 39.2, we can see that CSRNN algorithm outperforms ABC-BP, ABC-LM, and BPNN in-terms of CPU time, epochs, MSE, accuracy, and SD. The proposed CSRNN is converging within 14 epochs for the 4-5-1 network structure and again has 0 MSE, 100 % accuracy and 0 SD which is better than the other algorithms. While the other algorithm such as ABC-BP, ABC-LM, and BPNN still need more epoch and CPU time to achieve high accuracy and convergence to 0 MSE.

39.5 Conclusion

Nature inspired meta-heuristic algorithms provide derivative-free solution to optimize complex problems. A new meta-heuristic search algorithm, called cuckoo search (CS) is used to train RNN to achieve fast convergence rate, to minimize the training error and to increase the convergence accuracy. The proposed CSRNN algorithm is used to train network on the 2-bit XOR, and 4-Bit OR benchmark problems. The results show that the CSRNN is simple and generic for optimization problems and has better convergence rate, SD, and accuracy than the ABC-LM, ABC-BP and BPNN algorithms.

References

1. Narendra KS, Parthasarathy K (1990) Identification and control of dynamic systems using neural networks. J IEEE Trans Neural Networks 1(1):4–27
2. Chen S, Billings SA (1992) Neural networks for nonlinear dynamics system modelling and identification. J Int J Control 56(2):319–346
3. Su HT, McAvoy TJ, Werbos P (1992) Long-term predictions of chemical processes using recurrent neural networks: a parallel training approach. J Ind Eng Chem Res 31(5):1338–1352
4. Mastorocostas PA, Theocharis JB (2006) A stable learning algorithm for block-diagonal recurrent neural networks: application to the analysis of lung sounds. J. IEEE Trans Syst Man Cybern B Cybern 36(2):242–254

5. Rehman MZ, Nawi NM, Ghazali MI (2012) Predicting noise-induced hearing loss (NIHL) and hearing deterioration index (HDI) in Malaysian industrial workers using GDAM algorithm. J Eng Technol (JET) 3(1):179–197 (UTeM)
6. Li CJ (1995) Mechanical system modelling using recurrent neural networks via quasi-Newton learning methods. J Appl Math Modell 19:420–428
7. Chan LW, Szeto CC (1999) Training recurrent neural network with block diagonal approximated Levenberg marquardt algorithm. In: IJCNN'99, vol 3. pp 1521–1526
8. Rumelhart DE, Hinton GE, Williams RJ (1986) Learning internal representations by error propagation. In: Rumelhart DE, McClelland JL (eds) Parallel distributed processing: explorations in the microstructure of cognition, 45th edn. MIT Press, Cambridge
9. Warbos P (1993) The roots of backpropagation: from ordered derivatives to neural networks and political forecasting. Wiley, New York
10. Peng XG, Venayagamoorthy K, Corzin KA (2007) Combined training of recurrent neural networks with particle swarm optimization and backpropagation algorithms for impedance identification. In: Proceedings of the 2007 IEEE swarm intelligence symposium, pp 9–15
11. Juang CF (2004) A hybrid of genetic algorithm and particleswarm optimization for recurrent network design. J IEEE Trans Syst Man Cybern 34(2):997–1006
12. Yang XS, Deb S (2009) Cuckoo search via Lévy flights. In: Proceedings of world congress on nature and biologically inspired computing, India, pp 210–214
13. Yang XS, Deb S (2010) Engineering optimisation by cuckoo search. J Int J Math Modell Numer Optim 1(4):330–343
14. Tuba M, Subotic M, Stanarevic N (2011) Modified cuckoo search algorithm for unconstrained optimization problems. In: Proceedings of the European computing conference (ECC'11), pp 263–268
15. Tuba M, Subotic M, Stanarevic N (2012) Performance of a modified cuckoo search algorithm for unconstrained optimization problems faculty of computer science faculty of computer. Science 11(2):62–74
16. Yang XS, Deb S (2010) Engineering optimisation by cuckoo search. J Int J Math Modell Numer Optim 1(4):330–343
17. Chaowanawate K, Heednacram A (2012) Implementation of cuckoo search in RBF neural network for flood forecasting. In: Proceedings of 2012 fourth international conference on computational intelligence, communication systems and networks, pp 22–26
18. Elman JL (1990) Finding structure in time. J Cognitive Sci 14:179–211

Chapter 40
Using Time Proportionate Intensity Images with Non-linear Classifiers for Hand Gesture Recognition

Omar Ahmad, Basilio Bona, Muhammad Latif Anjum and Ikramullah Khosa

Abstract Gestures are signals that contain important spatiotemporal information. Understanding gestures is a trivial task for humans, but for machines it is a challenging task involving thousands of computations per video frame. This paper investigates an efficient hand gesture recognition technique which is based on time projections of the hand location. For recognition, non-linear classifiers, namely Support Vector Machines and Artificial Neural Networks, are tested. The proposed method performs much faster than the conventional Markov Model based gesture recognition techniques while achieving comparable recognition results.

Keywords Gesture recognition · Spatiotemporal segmentation · Computer vision · Machine learning · Classification · Human robot interaction

40.1 Introduction

Gestures are parts of human movement which contain certain information. They can convey certain meanings (like in Sign Language), or commands, or can be used to point to certain objects in the surroundings. From a computational point of

---

O. Ahmad (✉) · M. L. Anjum
Department of Mechanical and Aerospace Engineering (DIMEAS), Politecnico di Torino, Corso Duca degli Abruzzi 24, 10129 Turin, Italy
e-mail: omar.ahmad81@gmail.com

B. Bona
Department of Control and Computer Engineering (DAUIN), Politecnico di Torino, Corso Duca degli Abruzzi 24, 10129 Turin, Italy

I. Khosa
Department of Electronics and Telecommunications (DET), Politecnico di Torino, Corso Duca degli Abruzzi 24, 10129 Turin, Italy
e-mail: ikramullahkhosa@gmail.com

view they can be thought of as spatiotemporal signals that contain certain information which is vital for robots or machines to interact with their environments. A lot of research has been carried out in the recent years to develop techniques for gesture recognition [1]. From computer gaming to touch pads, and other such devices, we see a lot of applications of gesture recognition as results of this research.

As the computational power of the machines is rising with advances in IC manufacturing technology we see a lot of research activity in the field of Human Robot Interaction (HRI). The dream of making robots a household commodity is suddenly looking very much realizable in near future. Recent research is aimed at making the human-robot interaction natural and with as fewer constraints as possible. Humans interact with each other naturally, robustly identifying gestures as well. The robots have come a long way in understanding their environment, being now able to see, hear and feel to a certain limited extent, but a lot more is yet to be achieved. Understanding gestures robustly and naturally is a task that still remains an open area for research.

In this paper we present a gesture recognition technique that can robustly detect gestures without requiring any temporal segmentation beforehand. The proposed method performs the higher level tasks of gesture spotting, i.e., determining the start and end frames of the gestures (temporal segmentation) and classification, as well as low level tasks of spatial segmentation of the hand at each frame.

The algorithm starts by spatially segmenting the hand for every incoming frame by combining skin detection and motion calculation. This information is then sent to the Time Proportionate Intensity Accumulator unit which stores this information as an intensity image (TPI image). A high intensity on the TPI image corresponds to a recent impression and lower intensities correspond to previous impressions. Information from each new frame, after preprocessing is sent to the classifier unit which in turn classifies whether or not a gesture has just been performed. If a gesture is performed it is duly classified as being one of the gestures in the system vocabulary. Figure 40.1 is a block level representation of the proposed gesture recognition system.

A key aspect of the proposed approach is its very low computational cost. The algorithm can classify the gestures in real time mainly because it does not model the gestures as Markov Chains. Thus it does not have to match a Hidden Markov Model with a video query comprising of a series of frames.

The rest of the paper is organized as follows. Section 40.2 describes the recent research work in gesture recognition and how it relates to our work. Section 40.3 discusses the preprocessing step of spatial segmentation of hand and temporal projection. Section 40.4 is about the nonlinear classifiers used for gesture recognition and their offline training. In Sect. 40.5 we discuss the experimental setup, the video sets used and the results of classification, concluding with recommendations and remarks on future work in Sect. 40.6.

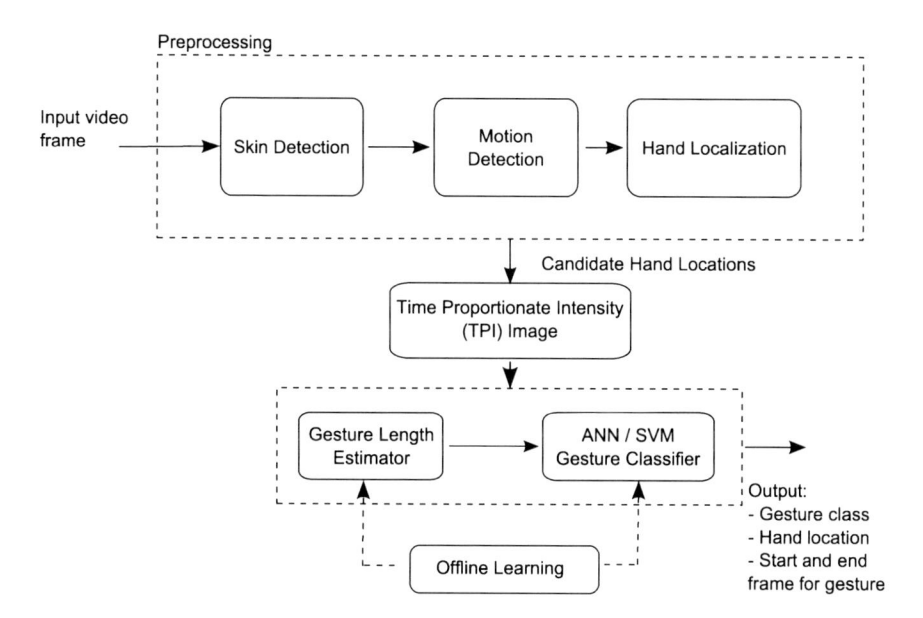

Fig. 40.1 Block diagram of the proposed gesture recognition system

40.2 Related Work

An important discriminating feature of the proposed algorithm from existing gesture recognition algorithms is that it does not require temporal segmentation of the gesture as preprocessing like other dynamic approaches [2–5]. In many gesture recognition algorithms [6, 7], spatial and temporal segmentation is done at a lower level and certain features of shape and velocity are extracted as preprocessing steps. These features are then passed to the classifier for recognition [6]. Recognition results deteriorate if these preprocessing segmentations fail. Although we perform a preprocessing step involving spatial segmentation of the hand, its failure in some frames does not make the next stages of the algorithm to fail. Ambiguities in hand segmentation causes some noise in the background in our algorithm but the nonlinear classifiers can handle such noise.

Pavlovic et al. [8], discuss the template matching based segmentation of the hand useful in determining the posture of the hand. They trained a hand shape determining classifier to detect hand posture and used skin color based classifier to spatially segment the hand. Aaron and James [9], discuss temporal templates for learning the history of motion. The proposed method is similar in the sense that it also captures the history of motion onto a spatial plane but differs in linear fading concept that we introduce.

Some algorithms like [3, 10, 11], extract global features from each frame like motion fields or use a transformed set of images such as intensity thresholded images or difference images as inputs to gesture recognition modules. These algorithms do not incorporate tolerance to background movements which can cause these algorithms to fail in noisy environments. We train our nonlinear classifiers for noisy environments as well thus it outperforms these other algorithms.

Most of the gesture recognition algorithms model the gestures as Markov Chains [12–14], with fixed or variable transition probabilities. The recognition of gestures in these algorithms becomes a problem of state by state aligning the query video sequence with the gesture model. This involves computations that increase exponentially with the gesture models in the vocabulary. To overcome this issue of time complexity these algorithms devise certain mechanisms to prune out certain hypotheses and rely on Dynamic Programming (DP) to reduce the computations. As we do not model gestures as Markov Chains our algorithm has to perform significantly less number of computations for classification.

The proposed method is similar to algorithms like [11] and [15], which model the gestures as rigid 3D patterns. These algorithms do not perform well if the gesturing speed is changed. On the other hand we learn the tolerances in gesturing speed from the training data and do not face these problems.

Finding the start and end frames in a gesture is referred to as *gesture spotting*. Algorithms can be divided in two categories based on the mechanisms they adopt for gesture spotting. One approach is to temporally segment the gestures before classification. This is usually achieved by inserting intervals between the gestures [16, 17]. The other approach indirectly performs the task of gesture segmentation based on results of certain cost functions over a window in time which slides over the temporal axis of the incoming video stream [7, 18, 19]. Our approach also lies in this second category where we find the start and end frames of a gesture during the classification.

Then there is the problem of sub gestures, i.e., when some gestures are part of other longer gestures. Classifiers usually either do not consider this possibility by imposing limitations on the gestures themselves [19] or they require additional looping over all the gestures in the vocabulary to determine these sub gestures. The proposed algorithm in its output not only classifies a gesture it also gives a confidence measure and expectancy of a super gesture.

40.3 Preprocessing and Hand Segmentation

In this section we describe the preprocessing steps performed on each frame before it is passed on to the classifier stage. First the spatial segmentation techniques to get the most probable hand locations are presented. Next discussed are the methods to incorporate multiple hand candidates and background noise removal. At the end of this section we describe in detail, the working of Time Proportionate Intensity Accumulator unit to get the projections of the temporal axis onto the spatial plain (TPI image) which in turn allows us to use existing techniques of feature based classifiers to recognize the gestures.

40.3.1 Hand Segmentation Based on Skin Pixel Estimation

First of all a skin likelihood image is computed using the point operation of Eq. (40.1). The mean μ_s, and variance $\sum_s$ from a generic skin model of [20] are used. Next the motion mask is calculated by taking the difference of the current frame and the previous frame. Hand likelihood image is obtained by applying the motion mask to the skin likelihood image.

$$p(x|skin\ pixel) = \frac{1}{(2\pi)^{\frac{1}{2}} |\sum_s|^{\frac{1}{2}}} exp\left(-\frac{1}{2}(x-\mu_s)^T \sum_s^{-1}(x-\mu_s)\right) \quad (40.1)$$

The hand likelihood image is filtered with a 3 × 3 order statistic median filter to remove background noise. This hand likelihood image is then passed onto the next stage of k-means clustering.

40.3.2 Hand Localization Using k-mean Clustering

This module takes as input the binary hand likelihood image. The white pixels in this image correspond to a high likelihood of hand location. The indices of all non-zero pixels are extracted from this image and are used to find K clusters. Each cluster corresponding to high probability of hand locations. The number of member points for each cluster determines the size of the impression made by this hand hypothesis on the TPI image in the next stage. This is approach is similar to the one used in [21] except that instead of using an integral image, and moving window, k-mean algorithm is used for more accurate hand localization.

The algorithm can be extended to accommodate two handed gestures for example and to make the algorithm robust. This also helps reduce the background noise due to moving distracters in the background. Figure 40.2 shows the results of outlier rejected hand segmentation.

40.3.3 Time Proportionate Intensity Projections

We get K candidate hand locations from the clustering unit. We place K blobs on the Time Proportionate Intensity Accumulator at the spatial locations received from the previous unit. We assign them maximum intensity. Upon processing of the next frame we receive further K blobs to be placed on the accumulator. Here we decrement the accumulator first by the fading factor α before placing the blobs. This way the TPI image periodically forgets or fades away the impressions made by previous frames, Fig. 40.3.

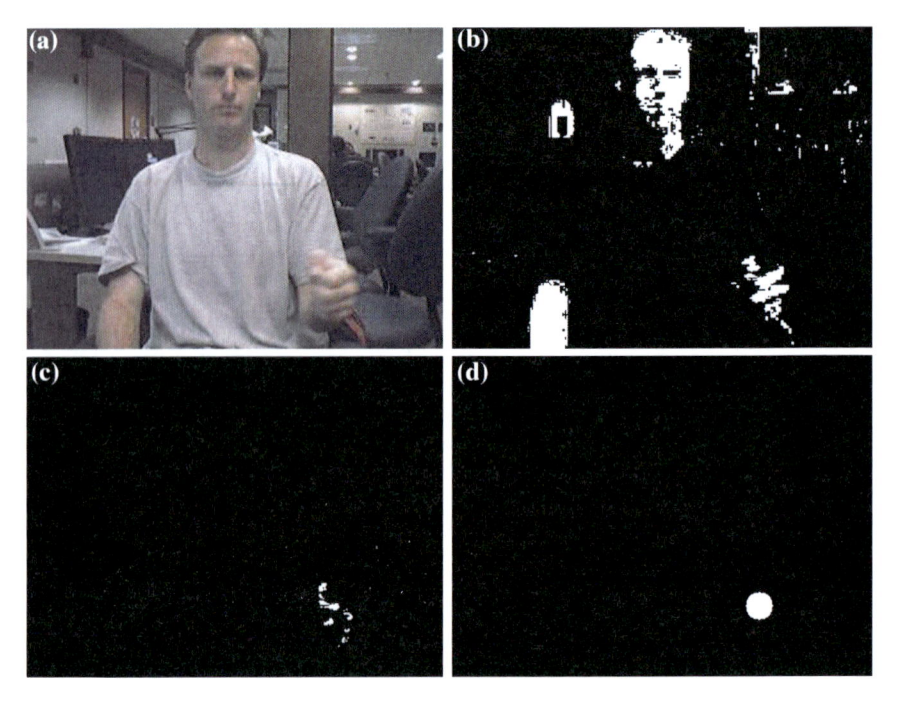

Fig. 40.2 Preprocessing of an incoming frame and corresponding impression onto the time proportionate intensity accumulator

Fig. 40.3 Example time proportionate intensity (TPI) accumulators from training sample videos (K = 1)

We have used a linear fade where the intensity of a pixel fades in the accumulator by a constant α after each new frame.

$$I_t = I_{t-1} - \alpha \tag{40.2}$$

To get training data, TPI image is thresholded (Eq. 40.3) to zero to forget all the information *n* frames ago thus temporally segmenting the gesture. The gesture length for each gesture in the training set videos is available in the ground truth files. Average gesture length is also learned from the training data for each gesture class.

$$threshold = max(I) - n\alpha \tag{40.3}$$

Fig. 40.4 Palm's Graffiti digits

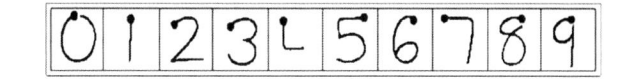

40.4 Spatiotemporal Matching

In this section the implementation of our gesture class learning method using nonlinear classifiers (Neural Networks and Support Vector Machines) is presented first, and then we discuss the classification mechanism used.

40.4.1 Model Learning and Non Linear Classifiers

We have used the video data sets by Athitsos [18] for experimentation. These video sets include a training set, an easy data set and a difficult data set. Each set contains 30 videos where signers make signs of the Palm's graffiti digits 0 to 9, Fig. 40.4. In each set of videos there are a total of 10 different signers with 3 videos from each signer. The signers wear colored gloves in the training data set only. All the videos are accompanied by ground truth text files which contain the temporal segmentation of the gestures. Meaning that start and end frames of all the gestures in each video are given. These data are useful for training and cross validation purposes.

40.4.2 Classification Using Support Vector Machines

Model learning phase includes feeding the TPI image state just after the gesture is completed to the SVM trainer. The TPI image is thresholded to forget all the information before the start of the current gesture. The m × n TPI image is down sampled to 25 × 25 image and then reshaped into 1 × 625 feature vector that along with the gesture label is used to train the SVM parameter vector theta θ. The optimization objective function for SVMs is a minimization problem given in Eq. (40.4). The kernel function we have used is the Gaussian kernel of Eq. (40.5).

$$\min_{\theta} C \sum_{i=1}^{m} \left[y^{(i)} cost_1 \left(\theta^T f^{(i)} \right) + \left(1 - y^{(i)} \right) cost_0 \left(\theta^T f^{(i)} \right) \right] + \frac{1}{2} \sum_{j=1}^{n} \theta_j^2 \quad (40.4)$$

$$f^{(i)} = k\left(x^{(i)}, l^{(i)}\right) = exp\left(-\gamma \|x^{(i)} - l^{(i)}\|^2\right), \quad \gamma = \frac{1}{2\sigma^2}, \text{ and } \gamma > 0 \quad (40.5)$$

The training data is divided into 11 classes. 10 classes for gestures 0 to 9 and one class for training examples where no gesture has been performed. Incomplete gestures and partially observed gestures are placed in this class. For the implementation of the SVMs we have used the library package LIBSVM [22].

For each incoming frame we get the current state of the TPI image as described in the previous sections. A window with this TPI image is passed to the classifier after each frame.

We use SVMs trained as multi-class classifiers. LIBSVM implements the multi-class classifiers using one against one approach. Therefore this TPI image is matched against all the models in the vocabulary. A positive match (based on majority voting) gives the gesture class as well as the temporal segmentation of the gesture as we now know the end frame of the gesture. The average time (in number of frames) for that particular gesture class was already learnt in the model learning stage.

Temporal segmentation can be improved if after initial classification we move the time window and recalculate the cost finding the frame corresponding to the minimum cost as the start frame. Of course this will be under the hypothesis that the cost function is convex with a single minimum.

40.4.3 Classification Using Artificial Neural Networks

Artificial Neural Network (ANN) is a computing system, composed of large number of highly interconnected units (called neurons) that emulate the organization and operation of biological nervous system. This is one of the most widely used technique in classification and pattern recognition problems [23].

For the learning of ANN, special training algorithms are developed based on the learning rules similar to learning mechanisms of biological systems. There are many types and architectures of neural networks, fundamentally depending on their learning mechanisms. For gesture recognition, we have chosen a three layer Multilayer Perceptron Neural Network (MLPNN) with back propagation training algorithm, consisting of one input, hidden and output layer each. Architecture of a typical MLPNN with three layers is shown in Fig. 40.5. The input layer of our network consists of 625 features and output layer has ten neuron; equal to the number of target classes. The number of nodes in the hidden layer has great influence on the performance of the network. An optimum number of nodes in the hidden layer are selected after trial and error in network performance. For the implementation of the ANNs we have used the OpenCV CvANN Class. For back propagation based training, the library uses the algorithm proposed in [24].

40.5 Experiments and Results

This section explains the experimental setup. Specifically we discuss the choice of data set of videos, the number of training examples, the validation data set and the test sets. Here we also present the results of classification and compare them with existing methods.

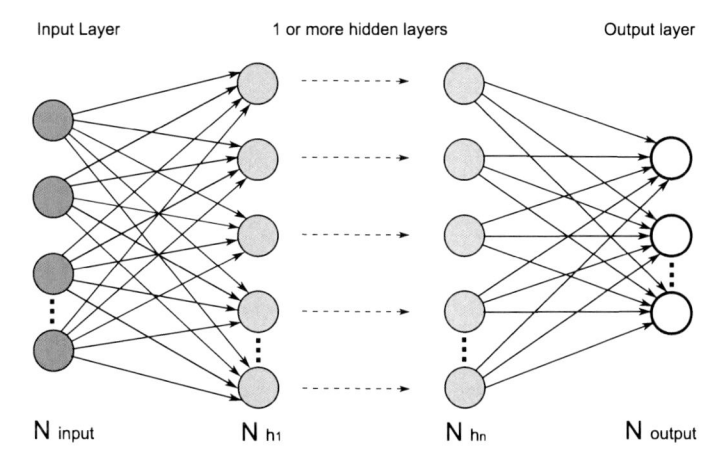

Fig. 40.5 Architecture of a multi-layered artificial neural network

40.5.1 Results Using Support Vector Machine

The training data set has only 30 videos, 3 from each of the 10 signers. Extracting a single TPI image per gesture class from every video leads to only 30 samples for training purposes. These are very few when training the SVMs to obtain good results. One way of increasing the samples is to extract multiple TPI images per gesture class from each video. This way multiple TPI images from above 80 % gesture completion can be extracted to increase the training examples. Following tables present the early stage results of the experimentation. As seen in Table 40.1, the accuracy of classification for the difficult data set is very poor. This is due to the background motion in the difficult data set. We are currently working to devise methods to incorporate multiple hand hypotheses with low time complexity.

40.5.2 Results Using Artificial Neural Networks

We have a total of 4,634 samples of hand gestures corresponding to ten target classes; 0–9. Sixty percent of the data representing each class is used for training, twenty percent is used for cross validation (CV) and the rest twenty percent is considered as test data. The network is trained with back propagation algorithm and a regularization parameter is also tuned to overcome the problem of over fitting as well as under fitting of data. The number of epochs is limited to 100. Using the best results of network on CV data, an optimum architecture of the network is settled with 196 nodes in the hidden layer. The network is then used to classify the test data. The overall network accuracy is presented in Table 40.2, and confusion matrix for test data is presented in Table 40.3.

Table 40.1 Class wise accuracy and false positives for easy (E) and difficult (D) data sets

Gesture	Accuracy percent		False positives		Confused with	
	E	D	E	D	E	D
0	74	34	5	19	6	6
1	81	73	3	6	7	9
2	91	56	2	14	4	7
3	78	43	4	15	2	2
4	85	67	3	9	2	2
5	81	48	3	13	8	8
6	88	62	2	10	0	0
7	87	69	2	9	1	2
8	76	33	5	20	5	5
9	77	35	5	19	7	1

Last column shows class label a gesture is confused with the most

Table 40.2 Accuracy accross the training, cross validation and the test data sets

Number of hidden layers	Training accuracy (%)	Cross validation accuracy (%)	Test accuracy (%)	Regularization parameter (λ)
182	99.89	92.12	96.24	0.005

Table 40.3 The confusion matrix for ANN

Actual class	Predicted class									
	1	2	3	4	5	6	7	8	9	10
1	102	0	0	0	0	0	0	0	0	0
2	0	107	0	0	0	0	0	0	0	0
3	0	0	78	0	14	0	0	0	0	0
4	0	0	0	82	0	0	0	0	0	0
5	0	0	18	0	78	0	0	0	0	0
6	0	0	0	0	0	95	0	0	0	0
7	0	0	0	0	0	0	80	0	0	0
8	0	0	0	0	0	0	0	90	0	0
9	0	0	0	0	0	0	0	0	99	0
10	0	0	0	0	0	0	0	0	0	86

40.6 Future Work and Conclusions

We presented a method of recognizing gestures with minimal computations thus making run-time gesture classification possible. Currently the algorithm is designed to recognize one-handed gestures. It can be modified to be able to accommodate two handed gestures or even other classes of gestures. Detection of right and left hands using the existing techniques of Viola and Jones [25] is possible. In this way two handed gestures can be recognized in run-time without having to compromise the speed.

One major drawback of the algorithm is the drop in accuracy of classification with movements in the background (tested on difficult data set). This is mainly due to the implementation decision of using one single reliable hand location, for impression on the TPI image. Although we do train our classifier with the background noise, thus increasing the accuracy in recognition other techniques need to be used to heuristically minimize the noise.

Hence the algorithm is suitable for time critical applications or for slower systems, which are incapable of running the traditional Markov Chains based algorithms successfully, and with fairly minimal background movements.

References

1. Poppe R (2010) A survey on vision-based human action recognition. Image Vis Comput 28(6):976–990. ISSN 0262-8856, doi:10.1016/j.imavis.2009.11.014
2. Corradini A (2001) Dynamic time warping for off-line recognition of a small gesture vocabulary. In: Proceedings of IEEE ICCV workshop recognition, analysis, and tracking of faces and gestures in real-time systems, pp 82–89
3. Darrell T, Pentland A (1993) Space-time gestures. In: Proceedings of IEEE conference computer vision and pattern recognition, pp 335–340
4. Cutler R, Turk M (1998) View-based interpretation of real-time optical flow for gesture recognition. In: Proceedings of third IEEE international conference on automatic face and gesture recognition, pp 416–421
5. Starner T, Weaver J, Pentland A (1998) Real-time American sign language recognition using desk and wearable computer based video. IEEE Trans Pattern Anal Mach Intell 20(12):1371–1375
6. Yang MH, Ahuja N, Tabb M (2002) Extraction of 2D motion trajectories and its application to hand gesture recognition. IEEE Trans Pattern Anal Mach Intell 24(8):1061–1074
7. Oka K, Sato Y, Koike H (2002) Real-time fingertip tracking and gesture recognition. IEEE Comput Graphics Appl 22(6):64–71
8. Pavlovic V, Sharma R, Huang T (1997) Visual interpretation of hand gestures for human-computer interaction: a review. IEEE Trans Pattern Anal Mach Intell 19(7):677–695
9. Bobick AF, Davis JW (2001) The recognition of human movement using temporal templates. IEEE Trans Pattern Anal Mach Intell (PAMI) 23(3):257267
10. Quattoni A, Wang S, Morency L-P, Collins M, Darrell T (2007) Hidden conditional random fields. IEEE Trans Pattern Anal Mach Intell 29(10):1848–1852
11. Gorelick L, Blank M, Shechtman E, Irani M, Basri R (2007) Actions as space-time shapes. IEEE Trans Pattern Anal Mach Intell 29(12):2247–2253
12. Chen F, Fu C, Huang C (2003) Hand gesture recognition using a real-time tracking method and hidden Markov models. Image Video Comput 21(8):745–758
13. Sato Y, Kobayashi T (2002) Extension of hidden Markov models to deal with multiple candidates of observations and its application to mobile-robot-oriented gesture recognition. In: Proceedings of 16th international conference on pattern recognition, vol II. pp 515–519
14. Alon J, Athitsos V, Yuan Q, Sclaroff S (2005) Simultaneous localization and recognition of dynamic hand gestures. Proceedings of IEEE workshop motion and video computing, vol II. pp 254–260

15. Ke Y, Sukthankar R, Hebert M (2005) Efficient visual event detection using volumetric features. In: Proceedings of 10th IEEE international conference on computer vision, vol 1. pp 166–173
16. Kang H, Lee C, Jung K (2004) Recognition-based gesture spotting in video games. Pattern Recogn Lett 25(15):1701–1714
17. Kahol K, Tripathi P, Panchanathan S (2004) Automated gesture segmentation from dance sequences. In: Proceedings of sixth IEEE international conference on automatic face and gesture recognition, pp 883–888
18. Alon J, Athitsos V, Yuan Q, Sclaroff S (2009) A unified framework for gesture recognition and spatiotemporal gesture segmentation. In: PAMI, September 2009. Video sets available at http://cs-people.bu.edu/athitsos/digits/
19. Lee H, Kim J (1999) An HMM-based threshold model approach for gesture recognition. IEEE Trans Pattern Anal Mach Intell 21(10):961–973
20. Jones M, Rehg J (2002) Statistical color models with application to skin detection. Intl J. Comput Vis 46(1):81–96
21. Alon J (2006) Spatiotemporal gesture segmentation. Ph.D. dissertation, technical report BU-CS-2006-024, Department of Computer Science, Boston University
22. Chang CC, Lin CJ (2011) LIBSVM: a library for support vector machines. ACM Trans Intell Syst Technol 2(27):1–27. Software available at http://www.csie.ntu.edu.tw/cjlin/libsvm
23. Zhang GP (2000) Neural networks for classification: a survey. IEEE Trans Syst Man Cybern C Cybern 30(4):451–462
24. LeCun Y, Bottou L, Orr GB, Müller KR (1998) Efficient backprop. In: LeCun Y, Bottou L, Orr GB, Muller KR (eds) Neural networks: tricks of the trade, vol 1524. Springer Lecture Notes in Computer Sciences. Springer, Berlin, pp 5–50
25. Viola P, Jones M (2001) Rapid object detection using a boosted cascade of simple features. In: Proceedings of IEEE conference on computer vision and pattern recognition, vol I. pp 511–518

Chapter 41
Applying a Multi-Agent Classifier System with a Novel Trust Measurement Method to Classifying Medical Data

Mohammed Falah Mohammed, Chee Peng Lim
and Umi Kalthum bt Ngah

Abstract In this paper, we present the application of a Multi-Agent Classifier System (MACS) to medical data classification tasks. The MACS model comprises a number of Fuzzy Min–Max (FMM) neural network classifiers as its agents. A trust measurement method is used to integrate the predictions from multiple agents, in order to improve the overall performance of the MACS model. An auction procedure based on the sealed bid is adopted for the MACS model in determining the winning agent. The effectiveness of the MACS model is evaluated using the Wisconsin Breast Cancer (WBC) benchmark problem and a real-world heart disease diagnosis problem. The results demonstrate that stable results are produced by the MACS model in undertaking medical data classification tasks.

Keywords Multi-agent classifier system · Fuzzy min–max neural network · Trust measurement · Medical data classification

41.1 Introduction

Multi-Agent System (MAS) technologies have recently gained significant attention amongst research community [20]. As an example, numerous MAS models have been applied to undertake medical diagnosis tasks [8, 17, 26]. In general, an

---

M. F. Mohammed · U. K. bt Ngah (✉)
Imaging and Computational Intelligence (ICI) Group School of Electrical and Electronic Engineering, Universiti Sains Malaysia, Nibong Tebal, Malaysia
e-mail: eeumi@eng.usm.my

M. F. Mohammed
e-mail: mofmah@yahoo.com

C. P. Lim
Centre for Intelligent Systems Research, Deakin University, Melbourne, Australia

MAS model consists of several independent agents, and they cooperate and collaborate with one another in solving a problem [25]. When several agents operate in a common environment, sharing resources and information among agents becomes important. Indeed, information sharing is one of the important factors affecting efficiency of an MAS model. In this study, the constituent agents in the MAS model consist of classifiers. Hence, the resulting system is known as a Multi-Agent Classifier System (MACS).

Numerous neural network models have been used for assisting medical diagnosis tasks [1, 2, 11, 23]. Among them, the Fuzzy Min Max (FMM) neural networks [21, 22] are useful for data classification and clustering problems. In this aspect, an FMM-based MACS model with the Trust-Negotiation-Communication (TNC) reasoning strategy was proposed to improve the classification accuracy rates [18, 19]. In the proposed TNC strategy, having an effective trust measurement is the key factor to ensure the success of the model. With the function of providing social control in effective interactions, trust is employed to reach an appropriate decision within the MACS model that encompasses multiple FMM agents. Indeed, trust not only is important for the TNC strategy, but is also an essential component that helps in enhancing the MAS performance through knowledge-sharing [4, 10].

A variety of trust and reputation methods are available in many fields [5, 13, 24]. Accordingly, an attempt has been made in our previous work to develop a novel trust measurement method known as Certified Belief in Strength (CBS) [12], which is based on reputation and strength. The CBS scheme enhances the performance of the MACS model by improving the accuracy rates of its constituent agents, i.e., the supervised FMM classifier [21]. In this study, the application of the MACS-CBS model is further evaluated using benchmark and real medical diagnosis problems.

The present paper is organized as follows. In Sects. 41.2 and 41.3, FMM and MACS-CBS are introduced. Section 41.4 presents the experimental study. Conclusions and suggestions for further work are presented in Sect. 41.5.

41.2 The FMM Neural Network Classifier

The supervised FMM neural network classifier was introduced by Simpson [21]. The network structure is built from hyperboxes. A hyperbox is defined by its minimum and maximum points, which are formed based on the input patterns. When a new input pattern is provided, the FMM learning algorithm is used to identify the best-matched hyperbox with respect to the input pattern, as shown in Eq. (41.1) [21].

$$B_j(A_h) = \frac{1}{2n} \sum_{i=1}^{n} \left[\max\left(0, 1 - \max\left(0, y \min\left(1, a_{hi} - w_{ji}\right)\right)\right) \right. \\ \left. + \max\left(0, 1 - \max\left(0, y \min\left(1, v_{ji} - a_{hi}\right)\right)\right) \right] \quad (41.1)$$

where B is the membership function, j is the hyperbox, $A_h = (a_{h1}, a_{h2}, \ldots, a_{hn}) \in I^n$ is the hth input pattern, $V_j = (v_{j1}, v_{j2}, \ldots, v_{jn})$ is the minimum point for B_j, $W_j = (w_{j1}, w_{j2}, \ldots, w_{jn})$ is the maximum point for B_j, and y is the sensitivity parameter that regulates how fast the membership value decreases as the distance between A_h and B_j increases. The membership function represents the degree to which an input pattern fits in the hyperbox. If the pattern does not belong to any hyperboxes, even with the expansion process in FMM, a new hyperbox is to be created to include the input pattern. In other words, FMM entails a dynamic network structure with an online learning capability whereby the number of hyperboxes can be increased whenever necessary; hence avoiding the problem of re-training as faced by many neural network models with an off-line learning capability [3, 14, 16]. Further details of FMM can be found in Simpson [21].

41.3 The MACS-CBS Model

The MACS-CBS model consists of an MACS model with two layers comprising the manager and the agents. Figure 41.1 shows the MACS-CBS structure. The manager chooses the winning agent, and is responsible for controlling the overall system. The agent layer consists of a number of agents (three FMM classifiers in this study), and the CBS trust measurement scheme is incorporated into each agent. Two important activities occur in MACS-CBS, as follows:

- *Training*: Since FMM learns online, different sequences of training data samples lead to the formation of different knowledge bases in individual FMM classifiers, which, in turn, produce different predictions for a given test pattern.
- *Prediction*: During the prediction step, the hyperbox accuracy rate indicates the degree of the belief element, which represents the knowledge of the agent. This belief element is updated during the test phase during the online classification stage.

To compute hyperbox accuracy, the number of samples predicted by each hyperbox and the associated number of correct predictions need to be determined. Hyperbox accuracy (in percentage) is computed with Eq. (41.2).

$$HA_i = \left(\frac{\sum_{j=1}^{n} CP_{ij}}{\sum_{j=1}^{n} (CP_{ij} + ICP_{ij})} \right) * 100 \qquad (41.2)$$

where HA is hyperbox accuracy, CP is the number of correct predictions, ICP is the number of incorrect predictions, i is the hyperbox number, and j is the number of input samples classified by hyperbox (i), where $j = (1, 2, \ldots, n)$.

Fig. 41.1 The structure of the MACS-CBS model

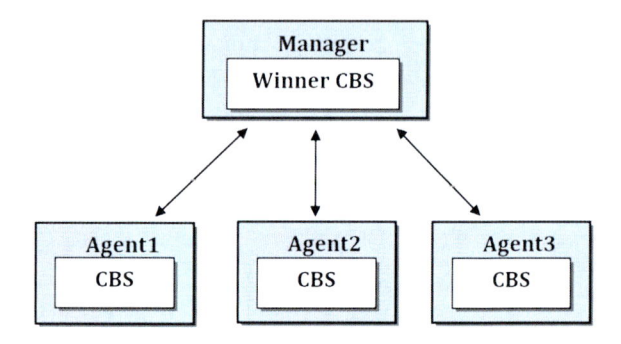

```
              ┌─────────────┐
              │   Manager   │
              │ Winner CBS  │
              └─────────────┘
              ↙      ↓      ↘
      ┌────────┐ ┌────────┐ ┌────────┐
      │ Agent1 │ │ Agent2 │ │ Agent3 │
      │  CBS   │ │  CBS   │ │  CBS   │
      └────────┘ └────────┘ └────────┘
```

The CBS trust measurement scheme is established based on the Bucket Brigade Algorithm (BBA) [7]. However, in the present CBS model, the bidding part of BBA is modified by incorporating HA_i (i.e., the belief degree), as shown in Eq. (41.2). Each agent participates in the auction by starting with a specific net worth (amount of money) defined by the manager. The net worth is called strength (S). Each agent then declares its maximum bid, which is proportional to its strength. The bid ratio (B) is affected strongly by the HA_i element which represents the state of knowledge of the agent, and is calculated based on the historical information of each agent. The information is taken during prediction, and is based on the accuracy rate of each hyperbox, i.e. Eq. (41.2).

All participating agents start bidding with an initial strength (i.e., $S = 100$ in this study), and all agents start their bids after receiving the test sample. It is found from many trials that a suitable value for the bid coefficient (C_{bid}) is 0.01. Each classifier places its bids in proportion to its strength as follows [7]:

$$B_k = C_{bid} S_k \tag{41.3}$$

where k is the agent number, $k = 1, 2, \ldots, n$. In this study, Eq. (41.3) is used as the reward and penalty scheme to update the strength, as embedded in Eq. (41.5). To compute the trust element, Eq. (41.3) is further modified as follows:

$$CBS_k = C_{bid}(S_k + HA_{ki}) \tag{41.4}$$

The agent that makes a correct prediction with the largest CBS (the highest trust value) is chosen as the winner by the manager, based on the sealed bid-first price auction strategy. Given the final decision, the strength of each agent is updated, by using the agent's bid as defined in Eq. (41.3). The updated value is positive if an agent gives a correct prediction, while it is negative in the case of an incorrect prediction. Equation (41.5) is used to update the strength [7].

$$S_i(t+1) = S_i(t) - P_i(t) - T_i(t) + R_i(t) \tag{41.5}$$

where P is penalty, T is tax, R is reward, i is the classifier index, and t is the time step. Both the strength and HA values are updated in each step; whereby the HA

element is updated according to the correct and incorrect predictions. The final test accuracy rate of the MACS-CBS model is computed as follows:

$$Test\ accuracy = \left(\frac{\sum_{j=1}^{n} CPTS_j}{\sum_{j=1}^{n} CPTS_j + ICPTS_j}\right) * 100 \qquad (41.6)$$

where *CPTS* and *ICPTS* are the numbers of correct and incorrect predicted test samples, respectively. Since test accuracy [Eq. (41.6)] is updated online for each test sample, it also serves as a tie-breaker to determine the winner in the case where all CBS values are equal. In this case, the agent with the highest test accuracy rate is selected as the final winner.

41.4 Experimental Study

Two cases studies were conducted to evaluate the applicability of MACS-CBS to medical data classification tasks. Firstly, the Wisconsin Breast Cancer (WBC) benchmark problem obtained from the UCI machine learning repository [15] was used. Then, a real medical diagnosis task was used. The overall results were obtained by using the bootstrap method [6], and the 95 % confidence levels of the averages were estimated.

41.4.1 Case Study I

The WBC data set was used to assess the robustness of the MACS-CBS model, whereby the results were compared with those published in the literature. In order to have a fair performance comparison, the experiments employed the same numbers of test samples and followed the same experimental procedures as in Hota [9]. A total of 75 % of the data samples were used for learning (384 samples for training and 140 samples for prediction), with the remaining data samples used for test. Tables 41.1 and 41.2 summarize the results from different methods published in Hota [9] and those from MACS-CBS, respectively.

As can be observed in Table 41.2, there was no overlap between the 95 % confidence intervals of MACS-CBS results and those from various methods reported in Hota [9], as well as FMM from this study. This indicates that MACS-CBS is statistically better (at the 95 % confidence level) than FMM and various models in Hota [9].

Table 41.1 Test accuracy rates of the WBC data set. The FMM and MACS-CBS results (in italics) are from this study, while the rest are from Hota [9]

Methods	Testing accuracy
EBPN	95.45
LVQ	92.04
CPN	93.75
CART	93.18
SVM	95.45
C5.0	92.61
Bayesian	93.18
Ensemble of Bayesian and C5.0	95.45
FMM	*96.57*
MACS-CBS	*97.99*

Table 41.2 The bootstrap averages of MACS-CBS

Value	Test accuracy (%)
Mean value	97.99
95 % Confidence level	(97.82–98.17)

41.4.2 Case Study II

In this case study, MACS-CBS was applied to tackle a real medical diagnosis task. A data set containing real medical records of Acute Coronary Syndrome (ACS) patients from a hospital in Malaysia was used. The data set contained 118 samples (patient records). After consultation with medical experts, a total of 16 features comprising physical symptoms (e.g. sweating, chest pain), background information (hypertension, smoking), and EEG signal representations, were used for the prediction of ACS. All the features, except one, were represented by 1 when the symptoms were present and by 0.5 when the symptoms were absent, while missing values were represented by 0. The only continuous feature (i.e., duration of pain) was normalized between 0 and 1. Given each input feature extracted from a patient record, the task was to provide a prediction of the diagnosis, i.e. whether the patient was suffering from ACS or otherwise.

To evaluate the performance of MACS-CBS, 80 % of data samples were used for training (60 % for training and 20 % for prediction), with the remaining (20 %) used for testing. Since the ACS data samples mainly contained three types of values, i.e. 0, 0.5, and 1, three levels of θ (hyperbox size) $0.0 < \theta < 0.5, \theta = 0.5, 0.5 < \theta < 1.0$ were of interest. As such, a series of runs with θ set to 0.4, 0.5, and 0.6 was conducted. The experiment was repeated 10 times for each step, and the overall results obtained by the bootstrap method, are shown in Table 41.3.

In general, MACS-CBS was able to yield better performances as compared with FMM. The main reason is that CBS is built from two useful elements. The first element is the hyperbox accuracy rate which represents the knowledge of the

Table 41.3 The overall results of FMM and EFMM for the ACS data Set

θ	MACS-CBS			FMM		
	Min	Mean	Max	Min	Mean	Max
0.4	76.66	81.30	87.08	55	62.53	70
0.5	83.75	85.02	86.25	57.50	65.06	75
0.6	73.75	85.24	86.66	76.66	81.30	87.08

agent, whereby it indicates the degree of the belief element, while the second element represents the strength (S). Both elements are useful indicators that help and enable MACS-CBS to make accurate predictions.

41.5 Conclusions

The application of the MACS-CBS model to medical data classification tasks has been evaluated. The CBS trust measurement, which is devised based on strength (money) and reputation (hyperbox accuracy) of each FMM agent, has been shown to be useful for the MACS model. Trust is built from meaningful elements that are linked with the FMM agents, allowing the CBS trust measurement to improve the performances of the overall MACS model. The first price method has been adopted for MACS to determine the winning agent. The MACS-CBS model has been evaluated using the WBC benchmark problem and a real ACS problem. The results have demonstrated the superior performances of MACS-CBS as compared with those from other existing models, in terms of accuracy and stability; hence ascertaining its usefulness in undertaking medical data classification tasks.

As the use of CBS with the original FMM agents has shown promising results, further work is focused on improving the performance of MACS-CBS by using other variants of FMM. It would be interesting to evaluate the resulting model using classification problems from different domains too.

Acknowledgments The authors gratefully acknowledge funding from the Fundamental Research Grant Scheme (No. PELECT/203/6711229) for supporting this project.

References

1. Al-Shayea QK (2011) Artificial neural networks in medical diagnosis. Int J Comput Sci Issues (IJCSI) 8(2):150–154
2. Amato F, López A, Peña-méndez EM, Vaňhara P, Hampl A, Havel J (2013) Artificial neural networks in medical diagnosis. J Appl Biomed 11:47–58
3. Benaim M, Samuelides M (1991) Arigorous result about the off-line learning approximation. International joint conference on neural networks (IJCNN), vol 2, p 979
4. Bentahar J, Khosravifar B (2008) Using trustworthy and referee agents to secure multi-agent systems. The 5th international conference on information technology: new generations (ITNG), pp 477–482

5. Boukerche A, Li X (2005) An agent-based trust and reputation management scheme for wireless sensor networks. IEEE global telecommunications conference (GLOBECOM), vol 3. p 5
6. Efron B (1979) Bootstrap methods: another look at the jackknife. Ann Stat 7(1):1–26
7. Goldberg DE (1989) Genetic algorithms in search, optimization and machine learning. Addison-Wesley, Boston, Longman Publishing Co., Inc, Harlow
8. Gupta S, Sarkar A, Pramanik I, Mukherje B (2012) Implementation scheme for online medical diagnosis system using multi agent system with JADE. Int J Sci Res Publ 2(6):1–6
9. Hota HS (2013) Diagnosis of breast cancer using intelligent techniques. Int J Emerg Sci Eng (IJESE) 1(3):45–53
10. Khosravifar B, Gomrokchi M, Bentahar J, Thiran P (2009) Maintenance-based trust for multi-agent systems. International conference on autonomous agents and multiagent systems, vol 2. pp 1017–1024
11. Mazurowski MA, Habas PA, Zurada JM, Lo JY, Baker JA, Tourassi GD (2008) Training neural network classifiers for medical decision making: the effects of imbalanced datasets on classification performance. Neural Netw 21(2–3):427–436
12. Mohammed MF, Lim CP, Quteishat A (2012) A novel trust measurement method based on certified belief in strength for a multi-agent classifier system. Neural Comput Appl. doi:10.1007/s00521-012-1245-2
13. Mui L, Mohtashemi M, Halberstadt A (2002) A computational model of trust and reputation. Hawaii international conference on system sciences (HICSS), pp 2431–2439
14. Nakashima T, Uenishi T, Narimoto Y (2010) Off-line learning of soccer formations from game logs. IEEE conference on world automation congress (WAC), pp 1–6
15. Newman DJ, Asuncion A, Hettich S et al (2011) UCI repository of machine learning databases. Accessed, available at http://archive.ics.uci.edu/ml/, Last visit on 25 May 2013 [Online]
16. Odeh SM, Khalil M (2011) Off-line signature verification and recognition: neural network approach. International symposium on innovations in intelligent systems and applications (INISTA), pp 34–38
17. Oprea M (2009) MEDICAL_MAS: an agent-based system for medical diagnosis medical diagnosisin. In: Iliadis L, Vlahavas I, Bramer M (eds) IFIP International federation for information processing, artificial intelligence applications and innovations III, vol 296. Springer, Boston, pp 225–232
18. Quteishat A, Lim CP, Saleh JM, Tweedale J, Jain LC (2011) A neural network-based multi-agent classifier system with a Bayesian formalism for trust measurement. Soft Comput 15(2):221–231
19. Quteishat A, Lim CP, Tweedale J, Jain LC (2009) A neural network-based multi-agent classifier system. Neurocomputing 72(7–9):1639–1647
20. Shemshadi A, Soroor J, Tarokh MJ (2008) An innovative framework for the new generation of SCORM 2004 conformant e-learning systems. International conference on information technology: new generations, pp 949–954
21. Simpson PK (1992) Fuzzy min–max neural networks. IEEE Trans Neural Netw 3(5):776–786
22. Simpson PK (1993) Fuzzy min–max neural networks—part 2: clustering. IEEE Trans Fuzzy Syst 1(1):32
23. Tan SC, Lim CP, Tan KS, Navarro JC (2009) An evolutionary artificial neural network for medical pattern classification, lecture notes in computer science on neural information processing: part II, vol 5864. pp 475–482
24. Wang P, Zhangz (2005) A computation trust model with trust network in multi-agent systems. International conference on active media technology (AMT), pp 389–392
25. Wu F, Zilberstein S, Chen X (2011) Online planning for multi-agent systems with bounded communication. Artif Intell 175(2):487–511
26. Yang Q, Shieh JS (2008) A multi-agent prototype system for medical diagnosis. 3rd international conference in intelligent system and knowledge engineering, vol 1. pp 1265–1270

Part IV
Electronic Design and Applications

Chapter 42
High-Speed Transmitter Designs for DDR3 SDRAM Memory Interfaces

Lim Zong Zheng, Mohd Tafir Mustaffa and Ch'ng Siew Sin

Abstract This work presents two high-speed transmitter designs for 2.4 Gbps Double Data Rate Generation 3 (DDR3) memory interfaces. The transmitters are designed using 45-nm CMOS process. Moreover, output slew rate of both transmitters is controlled at 4–6 V/ns, while their output impedance can be programmed between 20, 30 and 40 Ω, respectively. Output slew rate and impedance of both the transmitters are calibrated across process, voltage and temperature (PVT) variations. A comparison between Multi-Module Transmitter (MMTX) and Single-Module Transmitter (SMTX) is also presented with low pad capacitance and small far-end eye jitter as the main figures of merit.

Keywords DDR3 · Memory interface · Multi-module transmitter · Single-module transmitter

42.1 Introduction

The need for greater memory bandwidth to boost the computer performance has driven system memory evolution to Double Data Rate Synchronous Dynamic Read Access Memory (DDR SDRAM) technologies [1]. As operation frequency is increasing, Inter-Symbol Interference (ISI) has becomes significant that degraded

L. Z. Zheng (✉) · C. S. Sin
School of Electrical and Electronic Engineering, Universiti Sains Malaysia,
Engineering Campus 14300 Nibong Tebal, Penang, Malaysia
e-mail: zzlim2@hotmail.comzong.zheng.lim@intel.com

C. S. Sin
e-mail: cssin87@hotmail.com

M. T. Mustaffa
School of Electrical and Electronic Engineering,
Universiti Sains Malaysia, Nibong Tebal, Malaysia
e-mail: tafir@eng.usm.my

signal noise margin and increased signal jitter; thereby, causing data detection at receiver become very challenging [2, 3]. ISI can be illustrated as a symbol corrupted by another symbol traveling on the channel at an earlier time. When a stream of digital symbols are transmitted over a transmission channel, the output pad capacitance of the transmitter in combination with the skin-effect resistance in transmission channel can form a low-pass filter that attenuates high frequency components of a symbol; resulting ISI between successive symbols [4, 5]. Equalization techniques and pad capacitance reduction techniques have been proposed to overcome the ISI from transmission channel and transmitter structure, respectively [6–9]. Besides, ISI also can be resulted from signal reflection due to termination mismatch between transmission channel and transmitter [10]. Several impedance calibration techniques were proposed to control the transmitter impedance to match channel impedance; thereby, minimizing reflections and ISI across PVT variations [9–11]. This paper presents two types of transmitter design: MMTX and SMTX with impedance control technique to reduce ISI impacts. Both transmitters impedance can be programmed between 20, 30 and 40 Ω and their output slew rate can be controlled at 4–6 V/ns. This paper is organized as follows: Sect. 42.2 presents the design of SMTX and MMTX, Sect. 42.3 demonstrates the simulation results and comparison between both transmitters and Sect. 42.4 concludes this work.

42.2 Transmitter Design

42.2.1 MMTX Design

Figure 42.1a depicts the block diagram of MMTX with slew rate control. MMTX consists of six TX modules with each TX module decomposes into three TX segments. The reason of dividing a TX module into three TX segments is to achieve output slew rate control by means of driver segmentation. Each segment is activated in sequence through the control signals that delayed by delay cells with interval of ΔT [5]. Moreover, output impedance of each TX segment is calibrated at 360 Ω (by using ZQ Calibration Block in [8]) which gives an equivalent impedance of 120 Ω per TX module when all the three TX segments are activated. Therefore, output impedance value of 20, 30 and 40 Ω is possible by activating appropriate number of TX modules [12].

Additionally, Fig. 42.1b shows the structure of TX segment. It consists of an anchor leg and 4-bit binary weights dynamic legs. In each leg, there is additional transistor stacked into pull-up and pull-down branch that is biased by PBIAS and NBIAS, respectively. This is electrical over-stress protection on transistor gate to avoid any transistor operates at voltage higher than its sustainable voltage [13, 14]. Besides, in Fig. 42.2, Anchor legs are always turn on (depending on data) to provide impedance close to the transmission channel impedance, while dynamic

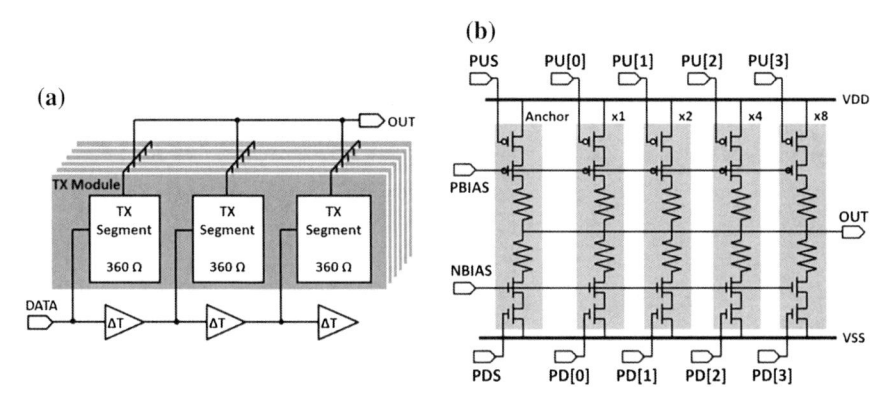

Fig. 42.1 **a** MMTX design with slew rate control. **b** Structure of one TX segment in MMTX

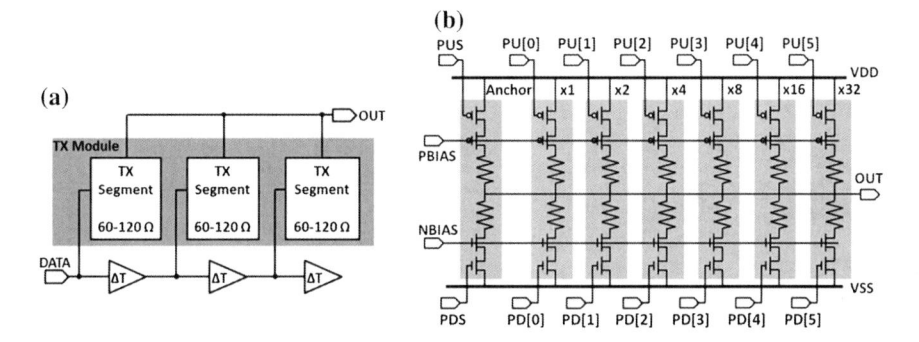

Fig. 42.2 **a** SMTX design with slew rate control. **b** Structure of one TX segment in SMTX

legs are turned-on to make the output impedance match to transmission channel impedance across PVT corners. The pull-up and pull-down branch in dynamic legs are controlled by ZQ Calibration Block through digital code adjustment using PU[3:0] and PD[3:0].

42.2.2 SMTX Design

The block diagram of SMTX with slew rate control is shown in Fig. 42.2a. SMTX consists of single TX module that is decomposed into three TX segments for slew rate control described in Sect. 42.2.1. Moreover, the impedance of each TX segment is controlled either at 60, 90 or 120 Ω by using Impedance Calibration Block (ICB) for an equivalent impedance of 20, 30 or 40 Ω per TX module when activating all three TX segments. Figure 42.2b shows the structure of TX segment

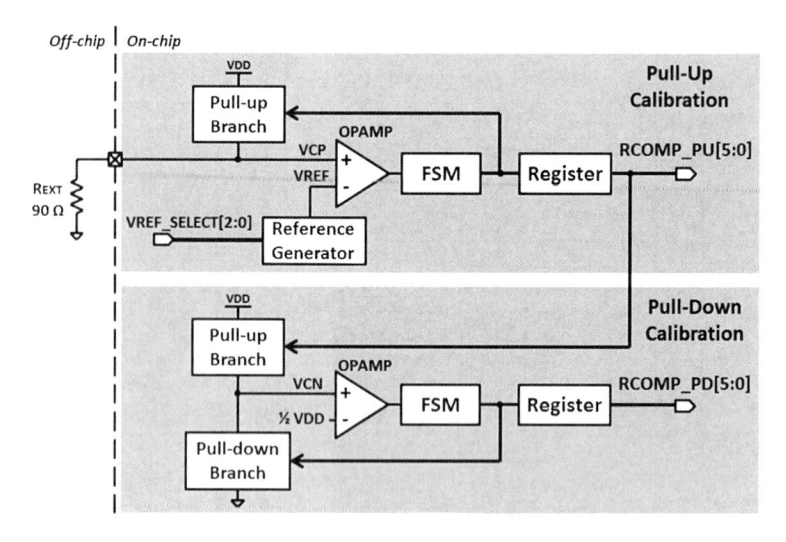

Fig. 42.3 Main components in ICB with reference generator

that is similar to Fig. 42.1b; but, it has 6-bit binary weights dynamic legs to reach output impedance range of 60–120 Ω through the digital code PU[5:0] and PD[5:0] adjusted by ICB.

Transmitter impedance is calibrated by using ICB shown in Fig. 42.3. Pull-up branch in Pull-up Calibration block is connected to a 90 Ω external resistor and is calibrated by comparing the output voltage VCP with reference voltage VREF. For a high-gain operational-amplifier, the potential at its positive terminal must equal to the potential at negative terminal which establishes Eq. (42.1) to show the relationship between VREF, VCP, external resistance R_{EXT} and pull-up resistance R_{PU_SEG} at steady state:

$$\text{VREF} = \text{VCP} = [R_{EXT}/(R_{EXT} + R_{PU\_SEG})] \times \text{VDD} \qquad (42.1)$$

Thus, by substituting R_{PU_SEG} with 60, 90 or 120 Ω, the desired VREF can be obtained. The VREF is generated by reference generator and can be varied through the VREF_SELECT[2:0] setting to calibrate the resistance of pull-up branch to the desired value at 60, 90 or 120 Ω per TX segment. Then, Finite State Machine (FSM) will adjust the six-bit dynamic legs in the pull-up branch until VCP matches VREF and the corresponding digital codes are stored in register. These codes are used to activate another pull-up branch that becomes reference resistance for the pull-down branch in Pull-down Calibration block. Similarly, pull-down branch is calibrated until its VCN is matched to reference voltage at $^1/_2$ VDD, which implies the resistance of pull-up branch is the same as the resistance of pull-down branch. Again, the digital codes of pull-down branch are stored in another register and both the pull-up and pull-down digital codes are ready to be delivered to the pre-driver of transmitters.

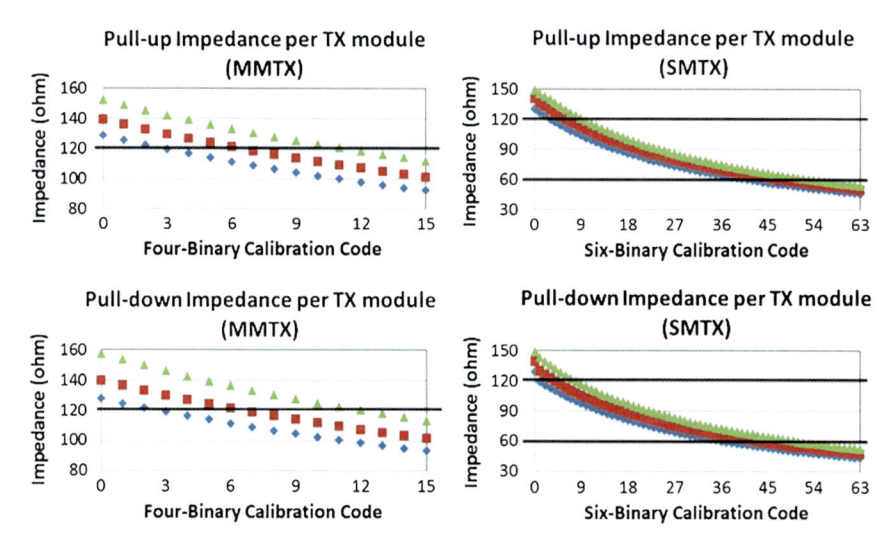

Fig. 42.4 Pull-up and pull-down impedance range per TX module in MMTX and SMTX

42.3 Simulation Results and Discussions

Both MMTX and SMTX are simulated at worst (slow transistor, 110 °C and VDD = 1.575 V), typical (typical transistor, 50 °C and VDD = 1.50 V) and best (fast transistor, 0 °C and VDD = 1.425 V) condition. In Fig. 42.4, pull-up and pull-down impedance per TX module in MMTX are sufficient to reach 120 Ω across worst (Green Δ), typical (Red □) and best (Blue ◇) condition. For SMTX, pull-up and pull-down impedance per TX module can also reach 60–120 Ω across worst, typical and best condition.

Besides, Fig. 42.5a shows the 2.4 Gbps far-end eye diagram of MMTX and SMTX with 4-inch memory channel model and 60 Ω center-tap termination at far-end [5]. For 40 Ω MMTX, the simulated eye height and jitter are 316.4 mV and 75.6 ps. Output impedance can be reduced from 40 to 20 Ω when higher eye height is needed. Thus, the eye height is improved to 447.8 mV for 20 Ω MMTX at the cost of larger jitter at 81.9 ps due to reflections caused by impedance mismatched with channel impedance. Besides, simulated eye height and jitter for 40 Ω SMTX in Fig. 42.5b are 436.2 mV and 52.1 ps, while 20 Ω SMTX has simulated eye height and jitter of 589.4 mV and 59.1 ps. It can be noticed that the eye opening in MMTX is smaller. This is because the pad capacitance in MMTX is larger due to larger TX module numbers as compared to single TX module in SMTX, although MMTX only use four-bit dynamic legs in TX segment of every TX module as compared to six-bit dynamic legs in SMTX. Large pad capacitance from transmitter in combination with skin-effect resistance in transmission channel can form a low-pass filter that attenuates high frequency components of a symbol,

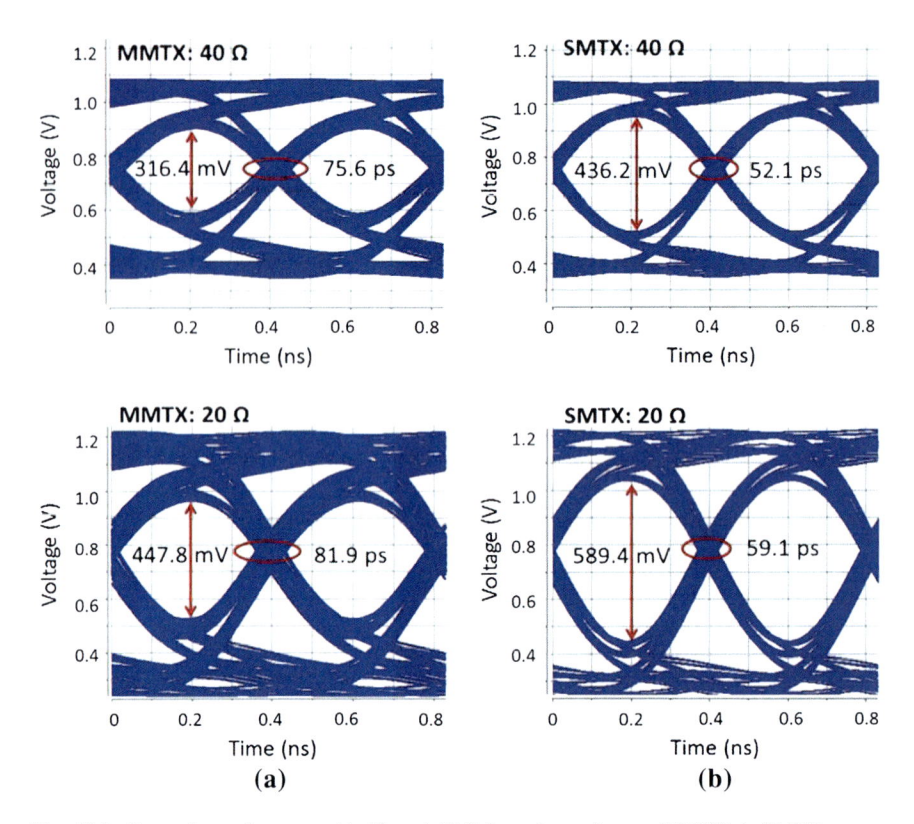

Fig. 42.5 Far-end eye diagram with 20 and 40 Ω impedance from **a** MMTX, **b** SMTX

thereby resulting more severe ISI between successive symbols. Thus, it was shown that SMTX is less sensitive to ISI through the controlled impedance and merit of low pad capacitance in its structure.

42.4 Conclusion

This paper presents two transmitter designs, MMTX and SMTX for 2.4 Gbps DDR3 memory interfaces. By using impedance control technique, the output impedance of both transmitters can be programmed between 20, 30 and 40 Ω across PVT variations. It was shown that SMTX has better far-end eye opening and is less sensitive to ISI impacts as compared to MMTX. This is the advantage of low pad capacitance in single module over multiple module structure, at the expense of slightly silicon area overhead due to the employment of reference generator in ICB block.

References

1. Jacob B, Ng SW, Wang DT (2007) Memory systems: cache, DRAM, disk. Morgan Kaufmann, San Francisco
2. Lee J, Lee S, Nam S (2010) Multi-slot main memory system for post DDR3. IEEE Trans Circuits Syst II: Express Briefs 57:334–338
3. Dally WJ, Poulton J (1998) Digital systems engineering. Cambridge University Press, Cambridge
4. Zhang L, Wilson JM, Bashirullah R, Luo L, Xu J, Franzon PD (2007) Voltage-mode driver preemphasis technique for on-chip global buses. IEEE Trans Very Large Scale Integr VLSI Syst 15:231–236
5. Lim, ZZ, Mustaffa MT, Navaratnam N (2012) A 2.4 Gbps transmitter with programmable de-emphasis scheme for DDR3 memory interface. In: IEEE international conference on 4th intelligent and advanced systems, Kuala Lumpur, vol 2. pp 713–718
6. Liu J, Lin X (2004) Equalization in high-speed communication systems. IEEE Circuits Syst Mag 4:4–17
7. Wong JKL, Hatamkhani H, Mansuri M, Yang KCK (2004) A 27-mW 3.6-Gb/s I/O transceiver. IEEE J Solid-State Circuits 39:602–612
8. Park C, Chung H, Lee YS, Kim J, Lee JJ, Chae MS, Jung DH, Choi SH, Seo SY, Park TS, Shin JH, Cho JH, Lee S, Song KW, Kim KH, Lee JB, Kim C, Cho SI (2006) A 512-Mb DDR3 SDRAM prototype with C_{IO} minimization and self-calibration techniques. IEEE J Solid-State Circuits 41:831–838
9. Kalyan G, Srinivas MB (2010) An efficient ODT calibration scheme for improved signal integrity in memory interface. In: IEEE Asia Pacific conference on circuits and systems, Kuala Lumpur, pp 1211–1214
10. Heidar D, Dessouky M, Ragaie HF (2007) Comparison of output drivers for high-speed serial links. In: IEEE international conference on microelectronics, Cairo, pp 329–332
11. Gabara TJ, Knauer SC (1992) Digitally adjustable resistors in CMOS for high-performance applications. IEEE J Solid-State Circuits 27:1176–1185
12. Ch'ng SS, Marzuki A, Tan SB (2012) Configurable output driver with programmable on-chip impedance supporting wide range data rates. In: IEEE international conference on 4th intelligent and advanced systems, Kuala Lumpur, vol 2. pp 801–806
13. Lim ZZ, Mustaffa MT (2011) An output driver with high-speed level shifter for 1.5 V applications using a 45 nm CMOS process. In: Electrical and electronic postgraduate colloquium, Pahang
14. Mishra NK, Jain M, Le P, Mukherjee S, Sendhil A, Amirkhany A (2011) An output structure for a bi-modal 6.4-Gbps GDDR5 and 2.4-Gbps DDR3 compatible memory interface. In: IEEE custom integrated circuits conference, San Jose, pp 1–4

Chapter 43
Wideband Low Noise Amplifier Design for 2.3–2.4 GHz WiMAX Application

Nor Zaihar Yahaya, Shahrul Yazid, Anwar Osman and Helmi Huzairi

Abstract This paper describes the design of a low noise amplifier using negative feedback topology for WiMAX 2.3–2.4 GHz base station application. The negative topology employs an enhancement mode feedback inductor and advanced GaAs PHEMT transistor which yields excellent performance on the critical RF parameters such as noise, gain and linearity. By using the negative feedback topology the design is compact and uses fewer components compared to other state-of-art wideband low noise amplifier designs. The low noise amplifier has been designed on a Rogers 4003C substrate PCB and produced a gain of 12 dB, noise less than 0.7 dB and IIP3 greater than 13 dBm. The design is also matched at both input and output for 50 Ω and proven to be unconditionally stable across the frequency range.

Keywords WiMAX · Wideband LNA · Negative feedback

N. Z. Yahaya
Electrical and Electronics Engineering Department, Universiti Teknologi Petronas, Seri Iskandar, Perak, Malaysia
e-mail: norzaihar_yahaya@petronas.com.my

S. Yazid · A. Osman (✉) · H. Huzairi
Spectre Solutions, Technical Support, George Town, Penang, Malaysia
e-mail: anwarfaizd@spectresolution.com

S. Yazid
e-mail: shahrulyazid@spectresolution.com; mohdshahrul@spectresolution.com

H. Huzairi
e-mail: mohdhelmi@spectresolution.com

43.1 Introduction

In WiMAX base station application, the receiver front end (RXFE) requires a good signal to noise (SNR) ratio and linearity performances. The low noise amplifier (LNA) is a critical part of the WiMAX RXFE, and must be able to provide a constant gain, low noise, high linearity and unconditionally stable in order to achieve the required SNR and linearity requirement of the RXFE [1]. To meet these requirements, however it requires design trade off and technology challenges for the LNA to achieve the optimum performances in terms of the RF parameters and design size. The negative feedback enhancement mode technology is chosen for the wideband LNA design due to its ability to produce low noise and high linearity performance while utilizing only a single supply. Designing the enhancement mode LNA without a feedback can produce a good performance LNA but difficult to achieve wideband with only a single amplifier. This is highlighted by Peng et al. [2], which describe the designing of good wideband LNA in the 2.5–5 GHz range utilizing the enhancement mode, albeit in a two-stage design topology. The difficulty in designing conventional LNA is that it requires multi-stage LNA and many discrete components to achieve wideband ability. A negative feedback drain-to-gate and/or source resistor as explained by Pozar [3] can produce a wideband range with only a single stage design. Pozar also highlighted the challenges in designing for achieving best noise and best gain (matching) LNA. To design for best gain, the LNA must be match for the conjugate of the input impedance, namely $Z_{in}*$, to get the optimum matching at input. Similarly for noise, the LNA must be match at the optimum source reflection for noise, Γ_{opt}, to achieve optimum noise. These two optimum points are not located on the same points in the Smitch Chart, and are usually far apart, making it difficult to get the optimum design for the LNA. Apart from achieving wideband performance, the source inductor feedback will also reduce the distance between these two optimum points, thus enabling the designer to better design for optimum LNA performance [3].

GaAs PHEMT transistor is selected for this design as it is potentially the best amplifier transistor for realizing low noise performance at microwave frequencies and is easily available from the market. A similar application but different topology design has been done by Mandeep et al. [4], which highlighted the benefit of using the GaAs PHEMT transistor for WiMAX application.

43.2 Negative Feedback Enhancement Mode LNA

The negative feedback technique is a method to design a wideband LNA while at the same time maintains a singular circuit and minimum PCB space and part counts. However, one disadvantage of this topology is the feedback circuit will degrade the noise and reduce the maximum available gain of the transistor [5].

Fig. 43.1 The amplifier with feedback series and shunt resistor

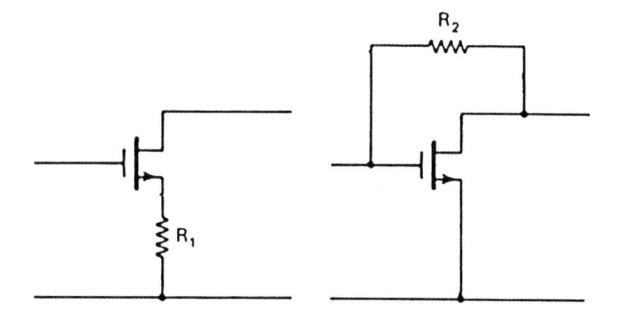

Fig. 43.2 The amplifier FET model with series source inductor

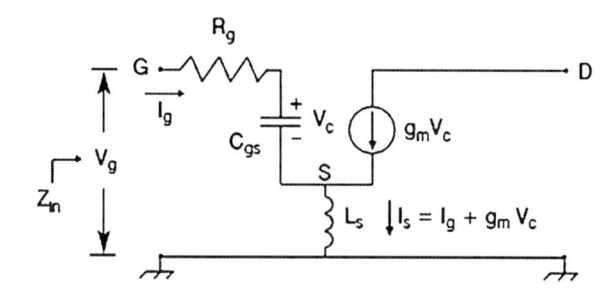

The negative feedback enhance mode is a technique which adds another series inductive element at the source of the amplifier to negate the degradation of the noise and the small signal gain of the LNA. This will make the negative feedback LNA to maintain the RF performance at the wideband frequency range and maintain low part count and PCB space as compared to other wideband topology, such as the balanced amplifier design, highlighted by Osman et al. [6]. A method of applying the negative feedback technique is by using the series and shunt resistor combination as shown in Fig. 43.1.

The addition of the drain to gate feedback resistor as proposed by Osman et al. [6] will enable the LNA to achieve wideband performance. However, the drawback of this design is the additional resistance at the input (gate) will increase the noise and reduce the overall gain. A method of enhancement mode adds a source inductance as a series feedback which reduces the noise and gain degradation of the LNA [7]. Figure 43.2 shows the small signal circuitry of the LNA with a source series inductor.

Typically, the inductance can be inserted by grounding the transistor through a short transmission line or by adding small value of inductor to the ground [7]. The adding of inductance moves the reflection coefficient required for an input conjugate match closer to Γ_{opt}. In Fig. 43.2, the voltage developed across the internal C_{gs} capacitor is

$$V_c = \frac{I_g}{SC_{gs}}, \qquad (43.1)$$

where S is the complex frequency variable

$$S = s + j\omega = j\omega. \tag{43.2}$$

Deriving the input impedance Z_{in} yields

$$Z_{in} = \frac{V_g}{I_g} = \frac{I_g R_g + V_c + I_s S L_s}{I_g}. \tag{43.3}$$

Substituting $\frac{I_g}{SC_{gs}}$ for V_c and $(I_g + g_m V_c)$ for I_s gives

$$Z_{in} = \frac{I_g R_g + I_g}{SC_{gs}} + \frac{(I_g + g_m V_c) S L_s}{I_g}. \tag{43.4}$$

From Eq. (43.4), substituting $\frac{I_g}{SC_{gs}}$ for V_c and dividing through by I_g leads to

$$Z_{in} = R_g + \frac{1}{SC_{gs}} + SL_s + g_m \frac{I_g}{SC_{gs}} \times \frac{SL_s}{I_g}. \tag{43.5}$$

Canceling out I_g in the last term yields the result

$$Z_{in} = R_g + g_m \frac{L_s}{C_{gs}} + S\left(L_s + \frac{1}{S2C_{gs}}\right). \tag{43.6}$$

Substituting $S = jw$ into the above equation gives the input impedance as a function of frequency. Hence the new transistor input impedance can be described as

$$Z_{in} = R_g + \frac{g_m L_s}{C_{gs}} + j\left(\omega L_s - \frac{1}{\omega C_{gs}}\right). \tag{43.7}$$

Based on the PHEMT capacitance and RF conductance/transconductance model explained by Wei [8], the term $\frac{g_m L_s}{C_{gs}}$ is equivalent to the FET output resistance based on the channel length modulation theorem. For simplicity this resistance is noted as R_o (positive value). Hence, Eq. (43.7) can be rewritten as

$$Z_{in} = R_g + R_o + j(Xl_s - Xc_{gs}). \tag{43.8}$$

In summary, Eq. (43.8) describes the input impedance of the FET with feedback. Without the feedback inductor, the impedance of FET without feedback can be written as

$$Z_{in} = R_g - jXc_{gs}. \tag{43.9}$$

It can be seen by Eq. (43.8) that the FET model with source inductor feedback adds resistive elements $(R_o + jXl_s)$ to the FET's input impedance, as compared with the non-feedback model, as shown from Eq. (43.9). By adding both a real resistive component R_o and a positive reactive component jXl_s, the source inductor feedback

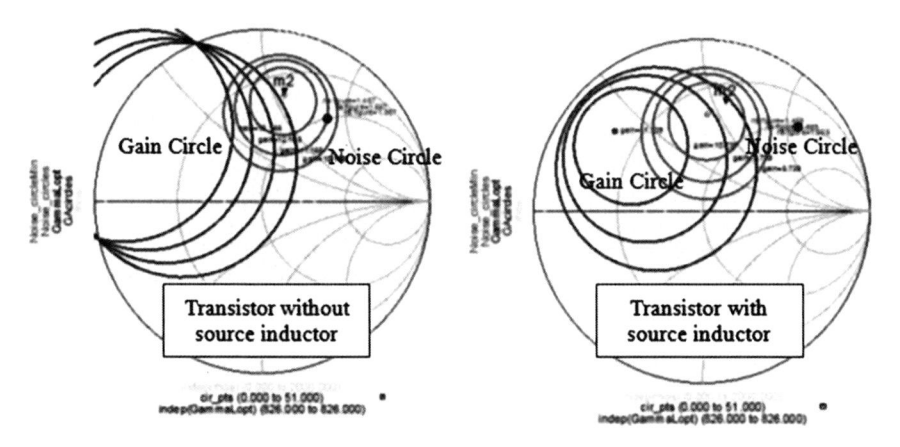

Fig. 43.3 Comparison between the Z_{in}^* and Γ_{opt} distance in a non-feedback and feedback FET

FET model moves the conjugate input impedance, Z_{in}^*, closer to the optimum reflection coefficient, Γ_{opt} [9]. As Γ_{opt} remains unchanged with the addition of source inductance, moving Z_{in}^* closer to Γ_{opt} will then reduce the distance between the noise match and the gain match optimum points in the smith chart. A graphical proof is shown in Fig. 43.3, in which a simple FET model is simulated with and without a source inductor (1 nH in this case). Hence, a better compromise between best gain and best noise of the amplifier design has been achieved in the feedback amplifier as opposed to the amplifier design without the feedback [9].

43.3 LNA Circuit Design

The LNA utilizes the Avago Technologies ATF-54143 high dynamic range, low noise enhancement mode PHEMT device as the main transistor [10]. The transistor has a SPICE model which is readily available from the manufacturer website. The schematic of the LNA is shown in Fig. 43.4. The transistor is biased at 70 mA with a single supply of 5 V, which was achieved using a voltage divider circuitry. DC block and RF choke are added to improve the isolation between the DC and RF path. The LNA are matched using lumped element as it is simple and save space. In order to achieve lowest noise as possible, the input stage of the amplifier is matched to Γ_{opt} (gamma optimum point). The negative feedback resistor is set at 10 kΩ while for the series feedback small amount of inductance is added to the source leading to better return loss while maintaining the noise near Fmin (lowest noise figure point). The design of balance LNA involves an additional step which is to design the coupler in addition to the biasing network. Optimization of the circuit is performed using the CAD software to tune the component values to meet the design requirement.

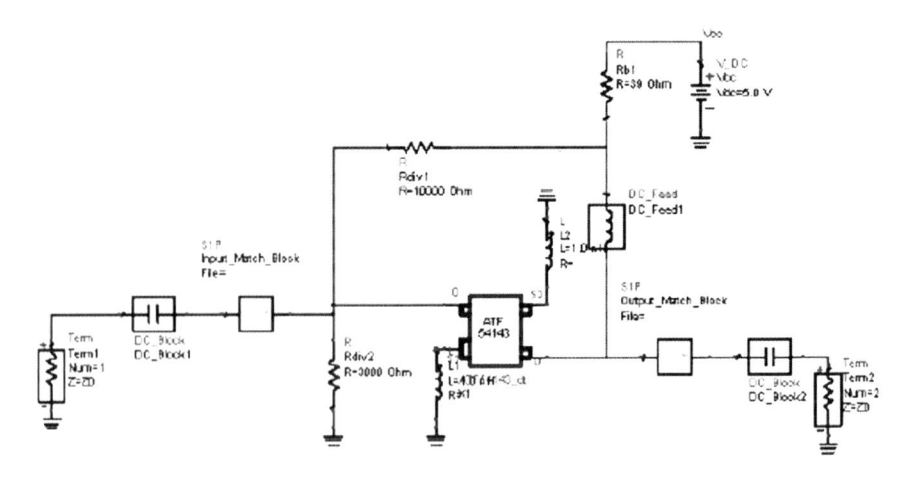

Fig. 43.4 The negative feedback LNA circuit

43.4 Method of Moments and PCB Design

After the LNA circuit of Fig. 43.4 has been finalized and simulated, the PCB RF transmission lines are then added to the schematic. A method of moments simulation have been conducted to simulate the RF PCB trace and also as a simulator for the PCB design. The method of moments simulation is a statistical estimation simulation based on the distributed parasitic components of the PCB board chosen for the circuit fabrication. This is an essential tool for the designers to estimate the final RF performance of the fabricated circuit. This simulation was done using Momentum RF simulator. The simulation is based on a standard Rogers 4003C PCB with dielectric constant of 4.78, thickness of 1.6 mm and metal thickness of 36 μm. After this simulation, the components value will be further tuned to compensate the parasitic effect of the PCB board. The resulting PCB design simulation is shown in Fig. 43.5. Once the circuit of Fig. 43.4 is implemented on PCB and has been optimized to achieve the desired specs, the PCB will then be fabricated and populated with the components and ready for the measurements.

43.5 Simulation Versus Measurement Result

The fabricated wideband LNA board is shown in Fig. 43.6. The circuit was supplied with a 5 V DC through the DC pins at the top of the board. The drain current was measured to be around 70 mA, which is proven to be consistent with the simulated performance. SMA connectors were attached at both RF input and output.

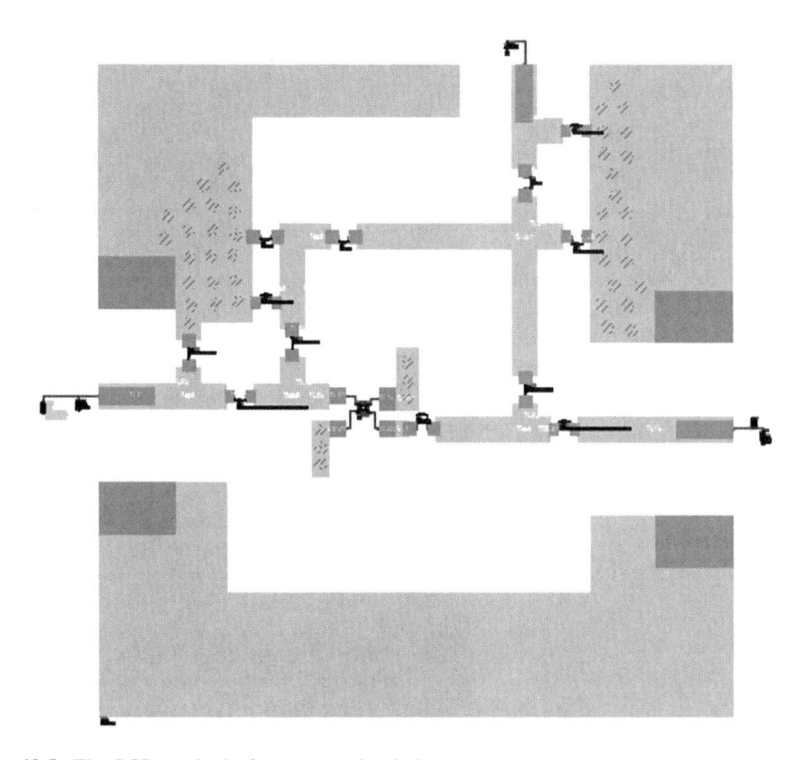

Fig. 43.5 The PCB method of moments simulation

Fig. 43.6 The fabricated LNA board

The fabricated board was measured using standard RF equipment (Vector Network Analyzers (VNA), Signal Generators, Spectrum Analyzer and Noise Figure Analyzer) to verify the performance metrics.

Fig. 43.7 The LNA simulated S-parameter results

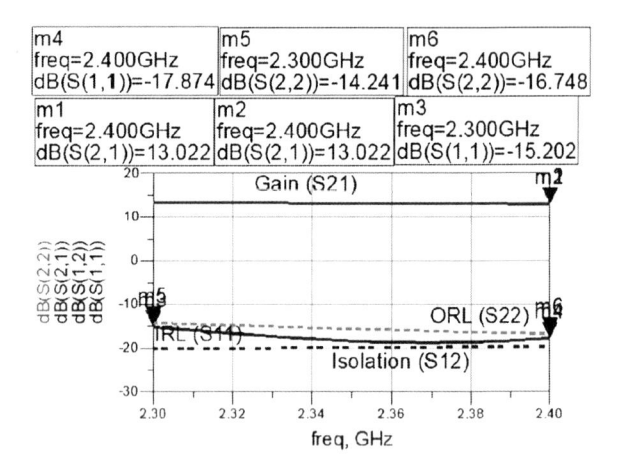

m4	m5	m6
freq=2.400GHz	freq=2.300GHz	freq=2.400GHz
dB(S(1,1))=-17.874	dB(S(2,2))=-14.241	dB(S(2,2))=-16.748
m1	m2	m3
freq=2.400GHz	freq=2.400GHz	freq=2.300GHz
dB(S(2,1))=13.022	dB(S(2,1))=13.022	dB(S(1,1))=-15.202

Fig. 43.8 The LNA measured S-parameter results

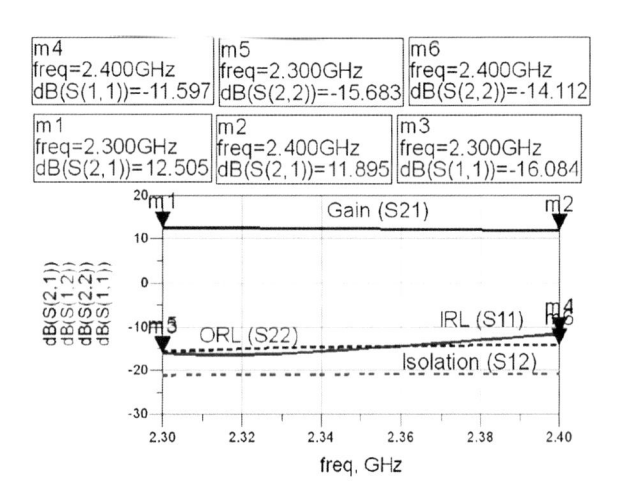

m4	m5	m6
freq=2.400GHz	freq=2.300GHz	freq=2.400GHz
dB(S(1,1))=-11.597	dB(S(2,2))=-15.683	dB(S(2,2))=-14.112
m1	m2	m3
freq=2.300GHz	freq=2.400GHz	freq=2.300GHz
dB(S(2,1))=12.505	dB(S(2,1))=11.895	dB(S(1,1))=-16.084

The simulated and measured S-parameter performances are shown in Figs. 43.7 and 43.8 respectively, while the simulated and measured NF performances are shown in Fig. 43.9. For the linearity test, Fig. 43.10 shows the simulated and measured IIP3 performance. For the IIP3 analysis, the separations between two fundamental signals are 1 MHz.

Looking at the results from Figs. 43.7, 43.8, 43.9 and 43.10, they can be observed that the simulated and measured performances are almost correlated to each other. For the return losses, the simulated input and output return loss (S_{11} and S_{22}) are better compared to the measured data at higher frequency whereas for the gain measurement, S_{21} is higher than the simulated data by 1 dB at higher frequency. This is because of the larger parasitic effect of the transistor package which has affected the simulation analysis. Parasitic components are more difficult to predict at higher frequencies (even with the SPICE model), thus explaining the slight differences of the small signal performances at the band. The difference in

Fig. 43.9 The LNA simulation versus measurement noise

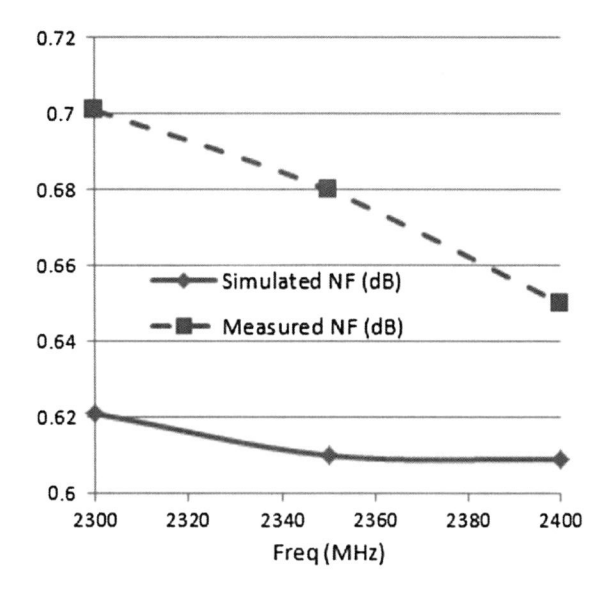

Fig. 43.10 The LNA simulation versus measurement IIP3

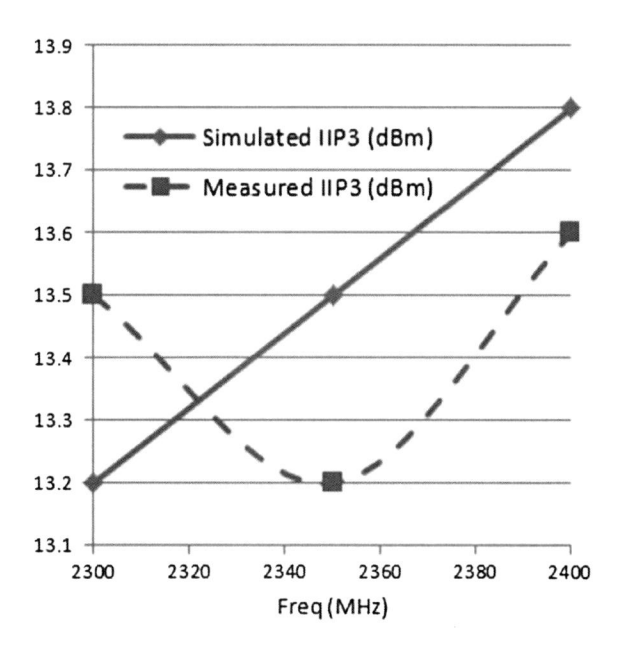

NF is expected as the environmental (temperature) effect and equipment uncertainties may contributed to the indifferences in performance. The IIP3 performance is much better correlated with differences less than 1 dB, this is because at higher power, the circuit is less affected by the environment and equipment uncertainties (smaller percentage of errors compared at lower power).

Table 43.1 Comparisons between measured results and other reported LNAs of similar work

References	Freq (GHz)	Gain (dB)	NF (dB)	IIP3 (dBm)
This work	2.3–2.4	12	0.7	13.2
Cimino et al. [11]	2.44	15.3	3.34	−10
Hsieh et al. [12]	2.6–6	13	2.5	−3.9
Wang et al. [13]	0.8–1.7	37	0.9	–
Chang et al. [14]	0.1–8	16	3.4–5.8	−9
Guan et al. [15]	2–10	10	4.6	–
Blaakmeer et al. [16]	2.5–4	10.6	4	−8

In addition, the K-factor of the design has been measured to be greater than 1 in both simulated circuit and the fabricated board. Hence, not only the design is able to achieve the desired responses, it is also unconditionally stable throughout the frequency bands. The performances are measured at a current consumption of 70 mA of drain current (I_{DS}) supplied by 5 V at the DC terminal. The measured results are summarized and compared with previous reports of LNA design in Table 43.1.

Based on the comparison of performance of Table 43.1, it can be noted that the measured LNA is much superior in the Stable Noise Figure performance across the 2.3–2.4 GHz frequencies. This has been achieved by the exercise of Eqs. (43.7) and (43.8) in which the addition of the feedback resistor and inductor move Z_{in}^* move closer to the Gamma Opt of the device (Γ_{opt}), while still maintaining a stable gain across the band.

43.6 Conclusion

The wideband LNA design, which uses the negative feedback enhancement mode topology, has been designed and developed. The design was fabricated and the measured with simulated results compared. The LNA exhibits a small signal gain of more than 12 dB, NF of less than 0.7 dB, IIP3 greater than 13 dBm and has been maintained to be unconditionally stable throughout the frequency of interest. The benefits of this feedback design are the consistency of its performances throughout the wideband frequency range, very space required for PCB fabrication and very few part counts due to only a single transistor is needed for the design. Overall, the performance of the wideband LNA for 2.3–2.4 GHz application has met the targeted performance.

Acknowledgements The authors would like to thank Universiti Teknologi PETRONAS Malaysia funded under UTP-STIRF internal grant no. 36/2011 for providing financial support towards publication of this work.

References

1. Osman AF, Noh NM (2012) Wideband LNA design for SDR radio using balanced amplifier topology. In: 4th Asia symposium on quality electronic design (ASQED), pp 86–90
2. Peng YY, Lu KJ, Sui WQ (2012) A low power 2.5–5 GHz low-noise amplifier using 0.5-μm GaAs pHEMT technology. J Semiconductors 33(10):105001-1–105001-4
3. Pozar DM (2001) Microwave and RF design of wireless systems, 3rd edn. Wiley, New Jersey
4. Mandeep JS, Abdullah H, Nitesh R (2010) A compact, balanced low noise amplifier for WiMAX base station applications. Microwave J 53:84–90
5. Henkes DD (1998) LNA design uses series feedback to achieve simultaneous low input VSWR and low noise. RF Des 10:26–32
6. Osman AF, Noh NM (2012) Wideband low noise amplifier design for software defined radio at 136–941 MHz. In: 4th international conference on intelligent and advanced system (ICIAS), pp 232–236
7. Gonzalez G (1984) Microwave transistor amplifiers analysis and design. Prentice Hall, New Jersey, pp 80–87
8. Wei CJ (2011) Capacitance and RF-conductance/transconductance look-up table based pHEMT model. In: Asia Pacific microwave conference proceedings (APMC) pp 1246–1249
9. Liao S (2001) Microwave circuit analysis and amplifier design. Prentice-Hall, Englewood Cliffs
10. Avago Technologies Datasheet (2008) ATF-54143 low noise enhancement mode pseudomorphic HEMT in a surface mount plastic package. Avago Technologies, USA, Tech. AV01-0602EN, Aug 5 2008
11. Cimino M, Lapuyade H, Deval Y, Taris T, Bégueret JB (2008) Design of a 0.9 V, 2.45 GHz self-testable and reliability-enhanced CMOS LNA. IEEE J Solid-State Circuits 43(5):1187–1194
12. Hsieh JY, Wang T, Lu SS (2009) Wideband low noise amplifier by LC load reusing technique. Electron Lett 45(25):1278–1280
13. Wang ZX, Ming YW, Hong PJ, Xin CJ, Bo XW, Chen S (2008) The design of SiGe HBT balanced broadband low noise amplifier. In: 2008 Beijing microwave and millimeter wave technology digest, pp 233–236
14. Chang TY, Chen JH, Rigge L, Lin JS (2008) A packaged and ESD protected and inductorless 0.1 to 8 GHz wideband CMOS LNA. IEEE Microwave Compon Lett 18(6):416–418
15. Guan X, Nguyen C (2006) Low power consumption and high gain CMOS distributed amplifiers using cascade of inductively coupled common source gain cells for UWB systems. IEEE Trans Microwave Theory Tech 54(8):3278–3283
16. Blaakmeer SC, Klumperink EAM, Leenaerts DWM, Bram N (2006) A wideband noise canceling CMOS LNA exploiting a transformer. In: IEEE radio frequency integrated circuits symposium digest, 2006, pp 137–140

Chapter 44
Development and Implementation of Synchronous DC–DC Buck Converter for Photovoltaic Power Generation

Muhammad Hafeez Mohamed Hariri, Norizah Mohamad and Syafrudin Masri

Abstract This paper presents the development and implementation of non-isolated synchronous DC–DC buck converter for photovoltaic power generation. The intention in designing synchronous DC–DC buck converter is to provide robust maximum power point tracker acting as an intermediate between the photovoltaic module and load. The dead-time interval and bootstrap technique are implemented in this topology. Experimental results demonstrate that the proposed prototype achieved an average of 80 % efficiency with maximum output power of 60 W at a constant input voltage of 20 V.

Keywords Buck converter · MPPT · Dead-time · Bootstrap circuits

44.1 Introduction

In recent decades, research on the use of photovoltaic as an alternative source of energy has become prominence in the field of electrical engineering [1]. A typical photovoltaic energy system compromises DC value at its output side. It is a promising technology and future energy demand in a sustainable and environmentally friendly way. Photovoltaic (PV) is made up of cells which are wired together into a module. A number of these modules can be arranged together to form PV array and produce a much higher voltage. The temperature surface area of

---

M. H. M. Hariri · N. Mohamad (✉) · S. Masri
School of Electrical and Electronics Engineering, Universiti Sains Malaysia,
Nibong Tebal, Penang, Malaysia
e-mail: norizah@eng.usm.my

M. H. M. Hariri
e-mail: aphieezsir@gmail.com

S. Masri
e-mail: syaf@eng.usm.my

a cell and the intensity of the light hitting the module determined the amount of total current produced by single PV module [2]. The amount of the maximum voltage and maximum total current depends on the arrangement of the modules. For instance, four BP 275F PV modules can reach up a maximum power of 300 W when connected in parallel and the value of its output voltage is high. Due to the load specification as well as the need to provide a lower voltage with high efficiency, a regulator or converter is required in most PV applications in order to step-up or step-down the output voltage that PV modules generate. In addition, PV modules require an intermediate maximum power point tracker (MPPT) as their I–V characteristics are nonlinear to harvest the maximum power available on PV panels [3]. The proposed topology in this paper is non-isolated synchronous DC–DC buck converter for battery charging application which it has the voltage output lower than the input. In order to verify the proposed design, the experimental work has been tested for 75 W.

44.2 Synchronous DC–DC Buck Converter

A DC–DC buck converter consists of two operating modes which are continuous conduction mode (CCM) and discontinuous conduction mode (DCM). Usually CCM technique is applied for efficient power conversion while the DCM is for a lower power and stand-by application [4]. For the CCM, the inductor current never reaches zero during one switching cycle. It has a smooth input current because an inductor is connected in series with the power source [5]. The operation of DC–DC buck converter can be divided into two modes where for mode 1 when switch, S is on and meanwhile for mode 2 when switch, S is off. The schematic diagram of synchronous DC–DC buck converter is illustrated in Fig. 44.1. The relationship between output voltage, V_O of buck converter and its input voltage, V_I is governed by Eq. (44.1) where k is defined as the duty cycle of the converter.

$$k = V_O/V_I \tag{44.1}$$

The roman numerals, I, II, III, IV, V, VI, VII and VIII denoted as current sensor block, 12 V voltage regulators, MOSFET drivers, bootstrap circuit, high-side MOSFET, low-side MOSFET, 2nd order low pass filter, and microcontroller block respectively. In designing a non-isolated synchronous DC–DC buck prototype, there are two issues that need to be considered regarding this arrangement. Firstly, high-side, S1 and low-side, S2 MOSFET switches should never conduct simultaneously or otherwise there will be a short-circuit between the power and ground which may destroy the device. Typically, a dead-time interval is inserted between their gate drive signals. On the other hand, its turn-on resistance is positively proportional to the gate-source voltage placed on it. The standard driven voltage of a MOSFET is 12 V which is straightforward to achieve for the low-side switch, S2. However, this is not the case with high-side MOSFET whose source is connected to

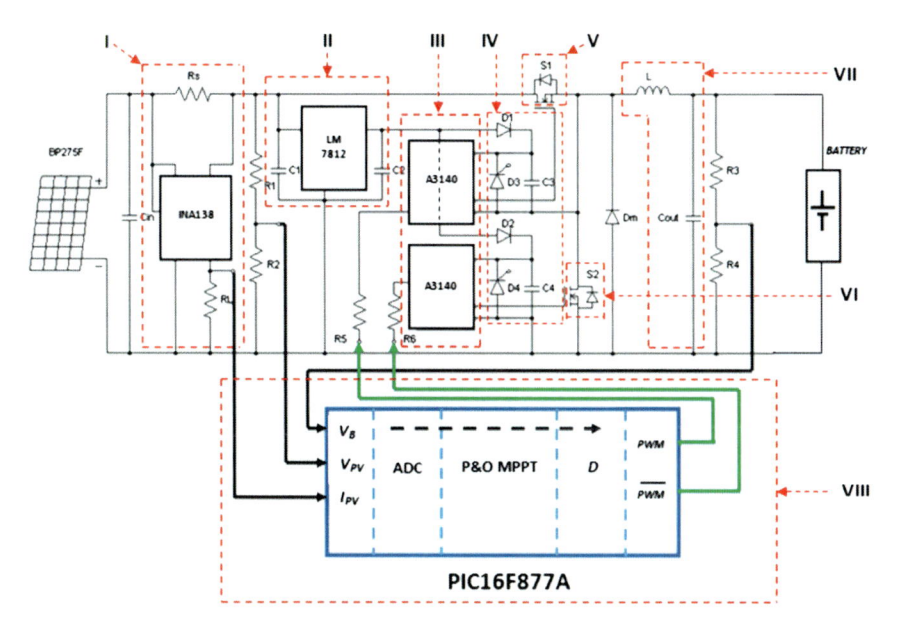

Fig. 44.1 Schematic diagram of synchronous DC–DC buck converter

the switch node. The source voltage will rise once the MOSFET is on. To ensure that the high-side MOSFET switch, S1 has a high enough gate-source voltage, bootstrap technology is introduced. Both of the two technologies are widely applied in a synchronous DC–DC buck driver [6].

44.2.1 Principle of Operation

The low-side driver is constructed to drive a floating high gate charge N-channel MOSFET, S1 as shown in Fig. 44.2. The gate voltage of the high-side switch, S1 is established by an external bootstrap supply circuit which contains of bootstrap diode, D3 and bootstrap capacitor, C3 connected between the power rails of the driver. The capacitor, C3 will provide a power source to the high-side MOSFET switch, S1. In Fig. 44.2, when the circuit is initiated, the bootstrap capacitor, C3 will be charge up through the bootstrap diode, D1 as S2 is turn-on to provide reference ground path to C3.

When the signal input goes high, the high-side driver will begin to turn-on the high-side MOSFET switch, S1 by injecting the charges of C3 into its input capacitance, Ciss in order to power up the Vgs reaches its threshold voltage resulting in conducting stage as illustrated in Fig. 44.3.

After completing the calculation, simulation and predicting the performance of the DC–DC buck converter for PV generation system using Matlab–Simulink, a hardware prototype of the PV system is built in order to clarify its performance. In

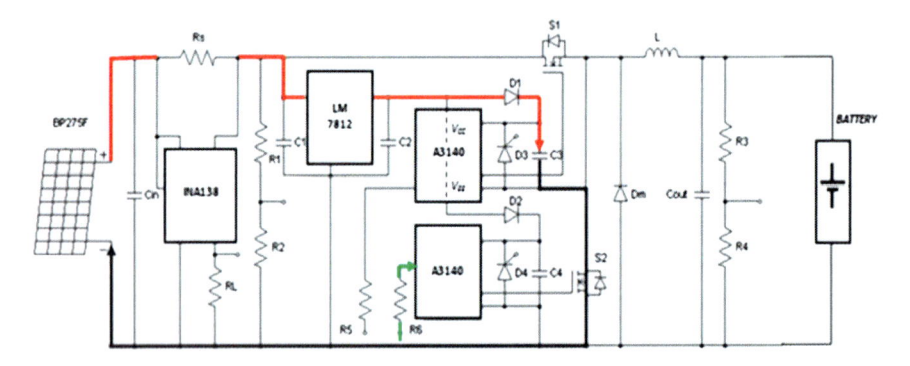

Fig. 44.2 Switch, S2 is turn-on

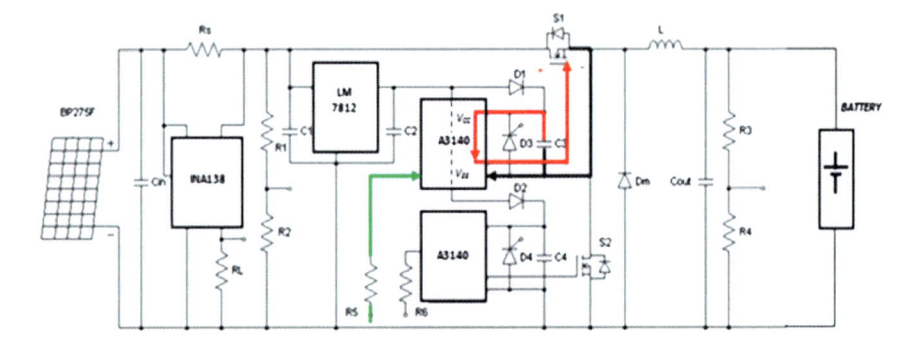

Fig. 44.3 Switch, S1 is turn-on

the later section, miscellaneous considerations of the prototype design will be discussed.

44.2.2 Selection of Components

The selection of components plays an important role in order for the PV system to operate at its full capacity. The design starts with the description of the BP275F solar module as the chosen PV module for the system.

44.2.2.1 PV Module

The BP275F solar module has V_{mpp} of 17 V and P_{mpp} of 75 W is selected in this research as it offers lots of advantages. Its mono-crystalline silicon cells, has high

Fig. 44.4 Experimental setup

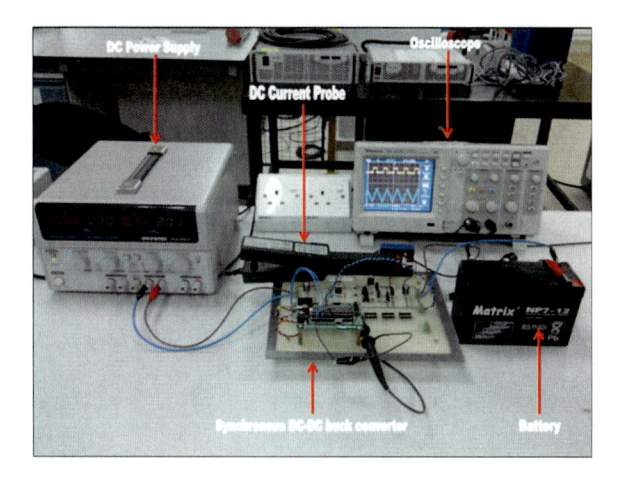

efficiency, highly resistant to water, abrasion, hail impact and other environmental factors. However, it was represented by DC supply source in laboratory conditions.

44.2.2.2 MOSFET and Drivers

The switch chosen for the prototype is the power MOSFETs IRF 540 N from the international rectifier. This device is rated at 100 V, 33 A. It drains current and has low on resistance, $R_{DS(on)}$ of 44 mΩ. In order to provide the control signal generated by the microcontroller to operate the power circuit, a driver is needed to connect between these two circuits. This driver boosted up the power of the control signal so that it is able to turn-on the power MOSFET. In this research, the driver used is HCPL-3140 from Avago Technologies.

44.2.2.3 Battery

The type of battery proposed in this research is a deep cycle battery. A deep cycle battery is a lead-acid battery designed to be regularly deeply discharged using most of its capacity. In this research, Matrix NP7-12 sealed lead rechargeable battery is used. Deep cycle battery is exclusively designed such that it can be discharged to low energy level and rapid recharge or cycle charged and discharged day after day for years.

In addition to that, other calculated converter parameters includes inductor with air gaps of 50 µH, free-wheeling diode, D_m MUR860, input and output capacitor of 220 µF, load, R of 4 Ω, frequency switching, f_s of 20 kHz. In order to verify the performance of the proposed non-isolated synchronous DC–DC buck converter, an experimental test is carried out. The experimental setup is displayed as in Fig. 44.4.

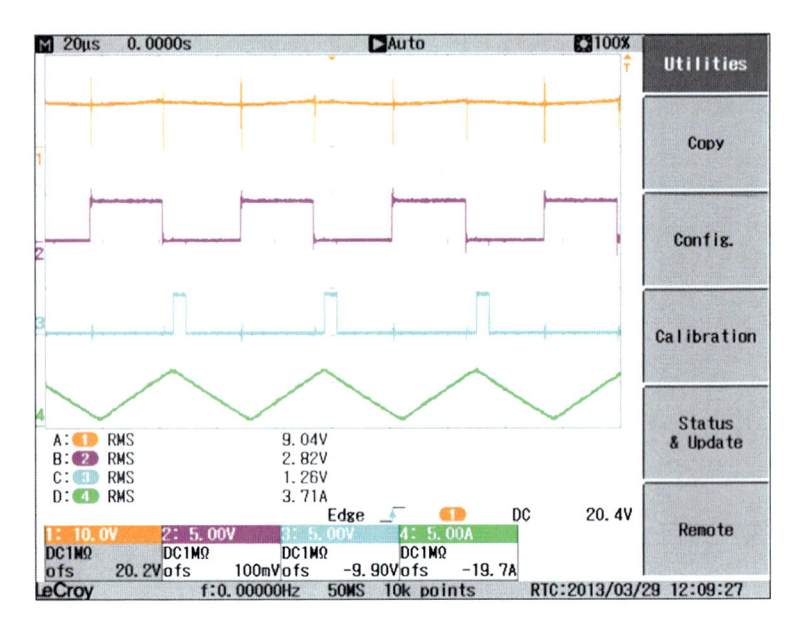

Fig. 44.5 Output waveform of non isolated synchronous DC–DC buck converter at k = 0.5

44.3 Result and Discussion

In this testing, the reading of input current, input voltage, output current and output voltage has been taken in order to calculate the input power and output power correspondingly and the efficiency, η is determined. The maximum recorded η is 80 % with the maximum output power of 60.50 W. Figure 44.5 shows the output waveform of DC–DC buck converter with $V_I = 20$ V, k = 0.5, and R = 4 Ω. Waveform labeled (1) indicated output voltage, V_O of buck converter, waveform labeled (2) represented PWM on the high-side driver while waveform labeled (3) correspond to PWM on low-side MOSFET driver. Subsequently, waveform labeled (4) denoted the value of inductor current, I_L. There are several factors that contributed to the degradation of the converter efficiency. There are losses, which exists in each component, passive elements, device elements, control circuit and others.

Moreover, as discussed before, dead-time is introduced between the switching signals as to avoid the short circuit condition. Dead-time of 3 µs is inserted after high-side switching signal is in off condition and before its goes to on state again as illustrated in Fig. 44.6.

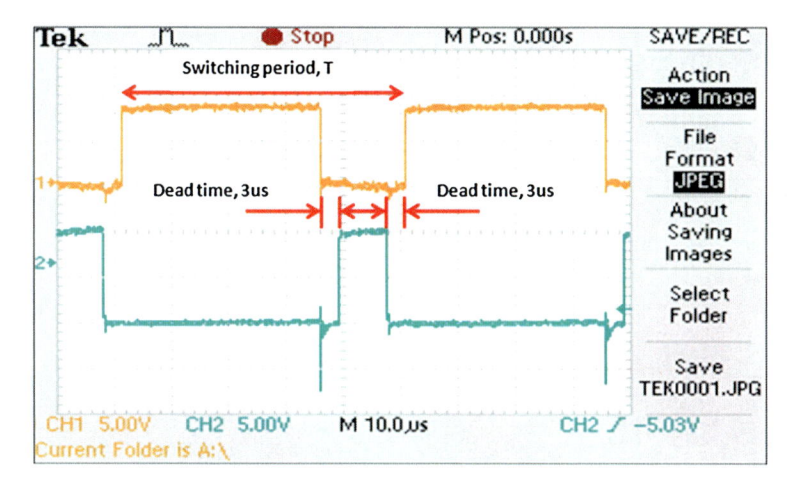

Fig. 44.6 Waveform of high-side and low-side switching signals

44.4 Conclusion

A proposed non-isolated synchronous DC–DC buck converter can be developed and implemented for photovoltaic power generation by considering the two issues regarding this arrangement. The introduction of dead-time interval as well as bootstrap technology could overcome certain limitations in this topology. The result has shown that the proposed design has achieved at an average of 80 % efficiency with maximum output power produced of 60 W at a constant input voltage of 20 V. Selection of right components with right parameters plays an important role in order to increase the converter's performance as well as its reliability.

Acknowledgments The authors would like to thank Universiti Sains Malaysia for the short term Grant # 304/PELECT/60310026. Under which this research was undertaken.

References

1. Bernardo PCM, Peixoto ZMA, Machado Neto LVB (2009) A high efficient micro-controlled buck converter with maximum power point tracking for photovoltaic systems. In: International Conference on Renewable Energies and Power Quality (ICREPQ'09), Spain, April 2009
2. Meksarik V, Masri S, Taib S, Hadzer CM (2004) Development of high efficiency boost converter for photovoltaic application. National Power & Energy Conference (PECon) 2004 Proceedings
3. Veerachary M (April 2011) Fourth-order buck converter for maximum power point tracking applications. IEEE Trans Aerosp Electron Syst 47(2)
4. Khader S (2011) Design and simulation of a chopper circuits energized by photovoltaic modules. In: 2011 IEEE GCC Conference and Exibition (GCC), United Arab Emirates, Feb 2011

5. Erickson RW, Maksimovic D (2005) Fundamental of power electronic. Kluwer Academic Publisher, University of Colorado, Colorado
6. Ming K, Wei Y, Wenhong L (2007) Design of a synchronous-rectified buck bootstrap MOSFET driver for voltage regulator module. In: ASIC, 2007. ASICON'07

Chapter 45
An Investigation of Raindrop Size in Raindrop Energy Harvesting Application via Photography and Image Processing Approach

Chin-Hong Wong, Joanne Neoh, Zuraini Dahari and Asrulnizam Abd Manaf

Abstract Renewable energy is gaining more attention as a new source of energy. Among all the renewable energies, kinetic energy is the most readily available energy source. One of potential kinetic energy to be utilized in energy harvesting is from the raindrop energy. The generated energy depends greatly on the size of raindrops. Therefore, it is important to study the raindrops characteristics prior for further investigation in the raindrop energy. This paper discusses the prediction of the size of raindrop based on simulated droplet using syringe pump. Edge detection segmentation method is applied to the image to detect the diameter of water droplet.

Keywords Renewable energy sources · Kinetic energy · Raindrop characteristics

---

C.-H. Wong (✉) · J. Neoh · Z. Dahari · A. A. Manaf
School of Electrical and Electronic Engineering, Universiti Sains Malaysia,
14300 Nibong Tebal, Pulau Pinang, Malaysia
e-mail: wch11_eee099@student.usm.my

J. Neoh
e-mail: jnws10524@student.usm.my

Z. Dahari
e-mail: zuraini@eng.usm.my

A. A. Manaf
e-mail: eeasrulnizam@eng.usm.my

45.1 Introduction

Due to the increasing concern about global warming, harvesting renewable energy sources has gained some researcher interests. Recently, there are a growing number of researches which study on the harvesting vibration energy from environment using piezoelectric technology recently [1, 2]. Nevertheless, only a limited number of researchers have explored research related to the vibration based raindrop energy harvesting application [3–6]. In raindrop energy harvesting application, the energy generated is greatly influenced by the size of raindrops and also its velocity. Therefore, it is important to study the raindrops characteristics in order to estimate the power generated by impact of water droplet on raindrop energy harvester. From Guigon et al. [4, 5] research, the raindrop energy harvester produced 73 µW from a single droplet with 3 mm diameter fall at the speed of 4.5 ms^{-1} impacted on a 25 µm thick, 3 mm wide and 10 cm long PVDF bridge piezoelectric mechanism. Whereas, it only produced 8 µW for a droplet with 1.6 mm diameter and fall at speed of 3.2 ms^{-1} impacted on the same structure.

To date, a few on-going researchers have conducted methods to measure raindrop size and fall speed. Based on the data collected by Gunn and Kinzer [7], the raindrop diameter is ranging from 0.1 to 5.8 mm. Most of them use high speed camera [8, 9] or laser-based optical sensor disdrometer [10], which are the easiest way to predict the droplet size and fall velocity. The data will be generated automatically during rain event which make it very convenience to use. However, the utilization of disdrometer is very costly. Another option is to apply the imaging method in order to estimate the raindrop size which is a cheaper option. In this paper, we present the prediction of the size of droplet via combination of photography and image processing approaches.

45.2 Methodology

Generally, there are two major steps which are firstly, the photography part and secondly, the image processing part. In order to generate a consistent and identical water droplet, a water droplet generator, NE 300 New Era Just Infusion Syringe pump was used. The syringe pump is used to adjust the injection flow rate. Various sizes of blunt needles which can be attached on a syringe were used to adjust the water droplet size. The size of the water droplet as shown in Fig. 45.1 can be calculated using expression.

$$D_{drop} = \sqrt[3]{(6D_{needle}\gamma)/(g\rho)} \qquad (45.1)$$

where D_{drop} is the diameter of the water droplet, D_{needle} is the external diameter of the blunt needle, g is the constant gravity, γ is the water surface stress at the liquid interface and ρ is the water density [4]. A needle with external diameter of 2.87 mm can generates a droplet with diameter of 5 mm, which is in the range of

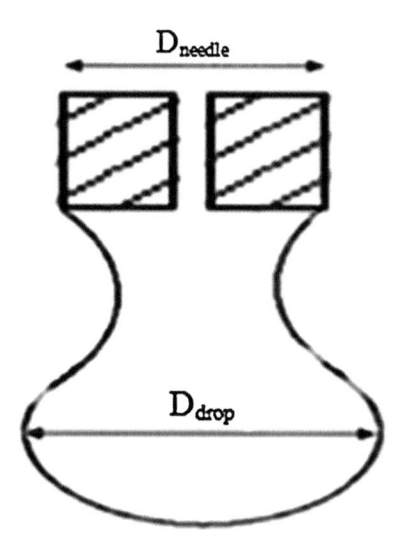

Fig. 45.1 Water droplet falling from a capillary

typical raindrop sizes, as mentioned in the previous section. With the aid of Eq. (45.1), the kinetic energy, *KE* of the falling raindrop can be calculated by using expression.

$$KE = (\rho \pi D_{drop} v^2)/12 \qquad (45.2)$$

where ρ is the water density, D_{drop} is the diameter of the water droplet and v is the droplet terminal velocity. In Eq. (45.2), the water droplet is assumed to have constant volume during drop, and spherical droplet.

A digital camera, Nikon J1 is used to record a high falling speed of water droplet. The camera is capable to record slow motion videos. The videos are recorded at 1,200 frames per second and play back at 30 frames per second. Figure 45.2 illustrates the experimental setup for the recording process.

The recorded video is then extracted into single gray scale images. The images are adjusted with the *imadjust* function from MATLAB image processing. The next step is to filter the image with a predefined 2-D filter. The filter used is an averaging filter with a vector of five rows and five columns. Sobel method of edge detection is used to calculate the threshold value to obtain a binary mask that contains the segmented droplet. The binary gradient mask shows lines of high contrast in the image, gaps can be seen in the lines surrounding the object in the gradient mask. These linear gaps will disappear if the Sobel image is dilated using linear structuring elements, which can be created with the *strel* function. However, there are still holes in the interior of the droplet. To fill these holes the *imfill* function is used. After that, the object is smoothening by eroding the image twice with a disk structuring element. The function *bwareaopen* is used to remove small objects in the image so that the image only contains the interested segmented droplet. Finally,

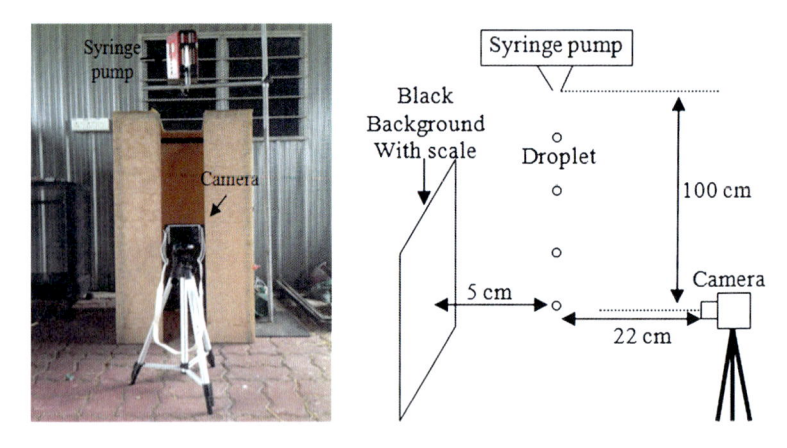

Fig. 45.2 Experimental setup for capturing droplet image

the droplet diameter is ready to be calculated. Using the *regionprops* function, which measures a set of properties for each connected component (object) in the binary image, the diameter of the droplet is obtained in pixels.

45.3 Results and Discussions

The water droplet was released from a syringe pump from tested heights of 0.5 and 1 m and the process of falling droplet was recorded by using Nikon J1 high speed camera. The experiments was repeated and recorded for five times for both heights.

From Fig. 45.3, it can be seen that the Nikon J1 camera is capable of capturing object with high fall velocity. However, it is unable to focus on the object clearly. The shape and the diameter of the droplet calculated have been observed as changing along the fall distance. Figure 45.3 shows the same droplet at three different positions. From these different positions, the calculated droplet diameter varied slightly. Therefore, an average of these values is taken as the final acquired value. The graphs illustrated in Fig. 45.4 show the comparison of the theoretical and MATLAB image processing calculation of droplet diameter. The details of the results are shown in Tables 45.1 and 45.2.

Figure 45.4 shows good correlations between theoretical and MATLAB image processing results. The MATLAB image processing results agree to the theoretical results. It is noticed that the highest percentage error occurred for the droplet generated by blunt needle with external diameter of 1.81 mm, which are 6.46 and 11.75 %, for both heights. This is due to the droplets in the images are too small and therefore it is more difficult to extract the exact size from the image. Bigger blunt needle size generated bigger droplets as a result it is easier to be segmented from image.

On the other hand, droplet released from 1.0 m height of has higher percentage error compared to the 0.5 m. As the released height of droplet increases, the falling

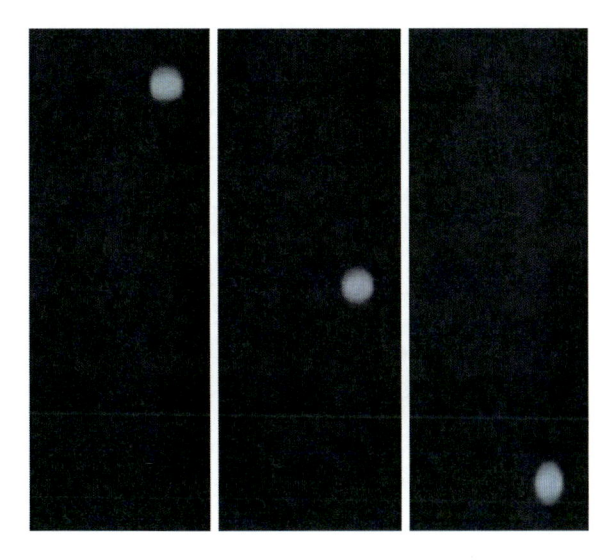

Fig. 45.3 Droplet fall along the vertical distance at different position

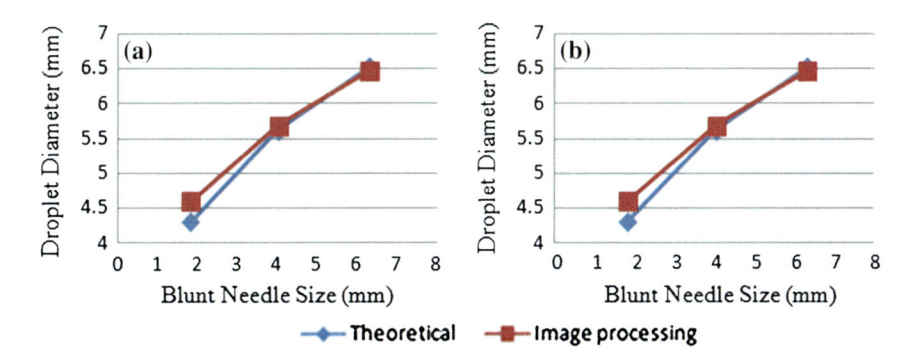

Fig. 45.4 Droplet diameter with respect to different blunt needle size released from **a** 0.5 m height, **b** 1 m height

Table 45.1 Comparison between theoretical and MATLAB results of droplet diameter released from 0.5 m height

Needle diameter (mm)	Theoretical results of droplet diameter (mm)	Imaging processing results of droplet diameter (mm)	Percentage error (%)
1.81	4.31	4.60	6.46
4.02	5.62	5.69	1.15
6.31	6.53	6.47	0.94

Table 45.2 Comparison between theoretical and MATLAB results of droplet diameter released from 1.0 m height

Needle diameter (mm)	Theoretical results of droplet diameter (mm)	Image processing results of droplet diameter (mm)	Percentage error (%)
1.81	4.31	4.85	11.75
4.02	5.62	5.92	5.12
6.31	6.53	6.46	1.17

droplets are more vulnerable to the wind, which makes the droplets do not fall back to the origin position.

Tables 45.1 and 45.2 showed that the percentage error of droplet diameter between theoretical and image processing is ranging from 0.94 to 11.75 %. The reasonable results showed that the proposed technique can be used to measure raindrops size.

45.4 Conclusion

Based on the findings in this paper, the droplet sizes are successfully predicted using the photography technique combined with the image processing technique. It is a cheaper way to measure droplet size suing photography and image processing approaches compared to laser-based optical sensor disdrometer. The results obtained from this algorithm agree to the theoretical results. The percentage error is ranging from 0.94 to 11.75 %. In this experiment, there are a few issues and challenges if it is to be tested in real application. In real rain event, raindrops size and shape are normally inconsistent while falling. It breaks up simply when it comes down due to air resistance or collision with their neighbours. It is difficult to calculate the kinetic energy of falling raindrop using Eq. (45.2) since it required constant raindrop size and shape. Besides, larger raindrops are usually impacted with splash. The splash of raindrop leads to significant energy loss and the amount is unpredictable. The delivered output of single droplet impact on a piezoelectric film is still small but the prospect for improvement looks positively inclined.

Acknowledgements The authors would like to express sincere appreciation for Fundamental Research Grant Scheme (FRGS), 203/PELECT/6071224.

References

1. Ramadass YK, Chandrakasan AP (2010) An efficient piezoelectric energy harvesting interface circuit using a bias-flip rectifier and shared inductor. IEEE J Solid-State Circuits 45:189–204
2. Sodano HA, Inman DJ, Park G (2004) A review of power harvesting from vibration using piezoelectric materials. Smart Mater Struct 36(3):197–205

3. Vatansever D, Hadimani RL, Shah T, Siores E (2011) An investigation of energy harvesting from renewable sources with PVDF and PZT. Smart Mater Struct 20(5):055019
4. Guigon R, Chaillout JJ, Jager T, Despesse G (2008) Harvesting raindrop energy: experimental study. Smart Mater Struct 17(1):015039
5. Guigon R, Chaillout JJ, Jager T, Despesse G (2008) Harvesting raindrop energy: theory. Smart Mater Struct 17(1):015038
6. Al Ahmad M, Jabbour GE (2012) Electronically droplet energy harvesting using piezoelectric cantilevers. Electron Lett 48(11):647–649
7. Gunn R, Kinzer GD (1949) The terminal velocity of fall for water droplets in stagnant air. J Meteorol 6:243–248
8. Salvador R, Bautista-Capetillo C, Burguete J, Zapata N, Serreta A, Playán E (2009) A photographic method for drop characterization in agricultural sprinklers. Irrig Sci 27:307–317
9. Miao Y, Hong H, Kim H (2011) Size and angle filter based rain removal in video for outdoor surveillance systems. In: Control conference (ASCC), 8thAsian, pp 1300–1304
10. Bloemink HI, Lanzinger E (2003) Precipitation type from the Thies disdrometer

Chapter 46
Latency Insertion Method with MNA Blocks Via Node Tearing

Patrick Goh

Abstract In this work, a method to interface simulations using the latency insertion method with blocks utilizing the modified nodal analysis method by a node tearing approach is presented. Using the proposed method, the insertion of fictitious latencies into blocks without latencies is avoided and thus the simulation of the traditional LIM is sped up.

Keywords Latency insertion method (LIM) · Modified nodal analysis (MNA) · Node tearing

46.1 Introduction

The shrinking trend of the semiconductor industry has resulted in an increase in the density of interconnects and the complexity of high-speed packages. As a result, signal integrity issues are becoming more prominent and can no longer be overlooked in the design of modern devices. In this aspect, the ability to accurately simulate large interconnect circuits in a reasonable amount of time is invaluable to capture the complicated electromagnetic behaviors of complex circuits. As a result, there is a constant need and push toward faster circuit simulation methods that are able to handle large circuits in a fraction of the time taken by conventional circuit simulators.

The latency insertion method (LIM) [1] has recently emerged as an efficient approach for performing fast simulations of very large circuits. The basic idea of LIM is to exploit the latencies in the circuit to solve the voltages and the currents explicitly at each time step. The resulting algorithm has been shown to exhibit a

P. Goh (✉)
School of Electrical and Electronic Engineering, Universiti Sains Malaysia,
14300 Nibong Tebal, Pulau Pinang, Malaysia
e-mail: eepatrick@eng.usm.my

linear numerical complexity with respect to the number of nodes [1], and thus is significantly faster than conventional matrix inversion-based methods. For example, results in [2] show that LIM is over 500× faster than HSPICE [3] for the simulations of large tightly coupled transmission lines. However, LIM has a disadvantage in that it requires latencies to be present at every node and branch in the circuit. When they are not present, small fictitious values are inserted to enable the method. This reduces both the accuracy and the efficiency of the method as LIM is only conditionally stable, whereby the maximum stable time step depends on the smallest latency in the circuit [4]. In this work, a method to combine LIM with a modified nodal analysis (MNA) circuit block via a node tearing approach [5] is proposed. While there have been previous work related to this topic, they are significantly different from the present work. For example, the work in [6] strips SPICE of its matrix solver engine and replaces it with LIM while [7] reformulates LIM using a block-matrix formulation. The present work, on the other hand, maintains both LIM and MNA blocks and works on interfacing the two components.

46.2 LIM Formulation

LIM can be applied to any arbitrary network, where it is assumed that through the use of Thévenin and Norton transformations, the branches and nodes of the circuit can be described by the general topologies shown in Fig. 46.1.

Each node is represented by a parallel combination of a current source, a conductance, and a capacitor to ground. The connection between two different nodes forms a branch and it is represented by a series combination of a voltage source, a resistor, and an inductor. In order to solve for the voltages and currents in the circuit, LIM discretizes the time variable whereby the voltages and currents are collated in half time steps. Specifically, the voltages are solved at half time steps while the currents are solved at full time steps. From Fig. 46.1, writing Kirchhoff's current law (KCL) at node i yields

$$C_i\left(\frac{V_i^{n+1/2} - V_i^{n-1/2}}{\Delta t}\right) + G_i\left(\frac{V_i^{n+1/2} + V_i^{n-1/2}}{2}\right) - H_i^n = -\sum_{k=1}^{M_i} I_{ik}^n \quad (46.1)$$

where the superscript n is the index of the current time step, Δt is the time step and M_i is the number of branches connected to node i. Solving for the unknown voltage yields

$$V_i^{n+1/2} = \left(\frac{C_i}{\Delta t} + \frac{G_i}{2}\right)^{-1}\left[\left(\frac{C_i}{\Delta t} - \frac{G_i}{2}\right)V_i^{n-1/2} + H_i^n - \sum_{k=1}^{M_i} I_{ik}^n\right] \quad (46.2)$$

for $i = 1, 2, \ldots, N_n$, where N_n is the number of nodes in the circuit.

Similarly, writing Kirchhoff's voltage law (KVL) at branch ij yields

Fig. 46.1 Node and branch topologies for LIM. *Left* node. *Right* branch

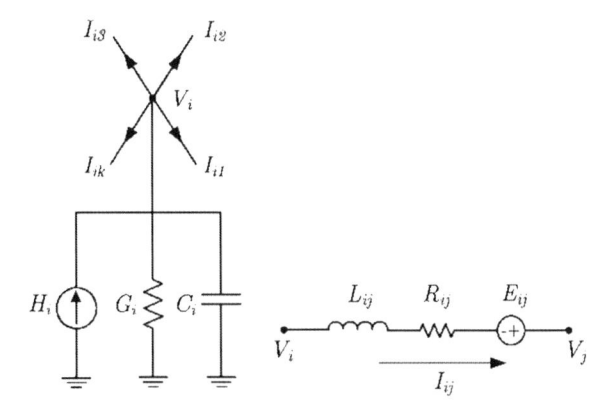

$$V_i^{n+1/2} - V_j^{n+1/2} = L_{ij}\left(\frac{I_{ij}^{n+1} - I_{ij}^n}{\Delta t}\right) + R_{ij}\left(\frac{I_{ij}^{n+1} + I_{ij}^n}{2}\right) - E_{ij}^{n+1/2} \qquad (46.3)$$

and solving for the unknown current yields

$$I_{ij}^{n+1} = \left(\frac{L_{ij}}{\Delta t} + \frac{R_{ij}}{2}\right)^{-1}\left[\left(\frac{L_{ij}}{\Delta t} - \frac{R_{ij}}{2}\right)I_{ij}^n + E_{ij}^{n+1/2} + V_i^{n+1/2} - V_j^{n+1/2}\right] \qquad (46.4)$$

Note that a semi-implicit formulation of (46.1)–(46.4) as detailed in [4] has been adopted. The computation of the node voltages and the branch currents are alternated as time progresses in a leapfrog manner. It is clear that the LIM algorithm relies on the latencies in the network to perform the leapfrog time stepping formulation. Thus, at every node, a capacitor to ground has to be present. If it is not, a small fictitious capacitor is inserted. Similarly, small fictitious inductors are inserted into branches without latencies. The values of these fictitious latencies must be made sufficiently small so as to not affect the accuracy of the overall simulation. However, having small fictitious latencies is detrimental to the overall simulation efficiency as LIM is only conditionally stable where the time step for a stable simulation is given by [4]

$$\Delta t < \sqrt{2}\min_{i=1}^{N_n}\left(\sqrt{\frac{C_i}{M_i}\min_{p=1}^{M_i}(L_{i,p})}\right) \qquad (46.5)$$

where M_i is the number of branches connected to node i and $L_{i,p}$ denotes the value of the pth inductor connected to node i. From (46.5), it can be seen that if the circuit contains portions of very small latencies (such as from fictitious latencies), the maximum stable time step for the entire simulation will decrease. This results in an increase in the number of computations that must be performed for a given simulation. Note that (46.5) is only valid for RLC or GLC circuits. For general RLGC circuits, the general criteria in [8] must be used. However, the observation above remains valid. In this work, a node tearing approach is applied to split the circuit into partitions with and without latency. LIM is then used to simulate the partition with latency while a traditional MNA is used to simulate the partition without latency. The result is a computational efficient method since no fictitious latency is inserted.

46.3 LIM-MNA Interface Via Node Tearing

Consider the circuit shown in Fig. 46.2 which contains sections with and without latency. The concept of the node tearing approach is summarized as follows. First the circuit is broken up at the nodes connecting the two sections. Voltage sources are then inserted on the side without latency to account for the response from the broken connection while current sources are inserted on the side with latency. This creates two separate partitions and at each time step, the voltage V_{LIM} is used to update the voltage source V_{MNA} while the current source I_{LIM} is updated using the current through the voltage source, I_{MNA}. Note that I_{MNA} is automatically solved for in an MNA formulation. In addition, at each time step, each partition is simulated using the two methods where LIM is used to simulate the partition with latency while MNA is used to simulate the partition without latency. No fictitious latency is inserted and as a result, the overall computational advantage of using LIM is retained. Numerical results will be presented in the next section.

46.4 Numerical Results

In this section, an example circuit is simulated to illustrate the method. Consider the circuit shown in Fig. 46.3 which consists of three transmission lines connected through a resistive connection. The circuit is driven by a voltage source V_{in} which is a single trapezoidal pulse with a magnitude of 1 V, rise and fall times of 1 ns, pulse width of 40 ns and an initial delay of 10 ns. The approach presented in the previous section is then applied to split the circuit into four distinct partitions as shown in Fig. 46.4 where each transmission line is modeled by 250 unit cells of a π-model RLC circuit where R = 0.01 Ω, L = 1 nH and C = 1 pF. The voltage at V_{out} is plotted in Fig. 46.5 along with the input voltage. For comparison, the same circuit is also simulated using the traditional LIM where fictitious latencies of $L_{fict} = 0.01$ nH were inserted into the branch without latencies to enable the method. The results are plotted in the same figure where no notable differences are observed in terms of accuracy. In terms of runtime, the stable time step for the traditional LIM method was limited to

$$\Delta t < \sqrt{2} \min_{i=1}^{N_n} \left(\sqrt{\frac{C_i}{M_i} \min_{p=1}^{M_i}(L_{i,p})} \right) = \sqrt{2} \left(\sqrt{\frac{1p}{3}(0.01n)} \right) = 2.58 \, \text{ps} \qquad (46.6)$$

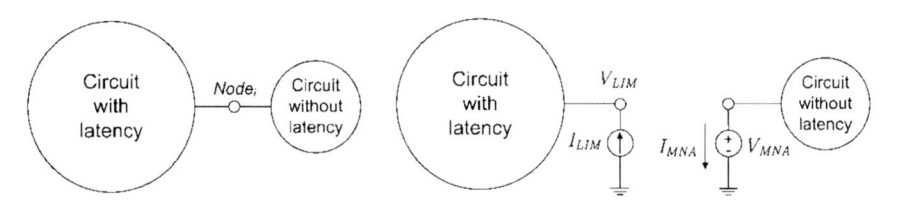

Fig. 46.2 Partitioning circuit via node tearing. *Left* original circuit. *Right* partitioned circuit

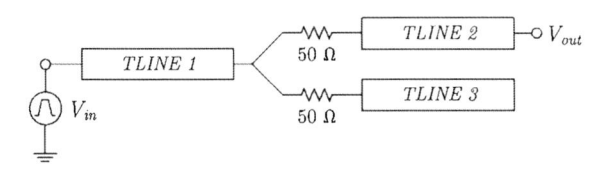

Fig. 46.3 Example interconnect circuit

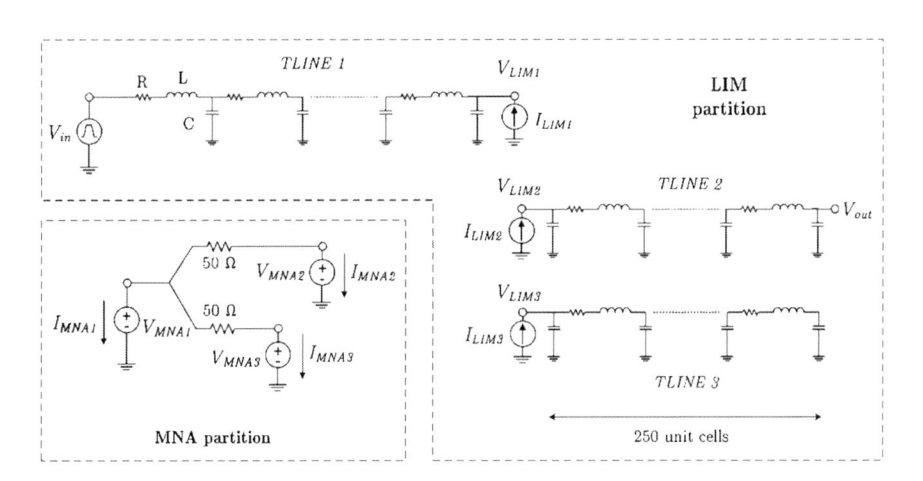

Fig. 46.4 Partitioned interconnect circuit

due to the fictitious latencies while the stable time step for the proposed method was limited to

$$\Delta t < \sqrt{2} \min_{i=1}^{N_n} \left(\sqrt{\frac{C_i}{M_i} \min_{p=1}^{M_i}(L_{i,p})} \right) = \sqrt{2} \left(\sqrt{\frac{1p}{3}(1n)} \right) = 25.8 \text{ ps} \qquad (46.7)$$

since no fictitious latencies where added. The time steps of 2.5 and 25 ps were used for the traditional LIM and the current proposed method respectively. This resulted in a runtime of 18.85 s for the traditional LIM and 1.87 s for the proposed method for a simulation interval of [0, 120 ns]. A speedup of about 10× is

Fig. 46.5 Simulation result for LIM and the proposed method (LIM-MNA)

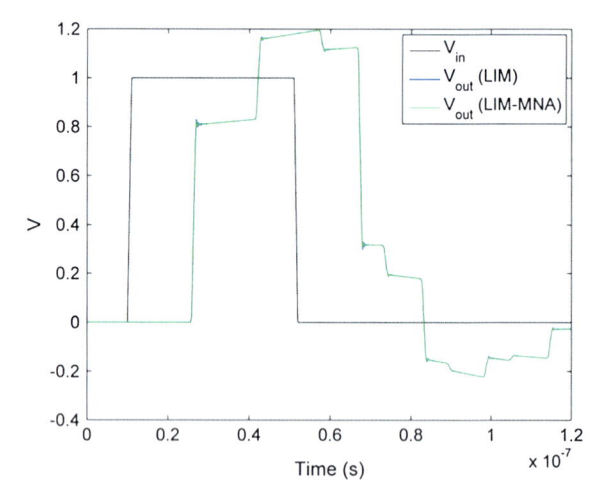

achieved in this example as the time step of the proposed method is not limited by the fictitious latencies. All simulations were performed in Matlab R2009a on an Intel Xeon 2.4 GHz computer with 4 GB of RAM.

46.5 Conclusion

In this work, a method to interface LIM simulations with MNA blocks via a node tearing approach has been presented. The proposed method avoids the insertion of fictitious latencies and thus allows for a larger time step which in turn speeds up the simulation. Future work on the subject will focus on parallelizing of the individual partitions.

References

1. Schutt-Ainé JE (2001) Latency insertion method (LIM) for the fast transient simulation of large networks. IEEE Trans Circuits Syst I: Fundam Theory Appl 48(1):81–89
2. Sekine T, Asai H (2011) Block-latency insertion method (block-LIM) for fast transient simulation of tightly coupled transmission lines. IEEE Trans Electromagn Compat 53(1):193–201
3. HSPICE Datasheet, Synopsys Inc., Mountain View, CA, 2010. [Online]. Available: http://www.synopsys.com
4. Lalgudi SN, Swaminathan M, Kretchmer Y (2008) On-chip power-grid simulation using latency insertion method. IEEE Trans Circuits Syst I 55(3):914–931
5. Pillage T, Rohrer R, Visweswariah C (1994) Electronic circuit and system simulation methods. McGraw-Hill, New York

6. Deng Z, Schutt-Ainé JE (2005) LIM-SPICE for the analysis of power distribution networks. In: Proceedings of 9th IEEE workshop signal propagation on interconnection. Garmisch-Partenkirchen, Germany, May 2005, pp 17–20
7. Kubota H, Tanji Y, Watanabe T, Asai H (2005) Generalized method of the time-domain circuit simulation based on LIM with MNA formulation. In: Proceedings of IEEE custom integrated circuits conference. San Jose, CA, Sept 2005, pp 282–292
8. Goh P, Schutt-Ainé JE, Klokotov D, Tan J, Liu P, Dai W, Al-Hawari F (2011) Partitioned latency insertion method with a generalized stability criteria. IEEE Trans Compon Packag Manuf Technol 1(9):1447–1455

Chapter 47
A Fully-Integrated Dual-Band Concurrent CMOS LNA for 2.45/5.25 GHz Applications

Hamidreza Ameri Eshghabadi, Mohd Tafir Mustaffa, Norlaili Mohd Noh, Asrulnizam Abd Manaf and Othman Sidek

Abstract This paper presents a fully-integrated dual-band concurrent CMOS low noise amplifier (LNA) which is implemented in Global Foundries 0.13 um RF CMOS technology. The LNA is designed to receive signals at 2.45 and 5.25 GHz frequency bands simultaneously. The concurrent operation of the proposed LNA makes it suitable for WLAN (802.11 b/g/a) applications. Based on post layout simulation results, this design achieved the power gain of 16 dB at 2.45 GHz and 11.7 dB at 5.25 GHz, the NF of 3 dB at 2.45 GHz and 4.3 dB at 5.25 GHz and the input and output return loss of −14 dB at the desired frequencies. Total power consumption of this design is only 6.5 mW.

Keywords Concurrent CMOS low noise amplifier · WLAN (802.11 b/g/a) · Power gain · NF · Input and output return loss

---

H. Ameri Eshghabadi (✉) · N. Mohd Noh · A. Abd Manaf
School of Electrical and Electronics Engineering, Universiti Sains Malaysia Engineering Campus, Nibong Tebal, Malaysia
e-mail: hae11_eee124@student.usm.my

N. Mohd Noh
e-mail: eelaili@eng.usm.my

A. Abd Manaf
e-mail: eeasrulnizam@eng.usm.my

O. Sidek
Collaborative Microelectronic Design Excellence Centre, Universiti Sains Malaysia Engineering Campus, Nibong Tebal, Malaysia
e-mail: othman@eng.usm.my

M. T. Mustaffa
School of Electrical and Electronic Engineering, Universiti Sains Malaysia,
Nibong Tebal, Malaysia
e-mail: tafir@eng.usm.my

Table 47.1 Wireless standards

Wireless standard	Frequency band (MHz)
WCDMA	2,110–2,170
WLAN (802.11 b/g)	2,400–2,483
WLAN (802.11 a)	5,150–5,825
Bluetooth	2,400–2,483

47.1 Introduction

Recent developments in wireless communication systems and the fast growth of its competitive market have resulted in spread range of wireless standards. Each standard adopted a specific range of bandwidth from the available spectrum. Table 47.1 shows some of the famous standards and their related bandwidths as well.

Accordingly, many RF receivers have been designed to perform in one or more frequency ranges. Standard receiver architectures, such as super-heterodyne and direct-conversion, accomplish high selectivity and sensitivity by narrow-band operation at a single input frequency [1]. These receivers suffer from the limitation on available bandwidth. Another approach is to use the conventional dual-band receivers. In conventional type, a signal path is allocated for each frequency band and a single-banded LNA is selected according to the operating band. This method leads to high implementation costs and large chip area [2].

Many wide-band LNAs have been introduced to operate at wide frequency range. These kinds of LNAs are more sensitive to out-of-band unwanted signals (blockers) due to transistor nonlinearity and the out of band blockers can severely degrade receiver's sensitivity [1].

Another approach is to use a multi-band concurrent LNA at the first stage of the receivers. It is capable to operate at multiple frequency bands (simultaneously) at any time. This approach resulted in significant decrease of the manufacturing costs due to the optimum area. Also the power consumption of the LNA is minimized comparing to the other methods [3].

This article presents a fully-integrated concurrent dual-band CMOS LNA which operates at 2.45 and 5.25 GHz for WLAN (IEEE 802.11 b/g/a) applications. Section 47.2 describes the design methodology of the proposed LNA, its topology and the theory of input and output matching circuits. Section 47.3 discusses the s-parameter results and noise performance based on post layout simulation. Finally Sect. 47.4 concludes this work and compares it to similar designs.

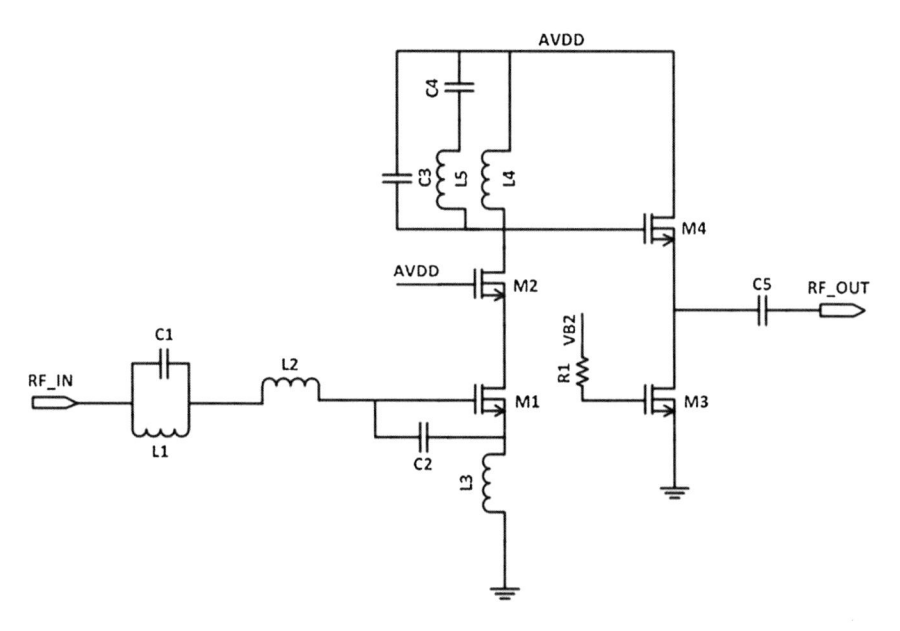

Fig. 47.1 Topology of concurrent dual-band CMOS LNA

47.2 Concurrent Dual-Band CMOS LNA Design

47.2.1 Topology

The topology of the proposed LNA is shown in Fig. 47.1. It is based on a combination of a cascode stage and a buffer stage added to the output. The cascode stage provides some advantages as well as decreasing the Miller Effect, increasing the isolation between the output and the input of the stage even added power gain of high frequency [3]. Also, the inductive source degeneration structure applied to the input circuit which helps to increase the stability of the LNA through the negative feedback at the expenses of negligible decrease in power gain [4]. A buffer circuit implemented at the output of the cascode to improve the output matching.

47.2.2 Input Matching

Figure 47.2 depicts the input matching circuit of concurrent LNA. The input circuit is made up of three inductors (L1, L2 and L3) and three capacitors [C1, C2 and Cgs1 (gate-source capacitance of M1)]. The input matching circuit is fully integrated using the on-chip inductors. Assuming R1, R2 and R3 as the resistances of L1, L2 and L3 inductors, as mentioned in [4] we can calculate the real and imaginary parts of the input impedance as:

Fig. 47.2 Input matching circuit

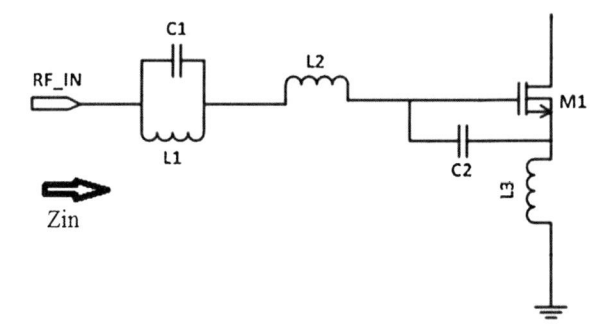

$$\operatorname{Re}(Zin) = \frac{gmL3}{C2 + Cgs1} + \frac{(\omega L1)^2/R1}{(1 - \omega^2 C1L1)^2 + (\omega L1/R1)^2} + (R2 + R3) \quad (47.1)$$

$$\operatorname{Im}(Zin) = \frac{\omega L1(1 - \omega^2 C1L1)}{(1 - \omega^2 C1L1)^2 + (\omega L1/R1)^2} + \omega(L2 + L3) - \frac{1}{\omega(C2 + Cgs1)} - \frac{gmR3}{\omega(C2 + Cgs1)} \quad (47.2)$$

The input impedance matching achieved by making the real part matched to 50 Ω [Eq. (47.1)] while eliminating the imaginary part [Eq. (47.2)]. To satisfy the matching conditions at both frequencies, L1, L2, C1, C2 and L3 need to be adjusted accordingly. L3 should be chosen to match the real part of Zin and C2 is implemented to enhance the specified frequency bandwidth. As the input matching circuit is fully integrated, it suffers from the low quality factor (Q) of on-chip inductors [5].

47.2.3 Output Matching

Figure 47.3 depicts the output matching circuit of the concurrent LNA before adding the output buffer. A series LC branch connected in parallel with a standard LC-tank to shape the output matching circuit. Neglecting the effects of the buffer load to the circuit, if we consider the non-ideal models for inductors and after massive calculations that mentioned in [4, 5], the output impedance is achieved as:

$$Zout = \frac{SL4\left(1 + S^2 L5 C4\left(1 - j/Q_L\right)\right)\left(1 - j/Q_L\right)}{1 + S^2(L4C3 + L5C4 + L4C4)\left(1 - j/Q_L\right) + S^4 L4 C3 L5 C4\left(1 - j/Q_L\right)^2} \quad (47.3)$$

Fig. 47.3 Output matching circuit

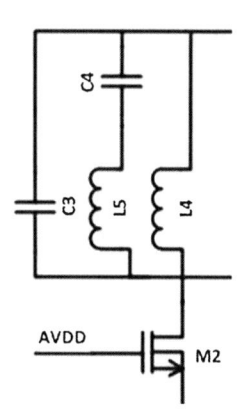

Making the imaginary part of the output impedance zero and after simplifying the achieved equations, the roots of the equation are:

$$\omega^2\left(1 + {1}/{QL^2}\right)\left(L5^2C4^2 + L4L5C4^2\right) - 2L5C4\omega^2 + 1 = 0 \quad (47.4)$$

$$\omega 1,2 = \frac{2L5C4 \pm \sqrt{(2L5C4)^2 - 4\left(1 + \frac{1}{QL^2}\right)\left(L5^2C4^2 + L5L4C4^2\right)}}{2\left(1 + \frac{1}{QL^2}\right)\left(L5^2C4^2 + L4L5C4^2\right)} \quad (47.5)$$

There are two desired resonance frequencies at 2.45 and 5.25 GHz that the concurrent output impedance should be matched to them. As mentioned in [4], here is a brief description on calculations that made to find the proper values for matching components.

Based on (47.4) we define two parameters P1 and S1 as:

$$P1 = \omega 1^2 \omega 2^2 = \frac{1}{\left(1 + {1}/{QL^2}\right)\left(L5^2C4^2 + L4L5C4^2\right)} \quad (47.6)$$

$$S1 = \omega 1^2 + \omega 2^2 = \frac{2L5C4}{\left(1 + {1}/{QL^2}\right)\left(L5^2C4^2 + L4L5C4^2\right)} \quad (47.7)$$

From (47.6) and (47.7) and considering the notch frequency as:

$$\omega 0 = \frac{\omega 1 + \omega 2}{2} \quad (47.8)$$

Equation (47.9) is defined as the assisting equation to calculate L5:

$$L5 = \frac{1}{2C4}\left(\frac{S1}{P1}\right) \quad (47.9)$$

Table 47.2 Values of used components

L1	1.5n	C1	2.2p
L2	5n	C2	75f
L3	1.82n	C3	150f
L4	4.8n	C4	450f
L5	5.1n	R1	25 k
M1	(75/0.13)u	M3	(35/0.15)u
M2	(35/0.13)u	M4	(35/0.15)u

Equations (47.4)–(47.9) are used to find the proper values of the components. QL (quality factor) of the inductors is provided in their model parameters from process design kit. Also, the real part of the Eq. (47.3) should be matched to 50 Ω as well. A buffer stage is added to the output of this circuit in order to improve the output matching of the LNA. Table 47.2 shows the values of both input and output matching circuits and other components.

47.3 Simulations Results

The fully-integrated concurrent CMOS LNA for 2.45/5.25 GHz applications was designed using the Global Foundries 0.13 um RF-CMOS process. Figure 47.4 depicts the s-parameter results based on post layout simulation. This design achieved the input matching (S11) of −19.08 dB at 2.45 GHz and −15.4 dB at 5.25 GHz and the power gain of 16 dB at 2.45 GHz and 11.67 dB at 5.25 GHz. Also the output matching parameter (S22) is −15.78 dB at 2.45 GHz and −15.15 dB at 5.25 GHz. There is a shift of frequency to the left side of S11 and S22 curves. This shift is produced by the parasitic capacitors which are more significant at high frequencies (5.25 GHz). The results are based on the optimization of worst case corner that influenced the shift of frequency. At lower frequency, the S11 is −19.08 dB that resulted in higher power gain compared to the upper frequency. C2 at the input matching circuit contributed to better input matching. Also the added buffer stage at the output of LNA, improved S22 significantly at higher frequencies.

Figure 47.5 shows the noise performance of the concurrent dual-band LNA. As the results show at both 2.45 and 5.25 GHz, the minimum noise figure (NFmin) and noise figure (NF) curves placed close to each other and this means the noise matching is acceptable for both bands. Figure 47.5 shows that, at 2.45 GHz, the NF is 3 dB and at 5.25 GHz, it is 4.3 dB. For the DC power consumption of this design, the first stage (cascode) is optimized to consume 3.2 mA and the buffer stage dissipates less than 2.2 mA under the supply voltage of 1.2 V. So the total power consumption of this LNA is less than 6.5 mW which considered as low-power design.

Fig. 47.4 S-parameter results

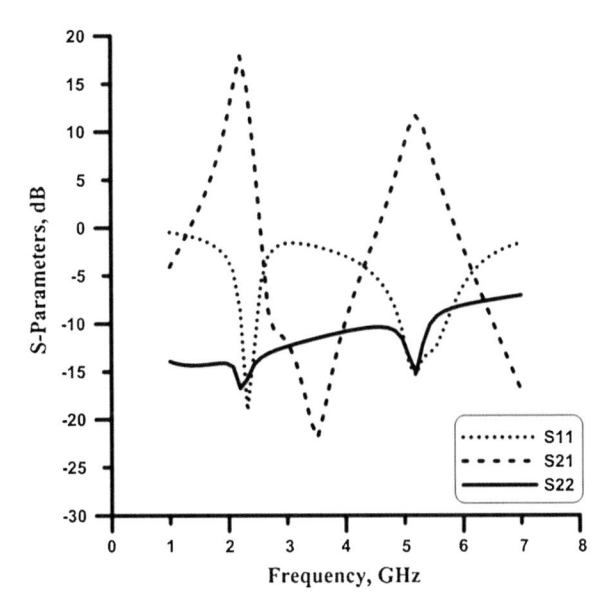

Fig. 47.5 Noise performance

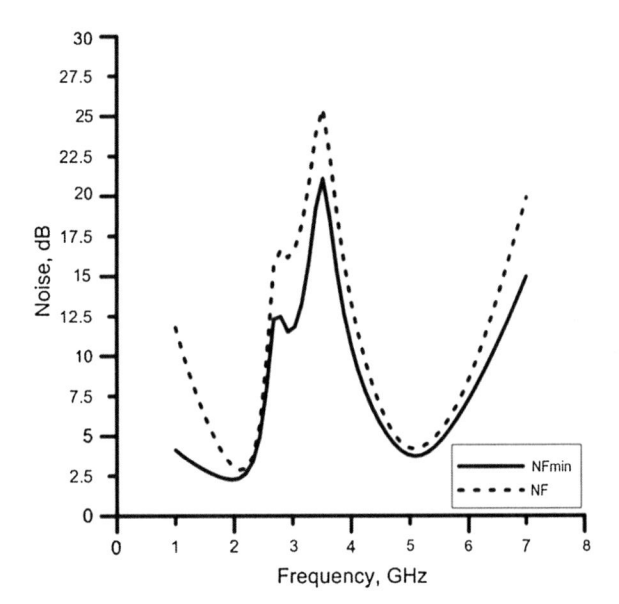

Table 47.3 Comparision analysis of related dual-band LNAs

Parameter	Hashemi and Hajimiri [1]	El-Gharniti et al. [2]	Kao et al. [3]	This work
Freq (GHz)	2.45/5.25	2.45/5.5	2.4/5.2	2.45/5.25
S11 (dB)	−25/−15	−24/−28	−10.4/−13.5	−19/−15.4
S22 (dB)	N/A	−11/−7.2	−12.5/−19.4	−15.8/−15.15
Gain (dB)	14/15.5	18.3/15.2	11.8/10	16/11.67
NF (dB)	2.3/4.5	2.26/3.2	3.9/3.7	3/4.3
Power (mW)	10	12	13.5	6.5
Technology	0.35um CMOS	SiGe–BICMOS	0.18um RF CMOS	0.13um RF CMOS

47.4 Conclusion

This article discussed a fully-integrated dual-band concurrent CMOS LNA for 2.45/5.25 GHz applications. The LNA has been designed using the Global Foundries 0.13 um RF CMOS technology. Based on the post layout simulation results, the LNA performs at desired bands simultaneously. Comparing the similar works as shown in Table 47.3, this design achieved acceptable results considering its lower power consumption. According to the additional buffer at the output stage, the design achieved better S22 results compared to [2] and [3] at lower frequencies and the input matching parameter is improved compared to [3] at both frequencies. The gain of the LNA improved at lower frequencies compared to [1] and [3]. At higher frequencies the gain is better than [3]. The noise performance of the design is improved at lower frequencies compared to [3] and also the high frequency noise performance is improved compared to [1]. Finally the proposed LNA dissipates less power compared to all the related designs mentioned in Table 47.3.

Acknowledgments Authors would like to thank Collaborative Microelectronic Design Excellence Centre (CEDEC) for supporting Cadence EDA tools software. Special acknowledgement to Universiti Sains Malaysia, short term research grant no. 304/PELECT/60311021 and the research university (RU) grant no. 1001/PELECT/814107 for the funding of this work.

References

1. Hashemi H, Hajimiri A (2002) Concurrent multiband low-noise amplifiers-theory, design, and applications. IEEE Trans Microwave Theory Tech 50(1):288–301
2. El-Gharniti O, Kerherve E, Begueret J-B, Belot D (2006) Concurrent dual-band low noise amplifier for 802.11 a/g wlan applications. In: 13th IEEE international conference on electronics, circuits and systems, 2006. ICECS'06, IEEE, pp 66–69
3. Kao C-Y, Chiang Y-T, Yang J-R (2008) A concurrent multi-band low-noise amplifier for wlan/wimax applications. In: IEEE international conference on electro/information technology, 2008. EIT 2008, IEEE, pp 514–517

4. Datta S, Datta K, Dutta A, Bhattacharyya TK (2010) Fully concurrent dual-band LNA operating in 900 MHz/2.4 GHz bands for multi-standard wireless receiver with sub-2dB noise figure. In: 2010 3rd international conference on emerging trends in engineering and technology (ICETET), IEEE, pp 731–734
5. Datta S, Dutta A, Bhattacharyya TK (2010) A gain boosted fully concurrent dual-band interstage matched LNA operating in 900 MHz/2.4 GHz with sub-2dB noise figure. In: 2010 IEEE international conference on communication control and computing technologies (ICCCCT), IEEE, pp 25–30

Chapter 48
Three Dimensional Through-the-Wall Imaging Using Ultrawideband (UWB) Sensors with Enhanced Delay-and-Sum Algorithm

N. S. N. Anwar and M. Z. Abdullah

Abstract In improving through-the-wall imaging technology, a 3-D imaging technique is developed combining enhanced delay-and-sum algorithm, EDAS and multistatic radar measurement. In this research, the scene was simulated using finite-difference time-domain, FDTD based software Meep, and then the image was reconstructed in Matlab and visualized using ParaView. A method to compensate the wall refraction in 3-D is developed and tested. The effect of freespace assumption on the 3-D images are presented and discussed. The result shows that 3-D imaging reveals more geometrical information of the target and EDAS had significantly increased the contrast and reduced the clutter.

Keywords Through-the-wall · Three dimensional · Imaging · Delay-and-sum · Wall refraction

48.1 Introduction

Through-the-wall imaging or in short TWI, has gained much interest in the last decade. This is mostly motivated by its potential applications in many critical areas either in civilian or military. The objective of TWI is to provide see-through-wall ability, and detect obscured objects in an enclosed structure. In military, the

---

N. S. N. Anwar (✉) · M. Z. Abdullah
School of Electrical and Electronic Engineering, Universiti Sains Malaysia,
Engineering Campus, 14300 Pulau Pinang, Malaysia
e-mail: syahrim@utem.edu.my

M. Z. Abdullah
e-mail: mza@usm.my

possibility to detect threats behind walls especially in urban warfare could increase situation awareness so that preemptive actions could be taken. In civilian area, the demand for TWI for example in airport and seaport security, disaster recovery and forensic discovery is also increasing as pointed out by Daniels [1].

The application of TWI can be divided based on the dimensions of the imaging space as explained by [2]. The simplest application is in 0-D, in which the system is considered as motion detector, 1-D as ranging, 2-D as a cross sectional view of the scene, and 3-D the volumetric representation of the scene. The 3-D imaging offers the possibility for object recognition, size determination, differentiation between clutter and real object, however at the expense of more computational cost. To take full advantages of TWI, this research was done in 3-D.

The biggest challenge in TWI arises from the effect of non-free space scattering of the wave which lead to attenuation, dispersion, multi-reflection, refraction and diffraction of the signal. The wall itself is accounted for the high insertion loss of the signal, since it has to pass through the wall twice. The signal will be distorted by the wall and suffers multiple reflections at the front and back surface of the wall. Due to these challenges, UWB is deemed as the best suited technology as it offers a good penetration property as well as compactness in time domain.

In this paper we investigate the effect of non-free space scattering on a Delay-And-Sum (DAS) based three-dimensional TWI. DAS is actually a free space imaging method, therefore some error and distortion are expected. The effect of the multiple reflections from the wall and the effect of the limited view imaging on the image are also studied.

48.2 Simulation Tools and Setup

In this work three main simulation tools were used starting with Meep, a free FDTD solver for electromagnetics [3]. The output from Meep is a 4-D Hierarchical Data Format, HDF which represents the volume vs simulation time, (Point$_x$ × Point$_y$ × Point$_z$ × T). In this work a scene with a dimension of 160 × 130 × 100 × 520 or equivalent to 208,000 points, with a size of memory of 845 MB was used. The HDF data will be imported into Matlab for image processing, which produces an output in form of Visualization Toolkit, VTK format that later will be imported into ParaView, an open source scientific 3-D visualization tool.

The primary work in Meep is to simulate cross-port transmission gains of unique transmitter–receiver pairs from 16 antennas, which produced 240 pairs of data. The antenna gap was set at 16 cm and arranged in a 4 × 4 array. The signal used was UWB Gaussian pulse with a central frequency of 2 GHz and bandwidth 4 GHz (Fig. 48.1).

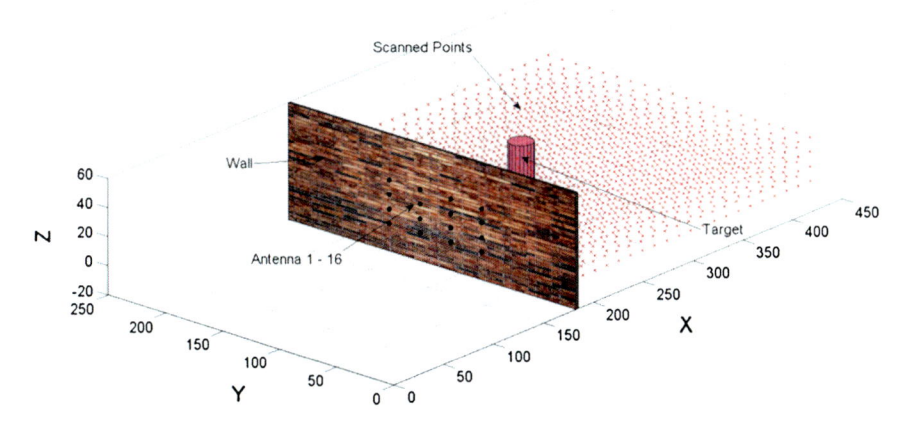

Fig. 48.1 The 3-D simulation scene

48.3 Enhanced Delay-and-Sum

Enhanced delay-and-sum, EDAS is an improvement of DAS, which is based on coherent summation of the reflected signal from different transmitter–receiver pairs. The number of transmitting antenna is denoted by M, and number of receiving antenna by N. Given the distance and the propagation speed of wave, the focusing delay, τ for each point, X can be calculated. Then, the energy reflected from each point is computed using DAS formula as follows:

$$\mathrm{En}(X) = \int_0^W \left[\sum_{m=1}^M \sum_{n=1}^N X(t-\tau) \right]^2 dt \qquad (48.1)$$

where W represents the window size.

In EDAS two more clutter suppression techniques which are (1) focusing quality weighting and (2) pairwise multiplication are applied. The formula for the focusing quality could be found in many literatures or as reported in here [4], and for pairwise multiplication in [5]. By applying focusing quality weighting, energy with low coherence factor will be weighted out before the summation. And through pairwise multiplication technique, more transmitter–receiver pair data was synthetically generated resulting in 32,640 pairs from originally 240 pairs for 16 antennas. The effectiveness of EDAS had been tested for breast cancer detection as reported in [6].

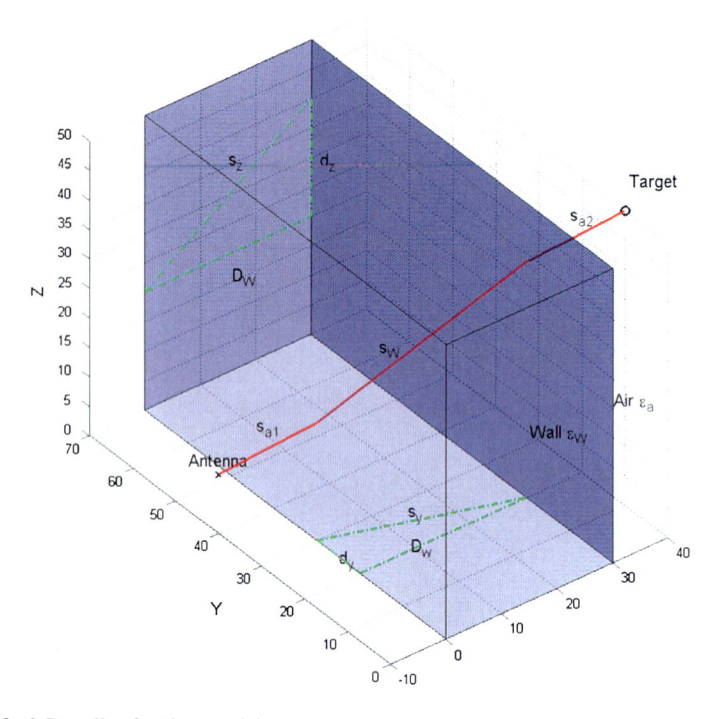

Fig. 48.2 3-D wall refraction model

48.4 The Wall Refraction Model in 3-D

Based on Eq. (48.1), the accuracy of calculation of the reflected energy is highly dependent on τ, which is directly related to time of arrival (TOA). In this case the TOA is the time taken by an electromagnetic signal to travel from a source to a target. For this reason, the signal's refraction path in the wall must be modelled correctly.

However unlike the positions of the antennas and the target points, the points of incident of the signal at the wall's surfaces are indeterminate. In a research, Aftana [7] pointed out that the distance between the antenna and the wall does not play any role in TOA, as long as the wall is in between the antenna and the target. Based on this finding, the refraction path could be modelled as in Fig. 48.2, by assuming the flight distance in air before the wall (s_{a1}) is equal to the flight distance in air after the wall (s_{a2}). The exact solution for TOA in xy plane would be as expressed in Eq. (48.2), assuming that the wall has a constant permittivity and thickness.

$$\text{TOA} = \frac{s_a}{c} + \frac{s_y}{V_w} = \frac{\sqrt{(H_x - D_w)^2 + (H_y - d_y)^2}}{c} + \frac{\sqrt{(D_w^2 + d_y^2)}}{V_w} \quad (48.2)$$

Fig. 48.3 The propagation of a point source UWB impulse at 5.3 ns after triggering

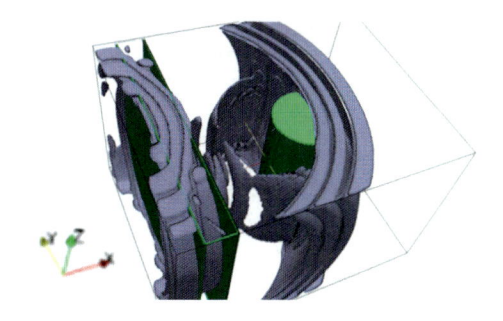

where s_a and s_y are the flight distances in air and wall respectively, c and v_w are the propagation speeds in air and wall respectively, H_x and H_y are the distances between the antenna and the target in x- and y-direction respectively, D_w is the wall thickness, and d_y is the component of s_y in y-direction.

Now the only unknown parameter in calculation of TOA is d_y. To calculate d_y, Eq. (48.2) can be differentiated with respect to d_y and equating the result to zero. This is based on Fermat's principle which dictates that an electromagnetic wave must follow the path with minimum time. Solving this is, however, a very time consuming procedure. A much faster way is to use the estimation derived by Thanh [8]. Mathematically,

$$d_y \approx \sqrt{\frac{\varepsilon_a}{\varepsilon_w}} \frac{H_y D_w}{H_x} \qquad (48.3)$$

This equation was implemented in data processing reported in this paper. It produced only negligible error up to an incident angle of 65°.

48.5 Results

Figure 48.3 shows the pulse propagation in the constructed scene which consisted of a brick wall with thickness 16 cm and a target cylinder with permittivity of 10, 20 cm radius, 60 cm height and located at 52 cm from the wall. The target is simulated in a very wide room by enclosing the volume of interest with an artificial absorbing layer or Perfectly Match Layers, PML. It can be seen from that the wavefront changes when passing through the wall. The reflection of the pulse from the target can also be clearly visualised.

Figure 48.4 shows the corresponding imaging results. The first four images show the vertical cross sectional view of the cylinder. The final two images show the 3-D image which is made-up from isosurfaces, or surface with same intensity value.

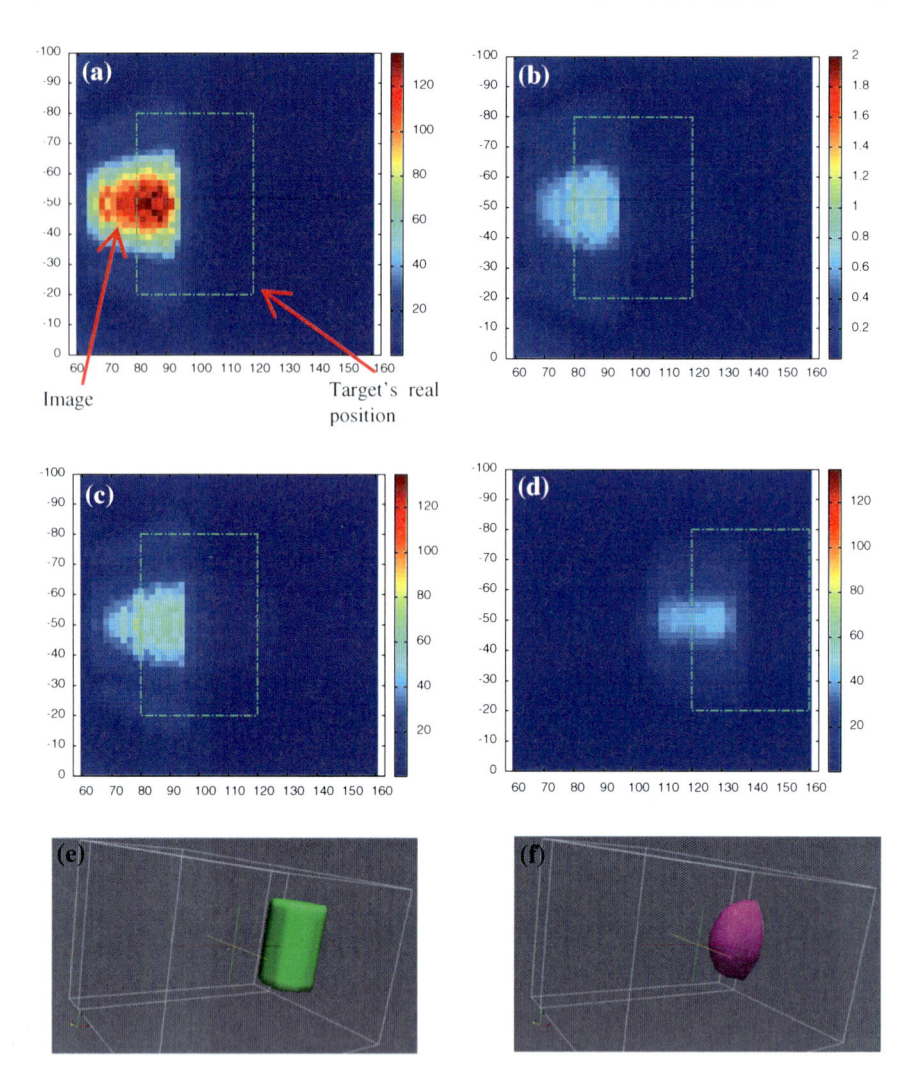

Fig. 48.4 a Vertical cross-sectional view of the cylinder, no wall, EDAS. b Vertical cross-sectional view, through-wall, 52 cm from wall, simple DAS. c Vertical cross sectional view, through wall, 52 cm from wall, EDAS. d Vertical cross-sectional view, through wall, 92 cm from wall, EDAS. e The cylinder at distance 92 cm from the wall. f 3D image of the cylinder using EDAS

48.6 Conclusion and Future Work

In conclusion the 3-D TWI offered a new perspective in interpreting images as compared to 2-D cross sectional view. A 3-D image shown Fig. 48.4f gives more geometrical information such as height and width of the target, even though it appears distorted due to the freespace assumption in DAS and limited viewing angles. With 3-D imaging, object recognition and risk elimination in real application would be possible. Closer examination of image in Fig. 48.4a–d reveals that EDAS had significantly increased the contrast and reduced the clutter. The agreement between the images for no wall and through-wall proves the correctness of the 3-D refraction model.

Migrating from 2-D to 3-D imaging comes with an extra computational and time cost as demonstrated by this work. FDTD calculation for one antenna took approximately 4 min of CPU time, resulting in a total time of 64 min for 16 antennas. Image reconstruction required another 30 min with DAS and 50 min with EDAS. All the computations were carried on a 32 bit desktop with Intel Core i7-3770 @3.4 GHz, 4 GB RAM. Further research includes validating these results using real experimental data and optimizing between speed and resolution.

Acknowledgments This work was supported by the Fundamental Research Grant Scheme (FRGS) grant 6071222.

References

1. Daniels DJ (2009) EM detection of concealed targets. Wiley, New York
2. Baranoski EJ (2008) Through-wall imaging: historical perspective and future directions. J Franklin Inst 345(6):556–569
3. Oskooi AF, Roundy D, Ibanescu M, Bermel P, Joannopoulos JD, Johnson SG (2010) Meep: a flexible free-software package for electromagnetic simulations by the FDTD method. Comput Phys Commun 181(3):687–702
4. Shum-Li W, Chen-Han C, Hsin-Chia Y, Yi-Hong C, Pai-Chi L (2007) Performance evaluation of coherence-based adaptive imaging using clinical breast data. IEEE Trans Ultrason Ferroelect Freq Control 54(8):1669–1679
5. Lim HB, Nhung NTT, Li EP, Thang ND (2008) Confocal microwave imaging for breast cancer detection delay-multiply-and-sum image reconstruction algorithm. IEEE Trans Biomed Eng 55(6):1697–1704
6. Tiang S (2012) Radar sensing featuring biconical antenna and enhanced delay and sum algorithm for early stage breast cancer detection. Prog Electromagnet Res B 46:299
7. Aftanas M (2009) Through wall imaging with UWB radar system. Technical University of Kosice
8. Thanh N, Van Kempen L, Savelyev T, Zhuge X, Aftanas M, Zaikov E, Drutarovsky M, Sahli H (2008) Comparison of basic inversion techniques for through-wall imaging using UWB radar. In: Proceeding IEEE 5th european radar conference, pp 140–143

Chapter 49
A Wideband LNA for Cognitive Radios

Chee Han Cheong, Norlaili Mohd. Noh and Harakrishnan Ramiah

Abstract This paper presents a wideband three-stage common-source (CS) low noise amplifier (LNA) with negative feedback for cognitive radio applications covering the range of 0.4–11.5 GHz. Designed in 0.13 μm CMOS process and through pre-layout simulations over the covered band, the minimum and maximum voltage gain are 11 dB and 14.3 dB respectively with noise figure of 3–4 dB. $S_{11} < -10.4$ dB for the covered band and IIP3 achieves a minimum of -2.4 dBm and a maximum of -1.2 dBm. The power consumption is 36 mW at 1.2 V.

Keywords Wideband LNAs · Cognitive radio · Three stage · Common-source · Negative feedback

49.1 Introduction

The explosive growth of number of wireless devices and data traffic are leading to congestion of the pre-allocated bands. However, certain bands are still underutilized given the time and location. Cognitive Radios (CRs) can be viewed as a potential solution to improve spectrum utilization by dynamically sensing and reconfigure itself to match the operating environment. This enables CRs to automatically detect vacant frequency bands and access them, while changing the transmission parameters to improve the overall efficiency of the radio communications [1–3].

C. H. Cheong (✉) · N. Mohd. Noh
School of Electrical and Electronic Engineering, Universiti Sains Malaysia,
Engineering Campus 14300 Nibong Tebal, Pulau Pinang, Malaysia
e-mail: chcheong1024@gmail.com

H. Ramiah
Department of Electrical Engineering, University of Malaya, Kuala Lumpur, Malaysia

Efforts have been made to expand the bandwidth of the LNA to open the way for more CR applications. The challenges may include a more stringent bound for linearity to tolerate interferers at any frequency bands (WCDMA, GPRS and GSM to name a few) in the bandwidth covered. Secondly, the broadband characteristics where CR receiver has to provide a relatively flat gain with adequate input return loss across these decades wide of frequency, thus traditional RF circuit techniques are struggling to comply [4].

The popular broadband LNA topology that employs noise cancellation technique is able to achieve a low noise figure but may face difficulties if the bandwidth is stretched wider as the noise figure escalates especially at higher frequencies [6, 8, 10]. An inductorless multi CS stages design with resistive feedback as demonstrated by [5] is able to achieve decades-wide bandwidth. This topology exploits the inductive effect possessed by the negative feedback so as to cancel the input capacitance to achieve a good input matching at low frequencies. As the design was implemented in 65 nm CMOS process, this would be an issue for processes having high terminal capacitance as it is too large for the inductive effect. In this work of the similar topology, in order to improve the input matching and overall performance in a 0.13 μm process, a gate inductor is added to generate a resonance at high frequencies.

Section 49.2 covers the design methodology of the proposed LNA. Section 49.3 presents the simulation results with analysis on the findings. Section 49.4 wraps up this paper with concluding remarks.

49.2 Proposed LNA

The concept in [5] is extended by adding a gate inductor, L_g and the schematic of the proposed wideband LNA is shown in Fig. 49.1. Component Z is replaced by a balun and the external capacitor, C_1 is for AC coupling. The circuit is biased by the output of each preceding cascaded common-source stage.

For high gain and low noise figure, the width of M_1 has to be relatively larger. Consequently, this introduces large gate-source capacitance, C_{gs} of ~30 fF. The total input capacitance can even reach a higher value if the Miller effect and capacitance of the succeeding stages are taken into account, which further upsets the S_{11}. By using the peaking inductor L_g which can be estimated by resonant frequency given by (49.1) and through a series of modeling, the value of L_g can be obtained at 10 GHz.

$$\omega_0 \approx \frac{1}{\sqrt{L_g C_{gs}}} \quad (49.1)$$

Achieving a better S_{11} could give room to optimize the feedback resistor, R_F where the gain and noise figure can be further improved. The parameters obtained are as shown in Table 49.1.

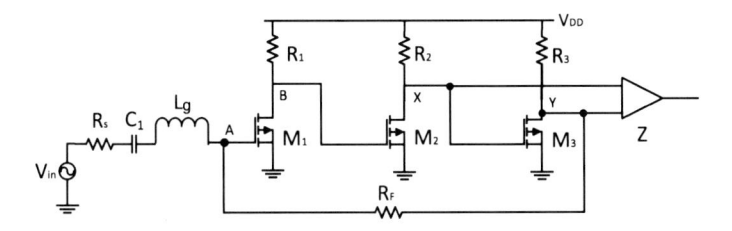

Fig. 49.1 Schematic of the proposed wideband LNA

Table 49.1 Parameters of the wideband LNA

Parameter	Value	Parameter	Value
M_1 (W/L)	110 μm/130 nm	R_F	350 Ω
M_2, M_3 (W/L)	66 μm/130 nm	L_g	1 nH
R_1, R_2, R_3	50 Ω	C_1	1 nF

49.3 Simulation Results

The S_{11} was simulated under three conditions, without L_g, with L_g and after optimization as shown in Fig. 49.2. Without the inductor, the S_{11} ranges from −48 dB at 0.4 GHz to −10 dB at 10.8 GHz. With the inductor, the bandwidth was extended to 14 GHz with $S_{11} < -10$ dB. After the optimization, S_{21} and NF were improved in the expense of S_{11} which achieved <-10 dB up until 11.5 GHz. Figures 49.3 and 49.4 show the results of S_{21} and NF after optimization. S_{21} achieved a maximum of 14.3 dB at lower frequencies and 11 dB at 11.5 GHz. Noise figure achieved a maximum of 4 dB at 0.4 and 11.5 GHz and a minimum of approximately 3 dB throughout the frequency range of 3–5.8 GHz. S_{12} and S_{22} achieved a relatively flat value of average −18.5 and −19.5 dB respectively. This confirms the statement in Sect. 49.2 where with the improvement in S_{11}, the S_{21} and NF had improved by approximately 2 and 1 dB, respectively. The results were obtained for the design without inductor from prior experiment. In other words, the bandwidth can be increased by up to 30 % with the introduction of gate inductor to the circuit, with the option to further improve both the S_{21} and NF by sacrificing part of the bandwidth.

The stability factor, K_f is plotted in Fig. 49.5. K_f is slightly more than 1.0 at 0.4 GHz and increases exponentially as frequency increases. The low stability factor is caused by the use of multiple stages where multiple poles were generated. In this case, the LNA may risk becoming unstable and oscillate at lower frequency. IIP3 and P1 dB are plotted in Fig. 49.6. P1 dB shows an increasing trend from 0.4 GHz with −18.2 dBm to 11.5 GHz with −15.6 dBm, while IIP3 achieved relatively flat trend with a minimum of −2.4 dBm at 0.4 GHz and −1.2 dBm at 11.5 GHz. Finally, Table 49.2 shows a comparison of the designs of other published broadband LNAs.

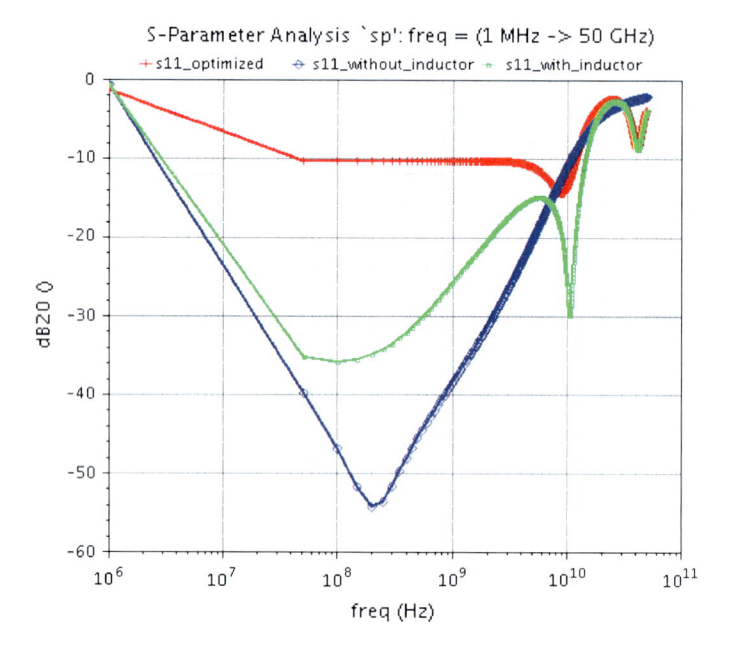

Fig. 49.2 Simulated S_{11}

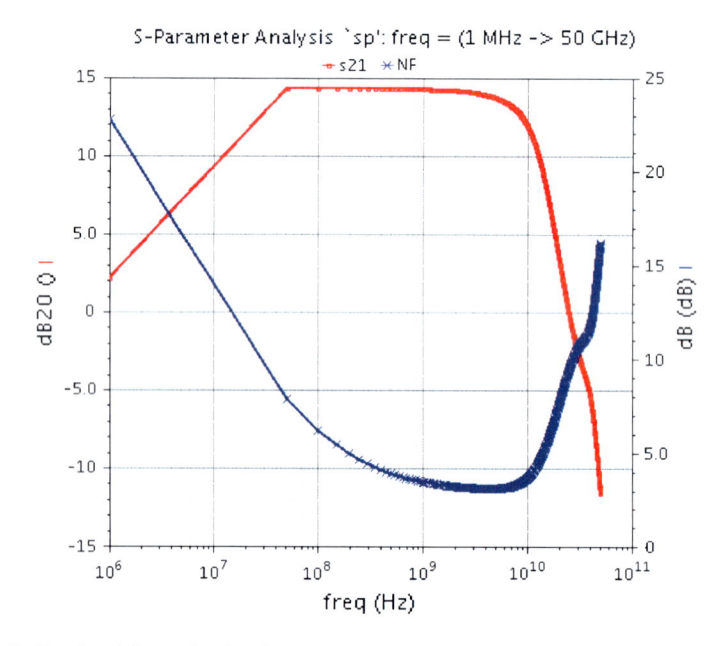

Fig. 49.3 Simulated S_{21} and noise figure

49 A Wideband LNA for Cognitive Radios

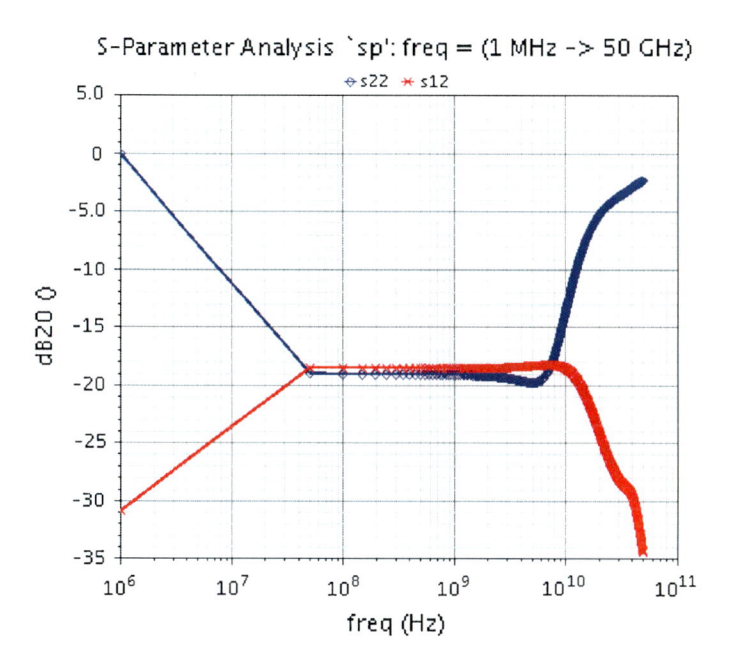

Fig. 49.4 Simulated S_{22} and S_{12}

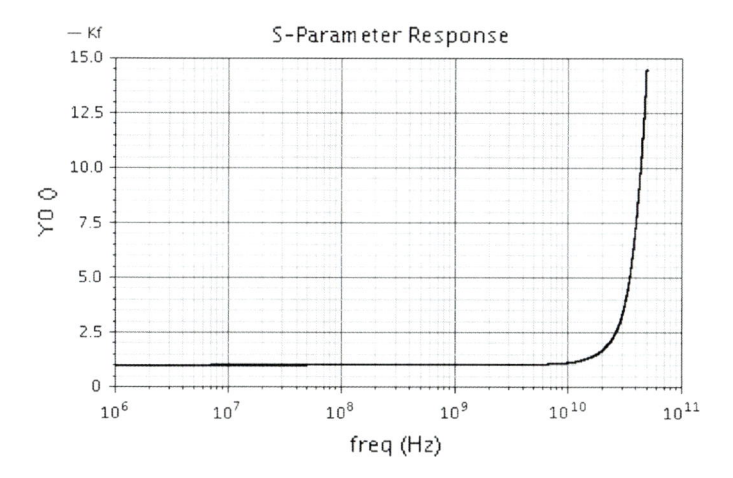

Fig. 49.5 Simulated K_f

Results from Table 49.2 show that this work is able to provide performance as expected from the 0.13 μm technology used. It is performing better than works by [9, 10] (which were designed using a lower technology). With the design from this work, it is obvious that the bandwidth is superior to the bandwidth of others and achieving good linearity too. Tradeoff of this topology is clearly on the power

Fig. 49.6 Simulated IP3 and P1 dB

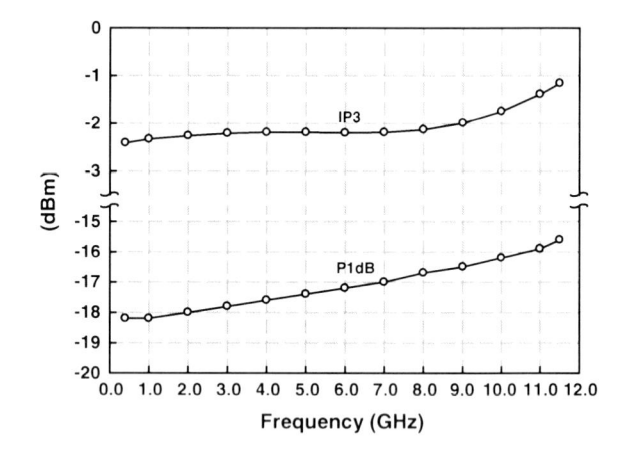

Table 49.2 Comparison of LNA performance

	[5]	[6]	[7]	[8][a]	[9]	[10]	This work[a]
CMOS Tech. (nm)	65	65	90	90	180	180	130
Bandwidth (GHz)	0.05–10	0.2–5.2	0.5–8.2	1–5	0.7–6.5	3–10.6	0.4–11.5
S_{11} (dB)	<−10	−12	−7	<−11	<−11	<−10	<−10.4
S_{21} (dB)	18–20	13–16	22–25	<11	12.5	<10	11–14.3
NF (dB)	2.9–5.9	2.9–3.5	1.9–2.6	<2	3.5–4.2	4.5–5	3–4
IP_3(min/max)	−11.2/−7	>0	−4/−16	2.4	−5	−6.2	−2.4/−1.2
Power (mW)	22	14	42	30	11	20	36

[a] Simulation results

consumption as the design is constructed of three CS stages. Overall performance is expected to perform slightly worse if this work is implemented in a lower technology. However, the power consumption can still be improved with the expense of linearity. Also, area of the LNA of this work will be slightly larger as opposed to the inductor-less design in [5] but the impact should be manageable as only one inductor is used in this work.

49.4 Conclusions

The wideband LNA presented shows an improved bandwidth and overall performance with the use of the gate inductor. In the bandwidth from 0.4 to 11.5 GHz, the minimum and maximum voltage gain achieved are 11 and 14.3 dB respectively with noise figure from 3 to 4 dB. $S_{11} < -10.4$ dB for the covered band and

IIP3 achieves a minimum of −2.4 dBm and a maximum of −1.2 dBm. The power consumption is 36 mW at 1.2 V. The bandwidth covered is able to serve frequency bands of UHF and SHF where applications may include TV broadcasting, WLAN, GPS, mobile phones, satellite, etc.

Acknowledgments The authors would like to thank Universiti Sains Malaysia, CEDEC and AISDE for the design tool and the many discussions with the IC design team. This research is supported by UM High Impact Research Grant UM.C/HIR/MOHE/ENG/51 from the Ministry of Higher Education Malaysia.

References

1. Haykin S (2005) Cognitive radio: brain-empowered wireless communications. IEEE J Sel Areas Commun 23:201–220
2. Kim JM, Sohn SH, Han N, Zheng G, Kim YM, Lee JK (2008) Cognitive radio software testbed using dual optimization in genetic algorithm. In: 3rd international conference on cognitive radio oriented wireless networks and communications, CrownCom, pp 1–6
3. Chen S, Newman TR, Evans JB, Wyglinski AM (2010) Genetic algorithm-based optimization for cognitive radio networks. Sarnoff Symposium, IEEE, pp 1–6
4. Razavi B (2009) Challenges in the design of cognitive radios (CICC), pp 391–398
5. Razavi B (2010) Cognitive radio design challenges and techniques. IEEE J Solid State Circuits 45:1542–1553
6. Blaakmeer S, Klumperink E, Nauta B, Leenaerts D (2007) An inductorless wideband balun-LNA in 65 nm CMOS with balanced output. In: 33rd European solid state circuits conference (ESSCIRC 2007), pp 364–367
7. Zhan J-H, Taylor SS (2006) A 5 GHz resistive-feedback CMOS LNA for low-cost multi-standard applications. IEEE international solid-state circuits conference (ISSCC 2006), Digest of technical papers, pp 721–730
8. Najari OE, Arnborg T, Alvandpour A (2010) Wideband inductorless LNA employing simultaneous 2nd and 3rd order distortion cancellation. In: NORCHIP, pp 1–4
9. Hu J, Zhu Y, Wu H (2008) An ultra-wideband resistive-feedback low-noise amplifier with noise cancellation in 0.18 μm digital CMOS. IEEE topical meeting on silicon monolithic integrated circuits in RF systems (SiRF 2008), pp 218–221
10. Liao C-F, Liu S-I (2007) A broadband noise-canceling CMOS LNA for 3.1–10.6-GHz UWB receivers. IEEE J Solid-State Circuits 42:329–339

Chapter 50
Direct Digital Synthesizer Based Clock Source for ADC Sampled System

Desmond Tung and Rosmiwati Mohd-Mokhtar

Abstract In this paper a Direct Digital Synthesizer (DDS) clock source application in sampled system is proposed. It is used to replace the analog Phase-Locked Loop (PLL) of a multi-formats test set that supports wireless mobile telecommunication. The DDS in this work operates from 1 µHz to 150 MHz with utilization range of 2.5–40 MHz. The rms jitter is less than 8 ps and peak jitter less than 20 ps for a reference clock signal of 20 MHz. From result, the design achieves phase noise less than −100 dBc/Hz at 1 kHz offset. It also demonstrates a total cost reduction of 61 % as compared to analog PLL implementation.

Keywords DDS · PLL · Clock source · Phase noise · Jitter

50.1 Introduction

In modern technology, frequency synthesizer is a fundamental part that can be found in wireless and telecommunication. The application includes mobile phone, wireless test set, Global Positioning System (GPS), walkie-talkie, wire or wireless network devices and many more. Mature products in the market utilize Phase-Locked Loop (PLL) frequency synthesizer to provide a referencing element either in Radio Frequency or digital application. The designs in old times used bulky passive components and rather expensive to achieve high performance.

D. Tung
Agilent Technologies, 11900 FTZ Bayan Lepas, Pulau Pinang, Malaysia
e-mail: desmondtch@gmail.com

R. Mohd-Mokhtar (✉)
School of Electrical and Electronic Engineering, Universiti Sains Malaysia,
Engineering Campus, 14300 Nibong Tebal, Pulau Pinang, Malaysia
e-mail: rosmiwati@ieee.org; eerosmiwati@eng.usm.my

Furthermore, most of the PLL design can only be used within a limited range of frequency [1].

Prior to that, Direct Digital Synthesizer (DDS) was introduced. DDS is a technique to generate sine wave with phase accumulator and digital-to-analog conversion. Since the operation of DDS is in digital, it offers a fast switching time compared to conventional analog PLL with excellent frequency resolution. Hence, it becomes key common requirement to number of industries [2]. DDS implementation varies from Field Programmable Gate Array (FPGA) to single chip solution. With FPGA, it enables migration in different implementations [3]. For mass production, single chip solution is preferred. Analog Devices who had been the leader in DDS market offers a single chip that fits customer's need. Through coding, the frequency, phase and amplitude can be adjusted. There are some that comes with modulation capabilities. Hence, the adaptive and flexibility of DDS makes it ideal for not only Radio Frequency but as many applications.

The purpose of this research work is to solve the Design for Manufacturability (DFM) issue in which the DDS will be used in replacing the analog PLL as a clock source of ADC system. The AD9852A from Analog Devices will be used for research investigation. A proper reconstruction filter will be designed to ensure the jitter and phase noise performance is met without jeopardizing the existing specification. In addition, the aim is also on reducing the cost and to improve the system performance so that its tunable frequency range can be increased. The ability of the proposed design in improving the system performance with reduced cost will be the main contribution of this work.

In brief, this paper will go as follow. Section 50.2 elaborates the DDS theory of operation. It will be followed by DDS clock generator implementation which available in Sect. 50.3. Section 50.4 will give result and analysis that have been conducted from the investigation. The magnitude and phase response, the phase noise performance and the outcome from jitter measurement will be demonstrated in this section. Finally, Sect. 50.5 concludes the paper.

50.2 DDS Theory of Operation

DDS can be used to generate variable frequency signal by controlling the rate of change of phase. It comprises of basic component like frequency multiplier (MUL), phase accumulator (PA), sine look-up table (LUT), digital-to-analog converter (DAC) and a low pass filter (LPF) as shown in Fig. 50.1.

A basic DDS takes in two inputs, an external reference clock and frequency tuning word, M. The frequency multiplier was added typically with ranging factor from 1 to 20 giving a possible of having low external clock frequency for synchronization. The reference frequency (f_{REF}) must satisfy both the system frequency (f_{SYS}) and output frequency (f_{OUT}) expressed as

50 Direct Digital Synthesizer Based Clock Source

Fig. 50.1 Basic block of DDS

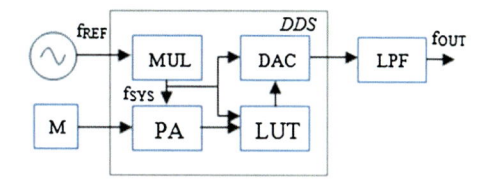

$$f_{OUT} \leq \frac{f_{SYS}}{2} \tag{50.1}$$

where

$$f_{SYS} = f_{REF} \times \text{MUL} \tag{50.2}$$

Frequency tuning word is essential to DDS implementation. It controls and steers how fast of that phase accumulator complete one revolution by interpreting them as a vector in digital phase wheel as shown in Fig. 50.2. A revolution of the vector around the circumference with a constant speed represents a sine wave [4]. Greater M will increase the speed of revolution and hence, increase the output frequency.

Ultimately the frequency resolution around the circumference is defined by the phase accumulator length. A 48 bits PA will generate fine frequency resolution of 1 µHz while 32 bits for 70 mHz at 300 MHz system clock. In other words, frequency resolution is

$$f_{OUTmin} = \Delta f = \frac{f_{SYS}}{2^N} \tag{50.3}$$

Therefore, the output frequency can be written as

$$f_{OUT} = \frac{M \cdot f_{SYS}}{2^N} \tag{50.4}$$

And the corresponding analog output signal will be

$$f(t) = A \sin\left(2\pi t \frac{M \cdot f_{SYS}}{2^N} + \theta\right) \tag{50.5}$$

The analog output being expressed with two other distinct variables, amplitude (A) and phase (θ). Both of them appeared to be adjusted in DDS chip provided by Analog Devices. It allows scalable amplitude to meet several of voltage thresholds such as transistor–transistor logic (TTL), Complementary metal–oxide–semiconductor (CMOS) stub-series terminated logic (SSTL) and more. On the other hand, the phase is adjustable to allow phase sensitive application. It is often applied in multiple microprocessor system to achieve synchronization and to avoid any data loss.

Fig. 50.2 Digital phase wheel

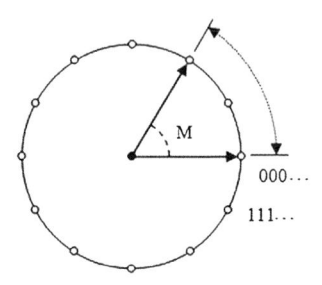

Output from phase accumulator is basically fed into a sinusoidal LUT. Typically ROM with hardcoded sine coordinate is being used. It converts the linear incremental digital words from PA into nonlinear digital words which represent the sine waveform. Its output will be supplied to the DAC dedicated to generate the analog output voltages. Due to the DAC that generated sine spectral components, it is necessary to ensure spectral purity. Another essential part of DDS implementation is the reconstruction of LPF. It helps to filter of the higher order spectral as well as ensure the output jitters reduction.

50.3 DDS Clock Generator Implementation

The DDS was implemented with both hardware and software. AD9852A comes in as a core device that generates the clock for system synchronization as well as sampling. To support multiple formats, variable frequency is inevitable. A fast switching time would be ideal for test and measurement industries. Accuracy is a critical measure to test and measurement company for performance and protocol analysis therefore, customer satisfaction.

In clock application, jitter and phase noise performance at the output are important. If the phase noise distribution is too large, the jitter might fall out of specification causing overall SNR degraded at the output. With this regards, phase noise and jitter can be used to define the specification of the intended clock design [5].

A wireless mobile test set fundamentally comes with receiver that utilizing down converted intermediate frequency (IF). The reason being the actual digital processing power is running at a lower speed that of the modulated radio frequency (RF) signal. Demodulation took place with the analog-to-digital converter (ADC) which requires number of sampling frequencies show in Table 50.1 to do the job. It occurs that the frequency resolution is important to a sample system where DDS played a main role on this (Fig. 50.3).

In test set, the performance optimization with enhanced RISC performance computing of PowerPC (PPC) serves as the main controller. It runs with real time operating system (RTOS), in this case, vxWorks. The DDS control module is

Table 50.1 Sampling frequency requirement

Formats	Sampling frequency (MHz)
GSM	18.95833
	19.9899
	13
	19
	20
TD-SCDMA	19.2
	15.36
	12.8
WCDMA	20.48
	20
	23.04
CDMA	16
	10
	14.7456
LTE	40

Fig. 50.3 DDS Clock generator in ADC sampled system

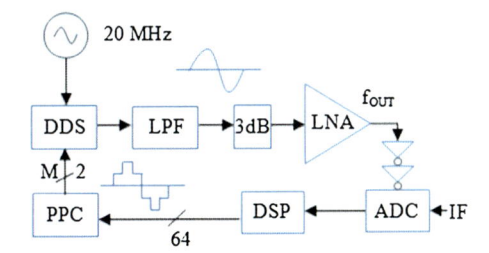

developed and bundled into the RTOS of operation. To save number of input output (IO), 2-wires operation was configured. Output of DDS is filtered by LPF and match through a 3 dB pad attenuator with a low noise amplifier (LNA) to increase the strength of signal. The sine to square conversion is done with two ultra-high speed (UHS) inverters. The seventh order elliptic reconstruction LPF is shown in Fig. 50.4.

To account for temperature and tolerance, a bandwidth of 42 MHz has been synthesized. The inductors were designed as ideal to synthesis value to achieve accuracy. LNA were used to suppress the amplification of noise at low output frequencies ranging from 2.5 to 5 MHz. This will widen the frequency range allows support towards more formats than original intended. However, the output performance of this design (Fig. 50.3) appears to be jittery. A rms jitter of 15 ps and peak jitter of 150 ps was measured from LNA and the inverters.

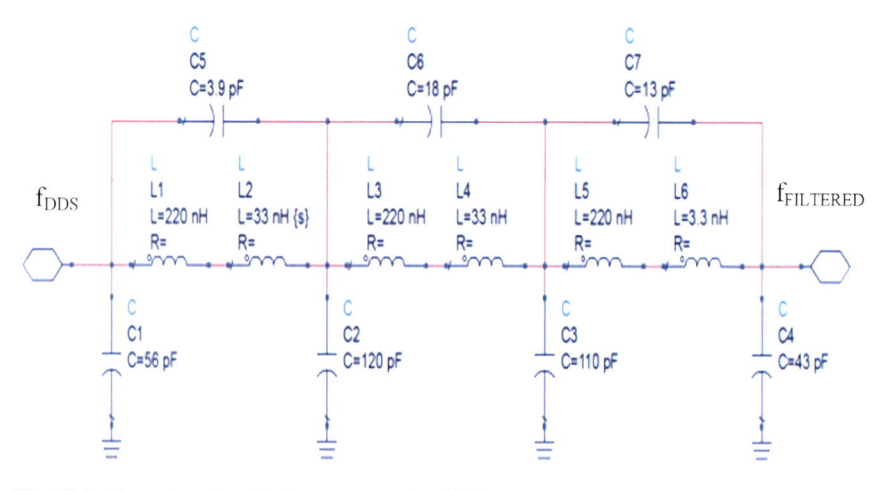

Fig. 50.4 Seventh order elliptic reconstruction LPF

Fig. 50.5 Improved DDS clock generator

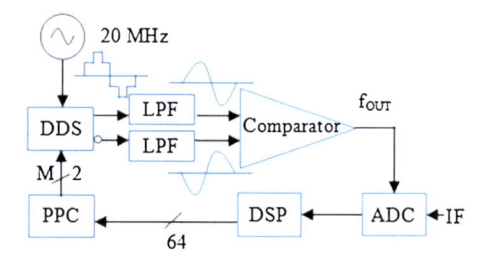

Therefore, an alternative approach being taken with the use of high speed comparator (Fig. 50.5). The specification of typical performance at about 4 ps is contributed by the comparator. Unlike LNA and inverters that has total jitter of 100 ps.

50.4 Result and Analysis

Characterization of reconstruction filter appears to be essential ensuring the output from DDS is reliable. The process took place with respect to temperature. A specification budget of operating temperature from 0 to 55 °C is guaranteed, hence, three temperatures were evaluated; those are 0, 25 and 55 °C.

Figure 50.6 shows LPF magnitude response with corner frequency of 41.6–42.18 MHz between 0 and 55 °C. The passband is not ideal as it would but with maximum voltage of 5 V, a −2.4 dB drop will reduce the output voltage to 3.7 V. With 2.4 V voltage threshold of UHS inverters, a noise margin of 1.3 V is sufficient. Within the passband, phase increases linearly as frequency increased. If

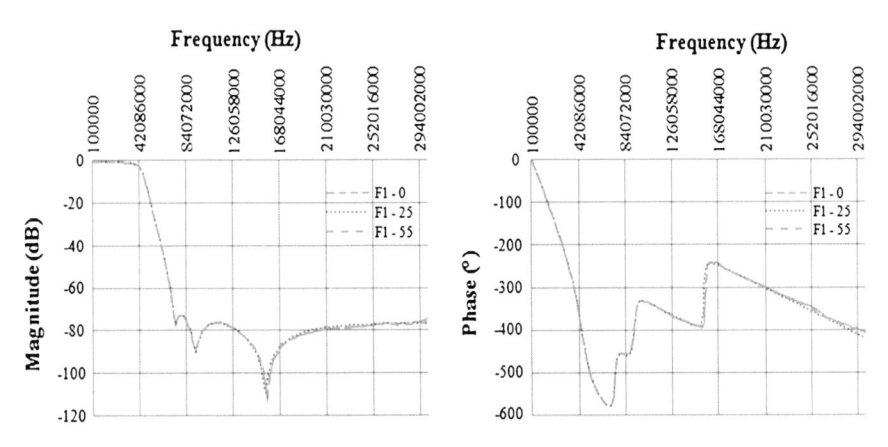

Fig. 50.6 LPF magnitude and phase response across temperatures

phase is critical in the system, AD9852 output can be adjusted programmatically to compensate the offset.

As shown in Fig. 50.7, the output frequency generated by DDS appeared to be precise. For GSM sampling frequency, the DDS was programmed to provide 18.95833 MHz. Measurement was made exactly at power level of 11.87 dBm. On the other hand, CDMA sampling frequency was programmed and measured with 14.7456 MHz at 11.93 dBm. The results show that DDS is capable to generate what an analog PLL does. Its output can be used to drive CMOS device without a problem.

As previously mentioned in Sect. 50.2, phase noise is important to be characterized as it is a function that defines the jitter in time domain [6, 7]. From Fig. 50.8, the phase noise at 1 kHz offset is measured at −101.6 dBc/Hz for GSM while −102.8 dBc/Hz for CDMA that is below −100 dBc/Hz. It shows that the DDS offers a better performance over the analog PLL.

The jitter measurement shown in Fig. 50.9 has also given promising performance. In GSM, the periodic rms jitter is measured at 6.75 ps while peak at 15.55 ps. For CDMA, since the frequency is low, the jitter appeared to be poorer with rms jitter of 7.38 ps and peak jitter of 19.79 ps.

Overall, the total cost of DDS implementation has reduced significantly. The new design that contain DDS chip AD9852, single-ended to differential converter and passive components accumulatively USD$25 while equivalent analog PLL that cost about USD$65 on the use of digital devices, VCOs and other components. There is a 61 % of cost reduction prior to the design. Comparison of initial design and alternative design is shown in Table 50.2.

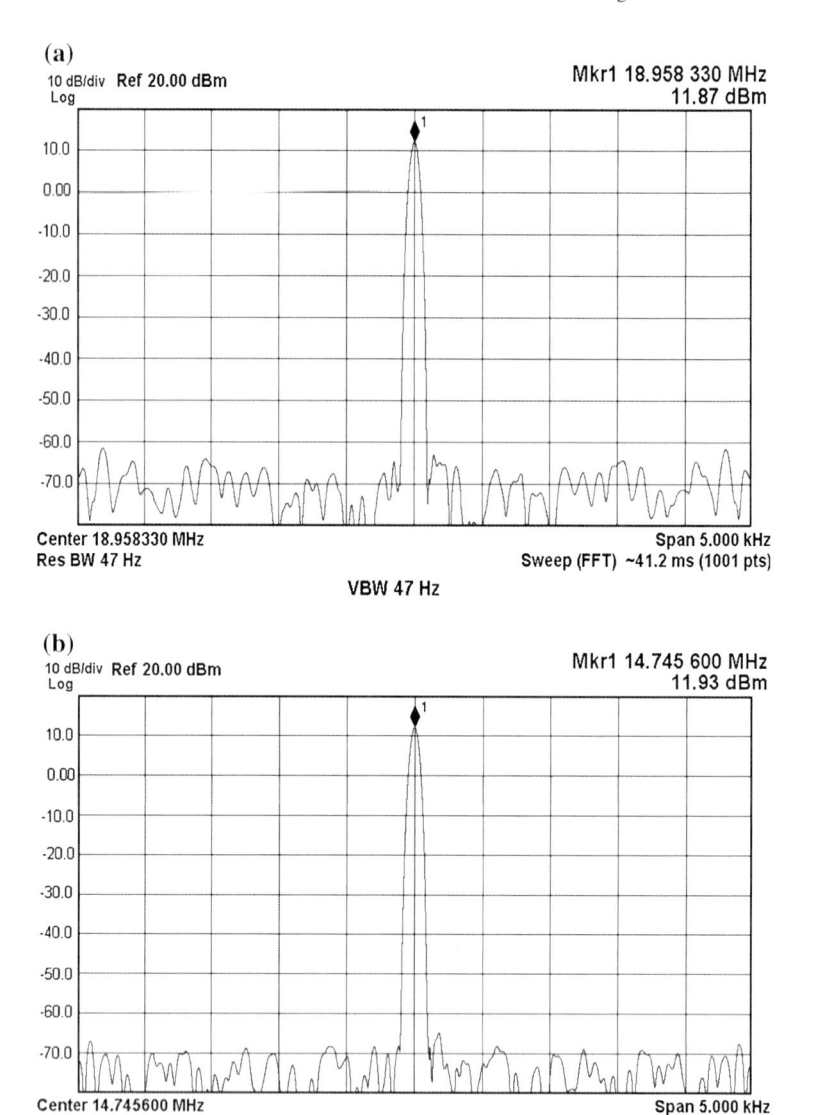

Fig. 50.7 Frequency accuracy of **a** GSM, **b** CDMA

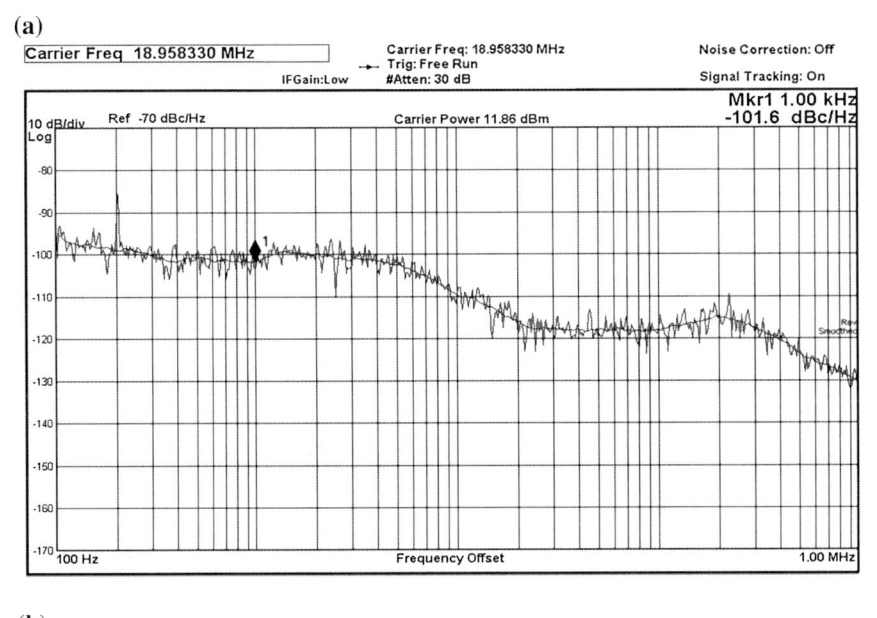

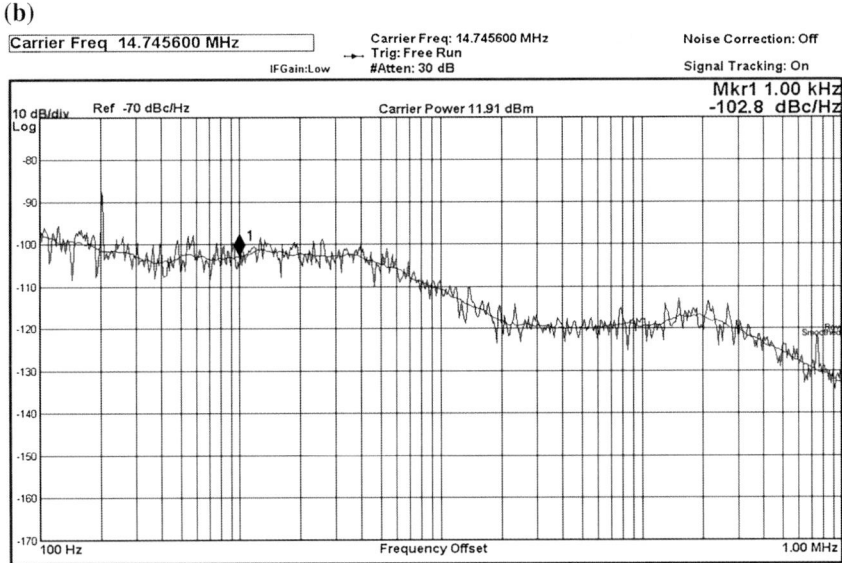

50 Direct Digital Synthesizer Based Clock Source

(a)

(b)

Fig. 50.8 Phase noise performance for **a** GSM, **b** CDMA

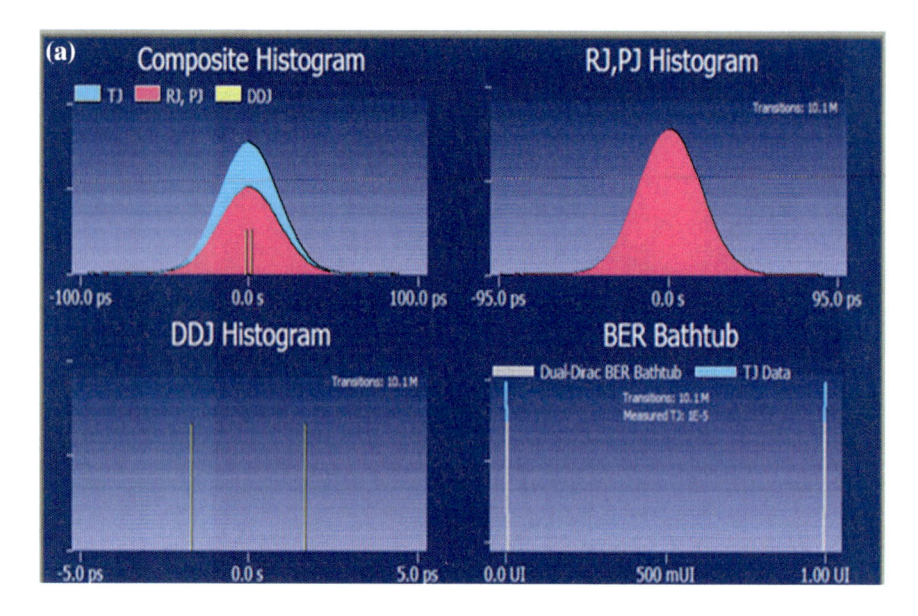

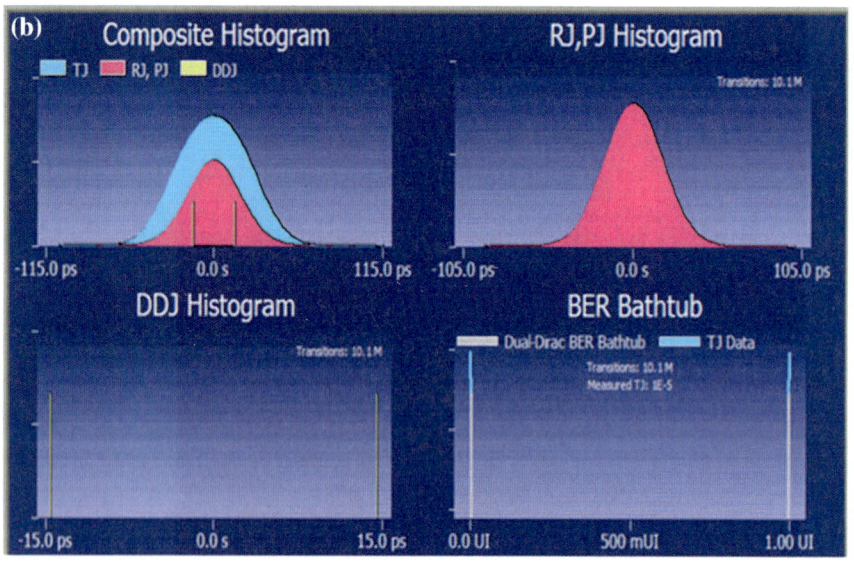

Fig. 50.9 Periodic jitter measurement **a** GSM, **b** CDMA

Table 50.2 Comparison between designs

Design	Power level (dBm)	Jitter (ps)		Cost (USD$)
		PJ_{RMS}	PJ_{PEAK}	
Initial	>5.3	<14	<110	65
Alternative	>11.5	<8	<20	25

50.5 Conclusion

The design has demonstrated an analog clock source replacement with DDS. It is ideal and provides an accurate and finest output frequency. Internal update clock was utilized since only one clock source is required. A total of ×15 multiplication was used to increase the system clock from reference clock of 20 to 300 MHz. This kept the first harmonic out from interest bandwidth.

Smallest frequency can go as low as 1 µHz. Actual maximum output frequency can go as high as 40 MHz as limited by the LPF. In terms of phase noise, at frequency offset of 1 kHz, a −100 dBc/Hz below has achieved. This ensures the output frequency is less jittery, less than 8 ps rms and 20 ps peak jitter and therefore, a clean and steady clock source. Finish product cost about USD$25 which is 61 % reduction compared to analog PLL implementation. For most of the industries, this will turn into revenue and increase the return over invested capital (ROIC).

Acknowledgments The authors would like to thank the Agilent Technologies and Universiti Sains Malaysia in supporting the carried out research investigation.

References

1. Kern P (2007) Direct digital synthesis enables digital PLLs. RF Des 3(33):26–30
2. Fang Y-Y, Chen X-J (2011) Design and simulation of DDS based on quartus II. In: IEEE international conference on computer science and automation engineering (CSAE), 10–12 June, pp 357–360
3. Stanford Research System (2013) Direct digital synthesis. Application Note, pp 1–5
4. Vankka J, Halonen KAI (2001) Direct digital synthesizer theory, design and application. Springer, London
5. Kester W (2008) Converting phase noise to time jitter. Analog Devices. MT-008 (Rev. A), 1–10
6. Yannick G, Enrico R (2012) Phase noise and amplitude noise in DDS. In: IEEE international conference on frequency control symposium (FCS), pp 1–6
7. Romashov VV, Romashova LV, Khramov KK (2011) Research of phase noise of direct digital synthesizers. In: International Siberian conference on control and communications (SIBCON2011), Krasnoyarsk, pp 168–171

Part V
Telecommunication Systems and Applications

Chapter 51
An Enhanced ITU-R 837-6 Rain Rate Prediction Model for Tropical Region

Folasade Abiola Semire, Rosmiwati Mohd-Mokhtar, Widad Ismail, Norizah Mohamad and J. S. Mandeep

Abstract This paper presents an enhanced rain rate model based on ITU-R 837-6 rain rate prediction. The improved model is derived from well behaved cumulative distribution function of over 100 rainfall rate samples from tropical region. The performance evaluation results show an improvement of -24.93 and $16.67\,\%$ in term of bias and root mean square of the relative estimation error respectively.

Keywords Rain rate · ITU-R-837-6 · TRMM · Rainfall database

F. A. Semire (✉) · R. Mohd-Mokhtar · W. Ismail · N. Mohamad
School of Electrical and Electronic Engineering, Universiti Sains Malaysia,
Engineering Campus, 14300 Nibong Tebal, Penang, Malaysia
e-mail: fasemire@lautech.edu.ng

R. Mohd-Mokhtar (✉)
e-mail: rosmiwati@ieee.org; eerosmiwati@eng.usm.my

W. Ismail
e-mail: eewidad@eng.usm.my

N. Mohamad
e-mail: norizah@eng.usm.my

F. A. Semire
Department of Electronic and Electrical Engineering, Ladoke Akintola University of Technology, Ogbomoso P.M.B 4000, Nigeria

J. S. Mandeep
Department of Electrical and Electronic Engineering, Universiti Kebangsaan Malaysia, 43600 Bangi, Selangor, Malaysia
e-mail: mandeep@eng.ukm.my

51.1 Introduction

The most difficult task faced by link design engineers is on how to prevent signal outage in any communication link operating above 10 GHz. Most rain attenuation prediction models require 1-min rainfall rate cumulative distribution function as one of the main parameters in the estimation of propagation impairment along the communication link [1–4]. However, rainfall rate measurements for short integration time are not typically available around the world most especially in the tropical and equatorial regions, where accurate rainfall measurements still remain a difficult task and most importantly in the remote areas [5–7]. Therefore, a prediction model is required to predict the local cumulative distribution of 1-min rainfall rate. The ITU-R P 837-6 rain rate prediction model is the globally accredited model for the prediction of 1-min rainfall cumulative function [8]. However, the model is developed based on scattered rainfall data around the world. The temperate rainfall database is adequately represented while tropical and equatorial regions are sparingly presented. For instance, Malaysia was represented with only one location in Kuala Lumpur. Therefore, rainfall measurements for over 100 stations are employed in the derivation of this enhanced model, couple with satellite radar rainfall derived from TRMM products [9].

51.2 Enhanced ITU-R 837-6 Rain Rate Model

The modelling of the enhanced rain rate prediction was based on 10 year rainfall data collected from Meteorological Departments of both Malaysia and Nigeria, and TRMM website for over 100 stations. The rainfall database was made up of 66 stations around Malaysia and 38 stations from Nigeria. The rainfall database of the two countries was combined because the two regions are located in the same latitudinal belts within the same tropical region. The rainfall distributions patterns around the region are closely related. From the reference rainfall database, an ample number of 72 locations were selected for modelling purposes while the remaining 32 stations are devoted for testing. The conditions for selection were largely based on well behaved shape of the curve with respect to the analytical model so as to obtain reasonable parameters from the fittings and geographical distribution of stations.

In this enhanced model, the analytical function proposed in ITU-R 837-6 was maintained. The expression is described as

$$P(R) = P_0 \cdot e^{-aR\frac{1+bR}{1+cR}} \qquad (51.1)$$

where

$$P_0 = P_{r6h} \cdot \left(1 - e^{-x\frac{M_s}{P_{r6h}}}\right) \qquad (51.2)$$

Table 51.1 ITU-R 837-6 and the new optimized coefficients

	ITU-R 837-6	Proposed coefficients for tropical region
a	1.09	0.395
x	0.0079	0.01033
y	21,797	50,192
z	26.02	14.429

$$b = \frac{M_c + M_s}{y \cdot P_0} = \frac{M_T}{y \cdot P_0} \qquad (51.3)$$

$$c = z \cdot b \qquad (51.4)$$

The coefficients of P_0, a, b and c were derived from the linear regression fittings of cumulative distribution function of the rainfall rate of each location. The probabilities of rainy 6-h periods P_{r6h} were derived from 3-hourly TRMM 3B42 V7 with resolution 0.25 by 0.25. The parameters to be enhanced are as given in the Eqs. (51.5)–(51.8). The unknown coefficients are derived from in situ data and linear regression fitting parameters ($a_{fit}, b_{fit}, c_{fit}, p_{0fit}$)

$$x = \frac{P_{r6h}}{M_s} \cdot \ln\left(1 - \frac{p_{0fit}}{P_{r6h}}\right) \qquad (51.5)$$

$$y = \frac{M_T}{b_{fit} \cdot p_{0fit}} \qquad (51.6)$$

$$z = \frac{c_{fit}}{b_{fit}} \qquad (51.7)$$

$$a = a_{fit} \qquad (51.8)$$

The weighted averages of all the derived parameters are obtained by taking an arithmetic average for the coefficient 'a' and a geometric average for coefficients 'x', 'y', and 'z'. The geometric average shows that the parameters are not independent of each other. The new coefficients obtained are shown in Table 51.1.

51.3 The Prediction Results of the Proposed Model

The result of rain rate cumulative distribution function of the proposed model is compared with measured data and ITU-R model. Results of the proposed model are closer to measured rain rate than ITU-R predictions. Figure 51.1 shows results for eight selected stations. The improved model gives a percentage error of ±10 % for time percentage from 1 to 0.001 % for the selected stations in Malaysia. Higher percentage error is recorded in some of the stations representing Nigeria. This is

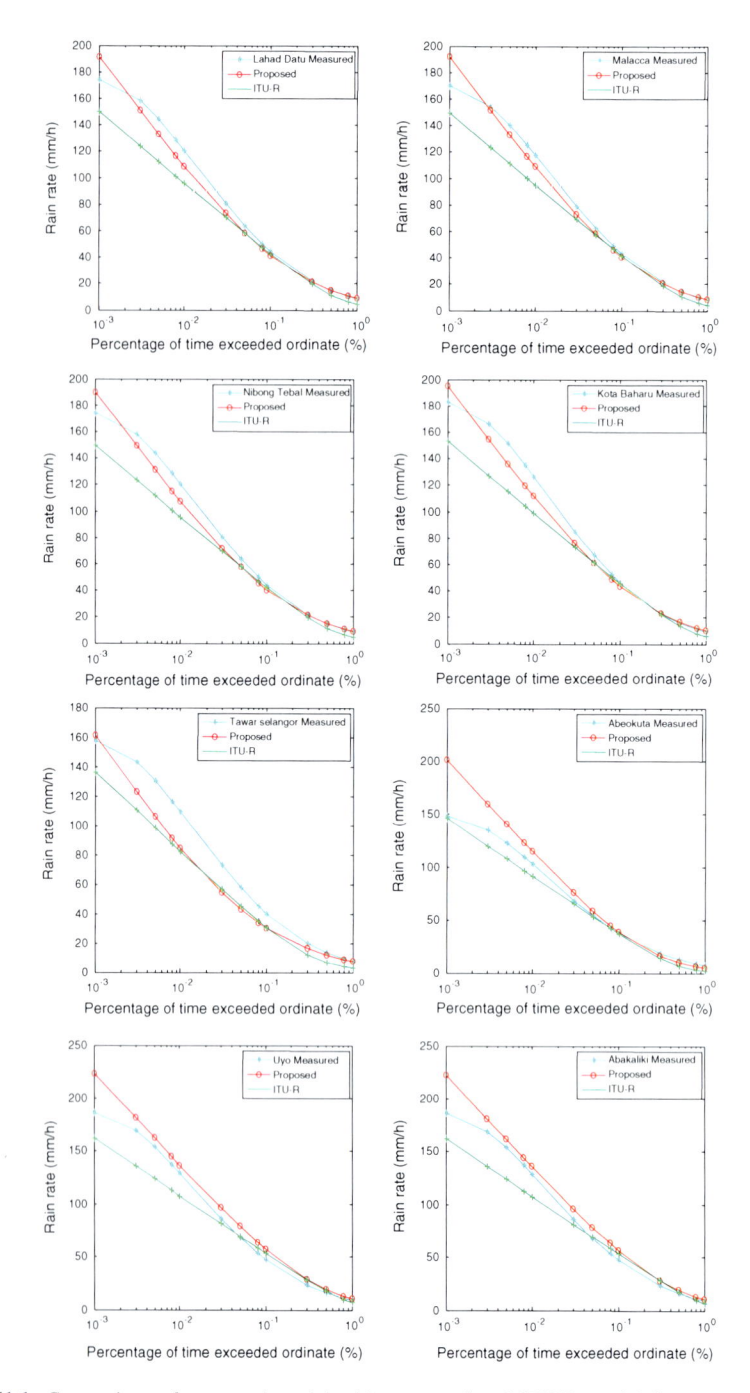

Fig. 51.1 Comparison of proposed model with measured and ITU-R model for some selected stations

evidenced in the cumulative distribution function of rainfall rate shown for Uyo and Abakaliki stations. The proposed model produced a slightly higher rain rate values as compared with the measured one. The percentage error recorded is slightly above $\pm 12\%$. In all the stations considered for testing, ITU-R 837-6 underestimates rain rate prediction in the area except for stations at Uyo and Abakaliki where rain rate is overestimated at higher time percentage from 1 to 0.1 %.

The reason for the closeness of the proposed model to measured rain rate from Malaysia may be attributed to the larger number of station database employed for model derivation compared to Nigeria and difference in rainfall distribution pattern observed due to topographic differences in the upper northern stations of Nigeria. Some of the results that are close to the proposed prediction are the ones that have similar distribution pattern with Malaysia within the same ITU-R zone P (Rainfall classification). The Northern parts of Nigeria are classified under zone N semi-arid region while Southern part are under zone P tropical wet region [10]. Therefore, the proposed model is most suitable for tropical wet and equatorial regions.

51.4 Performance Evaluation of the New Optimized Parameters Over ITU-R 837-6

The performance evaluation of the new constant coefficients derived from tropical region database over ITU-R 837-6 is evaluated based on three tests (Testing with all database, modelling database and testing database). The evaluation of the new model performance is tested using the following evaluation expressions

$$\varepsilon = \frac{R_{est}(p) - R_{meas}(p)}{R_{meas}(p)} \times 100\% \qquad (51.9)$$

$$E = \frac{\sum_{i=1}^{N} \sum_{j=1}^{n} \alpha_i \times \varepsilon}{\sum_{i=1}^{N} n_i \times \alpha_i} \qquad (51.10)$$

$$RMS = \frac{\sum_{i=1}^{N} \sum_{j=1}^{n} \alpha_i \times \varepsilon^2}{\sum_{i=1}^{N} n_i \times \alpha_i} \qquad (51.11)$$

$$\sigma = \sqrt{RMS^2 - E^2} \qquad (51.12)$$

As a result of differences in the number of years of observation of the data employed, a weighting factor as recommended by ITU-R is applied in order to account for the statistical stability of the experimental distribution. The weighting function is described as

$$\alpha_i = \frac{\sqrt{N}}{\sigma} \qquad (51.13)$$

Table 51.2 Performance evaluation results

	Mean error (%)	RMS (%)	STD (%)
Test result for all the sites in the database			
ITU-R 837-6	−36.359	39.508	15.458
Proposed model for tropical	−11.425	22.842	19.779
Test result for the testing sites in the database			
ITU-R 837-6	−37.071	38.561	10.618
Proposed model for tropical	−22.303	27.751	16.513
Test result for the modelling sites in the database			
ITU-R 837-6	−37.013	40.013	15.201
Proposed model for tropical	−12.553	24.309	20.817

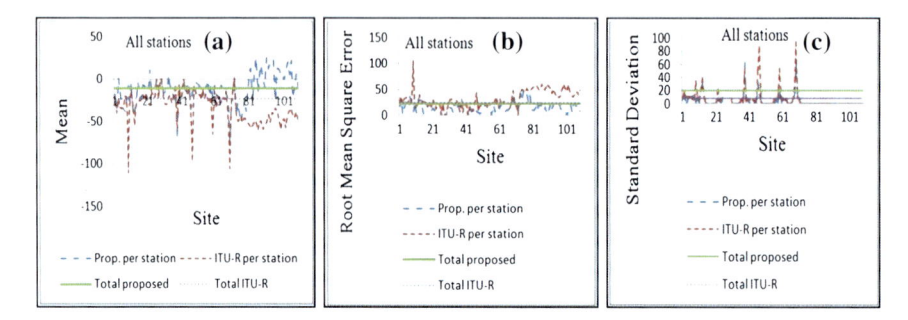

Fig. 51.2 Performance evaluation results for all sites. **a** Mean, **b** RMS, **c** STD

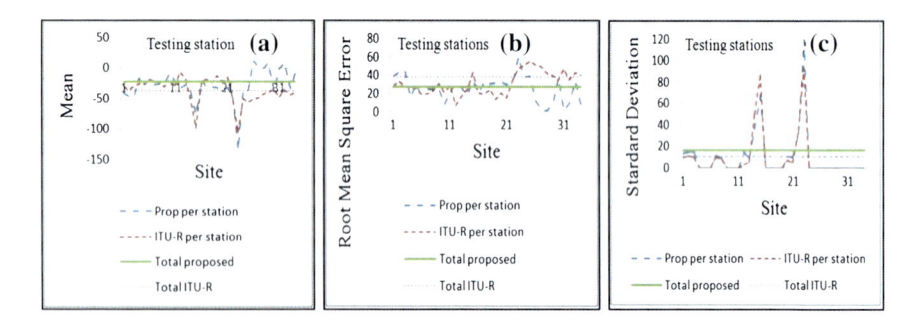

Fig. 51.3 Performance evaluation results for testing sites. **a** Mean, **b** RMS, **c** STD

with σ being the standard deviation of the year-to-year variability of rainfall and N is the number of years of measurements.

The results of the performance evaluation of the proposed model for the three categories of testing are as shown in Table 51.2. The weighted average error, root mean squares (RMS) and standard deviation (STD) are depicted in Figs. 51.2a–c, 51.3

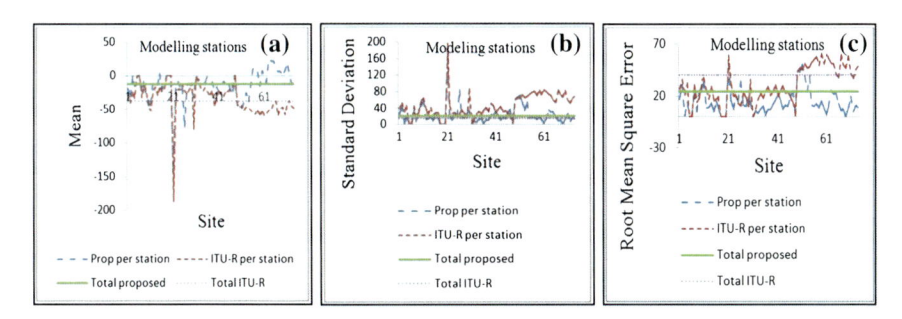

Fig. 51.4 Performance evaluation results for modelling sites. **a** Mean, **b** RMS, **c** STD

and 51.4a–c for all databases, testing database and modelling database respectively. The obtained results revealed that there is an improvement of 22.84 % (for the whole database), 27.75 % (for testing database) and 24.3 % (for modelling database) on the RMS of the estimated error over ITU-R 837-6 rain rate prediction model. Based on these results, the new enhanced ITU-R model is expected to perform better in tropical regions.

51.5 Conclusion

In this paper, an enhanced rain rate prediction model is presented. The new improved model followed the analytical function proposed in ITU-R 837-6. Well sampled rain rate cumulative distribution functions of over 100 stations from tropical regions were used. The results produced a new set of coefficients that are well fitted to rainfall distribution pattern in tropical regions. The results of the performance evaluation of the new improved rain rate prediction model over ITU-R 837-6 rain rate model, revealed that there was an improvement of 22.84 % (for the whole database), 27.75 % (for testing database) and 24.3 % (for modelling database) on the RMS of the estimated error.

Acknowledgments Authors would like to thank Universiti Sains Malaysia for the awarded RU-PRGS grant (1001/PELECT/8046022) to support this project.

References

1. Ippolito LJ (2008) Satellite communication systems engineering: atmospheric effects, satellite link design, and system performance. Wiley, London
2. Allnutt JE (1989) Satellite-to-ground radiowave propagation: theory, practice and system impact at frequencies above 1 GHz. Peter Perigrinus Ltd, London
3. Crane RK (2003) A local model for the prediction of rain-rate statistics for rain-attenuation models. IEEE Trans Antennas Propag 51(9):2260–2273

4. Ito C, Hosoya Y (1999) Worldwide 1 min rain rate distribution prediction method which uses thunderstorm ratio as regional climatic parameter. Electron Lett 35(18):1585–1587
5. Mandeep SJS, Hassan SIS, Ain MF, Ghani F, Kiyoshi I, Kenji T, Mitsuyoshi I (2006) Earth-to-space improved model for rain attenuation prediction at ku-band. Am J Appl Sci 3(8):1967–1969
6. Moupfouma F (1994) More about rainfall rate and their prediction for radio system engineering. IEE Proc 134-H(6):527–537
7. Semire FA, Mohd-Mokhtar R, Omotosho TV, Ismail W, Mohamad N, Mandeep JS (2012) Analysis of cumulative distribution function of 2-year rainfall measurements in Ogbomoso, Nigeria. Int J Appl Sci Eng 10(3):171–179
8. ITU-R 837-6 (2012) Characteristics of precipitation for propagation modelling. Telecommun Syst, Geneva
9. Simpson J, Adler RF, North GR (1996) A proposed tropical rainfall measuring mission (TRMM) satellite. Bull Amer Meteor Soc 69:278–295
10. Omotosho VT, Oluwafemi CO (2009) Impairment of radio wave signal by rainfall on fixed satellite service on earth-space path at 37 stations in Nigeria. J Atmos Solar-Terr Phys 77:830–840

Chapter 52
Study on the Effect of Dielectric Structure to the Cylindrical Dielectric Resonator Antenna (DRA)

Mohamadariff Othman, Mohd Fadzil Ain, Ubaid Ullah,
Seyi S. Olokede, Mohd Zaid Abdullah, Arjuna Marzuki,
Wan Fahmin Faiz Wan Ali, Zainal Arifin Ahmad, Julie Juliewatty
and Srimala Sreekantan

Abstract This paper presents the effect of variation of diameter and height of DR to the resonant frequency and radiation pattern of DRA. Analysis was carried out on the DR with diameter of 10.66 mm and height of 4.4 mm. The simulated and measured resonant frequencies were 3.67 and 3.77 GHz, respectively. The E-plane

---

M. Othman (✉) · M. F. Ain · U. Ullah · S. S. Olokede · M. Z. Abdullah · A. Marzuki ·
W. F. F. W. Ali · Z. A. Ahmad · J. Juliewatty · S. Sreekantan
School of Electrical and Electronic Engineering, Universiti Sains, Penang, Malaysia
e-mail: andikalusia83@yahoo.com

M. F. Ain
e-mail: mfadzil@eng.usm.my

U. Ullah
e-mail: xs2ubaid@yahoo.com

S. S. Olokede
e-mail: solokede@gmail.com

M. Z. Abdullah
e-mail: mza@eng.usm.my

A. Marzuki
e-mail: eemarzuki@eng.usm.my

W. F. F. W. Ali
e-mail: wanfahminfaiz87@gmail.com

Z. A. Ahmad
e-mail: zainal@eng.usm.my

J. Juliewatty
e-mail: srjuliewatty@eng.usm.my

S. Sreekantan
e-mail: srimala@eng.usm.my

radiation pattern of CDRA was bi-directional pattern. The analysis was based on the CST simulation which varied the thickness and height of DR with the same increment and decrement of 1 mm. The resonant frequency of CDRA shifts significantly when the height of DR is varied, in contrast to the variation of DR's diameter. The radiation pattern of antenna does not change when the dimension (height or diameter) of DR were altered.

Keywords Antenna · Dielectric material · Dielectric resonator · Dielectric resonator antenna

52.1 Introduction

Dielectric resonator antenna (DRA) is famously well known in the world of antenna for their appealing features of inherent hardness, lightweight, small in size as well as high efficiency due to absence of conductor loss and high flexibility to the various feeding method. It can also be formed into various shapes such as cylindrical, rectangular and ring [1–4]. Out of all the shapes, cylindrical and rectangular DRA are most commonly used in design [2, 5]. In cylindrical DRA (CDRA), it has aspect ratio of radius to height which can be manipulated to control the properties of antenna. A variety of modes in the CDRA can be excited to produce either broadside or omnidirectional radiation pattern [4]. In various papers, they discuss on the effect of dielectric resonator (DR) dimension on the microwave DR-based filter [6–8]. This later leads to the mechanical tuning screw method in controlling frequency of DR-based filter. However, there is less discussion on the effects of DR dimension to DRA. For instance, the application of stacked DR with different DR to perturb the permittivity of DRA [9, 10]. Variation of diameter as well as height of DR definitely alters the resonant frequency of DRA. Yet, it is interesting to observe whether either of DR height or diameter will have more significant influence on the resonant frequency and radiation pattern of DRA. This will pave the way for a simple method to control the resonant frequency of DRA especially during the prototyping stage. Hence, in this paper, analysis on the variation of the diameter and thickness of a cylindrical DRA is presented. The variation is carried out in equal increment and decrement for both the diameter and height. The influence of the both variation to the resonant frequency and pattern is investigated and compared.

52.2 Methodology

The DRA under investigation consists of cylindrical DR feed by simple 50 Ω microstripline. Figure 52.1 shows the structure of the CDRA. The DR was fabricated via conventional solid state reaction with permittivity of 55. The diameter

Fig. 52.1 Geometry of CDRA

of DR is 10.66 mm while its height is 4.4 mm. The characterization of permittivity of DR was made using Agilent 85070E Dielectric Probe Kit connected externally to the PNA-X Network Analyzer. Measurement on the S-parameter (S11) was done using PNA-X Network Analyzer. In order to predict the resonant frequency of CDRA, the simulation software of CST Microwave Studio was used. Later on, based on this particular CDRA, analysis on the variation of diameter and height was also done using CST Microwave Studio. The height of DR was varied from 2.4 to 6.4 mm with 1 mm increment. Whereas, the diameter of DR was changed form 8.66 to 12.66 mm with the same increment. The analysis was carried out on the resonant frequency and radiation pattern performances of CDRA. Comparison was made to reveals which variation of dimensions (height or diameter) of DR has much influence to CDRA.

52.3 Results and Discussion

Simulated and measured return loss for 10.66 mm diameter of CDRA is shown in the Fig. 52.1. Simulated return loss is -35.69 dB at 3.675 GHz. Measured return loss is -36.4 dB at the frequency higher than the simulated one at 3.77 GHz. The measured bandwidth is 201.5 MHz which is lower than the simulated bandwidth of 331 MHz. The dissimilarities occur because of inconsistencies value of permittivity and tan loss from the DR (Fig. 52.2).

As shown in Fig. 52.3, there are two curves representing the variation of the height and diameter of DR with the same differences of 1 mm each. As stated by [1], resonant frequency of DRA is proportional to the size, shape and dielectric constant of the DR. The frequency of DRA tends to decrease when the diameter and height of DR increases, and inversely proportional as the dimensions of height and diameter decrease. However, it can be clearly seen that the frequency shifting due to the height variation is much bigger compared to the variation of the diameter. When the height of DR reduces from 4.4 to 3.4 and 2.4 mm, the resonant frequencies of DRA shift to higher frequencies from 3.717 to 4.236 and 5.211 GHz, respectively. Comparing to the diameter's variation from 10.66 to 9.66

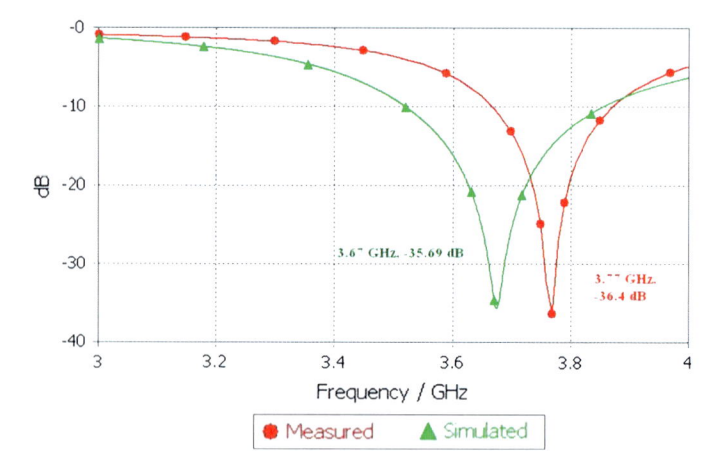

Fig. 52.2 Return loss of CCTO DRA for 10.66 mm diameter

Fig. 52.3 Variation of DR height and diameter versus frequency

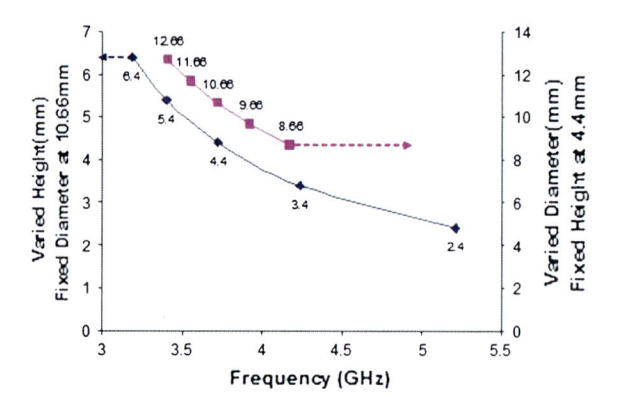

and 8.66 mm, the resonant frequency only increases to 3.922 and 4.175 GHz, respectively. Even when the diameter or height of DR is decreased, the frequency shifting occurs due to the height's variation is still larger. Additionally, based on the pattern of the curves, when the height of DR decreases (6.4–2.4 mm), the resonant frequency decreases exponentially whereas decreasing the diameter of DR (12.66–8.66 mm) reduces the frequency steadily.

The height of DR has more influence due to the electric field, E of excited HE mode which is perpendicular to the surface of the DRA. According to [8], the resonant frequency of any mode, in which the E is perpendicular to the DR's surface, is strongly dependent on the height of DR. As the height of DR changes, the E field is disturbed and resonant frequency shifting occurs.

Based on Figs. 52.4 and 52.5 the E-plane radiation pattern of CDRA is a bi-directional type pattern. These E-plane patterns were viewed in Φ (XY plane) at

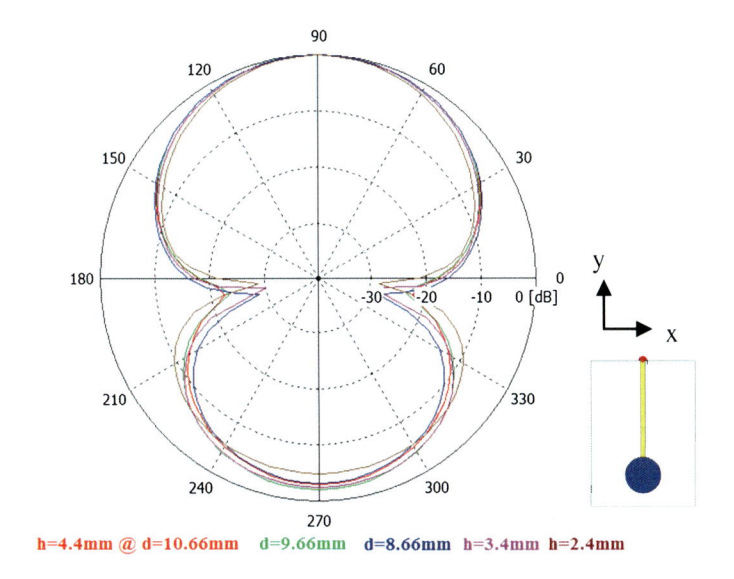

Fig. 52.4 E-plane radiation pattern of CDRA as the DR dimension decreases

Fig. 52.5 E-plane radiation pattern of CDRA as the DR dimension increases

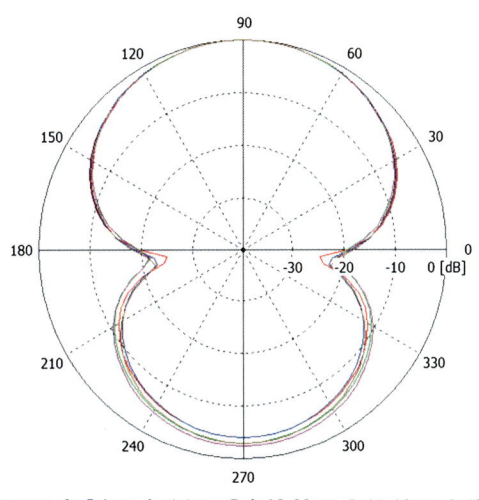

their respective resonant frequency. It is clear that bi-directional pattern does not suffer any side effect from the variation of diameter and height of DRA. Only when the height of DRA decreases to 3.4 and 2.4 mm, there are additional dips in 90° and 270°. Increasing the dimension of either the diameter or height of DR does not have any effect on the bi-directional type radiation pattern of this CDRA.

52.4 Conclusion

In conclusion, a simple CDRA design operating at 3.675 GHz was simulated and measured. The return loss shows good agreement between the simulation and measurement results. The variation of both the height and diameter does alter the resonant frequency of CDRA but not the bi-directional type radiation pattern. It indicates that by varying the height of CDRA, the frequency shifting occurs significantly as compared to the diameter. The maximum frequency shifting occurs from 3.675 to 5.211 GHz when the height is reduced to 2.4 mm. As a result, the height of CDRA is a better option in altering manually the resonant frequency of CDRA without any degradation on the radiation pattern.

Acknowledgments The works was supported by research university team grant (RUT) of 1001/PELECT/854004, postgraduate research grant scheme (PGRS) of 1001/PELECT/8044046 from Universiti Sains Malaysia and MyBrain scholarship from Malaysian ministry of higher education (Ref. No. KPT/B/831219045295).

References

1. Cuhaci M, Ittipoboon A, Petosa A, Roscoe D, Simons N (1996) Dielectric resonator antennas as an alternative technology for PCS applications. In: Canadian conference on electrical and computer engineering (CCECE'96). IEEE Press, Calgary, pp 867–870
2. Kishk AA (2003) Dielectric resonator antenna, a candidate for radar applications. In: IEEE radar conference. IEEE Press, Alabama, pp 258–264
3. Petosa A (2007) Dielectric resonator antenna handbook. Artech House, Bolton
4. Luk KM, Leung KW (2003) Dielectric resonator antenna. Research Studies Press Ltd, Hertfordshire
5. Mridula S, Pau B, Menon SK, Aanandan CK, Vasudevan K, Mohanan P, Bijumon PV, Sebastian MT (2006) Wideband rectangular dielectric resonator antenna for W-LAN applications. In: IEEE Antennas and Propagation Society International Symposium, IEEE Press, California, pp 1363–1366
6. Pratsiuk B, Tkachov D, Prokopenko Yu, Poplavko Yu (2010) Tunable dielectric resonator: design and parameters. In: 20th International crimean conference microwave and telecommunication technology, IEEE, Ukraine, pp 655–656
7. Poplavko Yu, Prokopenko Yu, Molchanov VI, Dogan A (2001) Frequency-tunable microwave dielectric resonator. IEEE Trans Microwave Theory Tech 49(6):1020–1026
8. Karp A, Herbert JS, Donald KW (1968) Circuit properties of microwave dielectric resonators. IEEE Trans Microwave Theory Tech 16(10):818–828
9. Sebastian MT, Jawahar IN, Mohanan P (2003) A novel method of tuning the properties of microwave dielectric resonators. Mater Sci Eng B97:258–264
10. Li L, Chen XM (2009) $Ba_2Ti_9O_{20}$–$Ba_{1.85}Sm_{4.1}Ti9O_{24}$ layered dielectric resonators with adjustable effective dielectric constant. Mater Lett 63:252–254

Chapter 53
Feed Coupling Comparative Assessment of Selected Microstrip Patch Antenna

Seyi Stephen Olokede, Mohd Fadzil Ain, Ubaid Ullah,
Arjuna Marzuki, Julie J. Mohammed, Srimala Sreekantan,
Sabar D. Hutagalung, Zainal A. Ahmad and Mohd Z. Abdullah

Abstract Feed coupling comparative assessment of some selected patch antennas is presented. Different available feed configurations were implemented ranging from the conventional microstrip feed line to coaxial feed probe, quarter wavelength feed, inset coupled, and a few available capacitively loaded coupled feed which are also optimized for maximal excitation, and hence, efficiency. Experimental results were understudy and a thorough comparative assessment summary table is presented. It was noted that the inset feed looks more promising.

Keywords Feed coupling · Comparative analysis · Maximal excitation · Performance assessment

---

S. S. Olokede (✉) · M. F. Ain · U. Ullah · A. Marzuki · M. Z. Abdullah
School of Electrical and Electronic Engineering, Universiti Sains Malaysia,
14300 Nibong Tebal, Penang, Malaysia
e-mail: solokede@gmail.com

M. F. Ain
e-mail: mfadzil@eng.usm.my

U. Ullah
e-mail: xs2ubaid@gmail.com

A. Marzuki
e-mail: eemarzuki@eng.usm.my

M. Z. Abdullah
e-mail: mza@eng.usm.my

J. J. Mohammed · S. Sreekantan · S. D. Hutagalung · Z. A. Ahmad
School of Material and Mineral Resources Engineering, Universiti Sains Malaysia,
14300 Nibong Tebal, Penang, Malaysia
e-mail: sreekantansrimala1974@gmail.com

S. D. Hutagalung
e-mail: mrsabar@eng.usm.my

Z. A. Ahmad
e-mail: zainal@eng.usm.my

53.1 Introduction

Antenna feed has always been a crucial requirements to determine its efficiency. Alternatively, the less efficient the antenna, the lower is its power utilization. Antenna impedance is the ratio of voltage to current at the antenna's terminals. The fundamental of antenna theory requires that the antenna be impedance matched to the feed. Otherwise, the antenna will not radiate. The extent to which this antenna radiates (the radiation efficiency) and also, its bandwidth depend largely but not solely on the feed mechanism. In [1], it was reported that an improved impedance match will ideally increase the bandwidth, the return loss, antenna size, and improved performance by reducing the excitation of unwanted modes of radiation. It is as a response to the severity of this implication that have led to the introduction of the concept of voltage standing wave ratio (VSWR), as a measure in order to determine the extent to which an antenna is matched. It is this same motive that has driven us to examine or rather understudy the assessment of feed coupling performance particularly for some selected patch antenna, with the view to determine their performance profile.

Therefore to accomplish this, these four objectives were set. The first one is to do a theoretical framework for these feeds in a bid to better understand them. The second is to design an antenna say for 5.8 GHz, and feed it with different feed mechanisms. When this is done, its performance and radiation characteristics are taken and documented. The third one is to critically compare these performance parameters with the view to determine the most efficient type, and finally, to draw our conclusion and suggest necessary recommendation. To that extent, Sect. 53.2 will be dedicated to the antenna theoretical framework, Sect. 53.3 to the antenna specifications and geometry, Sect. 53.4 to performance profile comparison, and finally, Sect. 53.5 will conclude our stand as regards feed coupling performance assessment.

53.2 Theoretical Framework

The input impedance is the impedance presented by the antenna at its terminals [2]. It is a complex function of frequency, and as such the frequency response of an antenna at its port is defined as the input impedance. Concisely, the input impedance Z_{in} can be stated as [3],

$$Z_{in}(f) = R_{in}(f) + jX_{in}(f) \tag{53.1}$$

where X_{in} is the energy stored in the near field of the antenna, f is the frequency of operation, and R_{in} is the energy used. This energy can be used in two ways namely: first for radiation, and secondly, as ohmic losses. Therefore, Eq. (53.2) states the energy utilization potential of a typical antenna [2].

$$R_{in} = R_{radiation} + R_{ohmic} \qquad (53.2)$$

Theoretically (and even in practice), antenna design engineers labour to ensure that $R_{rad} \gg R_{ohmic}$ as their design objective seeing that Rohmic is undesirable, where equation to evaluate the radiation resistance is stated in [4]. Effort to attain this will be determined by the efficiency of the feed employed. To this end, a proper study of types of feeds is imperative, with the view to assess their efficiency. In coaxial feed therefore, the probe position provides the impedance control in a similar manner to inserting the feed for an edge-fed patch [5]. Because of the direct contact between the feed transmission line and the radiating element, probe (coaxial) feeding is referred to as a direct contact excitation mechanism. The advantage of this technique is the direct coupling to the 50 Ω system and useful at lower frequency application where aperture-coupling may not be practical. The feed network, where antenna element may be located, is isolated from the radiating elements via a ground plane, and thus allows independent optimization of each layer. Of all the excitation methods, probe feeding is probably the most efficient because the feed mechanism is in direct contact with the radiating element and most of the feed network is isolated from the antenna, minimizing spurious radiation [6]. The microstrip feed line basically simplifies the problem of making many connections in a complex microwave circuits. Usually, it consists of a conductor, an insulator, and a flat plane called the ground plane.

The strip on the top is like the inner conductor, and the space in between is the insulator. Otherwise, the maximum amount of coupling is significantly reduced [7]. Notable but specific advantages include: ease of control of the level of the input impedance, relatively easy to model, simple transmission line models to estimates the input impedance performance of the antenna, several microwave transmission lines can be easily connected together by putting them on the substrate while the bottom serves as the grand plane, and finally, the feeding mechanism offers ease of integrations with the other microwave circuit. However, it has bandwidth and gain characteristics that are relatively narrow and moderate, poor surface wave efficiency, suffers from relatively high spurious feed radiation, and finally, because the feed network is not separated from the antenna, this causes the feed network to radiate. In a conventional microstrip-fed patch antenna, the input impedance is high at the edges and nose dive as it approaches the centre. Simultaneously, the current is low at the ends of a half-wave patch and increases in magnitude toward the centre. Therefore, the input impedance (Z_{in} = V/I) could be reduced if the patch is fed closer to the centre. An inset-fed method is an appropriate solution. Also, the voltage decreases in magnitude by the same amount of the increased current. Consequently, using (Z_{in} = V/I), the input impedance scale is [8],

$$Z_{in}(R) = Cos^2\left[\frac{\pi R}{t}\right] Z_{in}(0) \qquad (53.3)$$

where Z_{in} is at a point of the application of the feed to the assumed resonator. In coupled feed, the feed line is stopped just before the resonator. Consequently, a capacitive effect is introduced due to the presence of gap that existed between the patch and the feed. Therefore, the advantage of the feed is the effect of the introduced capacitance, which neutralizes the inductance added by the probe feed. A quarter wavelength transmission lines can be used to match the input impedance (Z_{in}) of an antenna to the transmission line (Z_0). When this is done, the input impedance is given as [8],

$$Z_{in} = Z_0 = \frac{Z_1^2}{Z_A} \qquad (53.4)$$

where Z_1 is the impedance of the quarter wavelength transmission line, and Z_A is the impedance of the antenna. The parameter Z_1 can be altered by changing the width of the quarter wavelength transmission line. This alteration in turn changes the input impedance Z_1. Basically, the wider the transmission line, the lower the impedance Z_0.

53.3 The Antenna Specification and Geometry

Figures 53.1 and 53.2 below show microstrip patch antennas designed for 5.8 GHz frequency operation. These antennas were fed in turn by different feed mechanism. The first one was excited by a coaxial feed probe, the second by a microstrip feed line, the third by an inset feed line, the fourth by an inset feed line forming a capacitively coupled, and finally, by a quarter wavelength transmission line. All the designs were understudy to specifically examine the efficiency of the feeds with respect to their return loss, bandwidth advantage, radiation characteristics. The simulation was done using full wave Finite-Integration Technique (FIT) CST microwave studio. The designs were later fabricated on a grounded Duriod RO4003C dielectric substrate with dielectric permittivity of 3.38 with a thickness of 0.813 mm, and of metal thickness of 0.035 mm.

53.4 Results and Discussions

Essentially, the conventional microstrip patch was excited with microstrip feed of width 1.898 mm. The simulated and measured return losses are as given in Fig. 53.3. The simulated return loss was -21.98 dB with a bandwidth of 68 MHz. The measured return loss was -18.22 dB with a bandwidth of 60 MHz and impedance of $50.01 - j1.21$ Ω. The simulated voltage standing wave ratio (VSWR) shown in Fig. 53.4 was 1.38:1 at a frequency of 5.78 GHz whereas the measured VSWR was 1.29:1 at a frequency of 5.8 GHz. The inset-fed Microstrip

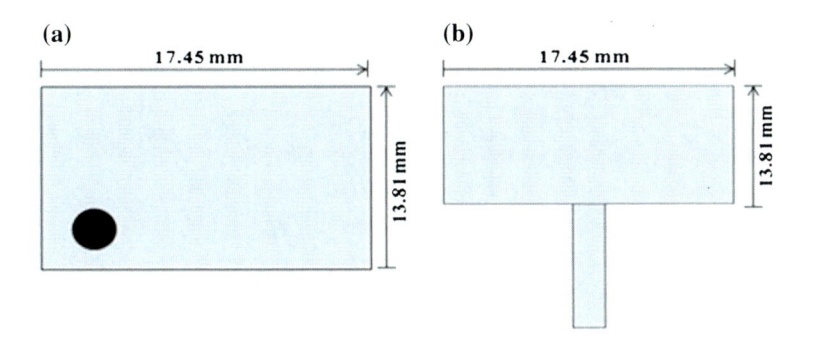

Fig. 53.1 Differently fed conventional microstrip patch antenna designs

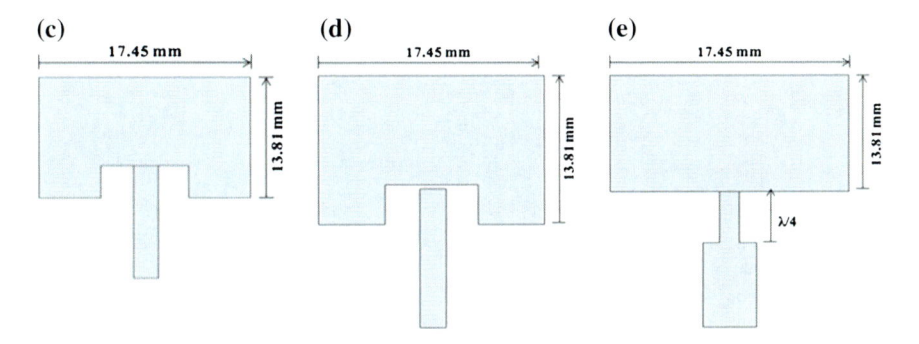

Fig. 53.2 Differently fed performance enhancement microstrip patch antenna designs

patch was designed and excited with Microstrip inset-fed (at an inset distance of R = $0.096\lambda_0$ mm) of width 1.898 mm. The simulated return loss was −18.05 dB at 5.8 GHz with a bandwidth of 169 MHz. The measured return loss was − 15.36 dB and the bandwidth was 145 MHz. The simulated and measured impedances were 51.18 − j1.91 Ω and 49.18 + j7.91 Ω respectively. The simulated voltage standing wave ratio (VSWR) was 1.57:1 whereas the measured VSWR was 1.48:1. Also, the inset-coupled Microstrip patch antenna was excited with Microstrip inset-fed (at an inset distance of R = $0.096\lambda_0$ mm) of width 1.898 mm. The simulated return loss was −14.55 dB at 5.8 GHz with a bandwidth of 139 MHz. The measured return loss was −13.98 dB with the bandwidth of 119 MHz and measured impedance of 48.51 + j12.22 Ω.

The simulated voltage standing wave ratio (VSWR) was 1.28:1 at the frequency of 5.808 GHz whereas the measured VSWR was 1.61:1. The quarter wavelength Microstrip patch antenna was excited with a Microstrip feed line of width 1.898 mm and a quarter wavelength width of 1.025 mm. The simulated return loss was −21.18 dB at 5.8 GHz with a bandwidth of 151 MHz The measured return loss was −19.24 dB at the same resonant frequency as obtained in the simulation but with the bandwidth of 130 MHz and measured impedance of 50.32 − j08.14 Ω.

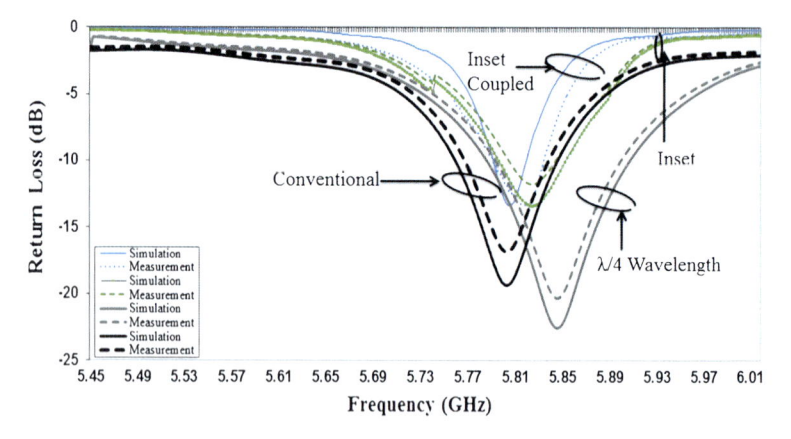

Fig. 53.3 Simulated and measured return loss of different MPAs

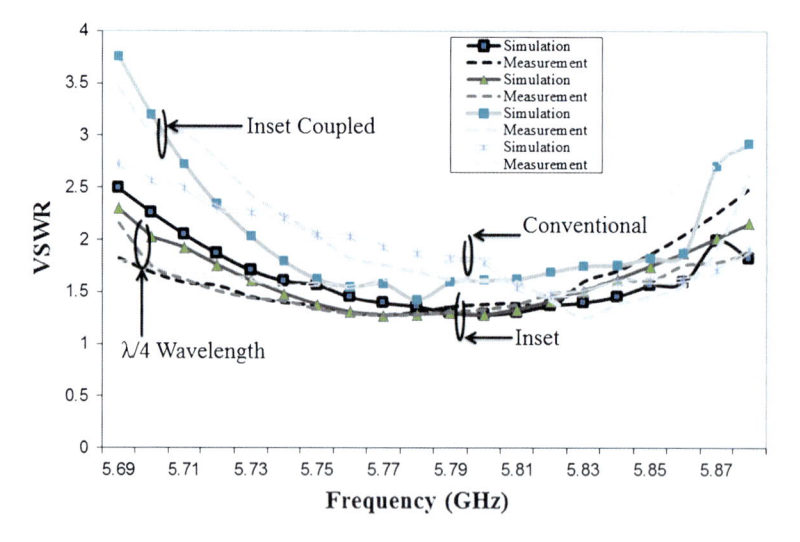

Fig. 53.4 Simulated and measured VSWR of different MPA

The simulated voltage standing wave ratio (VSWR) was 1.316:1 at the frequency of 5.8 GHz. The measured VSWR was the same with the simulation and both occur at the same frequency of 5.8 GHz.

The radiation pattern was as shown in Fig. 53.5a. The simulated gain was 6.49 dBi with main lobe magnitude of 2 dB, main lobe direction of 35° and angular width (3 dB) of 155.8° for E-plane. In H-plane, the main lobe magnitude was 3.8 dB, main lobe direction of 90°, a side lobe level of −0.9 dB and an angular width of 81.1°. In Fig. 53.5b, the radiation pattern of the inset-fed Microstrip patch antenna was depicted.

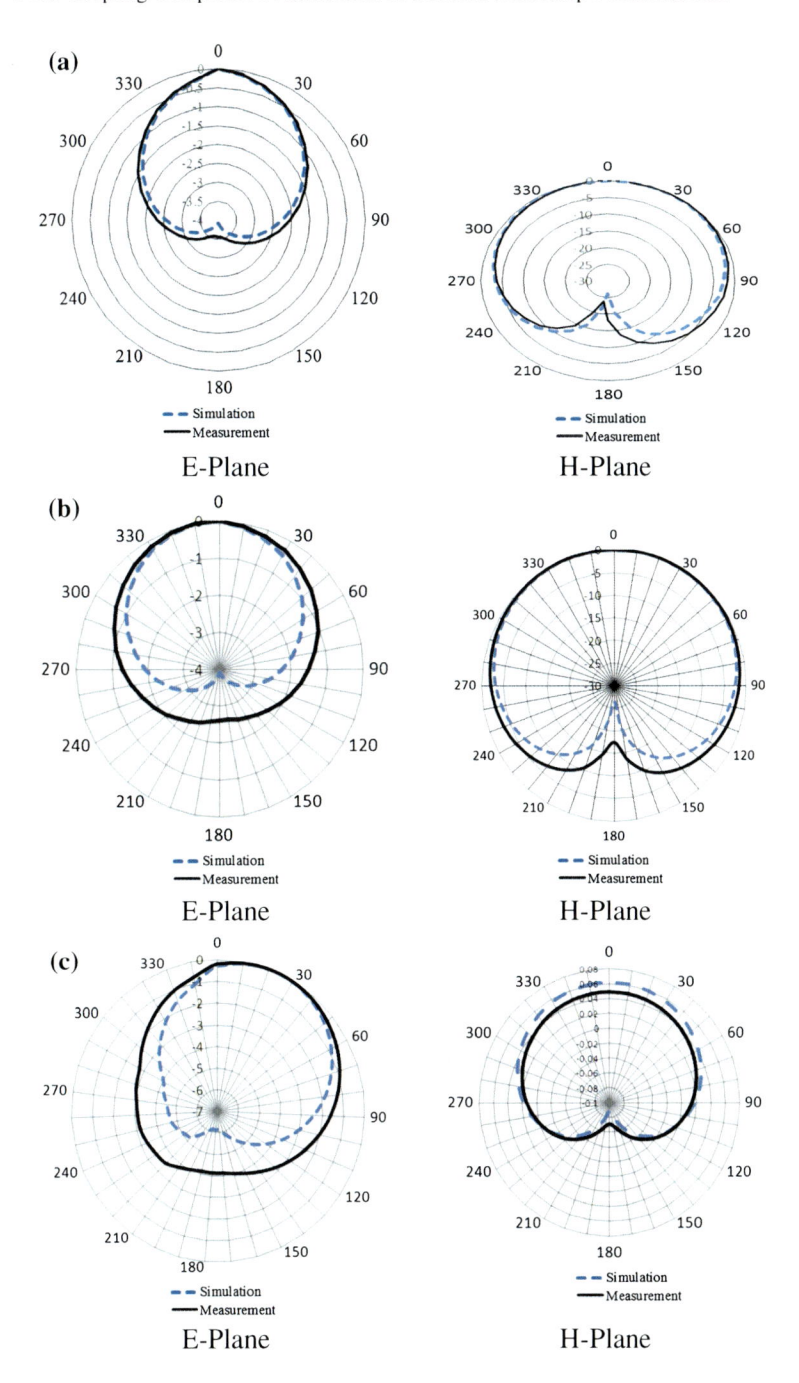

Fig. 53.5 Simulated radiation pattern of the microstrip patch enhancement antennas. **a** Coaxial fed, **b** Microstrip feed, **c** Inset-fed, **d** Inset-coupled, **e** $\lambda/4$ Wavelength

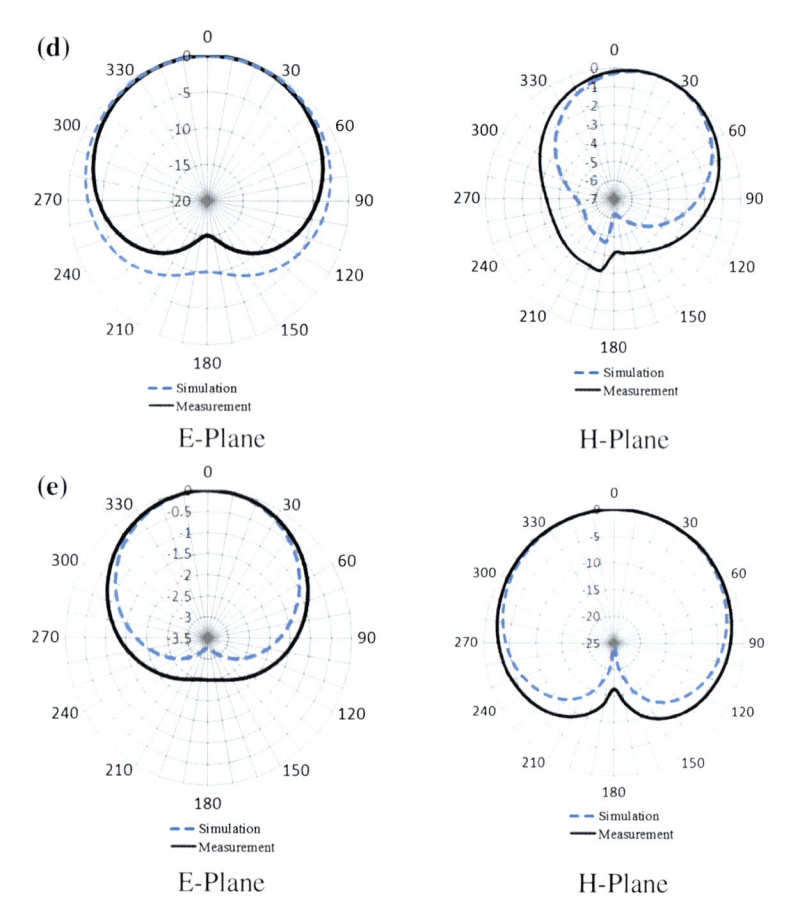

Fig. 53.5 (continued)

The simulated gain was 9.32 dBi with main lobe magnitude of 6.5 dB, main lobe direction of 10°, angular width (3 dB) of 76.5° and a side lobe level of −4.7 dB for E-plane. In H-plane, the main lobe magnitude was 6.2 dB, main lobe direction of 0°, and an angular width of 73.6°. Similarly, the radiation pattern of the inset-coupled Microstrip patch antenna was shown in Fig. 53.5b. The simulated gain was 8.66 dBi with the main lobe magnitude of 6.67 dB, main lobe direction of 6°, a side lobe of −5.2 dB and angular width at 3 dB of 69.9° for E-plane. For H-plane, the main lobe magnitude was 6.24 dB, main lobe direction of 10°, a side lobe level of −4.7 dB and an angular width of 66.2°. Finally, the radiation pattern of the quarter wavelength Microstrip patch antenna was as shown in Fig. 53.5d. The simulated gain was 6.004 dBi with main lobe magnitude of 3.6 dB, and main lobe direction of 90° for E-plane. For H-plane, the main lobe magnitude was 2.7 dB, main lobe direction of 30°, and an angular width of 141°.

Table 53.1 The performance profile assessment of some selected feeds

Antenna type	Size (mm)	Bandwidth (MHz)	Return loss (dB)	VSWR	Gain (dBl)	E-Plane side lobe (dB)	E-plane beamwidih (Degree)
Inset gap coupled patch	28 × 13.81	119	−13.98	1.61:1	8.66	−4.7	66.2
Coaxial fed patch	17.48 × 13.81	75	−17.18	1.115:1	8.19	−1.4	99.8
Microchip fed patch	28 × 13.81	60	−18.22	1.29:1	6.49	−0.9	81.1
Inset fed patch	27 × 13.81	145	−15.36	1.48:1	9.32	−4.7	73.6
Quartet wavelength patch	28 × 13.81	No	−19.24	1.316:1	6.0	−2.8	59.3

53.5 The Performance Comparative Assessment

Table 53.1 above shows the performance profile of various printed single element microstrip patch antenna with the view to compare their performance vis-à-vis the feed mechanism. The resonator itself is the same for all the antenna irrespective of the feed arrangements. Going by the table, it was obvious that coaxial fed presents a compact capability in terms of the size advantage with a good gain of about 8.19 dBi, only second to that of the Inset-coupled. The return loss of −17.18 dB (a convenient way of characterizing mismatch especially when the reflection is small) is much better than 10 dB.

However, it exhibits small bandwidth characteristics in the neighbourhood of about 75 MHz similar to that of a conventional microstrip line feed. Inset-coupled exhibited a bandwidth in excess of 119 MHz, a gain of 8.66 dBi and a marginal return loss of −13.98 dB with VSWR of 1.61:1. The same trend was also noticed in quarter wavelength transmission line feed with a good bandwidth of 130 MHz, gain of 6.0 dBi, a very good return loss of −19.24 dB and VSWR of 1.316:1. On a general note, the inset-fed demonstrated a bandwidth advantage of 140 MHz, a gain excess of 9.32 dBi, and reasonable return loss of −15.36 dB, and hence a VSWR of 1.48:1. Assessing these feeds on a 4-scale yardstick namely: gain, bandwidth, return loss and VSWR therefore, it was obvious that the inset-fed stand tall among its peers, as it has the best gain of 9.32 dBi, bandwidth of 145 MHz, a side lobe level of −4.7 dB. It also has reasonable return loss, VSWR, and beam width.

53.6 Conclusion

Performance profile comparison assessment of some selected microstrip antenna feeds were investigated and presented. These antennas were fed with different feed configurations and their performance recorded. In all cases, the antennas were designed, fabricated and measured under the same conditions. Detailed comparisons between all of the results were properly presented and in a tabular form for clarity. In most cases, the return loss characteristics indicated that the resonant frequency and impedance bandwidth were properly matched.

Acknowledgments Authors would like to acknowledge Universiti Sains Malaysia for research grant numbers 1001/PELECT/854004 RUT, and 1001/PELECT/814117 RUT.

References

1. Edwards TC, Steer MB, Edwards TC (2000) Foundations of interconnect and microstrip design. Wiley, Chichester
2. Balanis CA (2005) Antenna theory: analysis and design, 3rd edn. Wiley Interscience, New Jersey
3. Drabowitch S, Papiernik A, Griffiths HD, Encinas J (2005) Modern antennas, 2nd edn. Springer, New York
4. Johnson RC, Jasik H (1984) Antenna engineering handbook, 3rd edn. McGraw-Hill, New York
5. De-Kraus J, Marhefka RJ (2001) Antenna for all applications, 3rd edn. McGraw-Hill Science/Engineering/Math, New York
6. Stutzman WL, Thiele GA (1981) Antenna theory and design. Wiley, New York
7. Petosa A (2007) Dielectric resonator antenna handbook, Artech House, London
8. Antenna theory. http://www.antenna-theory.com

Chapter 54
Design of a 5.8 GHz Bandstop Filter Using Split Ring Resonator Array

Nor Muzlifah Mahyuddin and Nur Farah Syazwani Ab. Kadir

Abstract Nowadays, ISM band has been widely used by communication, military, medical and scientific department. The frequencies for this band are 2.4 GHz and 5.8 GHz. Heavy traffic has occurred in this frequency range due to high usage which makes noise easily to occur. Thus, in this work a 5.8 GHz bandstop filter has been designed to filter the noise that present at this frequency band using SRR array. It is designed using CST Microwave Studio software. Designing this filter also requires optimization process i.e. varying the different length of split gap and number of array used. Subsequently, the design is fabricated using Rogers 4003 as the substrate and is measured using Agilent PNA-X network analyzer. The result obtained satisfies a bandstop filter parameters such as insertion loss, return loss, and bandwidth. A 5.8 GHz SRR bandstop filter is successfully designed with 390 MHz bandwidth, insertion loss and return loss of −26.68 and −2.91 dB respectively.

Keywords Split-ring resonator · Bandstop filter · ISM band

54.1 Introduction

Nowadays, microwave filters are widely used in many RF and microwave design application. According to the filter specification, the filter will allow only particular band of frequencies. Subsequently, filter performs some functions which allow

---

N. M. Mahyuddin (✉) · N. F. S. Ab. Kadir
School of Electrical and Electronic Engineering, Engineering Campus, Universiti Sains Malaysia, 14300 Nibong Tebal, Penang, Malaysia
e-mail: ee.mnmuzlifah@eng.usm.my

N. F. S. Ab. Kadir
e-mail: farahwaney@gmail.com

H. A. Mat Sakim and M. T. Mustaffa (eds.), *The 8th International Conference on Robotic, Vision, Signal Processing & Power Applications*,
Lecture Notes in Electrical Engineering 291, DOI: 10.1007/978-981-4585-42-2_54,
© Springer Science+Business Media Singapore 2014

certain electrical frequencies to pass through, blocking certain frequencies from passing through the filter and completely blocking direct current electrical energy.

For this work, a band-stop filter is required to block any signal coming from the ISM band, specifically at 5.8 GHz. The ISM band is Industrial, Scientific and Medical radio bands, where the frequency bands of 2.4 and 5.8 GHz are considered unlicensed. As time goes by, there are more commercial products or devices operating within this band. More problems occur as more unwanted signals disrupting operation of other devices within this frequency band, thus the reason why band-stop filter is required for this purpose.

For low frequencies, lumped elements such as inductors and capacitors are used in filter design while transmission line sections and waveguide elements is used in filter design for microwave frequencies [1]. Additionally, microstrip filters also have been useful in new technology application such as military application, satellite application and mobile communication.

Recently, there have been high demands on designing microstrip filter using Split Ring Resonator (SRR) compared to other resonator. Using SRR in designing filter carries biggest advantage as it is smaller in size that lead to compact size filter. In addition, SRRs also provide easy fabrication, low radiation loss, high frequency stability and sharp selectivity in desired resonant frequency [2].

Therefore, in this paper, a square SRR array is designed for a 5.8 GHz bandstop filter which includes two optimization processes, i.e. varying the different length of split gap and number of array used. The design metric of interests are the centre frequency as well as the bandwidth, the insertion loss and the return loss. Several SRR designs are simulated using CST Microwave Studio and fabricated and measured using Agilent's PNA-X N5245 Network Analyzer.

54.2 SRR Band-Stop Filter Design

Essentially, SRR behave as a negative index metamaterial structure or also known as left handed materials (LHM). LHM is a material that has negative permittivity and permeability over a band of frequency. An example of metamaterial that carries LHM characteristic is SRR and metallic rods [3]. Characteristics of a negative index metamaterial structure are permittivity, $\varepsilon < 0$, permeability, $\mu < 0$ and index of refraction, n < 0. LFM was first proposed by Veselago [4]. His research was continued by Pendry after three decades. Pendry had successfully designed the SRR structures and his work was fabricated by Smith et al. in the year 2000 [5].

According to Mustafa K. Taher Al-Nuaimi and William G. Whittow [6], a microstrip line which is loaded with the SRR is a single negative medium that leads to exhibit a stop band characteristic. The resonance frequency obtained from designing a bandstop filter using SRR is much smaller compared to other resonators. This is due to the large distributed capacitance between the two rings. In designing the microstrip of the bandstop filter, SRR can only be etched at the upper substrate side near the transmission line [6]. To improve the coupling process, the distance between the transmission line and the ring should be as small as possible.

Magnetic field is generated when the microstrip transmission line is placed close together. As for example, if two arrays of SRR are placed at both side of the transmission line, a magnetic field line is induced by the transmission line across the SRR [7]. Electric Field, E is perpendicular the SRR plane which is at z-direction while magnetic field, H is perpendicular to the split gap of the SRR which is in y-direction.

54.2.1 Split-Ring Resonator Modeling

SRRs commonly known as metamaterial structure are resonators that have split in the opposite ends of each concentric annular ring. Usually the ring is designed in a pair which have small gap between them. Usually, it is made up of copper or gold [6]. Figure 54.1 shows the schematic diagram of a single split-ring resonator with its parameters. These parameters are tabulated in Table 54.1 indicating the equations used to calculate these parameters, which was derived by Saha, Siddiqui and Antar [8].

The resonant frequency and the bandwidth of the SRR can be determined by its total length and its physical width [9]. Other than that, according to Markos et al. the resonance frequency also depends on the thickness of the ring, inner diameter, and the split gap as well as the electrical permittivity of the board itself [10]. Additionally, the gaps of the inner and the outer rings are the main features in designing the SRR. It represents the capacitance of the rings while the inductance is represented by the area of both inner and outer rings.

Referring to Volkan Oznazl and Vakur B. Erturk [11], the microstrip line will have higher Q factor and act as a bandstop filters when it is loaded with the SRR. This also indicates that better performance will be realized when SRR is in array coupling to the microstrip transmission line, as shown in Fig. 54.2. Width of the transmission line is represented by w, b represent the gap between transmission line and SRR, a represent the gap between ring of SRR, g represent the split gap, c represent width of ring of SRR and lastly, d represent length of SRR. Values for the gap between transmission line and SRR, gap between ring of SRR, width of SRR ring and length of SRR are fixed (Table 54.2). The only parameters that are varied are the split gap as well as number of arrays.

The SRR array is modeled together with the RO4003C substrate and a 50 Ω microstrip line. The substrate has a thickness of 0.813 mm and a dielectric constant of 3.38. Figure 54.3 shows the side view of SRR modeling in CST MWS, as well as the dimension of a 50 Ω transmission line with a width of 1.8653 mm and length of 30 mm. For substrate in this design, it has been laminated with copper layer at both sides. The copper layer has an electrical conductivity of $5.9E + 007$ S/M and a height of 0.035 mm.

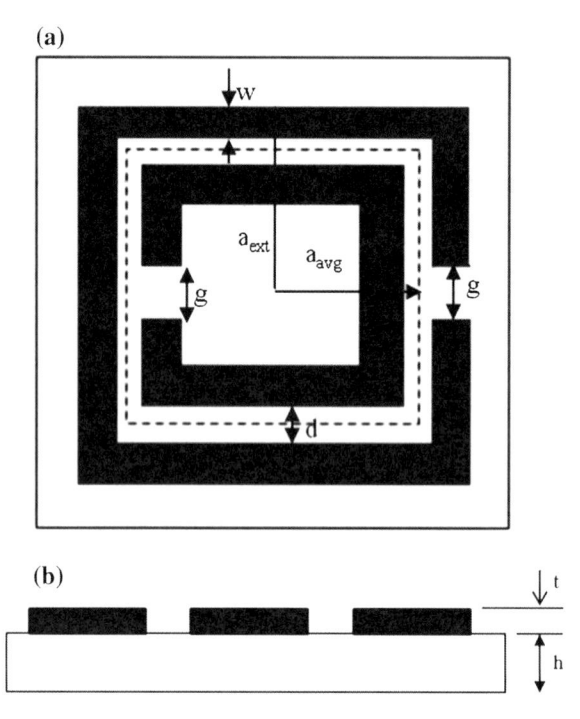

Fig. 54.1 Schematic view of **a** a square SRR formed with metallic strips of width, w, outer length, a_{ext}, and inner length, a_{avg} with inter ring spacing, d and split gap, g, **b** with thickness of metallic strip, t, printed on dielectric substrate having thickness, h

Table 54.1 SRR dimension and parameters [8]

Parameter	Definition
f_o	Resonant frequency
	$\dfrac{1}{2\Pi\sqrt{L_T C_{eq}}}$
L_T	Total equivalent inductance for a wire of rectangular cross section of a single SRR having finite length, l and width, w
	$0.0002l\left(2.303 \log_{10} \dfrac{4l}{w} - \gamma\right) \mu H$
	$l = \sim 8 a_{ext} - g$
γ	Constant for a wire loop of square qeometry, 2.853
C_{eq}	Total equivalent capacitance with series capacitance, C_o and gap capacitance, C_g
	$\dfrac{(C_o + C_g)}{2}$
C_o	$(4a_{avg} - g) C_{pul}$
a_{avg}	$a_{ext} - w - d/2$
C_{pul}	Capacitance per unit length
	$\sqrt{\varepsilon_r}/c_o Z_o$
Z_o	Characteristic impedance, 50 Ω
c_o	Velocity of light in free space, 3×10^8
C_g	$\dfrac{\varepsilon_o w t}{g}$

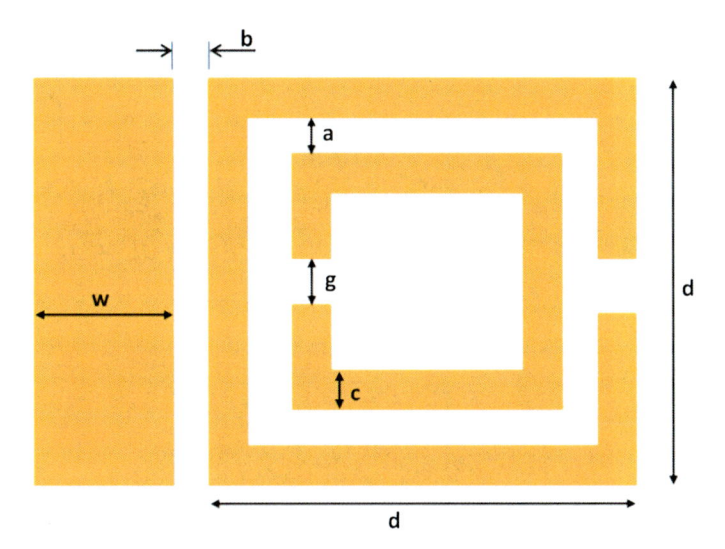

Fig. 54.2 Schematic view of a square SRR coupled to a microstrip line with its dimension parameters

Table 54.2 Dimension of SRR

Parameter	Dimension (mm)
Gap between transmission line and SRR, b	0.3
Gap between ring of SRR, a	0.3
Width of ring of SRR, c	0.5
Length of SRR, d	4.8

54.2.2 Optimization Process of SRR Band-Stop Filter Design

Optimization process is needed to make sure the design fulfill the requirement of the bandstop filter specification. In this work, the number of SRR array and the length of split gap are optimized. Firstly, one SRR array is designed with fixed split gap length at 0.5 mm to observe the simulation result for design with different number of array. After the simulated result is obtained, the result is analyzed to choose the best number of SRR array that satisfy the SRR bandstop filter specification.

The number of array is increased up to 5 arrays, starting with a single array SRR filter (Fig. 54.4a). The distance between the SRR arrays has been set as in Fig. 54.4, where 2 SRRs have been added on each side of the transmission line. According to Prof. Tzong-Lin Wu from Department of Electrical Engineering National Taiwan University in his lecture on Microwave Filter Design, the distance between each resonator is $\lambda g/4$. The wavelength, λg can be obtained by using Eq. (54.1) [12]. In order to acquire the best SRR array performance whilst still

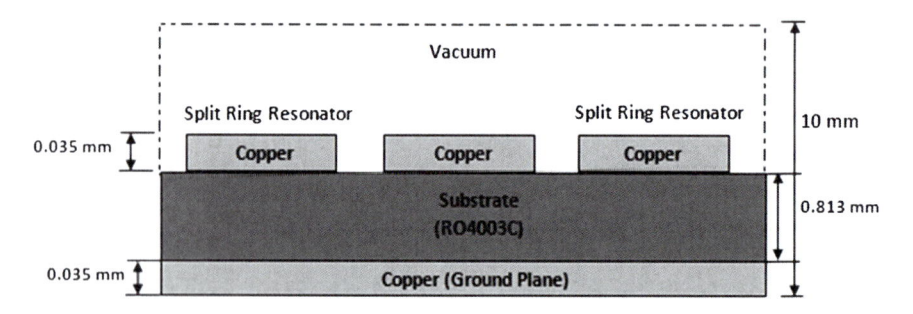

Fig. 54.3 Side view of SRR modeling with its parameters and dimension

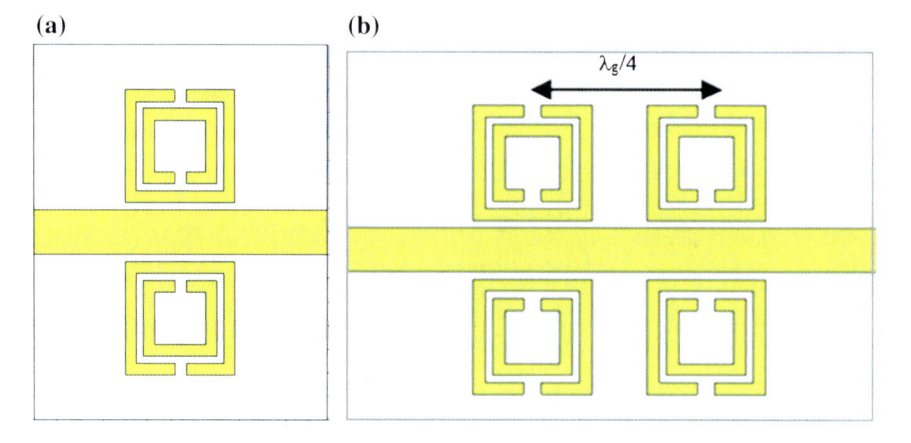

Fig. 54.4 Design of 2 SRR array at both side of the transmission line

maintaining the small and compact size, five different types of SRR array are designed.

$$\lambda_g = \frac{c}{\sqrt{\varepsilon_r f}}. \tag{54.1}$$

where
c   speed of light, 3×10^8 m/s
ε_r dielectric constant of a substance, 3.38
f   resonant frequency, 5.8 GHz

Finally, the best SRR bandstop filter design is obtained with optimized split-gap length and number of array. The final design is then fabricated and measured using Agilent PNA-X network analyzer, where its S-parameter results are analyzed.

54.3 Result and Discussion

The previous design process begins by a single SRR array using CST software to observe the behavior of the bandstop filter, and then increased gradually up to 5 array, taking account all the parameters that can influenced the performance such as the split gap length.

54.3.1 Optimization Results of SRR Band-Stop Filter Design

SRR Band-Stop Filter with Different Number of Array. The design of SRR bandstop filter is implemented by using different number of SRR array. The S11 parameter and S21 parameter results are tabulated in Table 54.3 to summarize the finding. From Table 54.3 below, the centre frequency is measured between 5.7 and 5.8 GHz for all different number of arrays. However, the most nearest to 5.8 GHz is the filter design with 5 SRR array and a single SRR array. Subsequently, the filter with 5 SRR array design has higher return loss and lower insertion loss compared to other array types. Therefore, for this first set of optimization, the 5 SRR array design is chosen.

SRR Band-Stop Filter with Different Split Gap Length. The split gap of SRR is the main feature in designing the SRR bandstop filter. Thus, to make sure the design fulfils the specification limit; the split gap is varied for three different lengths which are 0.5, 0.6 and 0.7 mm. This helps in analyzing which length of split gap satisfy the specification of the SRR bandstop filter for 5 array of SRR. In terms of centre frequency, the most nearest to 5.8 GHz is the filter design with split gap length of 0.7 mm. In addition, this design also has the best return loss performance, which is near to 0 dB. Even though the design with the split gap length of 0.7 mm has the worst insertion loss performance compared to the other designs, it is still acceptable as it still resides below -20 dB. Therefore, for this set of optimization, the SRR filter design with split gap length of 0.7 mm is chosen (Table 54.4).

54.3.2 Final Result of Optimized SRR Band-Stop Filter Design

The final design that successfully satisfies the bandstop filter design is the bandstop filter with 5 SRR array with split gap length of 0.7 mm. Figure 54.5 shows the layout and fabricated design of bandstop filter with 5 SRR array. The final design is simulated using CST Microwave Studio and measured using network analyzer where both simulated and measured results are compared as shown in Figs. 54.6

Table 54.3 Specification of SRR bandstop filter for different SRR array

Number of SRR array	Centre frequency (GHz)	S_{11} parameter (dB)	S_{21} parameter (dB)
1	5.8	−3.26	−9.36
2	5.69	−2.78	−30.7
3	5.67	−3.17	−18.97
4	5.68	−3.18	−26.33
5	5.7	−1.89	−60.91

Table 54.4 S-parameter results for different split gap length of SRR

Gap length, C_p (mm)	Centre frequency (GHz)	Return loss, S_{11} (dB)	Insertion loss, S_{21} (dB)
0.5	5.7	−3.48	−60.91
0.6	5.76	−2.71	−51.2
0.7	5.81	−1.89	−29.59

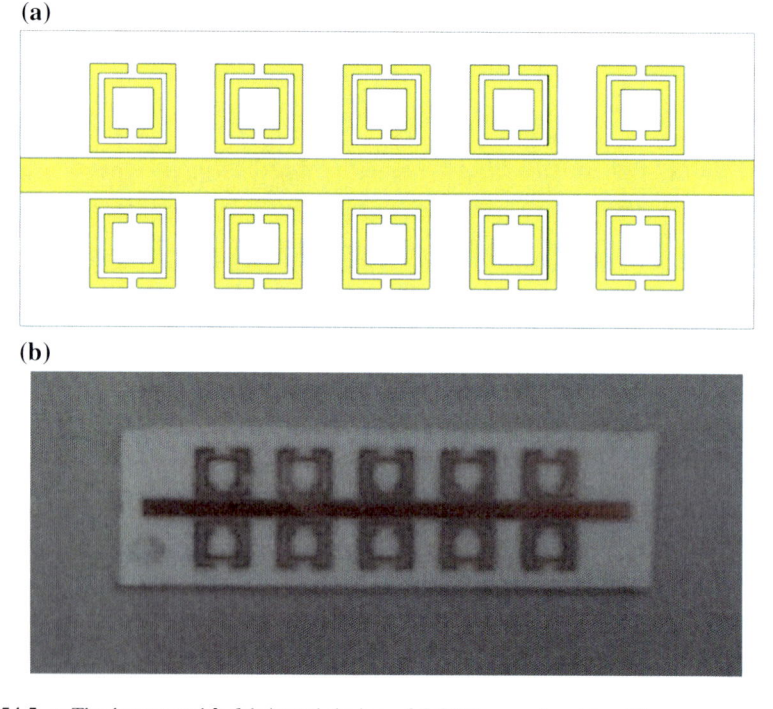

Fig. 54.5 **a** The layout and **b** fabricated design of 5 SRR array bandstop filter

and 54.7 indicating the centre frequency, insertion loss and return loss from both results.

The filter performance is simulated and measured at 4–7 GHz frequency range. The centre frequency is measured at 5.78 GHz where there is only 0.8 %

54 Design of a 5.8 GHz Bandstop Filter Using Split Ring Resonator Array

Fig. 54.6 The return loss of 5 SRR array bandstop filter

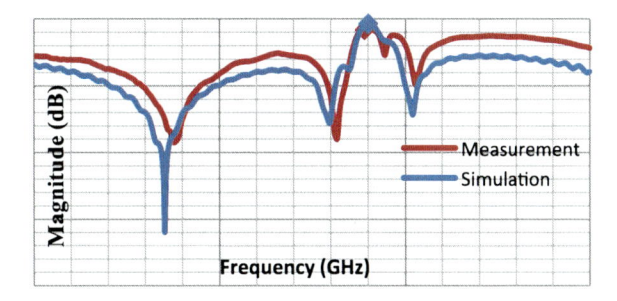

Fig. 54.7 The insertion loss of 5 SRR array bandstop filter

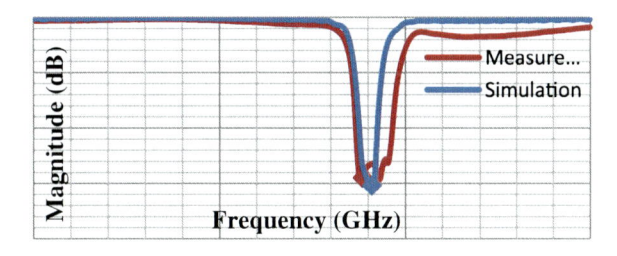

difference with the simulated one. In terms of bandwidth, the simulated result produces a bandwidth of 220 MHz whilst there is 44 % increase of bandwidth when measured, resulting in 390 MHz of bandwidth. However, both results show that the bandwidth for this filter design is slightly wider than the expected bandwidth of 5.8 GHz ISM band, which is at 150 MHz.

From previous simulated analysis, the insertion loss and return loss are measured at −29.59 and −1.89 dB respectively. However, when measured, the insertion loss decreases to −26.68 dB whilst the return loss is at −2.91 dB, with 10.9 and 35 % differences respectively. These differences may occur due to the losses that may present during fabrication process, soldering process, temperature of surrounding, the SMA connector used and testing apparatus which are cable and connector.

54.4 Conclusion

A bandstop filter incorporating SRR array is successfully designed at 5.8 GHz with a bandwidth of 390 MHz. The insertion loss obtained is more than −20 dB, which is at −26.68 dB and the return loss obtained is close to 0 dB, at −2.91 dB. This is acquired through two optimization processes, i.e. the length of split gap, g, and the number of SRR array are varied to find the best bandstop filter performance. From the results obtained, it can be concluded that the bandstop filter design with 5 SRR array with split gap of 0.7 mm meet the bandstop filter requirements at 5.8 GHz.

References

1. Mumtaz SH (2005) Spurious response elimination techniques for microwave low pass and band pass filter. Universiti Sains Malaysia
2. Jinse J, Choon Sik C, Lee JW, Jaeheung K, Tae H (2006) A low phase noise microwave oscillator using split ring resonator. In: 36th European microwave conference, pp 95–98
3. Pendry JB, Holdon AJ, Robbins DJ, Stewart WJ (1999) Magnetism from conductors and enhanced nonlinear phenomena. IEEE Trans Microw Theory Tech 47(11):2075–2084
4. Verelago VG (1968) The electrodynamics of substances with simultaneously negative values of ε and μ. Sov Phys Usp 10:509–514
5. Nornikman H, Ahmad BH, Abdul Aziz MZA, Malek MFBA, Imran H, Othman AR (2012) Study and simulation of an edge couple split ring resonator (ec-srr) on truncated pyramidal microwave absorber. Progress Electromagnet Res 127:319–334
6. Taher Al-Nuaimi MK, Whittow WG (2010) Compact microstrip band stop filter using SRR and CSSR: design, simulation and results. In: 2010 Proceedings of the 4th European conference on antennas and propagation (EuCAP), pp 1–5
7. Labidi M, Bel Hadj Tahar J, Choubani F (2011) New design of antenna array using left handed meta-material based on circular split ring resonator. In: 11th Mediterranean microwave symposium, pp 52–56
8. Saha C, Siddiqui JY, Antar YMM (2011) Square split ring resonator backed coplanar waveguide for filter applications. In: General assembly and scientific symposium, pp 1–4
9. Fallahzadeh S, Bahrami H, Tayarani M (2009) A novel dual-band bandstop waveguide filter using split ring resonators. Department of Electrical Engineering, Iran University of Science and Technology (IUST), vol 12, pp 133–139
10. Wu B, Li B, Su T, Liang CH (2006) Study on transmission characteristic of split-ring resonator defected ground structure. National Key Laboratory of Antennas and Microwave Technology Xidian University, vol 2, pp 710–714
11. Öznazl V, Erturk VB (2013) On the use of split-ring resonators and complementary split-ring resonators for novel printed microwave elements: simulations, experiments and discussions. Bilkent University
12. Pozar DM (2005) Microwave engineering. Wiley, Danvers

Chapter 55
Optic Flow Based Occlusion Analysis for Cell Division Detection

Sha Yu and Derek Molloy

Abstract The computer vision domain has seen increasing attention in the design of automated tools for cellular biology researchers. In addition to quantitative analysis on whole populations of cells, identification of the cell division events is another important topic. In this research, a novel fully automated image-based cell-division-detection approach is proposed. Differing from most of the existing approaches that exploit training-based or image-based segmentation methods, the main idea of the proposed approach is detecting cell divisions using a motion based occlusion analysis process. Testing has been performed on different types of cellular datasets, including fluorescence images and phase-contrast data, and it has confirmed the effectiveness of the proposed method.

Keywords Motion estimation · Optic flow · Cell division detection · Occlusion detection · Forward–backward motion consistency

Thanks the National Biophotonics and Imaging Platform Ireland (NBIPI).

55.1 Introduction

Within cellular biology research, recognition and identification of cell division events is an important task for monitoring the health and growth rate of a cell population. Existing work on cell division detection largely relies on training

---

S. Yu (✉) · D. Molloy
Centre for Image Processing and Analysis (CIPA), Dublin City University, Dublin 9, Ireland
e-mail: sha.yu3@mail.dcu.ie
URL: http://www.cipa.dcu.ie/

D. Molloy
e-mail: derek.molloy@dcu.ie

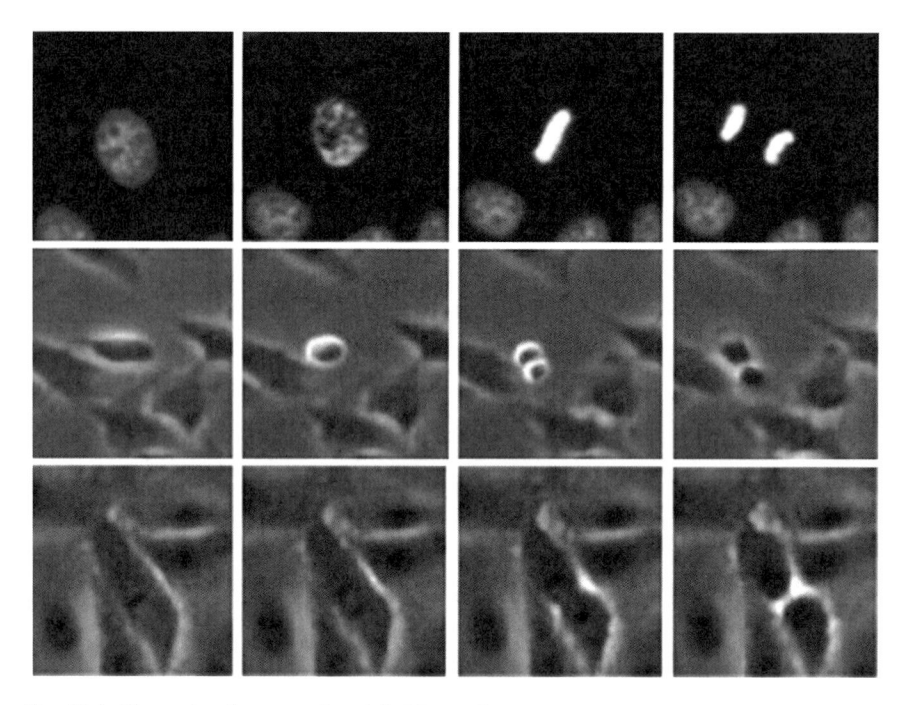

Fig. 55.1 Illustrating three samples of dividing cells displaying differing behaviours. The image details will be described in Sect. 55.4

based methods, or morphological segmentation: (1) In order to recognise cell division patterns, trained classifiers usually assume cells have particular appearance changes during the cell division processes. For example, cells will become more round in shape and abruptly display bright intensities. However, different types of cells may show large variances in appearances and behaviours (see Fig. 55.1). And, within the same cell type, the frequency of the images being taken, i.e. the time resolution of an image sequence, significantly affect the cell division pattern. Therefore, the training based approaches have relatively weak application generality; and, considering an offline pre-training is required, those methods are usually not suitable for online event detection. (2) Segmentation based approaches have difficulties in segmenting cells within low contrast images, i.e. where cells and the background have similar intensity distributions, and closely packed cells in dense environment. Fluorescence cellar images are relatively easier to segment due to the high image contrast. However, fluorescence-microscopy techniques usually require additional preparation such as staining, which may kill the cells. Therefore, a large amount of cellular datasets have been captured by non-fluorescent microscopy, such as phase-contrast microscopy. Within the obtained images, cells usually have vague boundaries. For this case, image-based segmentation and detection of interest regions is not trivial [1].

55.2 Related Work

Within the work of Padfield et al. [2], dividing cells are assumed to have intensity changing from bright to dark. Division events are recognised and tracked within an unified framework combining a level-set algorithm with a fast marching method, which has an associated cost function favoring objects with bright-to-dark changes during the image sequence. Nath et al. [3] also adopted a level-set method for the task of cell division detection and tracking. Level-set based methods, are acclaimed well suitable for identifying and tracking cell division due to their flexible topology [3, 4]. However, at least one important pre-condition should be satisfied: the mother cell in one image frame should have some overlapping image information with its children cells in the subsequent frame. This pre-condition requires the image sequence has sufficiently high time resolution.

Li et al. [4] proposed a simple cell-division-detection method by measuring a set of cell shape and appearance criterion. A nuclei division detection method, contributed by Li et al. [5], relies on a support vector machine (SVM) classifier, which is constructed based on manually selected cell features. Kanade et al. [6], also applied a trained classifier to recognising cell division events. Huh and Chen [1], presented a probabilistic model for division event detection. The model is obtained by a carefully designed learning process. The aforementioned approaches have the limitation of being applied to datasets that have been chosen by the authors.

Quelhas et al. [7], proposed an optic-flow based cell-division-detection method. Their approach is based on the difference of speeds of the cell division, i.e. they assume cell division is a much faster process. This method cannot guarantee a correct recognition of cell division when applied in a dataset with large cell speed variations, which is common in most of the cases.

For a detailed literature survey about cell division detection, please refer to the work of Huh and Chen.

55.3 Proposed Approach

At the heart of our approach is the assumption that where cell division happens, a group of pixels newly appearing/disappearing in one frame cannot be found in the adjacent frame within an image sequence. The most similar work to our approach is the one proposed by Quelhas et al. However, it worth noting that we assume neither particular changes about cell colour and shape, nor cell speed characteristics during divisions.

In this section, the key techniques chosen for the proposed approach will be introduced, including optic flow based motion estimation, and occlusion detection. After that, the major components of the overall system will be briefly explained.

55.3.1 Optic Flow Based Motion Estimation

Optic flow (OF) estimation computes approximate motion fields for time-varying image sequences. All OF approaches rely on the temporal conservation of some image elements, e.g. conservation of pixel intensity and/or gradient magnitude. Except for data conservation, another important assumption is that nearby pixels share similar motion, which is usually referred as the motion smoothness constraint. One classic formulation of OF techniques—the Horn-Schunk (HS) OF algorithm [8], can be defined as:

$$E(u, v) = E_D + \alpha E_S \tag{55.1}$$

$$E_D(u, v) = \sum_{(x,y) \in \Omega} (I_2(x + u_{x,y}, y + v_{x,y}) - I_1(x, y))^2 \tag{55.2}$$

$$E_S(u, v) = \sum_{(x,y) \in \Omega} (|u_{x,y} - u_{x+1,y}|^2 + |u_{x,y} - u_{x,y+1}|^2) \\ + \sum_{(x,y) \in \Omega} (|v_{x,y} - v_{x+1,y}|^2 + |v_{x,y} - v_{x,y+1}|^2) \tag{55.3}$$

for any one pixel (x, y) inside a frame I_1, a corresponding OF vector $(u_{x,y}, v_{x,y})$ is to be calculated, representing the pixel displacement between the frames I_1 and I_2. The data term E_D measures how much the intensity information is conserved during the pixel translocation. In practice, for robustly estimating objects with illumination change, gradient-magnitude based conservation can be exploited. E_D calculates the smoothness of nearby OF vectors. $\alpha > 0$ denotes a constant parameter controlling the balance between the terms E_D and E_S. In order to find the most probable displacement field, $E(u, v)$ is to be minimised, which is usually achieved by Euler–Lagrange equations [9].

Within our work, we choose the HS based optic flow algorithm, featured by having both data conservation and motion smoothness terms, and resulting in a dense OF field (u, v). Other motion estimation algorithms such as the Lucas–Kanade (LK) [10] based ones rely on texture features, and thus lead to sparse flow fields. Demons based image registration methods [11], which are firstly introduced for estimating non-rigid motions, lack smoothness assumptions and usually suffer from relatively high rate of motion estimation noise.

One problem with the HS based optic flow is the estimation of large displacements. The chosen OF algorithm is embedded within the coarse-to-fine framework. Within such a framework, the main components of the OF vectors are first estimated between a down-sampled image pairs. Then, the higher resolution versions of the same images will refine the flow estimation progressively.

55.3.2 Occlusion Detection

Given two subsequent image frames, occlusions that happen in motion estimation are caused by: that pixels inside one of two frames fail to be found inside the other, which is due to an object appearing/disappearing, for example an object being shielded by other object; that pixels are mismatched between two frames, because of shade and/or illumination change, etc. In this work, occlusions caused by newly incoming cells or cells exiting the image field will not be considered. This can be easily achieved by ignoring occlusion regions detected near image borders. Motion estimation errors caused by illumination changes can be largely reduced by adding a gradient-magnitude data conservation into Eq. (55.2).

The chosen occlusion detection method checks the mutual consistency between the forward and backward optic flow fields [9, 12]. Based on this scheme, an occlusion map O_1 is defined in this way: for a match found between two pixels (x, x') according to the backward OF field (the two pixels are respectively inside I_2 and I_1), if using the forward flow field x' is not matched to x, the pixel x will be flagged as an occluded pixel, i.e. $O_1(x)$ will be set as 1. Flow vectors for off-grid pixels are obtained by interpolation. In addition, a second occlusion map O_2 is also constructed, defined as

$$O_2(\mathbf{x}) = \begin{cases} 0 & \text{if } \exists (\mathbf{x}' + \mathbf{v}'_{x'}) \approx \mathbf{x}, \\ 1 & \text{otherwise.} \end{cases} \quad (55.4)$$

where $\mathbf{v}'$ represents the forward OF field. The symbol $\approx$ is for checking if $\mathbf{x}$ has been matched by any $\mathbf{x}'$ according to the forward flow.

In theory, O_1 and O_2 should have similar results, however, because of the interpolation process required for O_1, there ar e some occlusions not detected inside O_1 but appear in O_2. Therefore, the final occlusion map is a combined binary field: $O = O_1 \cup O_2$. In practice, a gray-value occlusion map can be obtained, taking fractional values, ranging from 0 to 1, in both of the occlusion maps.

55.3.3 System Design

This paper implements an automated cell-division-detection system, which consists of four major components:

- Image based motion estimation. For every two adjacent frames within the image sequences, a forward–backward optic flow estimation, within the coarse-to-fine framework, will be applied.
- Motion based occlusion detection. This process results in a binary or gray-value map indicating where the occlusions happen.

- Morphology operations. Two morphology operations will be conducted in order: a morphology erosion for removing scattered noises in the estimated occlusion map, and a morphology close for connecting nearby small occlusion regions.
- Cell division map output. This process calculates region centroids within the binary occlusion map. The regions that have their area sizes less then a threshold τ will be ignored. τ can be manually or automatically selected according to the average cell size in the tested data. For a gray-value occlusion map, a simple peak detection method or the mean-shift algorithm [13] can be used to detect the cell division positions.

55.4 Evaluation and Discussion

To demonstrate the validity of the proposed approach, experiments on several representative datasets have been conducted, including: Henrietta Lacks (Hela)—the donor of a cervical cancer cell dataset, captured under fluorescence microscopy, and phase contrast image sequences of C2C12 myoblastic stem cells, wound healing (WH) cells (taken from [3]), and Madin Darby Canine Kidney Epithelial cells (MDCK).

In regard to the related parameters adjustment: (1) The morphology operation parameters. Both of the morphology operations take disk-shaped structuring elements with radius as 1; (2) The parameter τ. Assuming the average cell size is a_c, τ can be set as $\gamma \times a_c$ with γ a positive fractional value less than 1; (3) The pyramid-level number l for the coarse-to-fine framework. With experiments, it is noticed that occlusions also happen in the regions where cells have large displacements. Therefore, the value l should be accordingly adjusted for different cellular datasets.

The cell-division-detection results returned by the proposed approach are compared against the manually annotated data. Three values will be recorded for each image sequences, the total number of dividing cells manually annotated, the correct number of automatically detected division events (for each dividing cell, multiple detection on the same cell will be counted just once), the number of non-division cells detected by the proposed method.

Figure 55.2 illustrates cell division events that are detected within the sample images. The statistical results about detection accuracy[1] and recall[2] values are listed in Table 55.1. High accuracies and recalls have been achieved within the HeLa and C2C12 image sequences. There are relatively lower detection accuracies for the WH and MDCK datasets. The main reason is that a considerable number of cell death events (within WH and MDCK) occur and there is much cell debris (in

---

[1] The rate of the correct automatically detected division number out of the total automatically detected number.

[2] The rate of the correct automatically detected division number out of the total manually annotated number.

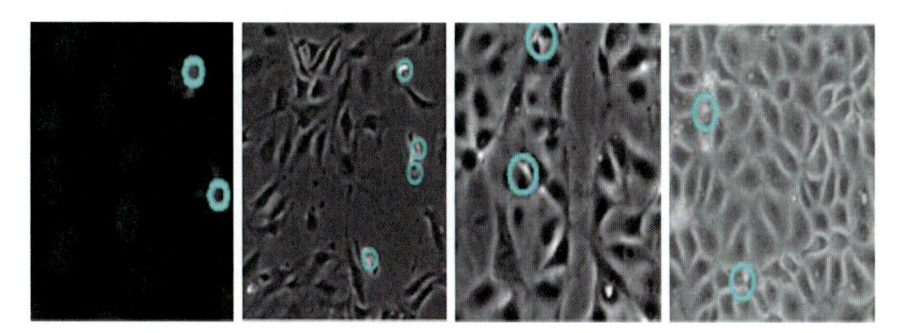

Fig. 55.2 Cell divisions being detected in the *circle-marked regions*

Table 55.1 Quantitative results for cell division detections

Data	No. of frame	Total division events (manually annotated)	Correct division detected	Non-division detected	Recall (%)	Accuracy (%)
HeLa	54	69	58	10	84.1	85.3
C2C12	200	98	89	11	90.8	89.0
WH	104	21	19	11	90.5	63.3
MDCK	150	51	42	24	82.4	63.6

MDCK), and therefore this affects the detection accuracies. However, the mean recall value of the four tests is 87.0 %, which reflects that most of the division events have been successfully identified.

We also add some clarifications, taking account of the comments of the reviewers, on the performance of the proposed approach under several challenging conditions: (1) In the case of a large spatial location change during a cell division process. This situation is reported in both of the HeLa and the C2C12 datasets, which have relatively lower time resolutions than the other considered datasets. In the HeLa case, two newly born cells can have large location deviations from their motion cell's spatial location. The proposed motion estimation method will match the mother cell region into one or both of the two children cells. So, the motion occlusion can be effectively detected by the proposed method. For the C2C12 dataset, large location changes usually take place when a mother cell has not started splitting itself. This case will be detected as a large motion, and therefore will not affect the robustness of the proposed approach for cell division detection. (2) OF based motion estimation can be tolerant to a small ratio of a cell region appearing/disappearing due to cell growing/shrinking. So, our approach works better in detecting occlusions caused by a relatively larger ratio of a cell region being occluded. By observation, we find division cells indeed have much larger ratios of region occlusion than growing/shrinking cells. (3) Distinguishing cell overlapping and cell division events are currently out of the scope. Furthermore, because the image sequences being tested are taken from in vitro cells that are cultured within considerably thin containers, so the rate of cells that heavily overlap with each other is small within this kind of cellular datasets. (4) The proposed approach is robust in the presence of scattered noise, due to the motion regularisation scheme embedded within the selected dense motion estimation algorithm.

To summarise, the generalised approach proposed in this paper is achieving good performance levels when it is considered that it is being tested under challenging conditions, where the cell division behaviours are very different in each of the four image sequences.

55.5 Conclusion and Future Work

In this paper, we have described a novel cell-division-detection approach. The proposed approach adopts a motion based occlusion detection strategy, which is quite different from the existing segmentation or training based methods. The major contribution of this work is that the proposed approach can successfully detect dividing cells with a variety of division behaviours. This has been proven by our experimentation on four different types of image sequences.

For future work, our major plan will be to combine the proposed approach with existing cell tracking techniques, so as to extend this cell-division-detection method into an unified tracking framework.

Acknowledgments This research was supported by the National Biophotonics Imaging Platform (NBIP) Ireland funded under the Higher Education Authority PRTLI Cycle 4, co-funded by the Irish Government and the European Union—Investing in your future.

References

1. Huh S, Chen M (2011) Detection of mitosis within a stem cell population of high cell confluence in phase-contrast microscopy images. In: IEEE computer vision and pattern recognition (CVPR'11), pp 1033–1040
2. Padfield DR, Rittscher J, Thomas N, Roysam B (2009) Spatio-temporal cell cycle phase analysis using level sets and fast marching methods. Med Image Anal 13(1):143–155
3. Bunyak F, Palaniappan K, Nath SK, Baskin TI, Dong G (2006) Quantitative cell motility for in vitro wound healing using level set-based active contour tracking. In: Proceedings of the 3rd IEEE international symposium biomedical imaging (ISBI), 1040–1043 April 2006
4. Li K, Miller ED, Chen M, Kanade T, Weiss LE, Campbell PG (2008) Cell population tracking and lineage construction with spatiotemporal context. Med Image Anal 12(5):546–566
5. Li F, Zhou X, Ma J, Wong ST (2010) Multiple nuclei tracking using integer programming for quantitative cancer cell cycle analysis. IEEE Trans Med Imaging 29(1):96–105
6. Kanade T, Yin Z, Bise R, Huh S, Eom SE, Sandbothe M, Chen M (2011) Cell image analysis: algorithms, system and applications. In: IEEE workshop on applications of computer vision (WACV)

7. Quelhas P, Mendona A, Campilho A (2010) Optical flow based *Arabidopsis thaliana* root meristem cell division detection. Lect Notes Comput Sci 6112:217–226
8. Horn BKP, Schunk BG (1981) Determining optical flow. Artif Intell 17:185–203
9. Brox T, Bruhn A, Papenberg N, Weickert J (2004) High accuracy optical flow estimation based on a theory for warping. In: Computer vision—proceedings of 8th European conference on computer vision, 2004
10. Lucas BD, Kanade T (1981) An iterative image registration technique with an application to stereo vision. In: IJCAI, vol 81, pp 674–679
11. Thirion JP (1998) Image matching as a diffusion process: an analogy with Maxwell's demons. Med Image Anal 2(3):243–260
12. Alvarez L, Deriche R, Papadopoulo T, Sanchez J (2002) Symmetrical dense optical flow estimation with occlusion detection. In: European conference on computer vision. Springer, pp 721–735
13. Cheng Y (1995) Mean shift, mode seeking, and clustering. IEEE Trans Pattern Anal Mach Intell 17:790–799

Part VI
Power System and Industrial Applications

Chapter 56
Design of Coreless PCB Transformer for DC/DC Converter Applications

Mohammadali Hashemi, Mohd Fadzil Ain and Majid Rafiee

Abstract This study offered a compact planar, low cost and high density power converter for applications. A fly-back DC/DC converter using a new coreless PCB transformer which operates in MHz band is simulated and fabricated. This is the fact that the size of all passive components such as capacitors, inductors or transformers will be reduced by increasing the frequency. In the proposed transformer, appropriate switching components with high power dissipation level in the vital current/voltage range are chosen so that the efficiency and power of aforementioned converter are improved. The fly-back DC/DC converter uses zero voltage switching (ZVS) technique to reduce switching losses. The assessment of the input and output displays efficiency in the range of 78 %.

Keywords Coreless PCB transformer · Zero voltage switching (ZVS) · Fly-back converter · Energy efficiency · High frequency region

56.1 Introduction

The conventional linear power supplies use a bulky power transformer to prepare isolation and lessen voltage from AC sources. The working frequency of the power transformer used in this case is at about 50/60 Hz due to which its weight is heavy in and its size is large. The linear power supply is considered as an ineffective

---

M. Hashemi (✉) · M. F. Ain · M. Rafiee
School of Electrical and Electronic Engineering, Universiti Sains Malaysia,
14300 Nibong Tebal, Pinang, Malaysia
e-mail: alihashemi2005@gmail.com

M. F. Ain
e-mail: mohdfadzilain@yahoo.com.my

M. Rafiee
e-mail: rafiee6@gmail.com

method of transmitting the signal/power from the primary side to the loads owing to the losses caused by series pass element and bulky transformer. For decreasing losses in the large transformer and converter, there is a rising need for offering a high frequency transformer to lessen the converter size and weight. Coreless printed circuit board (PCB) transformer includes of spiral windings which printed on the PCB laminate. Coreless PCB transformers do not have restrictions connected with magnetic cores, like frequency restriction, core losses and magnetic saturation [1]. In modern uses, features like power density and transient response are important in the DC/DC power converters function. With the enhancement of switching frequency of the operating converter the size of the passive components like transformers, inductors and capacitors get lessened [2]. The fly-back converter is the most frequently used SMPS circuit in low power uses. As, in converter the switching frequency is amplified, the power MOSFET's switching losses increase which brings about a reduced hard switched power converter efficiency accordingly high electromagnetic interference (EMI) [3]. Besides, with the assist of the soft switching tactics like zero current switching (ZCS) and zero voltage switching (ZVS), the electromagnetic interfering would be reduced [4, 5]. By using such approaches, each switch cycle renders a quantized 'packet' of energy to the converter output, and switch turn-off and turn-on happens at zero current and voltage, results in an fundamentally lossless switch [6]. In this study, design and real application of high frequency DC/DC fly-back converter employing coreless PCB step-down transformer is discussed.

56.2 Description of Coreless PCB Transformer

In this research, three circular spirals as primary and secondary winding which are printed on the two-layer of the FR-4 PCB laminate are employed. There are two primary windings (P1 and P2) in two dissimilar layers that they are linked in series by using via hole. In this case, to achieve better coupling [7] and lessening the leakage inductances, one secondary winding (S) is inserted between the primary windings on the reverse side of the one doubled-side PCB laminate. The number of turns is 24 in all P1, P2 and S consequently the total number of primary turns is 48. The formation of the suggested PCB transformer is illustrated in Fig. 56.1. The geometrical parameters have a significant position in the transformer function [8]. The electrical parameters like resistance, capacitance and inductance rely on geometrical factors. The coils dimensions, the PCB outline, the copper thickness and the track width should be chosen concerning the utmost allowable power losses. The geometrical parameters of coreless PCB step-down transformer are shown in Table 56.1.

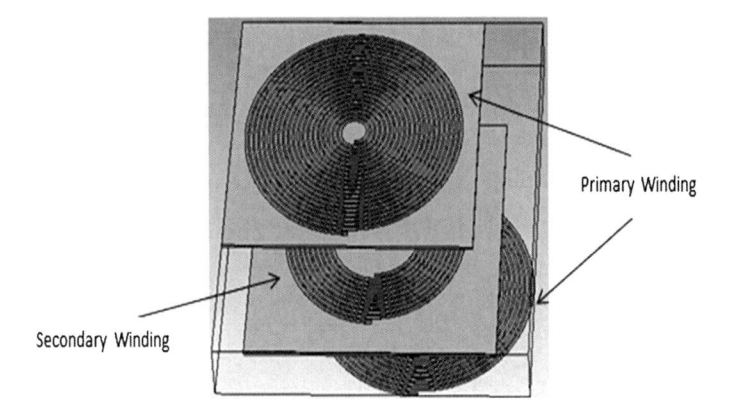

Fig. 56.1 Formation of the suggested PCB transformer

Table 56.1 Transformer geometrical factors

S. No	Factors	Value
1	No. of primary/secondary turns	24/24
2	Track division (mm)	0.4
3	Conductor width (mm)	0.6
4	Outermost diameter (mm)	36
5	Conductor height (μm)	35
6	Board thickness (mm)	1.6
7	Air space between two layers PCB (mm)	0.4

56.3 Design and Practical Application of Fly-Back Power Converter

This section explains a preferred fly-back power converter works at high frequency. This topology is preferred comparing to the other topologies for its high speed. It is significant to choose the appropriate components for working in high frequency range. All the parameters have been discussed in [7] in details. In this circuit, the ZXMN10A11G is regarded as a power MOSFET which has drain-source voltage of 100 V with an on-state resistance 'Rds–on' of 350 mΩ at 'Vgs' of 10 V. The MOSFET driver (LM5111-1 M) has a vital function in the drive power MOSFET.

The features like rise, fall and typical propagation delay of MOSFET driver are 14, 12 and 25 ns at 2 nF load respectively. Besides, a Schottky diode (STPS15L45CB) which has a forward current rating of 15 A with a blocking capacity of 45 V was used owing to low voltage drop of 0.46 V. The offered tools are located in the circuit as illustrated in Fig. 56.2. The coreless PCB transformer works very well in MHz band and could be planned for that aim [9, 10]. Since the converter switching operates in high frequency, the size of the filters become

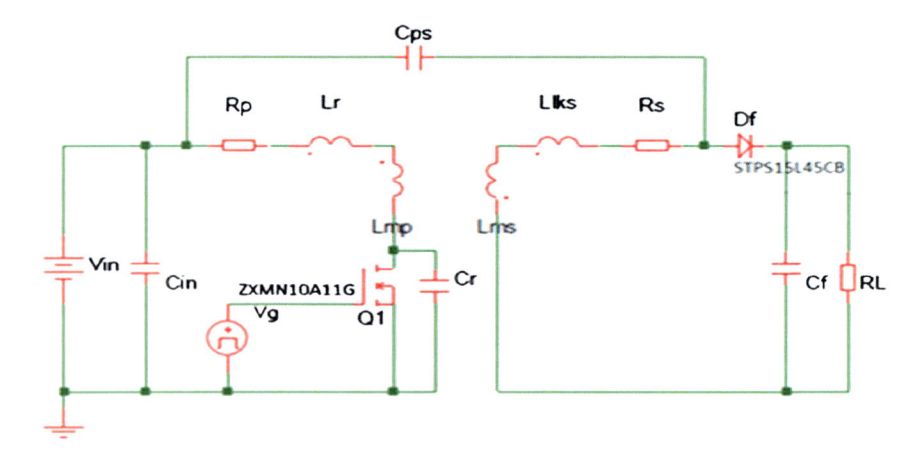

Fig. 56.2 The preferred fly-back DC/DC converter using coreless PCB step-down transformer

decreased so that the proposed design is smaller comparing to the previous designs in [9] and [10]. The filter capacitor has to be planned rooted in fly-back topology situations like load current and allowable ripple magnitude. In this circuit, 10 μF capacitor and 1 μF have located on input and output parts respectively. The output capacitor of the MOSFET could be employed in soft switching conditions. In this case, one small value of capacitor like 6 pF added to soft switching conditions.

56.4 Simulation and Experimental Results

For assessing the function and reactions of the suggested power converter, a replica of power converter was planned by SIMetrix simulation software. In this case, the planned fly-back DC/DC converter simulated and checked at nominal input voltage of 32 V and a load resistance of 30 Ω. The output voltage (V_O), the gate-source voltage (V_g) provided for the MOSFET, the drain-source voltage crosswise switch (V_d), and the mean current flowing through MOSFET (I_m) at 2.3 MHz are graphically illustrated in details in Fig. 56.3. It shows the impact of zero voltage switching (ZVS) condition on the output voltage (V_O) and the mean current flowing through MOSFET (I_m). Besides, it is illustrated that the converter output voltage is highest at a frequency of 2.3 MHz. In accordance with the attained outcomes from simulation procedure, a sample of the suggested circuit is made and checked. The converter energy efficiency is specified by (56.1) [10]:

$$Efficiency = \frac{P_{out}}{P_{in}} = \frac{V_{out}^2/R_l}{V_{in}I_{in}} \quad (56.1)$$

The energy efficiency of the converter in both soft and hard switched conditions at a load resistance of 30 Ω in the dissimilar input voltage V_{in} (10–32 V) is

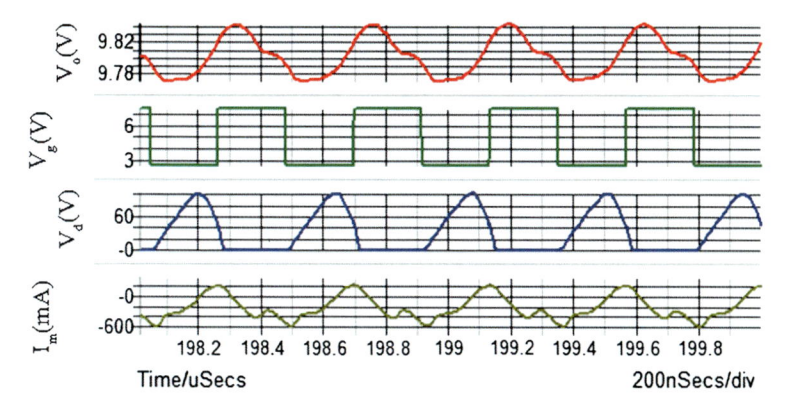

Fig. 56.3 Waveforms of V_o, V_g, V_d and I_m of fly-back converter at 2.3 MHz

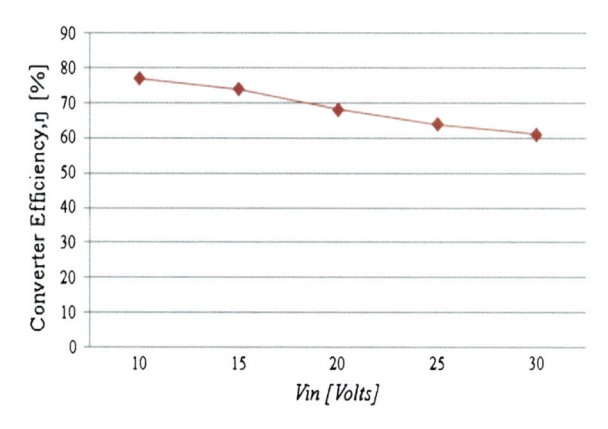

Fig. 56.4 Fly-back converter energy efficiency for V_{in} variations

illustrated in Fig. 56.4. In this case, the 50 % duty cycle is regarded. By using soft switching techniques for example ZVS, the converter energy efficiency can be developed. It is revealed that the energy efficiency power converter is uppermost at nominal input voltage of 10 V and it has lessened at higher input voltages due to the increased switching losses of the power converter.

56.5 Conclusions

The fly-back converter energy efficiency applying coreless PCB step-down transformer is measured to be almost 78 % of the supposed input voltage of 10 V at switching frequency of 700 kHz. At the utmost input voltage of 30 V, the output power of the fly-back converter is roughly 1.24 W. It could be seen that the energy efficiency is decreased at higher input voltages due to the increased switching losses. In this case, the conduction losses, the gate power consumption and the switching losses could be regarded as major losses. The results of the converter using coreless PCB transformer shows high energy efficiency and a suitable power quality for both output voltage and current. In addition, in this type of transformer, the greater part of losses is linked with copper losses owing to no core existence.

Acknowledgments Authors would like to thank Universiti Sains Malaysia Research University Grant No. 1001.PELECT. 814117 for supporting this research.

References

1. Tang S, Hui S, Chung HSH (2000) Coreless planar printed-circuit-board (PCB) transformers-a fundamental concept for signal and energy transfer. IEEE Trans Power Electron 15:931–941
2. Kotte HB, Ambatipudi R, Bertilsson K (2011) High speed series resonant converter (SRC) using multilayered coreless printed circuit board (PCB) step-down power transformer. IEEE, pp 1–9
3. Majid A et al (2012) Design and implementation of EMI filter for high frequency (MHz) power converters. In: 2012 International symposium on electromagnetic compatibility (EMC EUROPE), pp 1–4
4. Hwang SM, Ahn TY (2001) A ZVS forward DC–DC converter using coreless PCB transformer and inductor
5. Chuang YC, Ke YL (2007) A novel high-efficiency battery charger with a buck zero-voltage-switching resonant converter. IEEE Trans Energy Convers 22:848–854
6. Marchetti R (2007) Vicor corp—comparing DC/DC converters' noise-related performance. EDN Networks
7. Kotte HB, Ambatipudi R, Bertilsson K (2011) High speed cascade flyback converter using multilayered coreless printed circuit board (PCB) step-down power transformer. IEEE, pp 1856–1862
8. Meyer P et al (2012) FEM modeling of skin and proximity effects for coreless transformers. In: 2012 15th International conference on electrical machines and systems (ICEMS), pp 1–6
9. Dallago E, Passoni M, Venchi G (2004) Design and optimization of a high insulation voltage DC/DC power supply with coreless PCB transformer. IEEE 2:596–601
10. Bouabana A, Sourkounis C (2010) Design and analysis of a coreless flyback converter with a planar printed-circuit-board transformer. IEEE, pp 557–563

Chapter 57
Vector Control of Induction Motor Using Neural Network

Azuwien Aida Bohari, Wahyu Mulyo Utomo, Zainal Alam Haron, Nooradzianie Muhd. Zin, Sy Yi Sim and Roslina Mat Ariff

Abstract This paper deals with field oriented control of induction motor drive system with an online neural network for speed control. The field oriented control used to decoupling the flux and torque in order to get the performance as well as direct current motor. The online neural network is designed to maintain the output speed variation. To verify the effectiveness of the proposed method, a simulation model was developed. The result shows that the performance of transient response is improved in term of overshoot and settling time by using neural network field oriented control system. It is concluded that neural network based field oriented control schemes of induction motor drive is more effective to replace the conventional proportional integral derivative based field oriented control technique.

Keywords Induction motor · Field oriented control · Neural network controller

57.1 Introduction

Induction motor (IM) operated in variable speed alternating current (AC) drives with the employment of vector control (VC) method is expand in recent years to achieve better performances set by direct current (DC) drives. In order to provide good performance in dynamic response, decoupling between the torque and flux is highly recommends. The translation of coordinate from the fixed references stator frame to the frame of rotating synchronous is implied by the VC [1]. High dynamic performance in IM can be achieved by means of field oriented control (FOC) where it provides a suitable mathematical description of three phases IM.

A. A. Bohari (✉) · W. M. Utomo · Z. A. Haron · N. Muhd. Zin · S. Y. Sim · R. M. Ariff
Faculty of Electrical and Electronic Engineering, Universiti Tun Hussein Onn Malaysia, 86400 Parit Raja, Batu Pahat, Johor, Malaysia
e-mail: azuwienaidabohari@yahoo.com

By applying the control techniques such as adaptive control [2], a good performance can be achieved with parameter sensitive property. The neural network (NN) has capability for tolerance, to miss data, to fault, and to carry out in noise environment [3]. In order to identify and control nonlinear dynamic systems and nonlinear parameter estimation, the use of NN has been proposed [4, 5]. The NN model adapts to variation quickly in the nonlinear behavior of the system [6]. The NN are excellent estimators in non-linear system because it does not use the mathematical model of the system [7]. An estimation of the stator flux and trained to map the nonlinear behavior of a rotor flux is performed on the proposed NN [8] which presents robust neural controller and against parameter variation.

This study has been focus on developing the speed tracking control strategies in order to obtain an approach on modeling of the IM drives system for vector control purposes focusing on FOC. Intelligent controller which is NN controller are implemented and the space vector pulse with modulation (SVPWM) is used to generate pulse for switching three phase inverter to provide a higher voltage to the motor with lower overshoot and settling time to improve the overall performance of the motor.

57.2 Proposed NNFOC of IM Drive System

The controller is designed to compare between the references and the actual speed of IM to generate signal of quadrature current so that the desired speed will be reached. IM based proportional integral derivatives (PID) controller will be compared with the neural network field oriented control (NNFOC) by varying the speed reference. The block diagram of the proposed NNFOC of IM drive system is shown in Fig. 57.1.

57.3 Structure of NNFOC

The input and output neuron number of the NN control system will be based on the input signal and output signal numbers [9, 10]. In order to develop the NNFOC a multilayer perception with certain structure was proposed as shown in Fig. 57.2.

The controller consists of input layer, hidden layer and output layer. Based on number of the neuron in the layers, the NNFOC is defined as a 1-5-1 network structure. The first neuron of the output layer is used as a torque reference signal $(a_1^2 = m_f)$. The connections weight parameter between jth and ith neuron at mth layer is given by w_{ij}^m, while bias parameter of this layer at ith neuron is given by b_i^m. Transfer function of the network at ith neuron in mth layer is defined by:

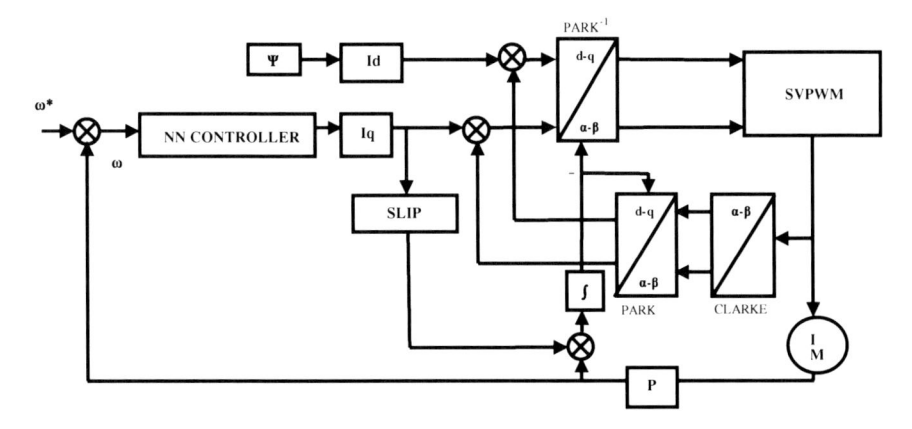

57 Vector Control of Induction Motor Using Neural Network

Fig. 57.1 The proposed NNFOC diagram

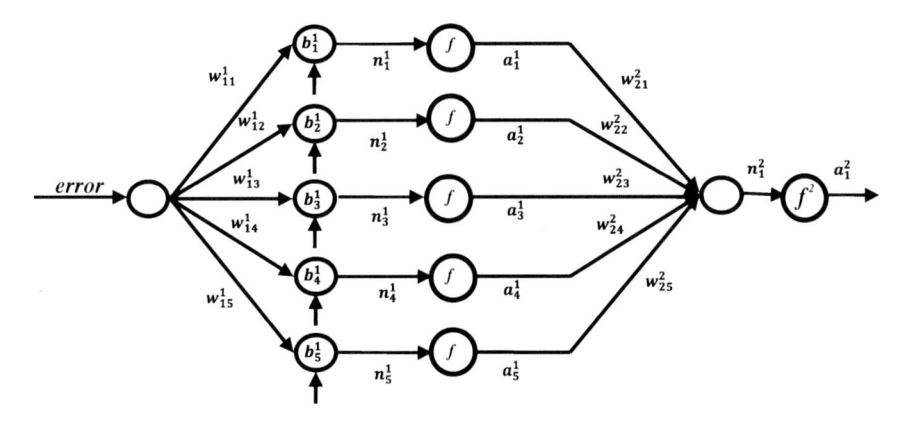

Fig. 57.2 NNFOC structure

$$n_i^m = \sum_{j=1}^{S^{m-1}} w_{ij}^m a_j^{m-1} + b_i^m \cdot z \quad (57.1)$$

The output function of neuron at *m*th layer is given by:

$$a_i^m = f^m(n_i^m) \quad (57.2)$$

where f is activation function of the neuron. In this design the activation function of the output layer is unity and for the hidden layer is a tangent hyperbolic function given by:

$$f^m(n_i^m) = \frac{2}{1 + e^{-2n_i^m}} - 1. \quad (57.3)$$

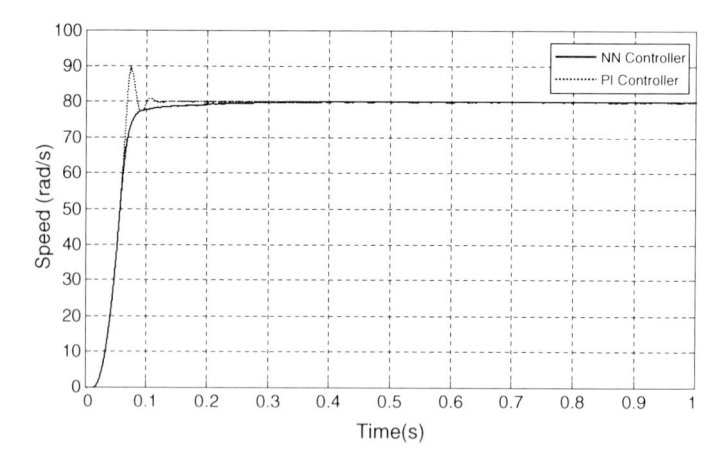

Fig. 57.3 Start-up response for PIDFOC and NNFOC controller

Updating of the connection weight and bias parameters are given by:

$$w_{ij}^m(k+1) = w_{ij}^m(k) - \alpha \frac{\partial F(k)}{\partial w_{ij}^m} \quad (57.4)$$

$$b_i^m(k+1) = b_i^m(k) - \alpha \frac{\partial F(k)}{\partial b_i^m} \quad (57.5)$$

where k is sampling time, α is learning rate, and F performance index function of the network.

57.4 Simulation and Results

Simulation was constructed to investigate the performance of NNFOC of IM. The dynamic model of a three-phase IM, SVPWM and NN controller model have been developed by using MATLAB/Simulink software in this section. The parameters for the motor are 240 V, 50 Hz, Poles = 4, Rs = 0.3 Ω, Rr = 0.25 Ω, Ls = 0.0415 H, Lr = 0.0412 H, Lm = 0.0403 H, and J = 0.1 kg-m².

The effectiveness of the proposed NNFOC is verify with the comparison of simulation results between conventional proportional integral derivative field oriented control (PIDFOC) and NNFOC system. With the same speed and load torque reference, the simulations of both methods are run simultaneously. Start-up response from time 0–1 s are shown in Fig. 57.3.

Figure 57.3 shows the start-up response for IM drive system. The NNFOC have a start-up response improved compared to PIDFOC system. The results show the overshoot of transient response for NNFOC is reduce and improved the settling

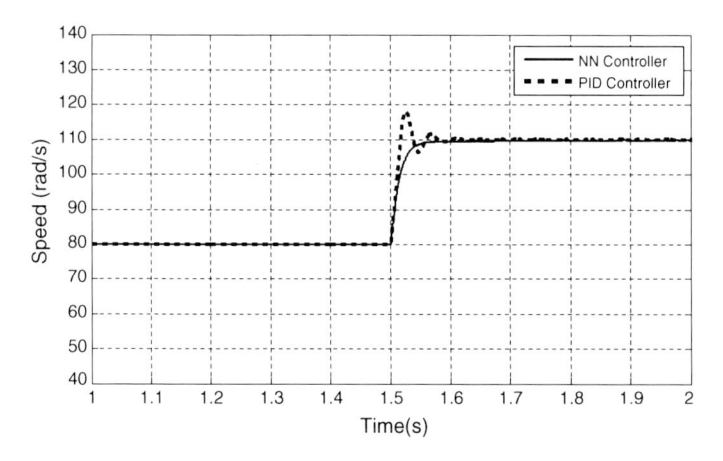

Fig. 57.4 Step-up response for conventional PIDFOC and NNFOC controller

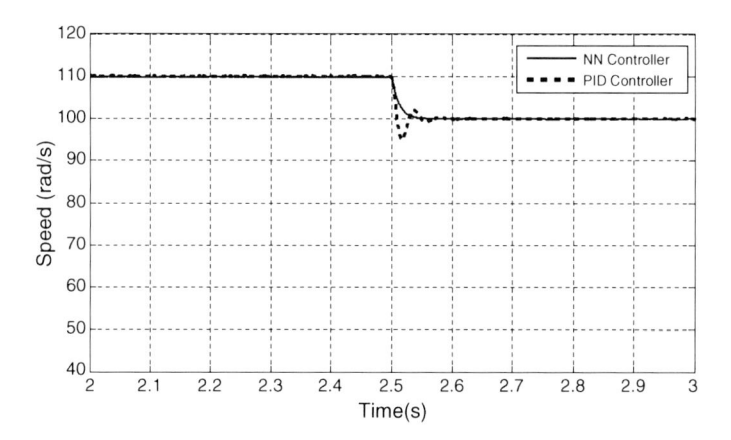

Fig. 57.5 Step-down response for PIDFOC and NNFOC controller

time. The step-up response for both PIDFOC and NNFOC system is shown in Fig. 57.4.

The improvement of the step-up response by great reduce in overshoot for NNFOC as well as the settling time is shown in the result in Fig. 57.4. The simulation testing is carry on by vary the speed reference from 110 to 100 rad/s with a constant load is shown in Fig. 57.5.

Regarding to the Fig. 57.5 the settling time and the overshoot is also greatly reduced by applying the proposed neural technique thus; improve the system performance at the same time.

57.5 Conclusion

This paper presents a comparative study on speed response of NNFOC and PIDFOC system. The performances differences due to both NN and PID controller were examined. Both controllers are tested for reference speed variation with constant load. The results show that the performance of transient response is improved in term of overshoot and settling time by using NNFOC system. It is concluded that NN based FOC schemes of IM drive is more effective and, hence, found to be a suitable replacement of the conventional PID controller. The NNFOC of IM drive system that has been proposed is useful and able to improve the performance.

Acknowledgments The author would like to gratitude University Tun Hussein Onn Malaysia for any valuable supports during conducting this research and in preparing this manuscript.

References

1. Goyat S, Ahuja RK (2012) Speed control of IM using vector or field oriented control. Int J Adv Eng Technol 4(1):475–482
2. Zuben V, Netto FJ, Bim MLA, Szajner E (1994) Adaptive vector control of a three-phase IM using neural networks. IEEE world congress on IEEE international conference of computational intelligence, vol 6, pp 3750–3755
3. Ba-Razzouk A, Cheriti A, Olivier G (1997) A neural network based field oriented control scheme for induction motor. Industry applications conference of 32nd IAS annual meeting, IAS on conference record of the IEEE, vol 2, pp 804–811
4. Wishart MT, Harley RG (1995) Identification and control of induction machines using artificial NNs. IEEE Trans Ind Appl 31(3):612–619
5. Narendra KS, Parthasarathy K (1990) Identification and control of dynamical systems using neural network. IEEE Trans Neural Network 1(1):4–27
6. Yang HT, Huang KY, Huang CL (1993) An ANN based identification and control approach for field-oriented induction Moto. In: 2nd international conference on advances in power system control, operation and management, APSCOM-93, vol 2, pp 744–750
7. Sivasubramaniam A (2009) Application of neural network structure in voltage vector selection of direct torque control induction motor. Int J Appl Eng Res 4(6):903
8. Ba-Razzouk A, Cheriti A, Olivier G, Sicard P (1995) Field oriented control of IMs using neural network decouplers. In: Proceedings of the IEEE Industrial electronics, control, and instrumentation
9. Yatim AHM, Utomo WM (2007) Neural network efficiency optimization control of a variable speed compressor motor drive. In: IEEE international electric machines and drives conference, pp 1716–1720
10. Yatim AHM, Utomo WM (2004) On-line optimal control of variable speed compressor motor drive using neural control model. In: PECon proceedings conference, pp 83–87

Chapter 58
Autonomous Dual Axis Solar Tracking System Using Optical Sensor and Sun Trajectory

Wong Yoong Wai Adrian, Vickneswari Durairajah and Suresh Gobee

Abstract In our daily life, the need of energy increases each and every day. However, the amount of natural resources on the earth is limited so this makes renewable energy rapidly gaining importance as an energy resource. Solar power is one of the most popular energy sources that available and can be converted into electricity by using solar panels. For solar panels to produce maximum output power, the incidence angle of the sunlight needs to be constantly perpendicular to the solar panel. However, most of the solar panels that used by the users in nowadays are in static direction. As the sun's position changes, low output power will be generated. In this paper, a two axis solar tracking system was proposed to keeps the solar panel perpendicular to sunlight by using sunflower tracking and path calculation modes. Additionally, a friendly graphic user interface (GUI) system also has been built for monitoring the performance of the system. In order to validate the design, a scaled down dual axis solar tracker was designed, built and tested. As results, the built system had been calculated to have an energy gain of 58.59 and 59.24 % compared to a stationary solar panel during sunny and random weather condition respectively.

Keywords Two axis solar tracking system · Sunflower tracking mode · Path calculation mode · GUI for monitoring the performance of the system

---

W. Y. W. Adrian · V. Durairajah (✉) · S. Gobee
Asia Pacific University, Kuala Lumpur, Malaysia
e-mail: vicky_nesa@apu.edu.my

W. Y. W. Adrian
e-mail: Adrian_wong507@hotmail.com

S. Gobee
e-mail: suresh.gobee@apu.edu.my

58.1 Introduction

Electricity is an important energy source for human beings that live in 21st century. Electrical generation is commonly provided by fossil fuels but the amount of fossil fuels on the earth is limited and it is not environmental friendly so this makes renewable energy rapidly gaining importance as an energy resource [1]. Solar power is one of the most popular and ideal renewable energy sources because it is clean, unlimited, economical and universal. Solar energy can be converted into direct current (DC) electrical energy through solar panel and the amount of DC electrical energy that generated is directly proportional to the angle and the intensity of the sunlight that falls on the panels [2]. In recent years, many residential around the world started to use solar panels for electricity generation. For the solar panels to produce maximum output power, the incidence angle of the sunlight needs to be constantly perpendicular to the solar panel. However, most of the solar panels in residence areas are mounts statically aligned on a roof and not facing towards the sun perpendicularly throughout the day. As the sun's position changes, low output power will be generated by the solar panels [3]. Many researches were conducted to increase the efficiency of solar panels and one of the methods is by employing a tracking system. Generally, there are two fundamental types of solar tracking systems. They are single axis and dual axis solar tracking system. A dual axis tracker usually will be having higher efficiency compare to the single axis tracker. This is because the dual axis tracker able to maintains the panel surface to the optimal during the day to collect maximum energy from the sun [4]. Besides that, tracking control method that used in solar tracker can also be divided into open loop and closed loop tracking. Both tracking methods will be having their individual limitations. Therefore, in this paper, a two axis solar tracking system was proposed to keeps the solar panel perpendicular to sunlight by using sunflower tracking (Closed loop) and path calculation (Open loop) modes to eliminate the limitations from previous conventional designs. In this presented tracking system, the tracker is able to select different modes for operating in different weather conditions to improve the tracking precision, the power generation efficiency of photovoltaic system and reduce the energy consumption of tracking system. In sunny condition, sunflower tracking mode will be employed to track the movement of sun by using optical sensors (positioning sensors). Meanwhile, in cloudy condition, path calculation mode will be used to track the sun. Lastly, a hibernation mode will also be integrated in the tracking system and to be used in sunless weather condition. Additionally, a monitoring system will also be used to monitor the performance of the tracking system according to the weather conditions. This paper will be structured as follows: In Sect. 58.2 the conceptual development; In Sect. 58.3 hardware design of the tracking system; In Sect. 58.4 the solar tracking software designed for tracking and monitoring systems; In Sect. 58.5 testing of the design. In Sect. 58.6 the results of simulation experiments will be analysed and In Sect. 58.7 is conclusion.

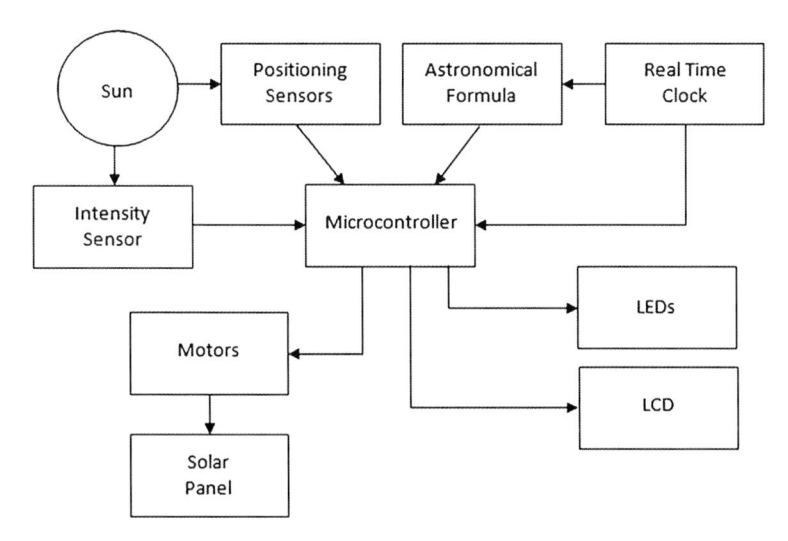

Fig. 58.1 Block diagram of solar tracking system

58.2 Concept Development

58.2.1 Solar Tracking System

In order to improve the efficiency of the solar panel, a conceptual dual axis tracker that contains of three modes which suitable for three different kinds of weather condition were proposed. Figure 58.1 shows the block diagram of the proposed conceptual tracking system.

The presented dual axis tracker is contains of sunflower tracking mode, path calculation tracking mode and hibernation mode. At the beginning, the system will determine the light intensity in the surrounded environment by using the intensity sensor for selecting the appropriate tracking mode. In sunny day, sunflower tracking mode will be selected and the positioning sensors will be used for sensing the sun direction and directing the motors to the desired position. Meanwhile, path calculation tracking mode will be selected in cloudy day. The rotation angles for the motors to direct the solar panel to face the sun will be based on astronomical formula. During sunless condition such as night and raining day, the system will be hibernated. In addition, this system also contain of three different colour of LEDs for representing the modes selection in different weather conditions and a LCD for displaying the current real time and the rotation angles of both motors.

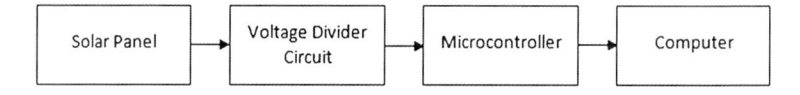

Fig. 58.2 Block diagram of monitoring system

Fig. 58.3 Arrangement of LDRs in four quadrants

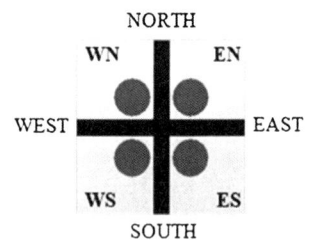

58.2.2 Monitoring System

Besides that, another microcontroller is also been used in this system for data collection purpose. This monitoring system (Graphic User Interface system) is built for monitoring and collecting the voltage that generated by the solar panel throughout the day. These resulted data will be used to calculate the efficiency of the tracking system compared to a stationary solar panel. The block diagram of the monitoring system is shown in Fig. 58.2.

58.3 System Hardware Design

In the real setup, a base with four legs is used as the main structure of the tracking system which supports the weight of the upper portion of the tracker. Besides that, a "Cross" shape of four quadrants structure as shown in Fig. 58.3 has been design for the placement of positioning sensors (LDRs) to improve the tracking efficiency in sunny condition. An LDR is a variable resistor whose output value changes with the change of incident light intensity. When the incidence angle of the sun is perpendicular to the positioning sensors, each sensor will be generating the same amount of output voltage to cause the system to stop rotating because of the sunlight is equally distributed on each sensor. When the sun moves, the distribution of the sunlight on each sensor is changed and its output voltages will also change accordingly too. Therefore, this principle has been utilised to position the solar panel for facing the sun.

In this design, LDR's WN + EN and WS + ES work as the altitude sensors that track the sun in North-South direction. Meanwhile, LDR's WN + WS and EN + ES work as the azimuth sensors to track the sun in East–West direction.

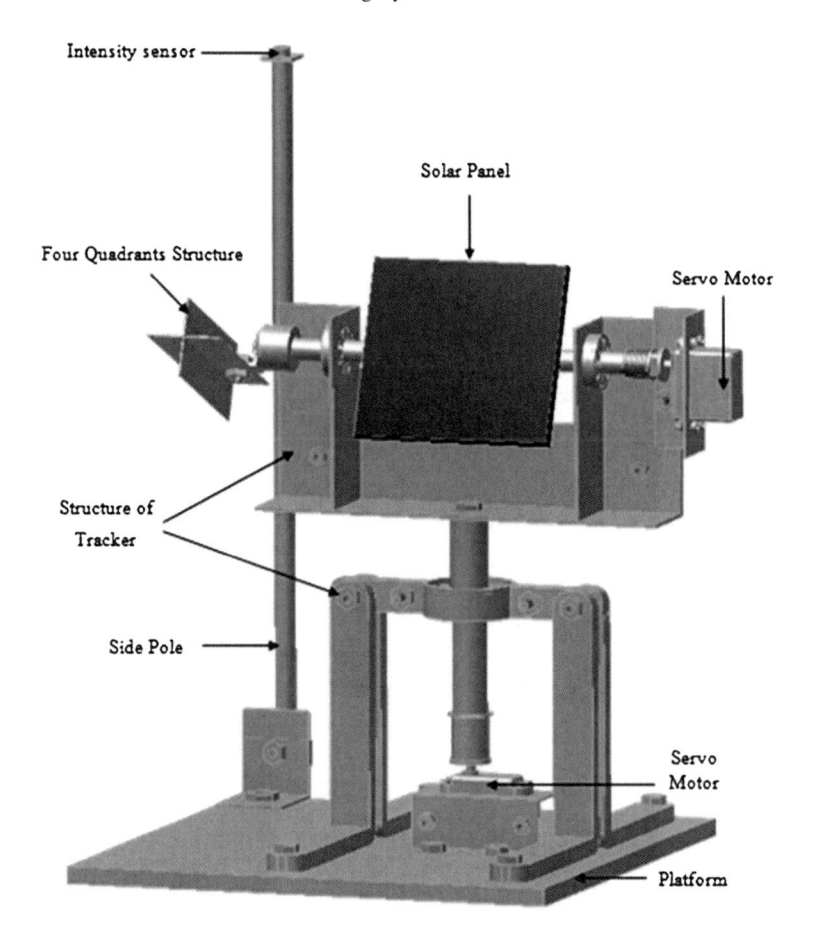

Fig. 58.4 Mechanical design of solar tracker

Therefore, it provides wider measurement range. The movement of the sun in the directions of x-axis and y-axis shows the changes of sun azimuth and altitude angles respectively. Furthermore, an intensity sensor was also mounted on top of a side pole so the light intensity of the environment can be determined accurately without affecting by shadows. In order to fully track the sun from sunrise to sunset, a dual axis motion mechanism which consists of azimuth rotation axis perpendicular to the horizon plane and altitude rotation axis which parallel to the horizon plane has been employed. This solar tracker is design to rotate around the horizontal axis from 0° to 180° for the East–West tracking of the sun and rotate around the vertical axis from 0° to 90° for the North–South tracking of the sun. The 3D mechanical design of the solar tracker is shown in Fig. 58.4.

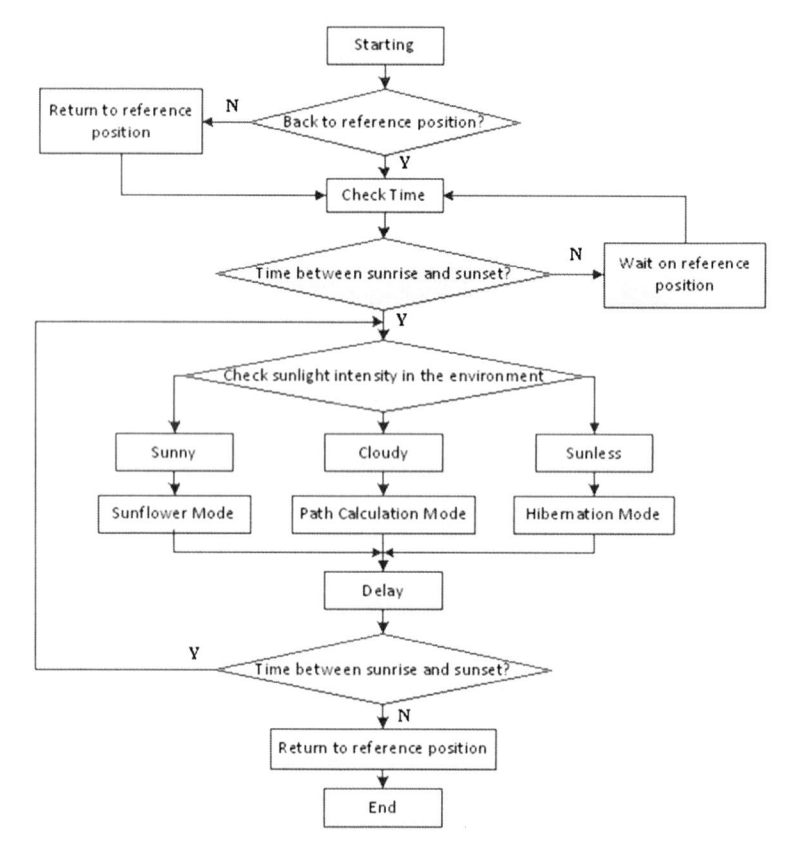

Fig. 58.5 Flowchart of solar tracking system

58.4 System Software Design

58.4.1 Solar Tracking Control System

Tracking control method involves of open loop and closed loop tracking. Open loop tracking calculates the solar altitude angle and azimuth angle of the sun position to track the sun. Meanwhile, closed loop tracking is using optical sensors to track the sun according to illumination with high precision. In this proposed system, the tracking system will be utilising only one tracking method each time and the methods will be selected based on the sunlight intensity in the environment by using an optical sensor (intensity sensor). According to the time, the system also controls the tracker back to the reference position after sunset. The flowchart of the solar tracking system is shown in Fig. 58.5.

The details of the presented dual axis solar tracking system are discussed below:

1. During sunny condition:
   - Sunflower mode (closed loop tracking method) will be selected.
   - This mode will be using four LDR sensors to track the movement of sun.
2. During cloudy condition:
   - Path calculation mode (open loop tracking method) will be selected.
   - This mode will be using real time clock and astronomical formula to calculate the solar azimuth angle and altitude angle of the sun position to track the sun [5].
   - The sun position (altitude and azimuth angles) can be determine by using Eqs. (58.1)–(58.3):

$$\sin \theta_z = (\cos \phi \cdot \sin \delta) + (\cos \phi \cdot \cos \delta \cdot \cos \omega) \quad (58.1)$$

$$\sin \theta_A = \frac{\cos \delta \sin \omega}{\cos \theta_z} \quad (58.2)$$

$$\delta = 23.45 \sin \left[\frac{360}{365} (284 + N) \right] \quad (58.3)$$

where:
- θ_z Altitude angle of the system.
- θ_A Azimuth angle of the system.
- ϕ Latitude of the area.
- ω Hour angle (15°/h) where noon is 0°.
- δ Solar declination.
- N Day of the year, where 1 is 1st of January.

3. During sunless/night condition:
   - Hibernation mode will be selected.
   - Both tracking modes are set to be in rest for saving energy.

58.4.2 Monitoring System

Besides that, a LabVIEW monitoring system (Graphic User Interface system) has been built for monitoring and collecting the voltage that generated by the solar panel throughout the day in this project. This system will be collecting one data in each two second in real time into a table and then saved into text file. In addition, the monitoring system also designs to plot the collected data (voltage and power) for each day into graph to compare the performance of solar panel between stationary and design dual axis tracking system. The flowchart for the monitoring system is shown in Fig. 58.6.

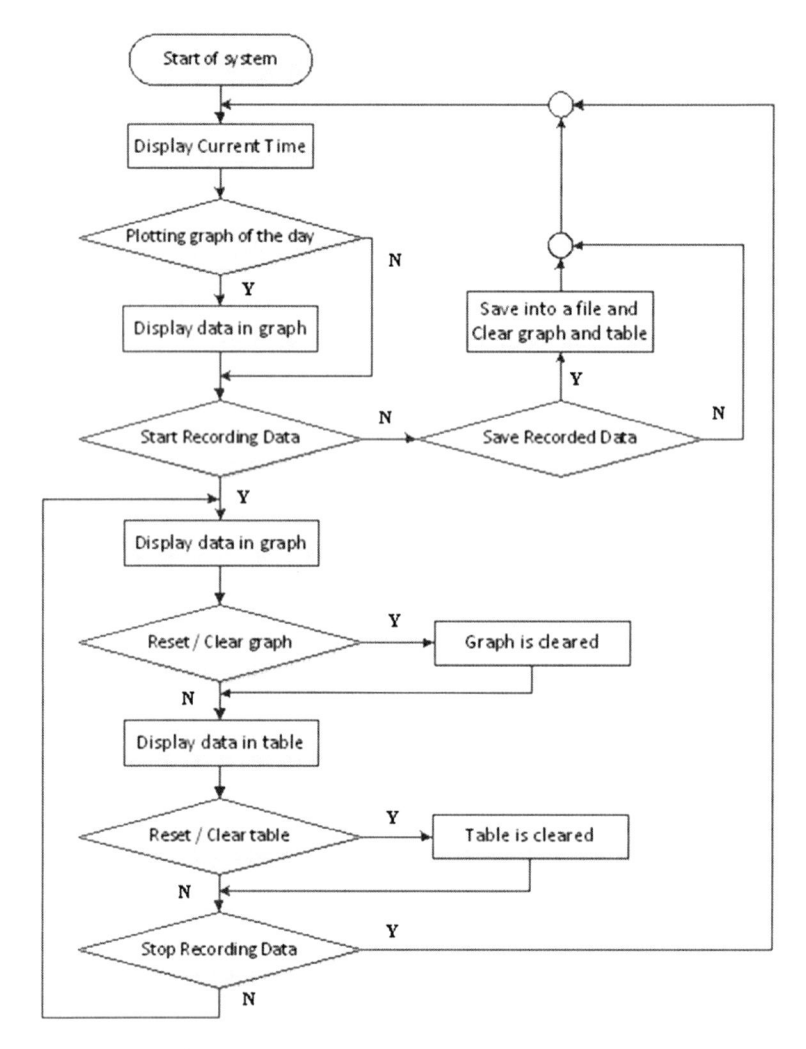

Fig. 58.6 Flowchart of monitoring system

In main display of the monitoring system, there is a "Start Recording Button" for user to begin data collection process. When the system starts to collect data, the voltage that generate by the solar panel will be recorded. Current and power that generated by the solar panel will be calculated by the system itself through the collected voltage data with some electrical formulas. The collected voltage, current and power data will be displayed in their individual real time graph. The data collection process will be continued until a "Pause" button was press by the user. The monitoring system also designs to plot the daily collected data (voltage and power) which had been saved into a waveform graph and chart by pressing a "Plotting the entire collected data" button. The main display of the monitoring system is shown in Fig. 58.7.

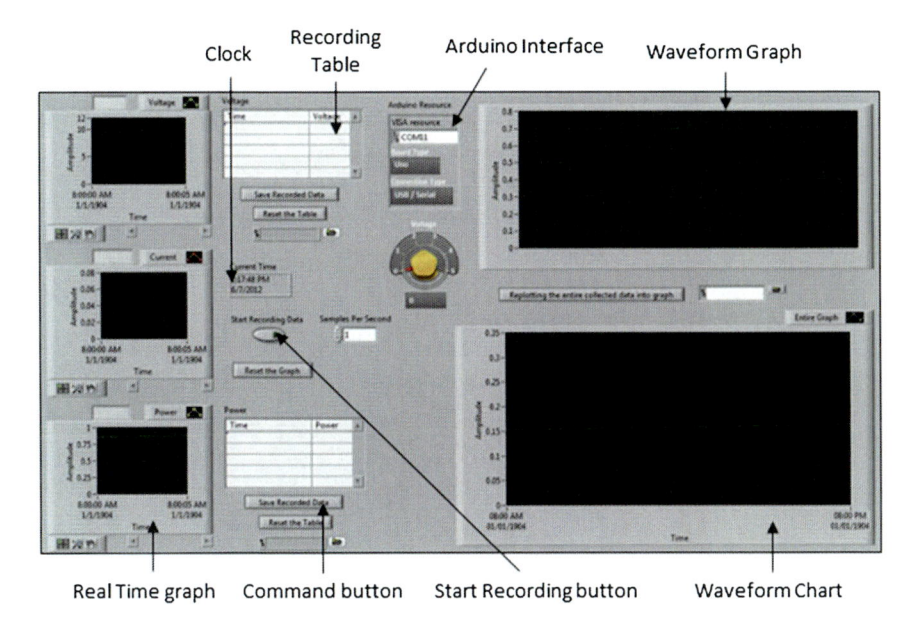

Fig. 58.7 Main display of monitoring (GUI) system

58.5 Testing of Design

The prototype of the solar tracking system will be tested in real life and output voltage that generated by the solar panel will be collected by the monitoring system. In addition, a stationary solar system also will be tested simultaneously under the sun by placing the panel inclined according to the latitude angle 30°. Both systems will be tested from morning 7 a.m. to evening 6 p.m. under two conditions which are the sunny weather condition and random weather condition in Kuala Lumpur, Malaysia. In sunny weather condition, optimum power that generated by both systems can be analysed. Meanwhile, in random weather condition, the performance of both systems under sunny, cloudy or sunless condition can be analysed. Finally, these collected data can be used to plot the "Power" against "Time" graph for comparing the efficiency between the tracking system and the stationary solar panel.

58.6 Experimental Results and Conclusion

58.6.1 During Sunny Weather Condition

The data of power which generated by the static solar panel and solar tracking system is tabulated in Table 58.1. Besides that, the power characteristic curve of static solar panel and solar tracking system is shown in Figs. 58.8 and 58.9 respectively.

Table 58.1 Power data for stationary system and solar tracking system for sunny weather condition

Local time (H)	Power (W)	
	Stationary system	Solar tracking system
7.00 a.m.	0	0
7.30 a.m.	0.000618	0.00061
8 00 a.m.	0.010387	0.02451
8.30 a.m.	0.031162	0.090162
9.00 a.m.	0.061456	0.209262
9.30 a.m.	0.168083	0.318218
10.00 a.m.	0.309867	0.399867
10.30 a.m.	0.409024	0.517391
11.00 a.m.	0.467391	0.528717
11.30 a.m.	0.51974	0.598717
12.00 p.m.	0.507198	0.629509
12.30 p.m.	0.508187	0.617063
1.00 p.m.	0.478367	0.581145
1.30 p.m.	0.418367	0.568735
2.00 p.m.	0.312648	0.523144
2.30 p.m.	0.244765	0.514377
3.00 p.m.	0.136213	0.433319
3.30 p.m.	0.075188	0.373175
4.00 p.m.	0.040586	0.294012
4.30 p.m.	0.033069	0.185331
5.00 p.m.	0.027125	0.083625
5.30 p.m.	0.007468	0.048777
6.00 p.m.	0.004529	0.027487
Total power collection	4.771438	7.567153

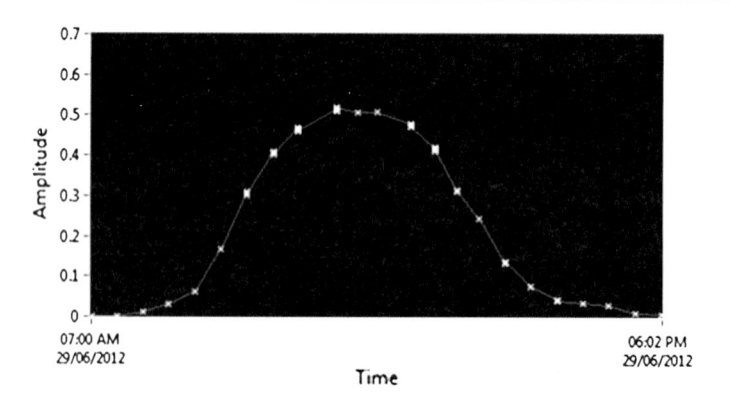

Fig. 58.8 Power characteristic curve of static solar panel for sunny weather condition

Based on the data in Table 58.1, the total power collection for stationary system is 4.771438 W. Meanwhile, the total power collection for solar tracking system is 7.567153 W. The efficiency of the solar tracking system can be calculated by using the Eq. (58.4):

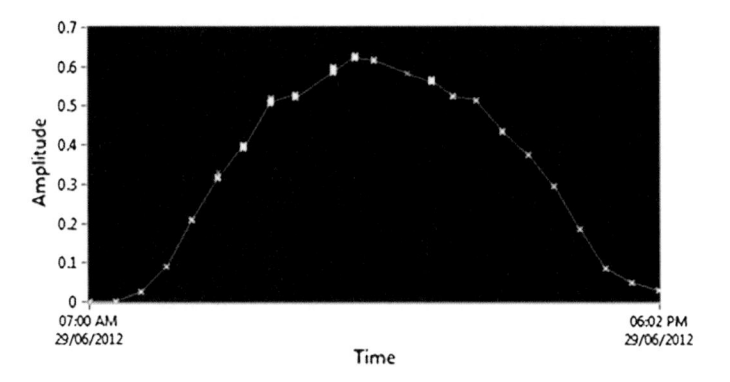

Fig. 58.9 Power characteristic curve of solar tracking system for sunny weather condition

$$E = \frac{Power_{Tracking} - Power_{Stationary}}{Power_{Stationary}} \times 100\,\%$$
$$= \frac{7.567153 - 4.771438}{4.771438} \times 100\,\% \qquad (58.4)$$
$$= 58.59\,\%$$

It seems that the efficiency of the solar tracking system is 58.59 % better than the stationary system. Therefore, it can be concluded that the solar tracking system is able to receive more sunlight and consequently generate more power as compared to stationary system on sunny weather condition.

58.6.2 During Random Weather Condition

The data of power which generated by the static solar panel and solar tracking system is tabulated in Table 58.2. Additionally, the power characteristic curve of static solar panel and solar tracking system is shown in Figs. 58.10 and 58.11 respectively.

Based on the data in Table 58.2, the total power collection for stationary system is 3.901788 W. Meanwhile, the total power collection for solar tracking system is 6.21319 W. The efficiency of the solar tracking system can be calculated by using the Eq. (58.5):

$$E = \frac{Power_{Tracking} - Power_{Stationary}}{Power_{Stationary}} \times 100\,\%$$
$$= \frac{6.21319 - 3.901788}{3.901788} \times 100\,\% \qquad (58.5)$$
$$= 59.24\,\%$$

Table 58.2 Power data for stationary system and solar tracking system for random weather condition

Local time (H)	Power (W)	
	Stationary system	Solar tracking system
7.00 a.m.	0	0
7.30 a.m.	0.000648	0.000748
8.00 a.m.	0.01051	0.025887
8.30 a.m.	0.032608	0.086162
9.00 a.m.	0.020884	0.056888
9.30 a.m.	0.155218	0.289984
10.00 a.m.	0.236509	0.344867
10.30 a.m.	0.368289	0.449585
11.00 a.m.	0.437391	0.515064
11.30 a.m.	0.219695	0.35374
12.00 p.m.	0.485914	0.579509
12.30 p.m.	0.476187	0.57794
1.00 p.m.	0.449671	0.545096
1.30 p.m.	0.095343	0.207798
2.00 p.m.	0.324364	0.426648
2.30 p.m.	0.224377	0.484377
3.00 p.m.	0.164516	0.455327
3.30 p.m.	0.089388	0.327384
4.00 p.m.	0.053683	0.264012
4.30 p.m.	0.015894	0.077294
5.00 p.m.	0.027115	0.091625
5.30 p.m.	0.009467	0.046768
6.00 p.m.	0.004117	0.006487
Total power collection	3.901788	6.21319

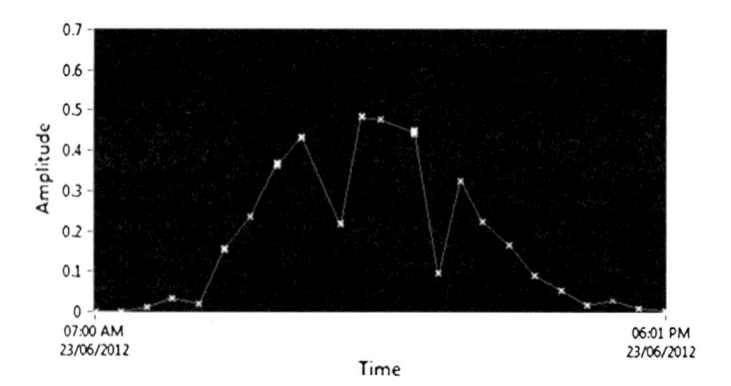

Fig. 58.10 Power characteristic curve of static solar panel for random weather condition

It seems that the efficiency of the solar tracking system is 59.24 % better than the stationary system. Therefore, it can be concluded that the solar tracking system is able to receive more sunlight and consequently generate more power as compared to stationary system on random weather condition.

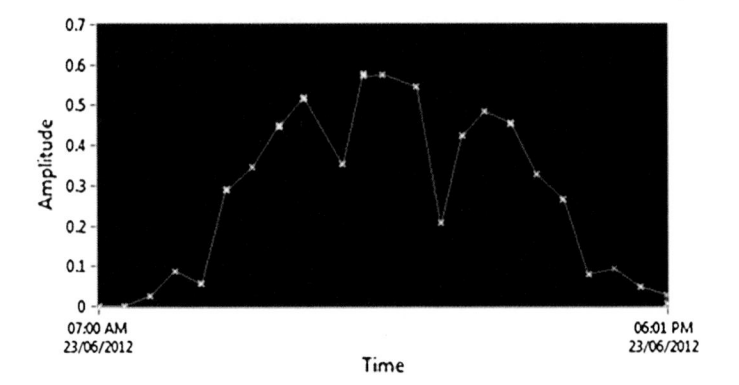

Fig. 58.11 Power characteristic curve of solar tracking system for random weather condition

58.7 Conclusion

Based on the experimental results, the dual axis solar tracking is able to deliver a maximal power from solar panel and having the best efficiency of conversion compared to a static solar panel with fix inclination under the climatic and geographical conditions of the testing site. Additionally, it also proves that the tracking system is characterized by the following advantages:

- The solar tracking system is able to follow the sun according to its movement with two tracking modes.
- The tracking system plays an important role especially in the morning and evening for generating power when the static solar panel hardly receive direct sunlight.

Through experimental testing, the obtained results show that the designed solar tracking system has approximate 60 % energy gain over a fixed solar panel setup during sunny and random weather conditions assuming the solar panel mounted on the solar tracker and the stationary solar system are a similar 1W panel. Therefore, it can be concluded that the Autonomous Dual Axis Solar Tracking System using Optical Sensor and Sun Trajectory built in this project shows apparent benefits over the widely use stationary solar system. In addition, the designed system also has wide suitability which can be applied in all types of solar panel for reducing the overall cost of photovoltaic power generation system and setting up the environmental protection economy society.

References

1. RenewableEnergyWorld.com. http://www.renewableenergyworld.com/rea/tech/home
2. Solar Technologies. http://www.solartechnologies.com/cm/About-Solar-Power/How-Does-Solar-Power-Work.html

3. Shayani RA, Oliveira MAG (2006) Global performance measurement of a stand-alone photovoltaic system. IEEE, pp 1–6
4. Barsoum N (2006) Implementation of a prototype for a traditional solar tracking system. IEEE, pp 23–30
5. Hanieh AA (2011) Solar photovoltaic panels tracking system. IEEE, pp 30–37

Chapter 59
Image Acquisition System for Boiler Header Inspection Robot

Dickson Neoh Tze How, Khairul Salleh Mohamed Sahari,
Adzly Anuar, Mohd Zafri Baharuddin,
Muhammad Fahmi Abdul Ghani and Mohd Azwan Aziz

Abstract This paper aims to present development of an image acquisition in the task of power plant boiler header inspection. Among the content of this paper includes hardware selection and software selection and integration of both systems to form a complete image acquisition system. The real challenge behind the project is the miniature scale of the robot that will be used to carry the image acquisition system into the boiler header, since a boiler header entrance is normally only about 80 mm in diameter. Hardware system needs to be small enough to fit into the boiler header entrance. This paper also discusses the results from testing in Stesen Janakuasa Sultan Ismail, Paka as a verification of the workability of the designed system.

Keywords Image acquisition · Boiler header inspection · Visual inspection

---

D. N. T. How (✉) · K. S. M. Sahari · AdzlyAnuar · M. Z. Baharuddin · M. F. A. Ghani
Centre for Advanced Mechatronics and Robotics, Universiti Tenaga Nasional,
Jalan IKRAM-Uniten, 43000 Kajang, Selangor, Malaysia
e-mail: dickson.neoh@gmail.com; dicksonn@uniten.edu.my

K. S. M. Sahari
e-mail: khairuls@uniten.edu.my

A. Anuar
e-mail: adzly@uniten.edu.my

M. F. A. Ghani
e-mail: fahmi.gie@gmail.com

M. A. Aziz
TNB Research Sdn Bhd Lorong Ayer Itam, Kawasan Institusi Penyelidikan,
43000 Kajang, Selangor, Malaysia

59.1 Introduction

In the world of autonomous robotics, guidance for mobile robots to ensure collision free travel path has become essential. Mobile robots need to have the capability to efficiently access the situation around to make critical decisions ensuring its own survival. Visual image acquisition is one of the techniques used to acquire and sense the environment of the mobile robot. Blais et al. [1] pointed out that environment sensing serve five fundamentally important roles are collision avoidance, detection of the limits of free space, position estimation, environment description detection of unknown objects and scene identification. For the role of environment description and scene identification, the visual image acquisition system will be able to acquire the image and identify the image via image processing whether or not the image acquired is the object or 'scene' that is expected by the robot i.e. cracks on the boiler header wall.

For boiler header maintenance, it has also become a necessity for occasional inspection to be carried out to maintain the condition of the boiler headers. Liu et al. [2] pointed out that failures that occur in piping environment are affected by static and dynamic factors. Static factors include pipe material, size, age, etc. while dynamic factors pressure, humidity, etc. Newman and Jain defined the task of inspection as follows: Inspection is the process of determining if a product deviates from a given set of specifications [3].

This version of development is a continuation of the first stage development in [4]. In [4], Sahari et al. had utilized the image acquisition device of a borescope. The probe of the borescope is mounted on a robot that carries it into the boiler header. The image is captured by an operator that controls the borescope while the robot traverses deeper into the boiler header. The disadvantage of this method is that the navigational and image acquisition system is not integrated, therefore; it requires more operators to carry out the inspection process. Integrating both the systems not only reduces the operators needed for the inspection process but also eases the task of automating the inspection robot in upcoming developments.

59.2 Imaging Device Selection for Image Acquisition

Imaging device selection is the first step in constructing the image acquisition system for the inspection robot. Before deciding on the camera module that will be used, a study has been done on inspection robots and its camera modules. In [5], Lim et al. who researches on indoor pipeline inspection robot utilizes a camera module that consists of an analogue CMOS camera. The camera utilizes the analogue composite signal for long distance transmission (15 m) of the images to a PC. The camera resolution is 250,920 pixels. In [6], Suzumari and team utilize an analogue CCD camera with resolution of 410,000 pixels. In [7], Ohya et al. utilizes

59 Image Acquisition System for Boiler Header Inspection Robot

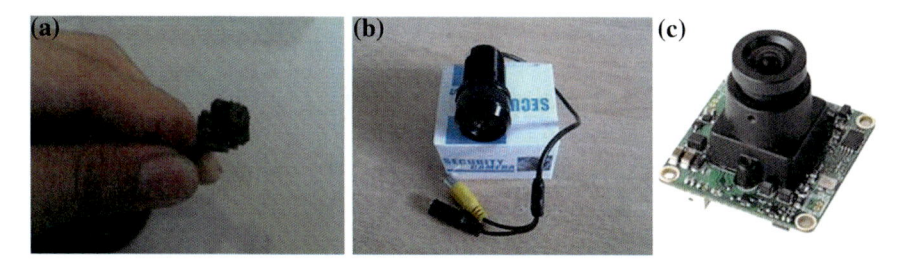

Fig. 59.1 Types of analogue camera tested. **a** CMOS spy camera. **b** Vehicle reverse camera. **c** Closed circuit television board level camera

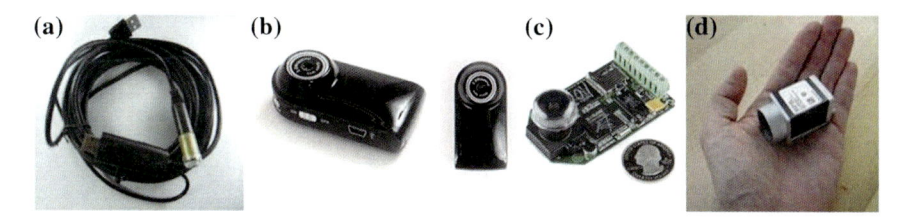

Fig. 59.2 Types of digital camera tested. **a** USB snake camera with LED. **b** High definition web camera. **c** High definition digital board level camera. **d** High definition network camera

analogue CCD camera with resolution of 380,000 pixels. The common point of these inspection robots is that they all utilize analogue cameras with resolution between 200,000 and 410,000 pixels.

Devices from various range and types are explored to ensure an accurate selection of the imaging device. The two classes of camera are tested before selecting the best candidate for the purpose of this project. The two classes of camera are the analogue camera (Fig. 59.1) and digital camera (Fig. 59.2).

Table 59.1 shows the summary of the types of cameras and its basic specification to undergo evaluation for camera selection for this project. The camera module that is selected for this research project is the CMOS analogue spy camera. The resolution of the camera is 480×640 pixels (307,200 pixels). The size of the camera is 15 mm $\times$ 15 mm $\times$ 20 mm. The main reason behind the selection of this imaging device is the size factor. This camera module offers the smallest size among all the candidates.

From the literature review, it is concluded that the main advantage of digital camera over their analog counterpart is the resolution. Digital cameras can provide far greater resolution compared to analog cameras. The data are also packaged in a digital format ready to be fed to a PC. However, digital cameras have a major disadvantage in terms of the size. Analog cameras are packaged in sizes far smaller than that of the digital cameras. Since the constraint of size is more crucial for this development, the advantage of miniature size weighs over smaller resolution. The subsequent chapters cover the integration of analog cameras with software in the PC.

Table 59.1 Camera type and basic specifications

Camera	Type	Resolution (in pixels)	Size (in mm)	Advantage	Disadvantage
CMOS spy camera	Analogue	307,200	15 × 15 × 20	Small in size	Moderate resolution
Vehicle reverse camera	Analogue	250,920	30 × 30 × 42	Wide view angle	Big in size
CCTV board level camera	Analogue	307,200	40 × 40 × 30	Moderate resolution	Big in size
USB snake camera	Digital	307,200	20 × 20 × 40	Long cable for transferring to PC	Big in size
HD web camera	Digital	921,600	50 × 30 × 30	Very high resolution	Big in size
HD digital board level camera	Digital	9,000,000	65 × 40 × 25	Ultra high resolution	Big in size
HD network camera	Digital	1,251,936	29 × 29 × 55.5	Ultra high resolution	Big in size

59.3 Image Acquisition Software and GUI

As for the software portion of the image acquisition system, C# programming language is used as a platform to create a friendly Graphical User Interface (GUI) for users to interact. The flexibility and universality of the C# programming language enables developers of the image acquisition software to adopt various libraries to acquire image from the camera hardware. The library that is being utilized in this project is named as "Touchless Library". The library is adopted because of the flexibility of open source programming. This library enables the image to be viewed from the PC via the GUI. The library enables the user to view the image from the camera hardware and to capture any frame if needed.

The best candidate selected for the application in the project is by utilizing the analogue CMOS spy camera. The output of the analog camera is in video composite signal. Therefore, there is a need of a hardware to digitize the analog signal before channeling it into the PC. The solution of this task would be to use a USB frame grabber. This device grabs the frame from the analog signal and converts it into digital format and relays the picture through the USB port.

In the PC, a GUI is constructed for the convenience of viewing the camera live feed and capturing the image in the boiler header. Figure 59.3 shows the screen capture of the GUI that is constructed in the PC. As can be seen from the GUI the camera's view in the boiler header is visible from the PC for the purpose of inspection and image acquisition.

59 Image Acquisition System for Boiler Header Inspection Robot   525

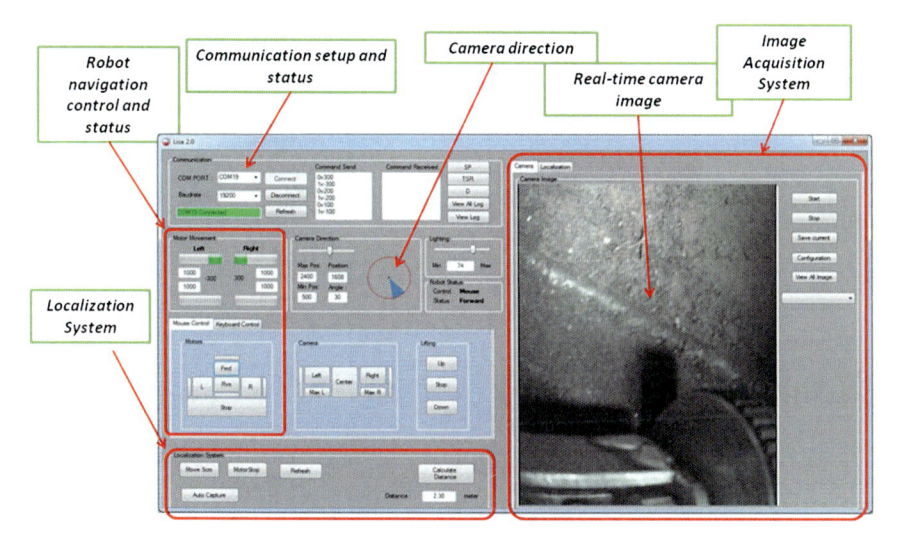

Fig. 59.3 GUI in the PC for grabbing picture from the capture card

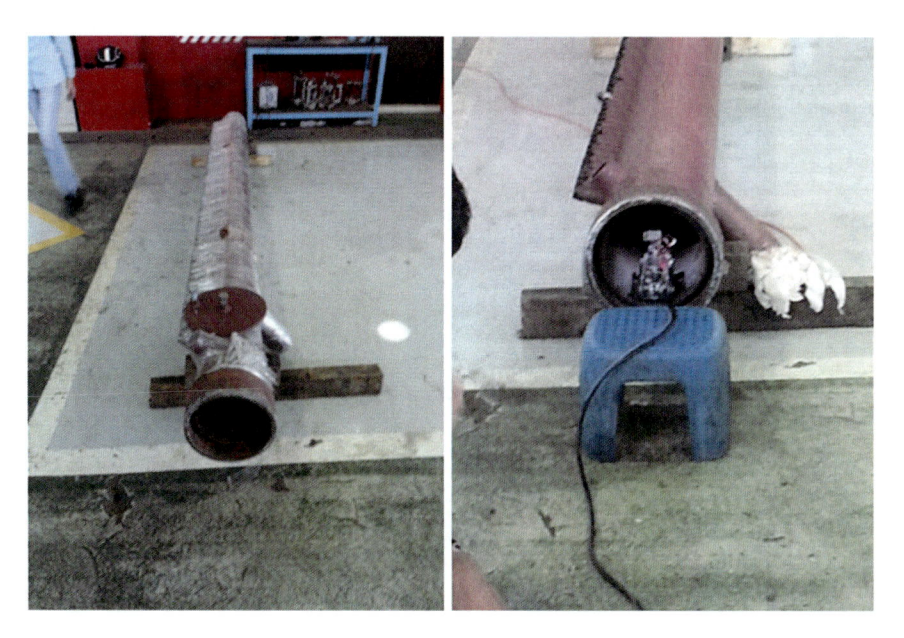

Fig. 59.4 Robot with the image acquisition system is tested on a real specimen of the boiler header in SJSI to verify the results tested previously in the lab

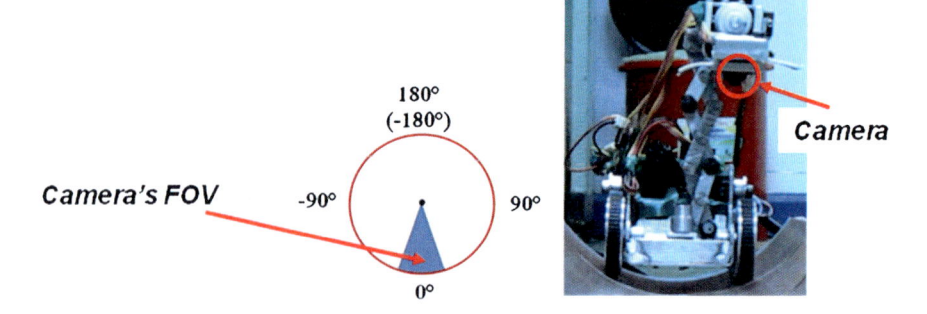

Fig. 59.5 Comparison image showing the Field of View (FOV) as seen in the GUI and on the actual robot

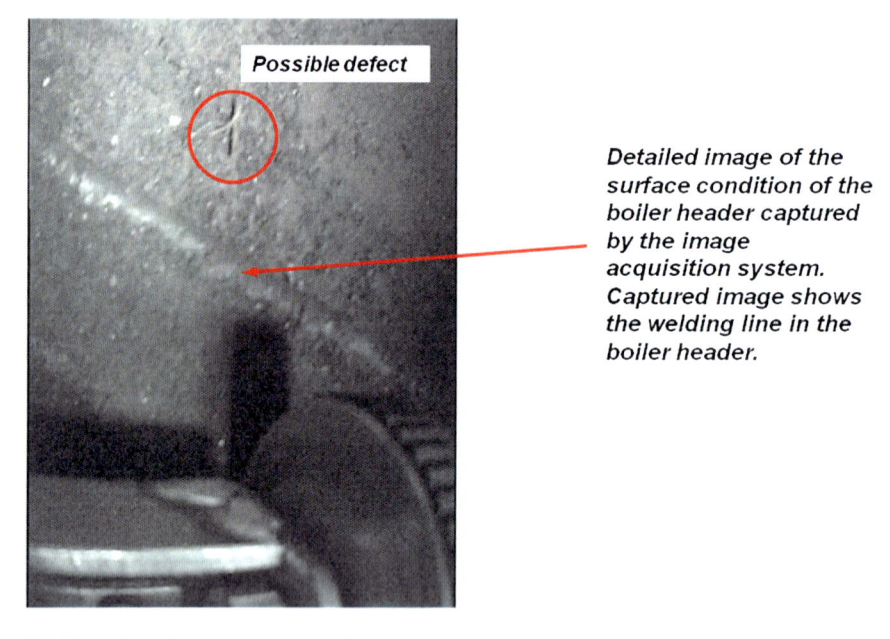

Fig. 59.6 Sample picture taken inside the actual boiler header indicating a possible defect in the boiler header

59.4 Results and Analysis

A series of tests have been carried out in the lab. A visit to the Stesen Janakuasa Sultan Ismail (SJSI), Paka, Terengganu was also organized to proceed with extensive testing and verification of the lab test results. In SJSI, a real specimen of a boiler header is taken out of the power plant and placed in the lab for testing. Figure 59.4 shows the robot mounted with the image acquisition system at the SJSI testing lab.

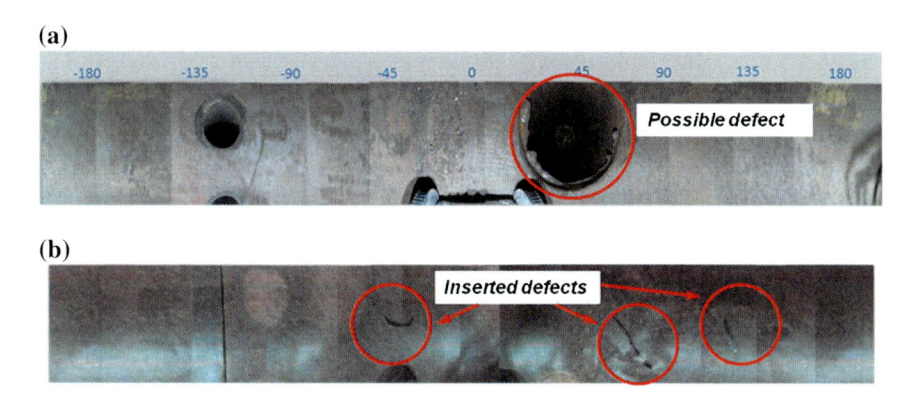

Fig. 59.7 Highlighted in the picture above includes the possible defect on the boiler header. **a** shows the combined individual pictures taken inside the boiler header with possible defects indication. **b** shows the combined images taken in the test-rig with inserted defects

Images in the boiler header are taken as the robot moves along in the boiler header. The aim of the image acquisition system is to acquire image from the inside of the boiler header to inspect for any visible defects on the surface of the boiler header wall. The series of pictures are visible from the GUI and are taken in a circular sequence to get a 360° view of the boiler header pipe. The image capturing in the pipe can be set to autonomous mode where the robot will take the picture in circular sequence every 5 cm of robot movement in the pipe. This method of image acquisition will ensure that every bit of the boiler header section is taken and an analysis of the images can be done by expert to determine if there are any defects in the boiler header. Figure 59.5 show the FOV as seen on the GUI and robot.

Apart from that, an overall mapping of the boiler header can also be done by combining the images that is taken in sequential manner into a strip of flatten images. Figure 59.6 shows a sample of the picture taken in a boiler header indicating a possible defect inside. Figure 59.7 shows a strip of images as combination of a series of individual images. Also highlighted are possible defects on the boiler header and test-rig.

59.5 Conclusion

This paper has presented briefly on a method of setting up an image acquisition system for an inspection robot. The main challenges in setting up the system includes the size constraint, clarity of image and the software to be used to tap the frames from the image acquisition hardware. The test in SJSI, Paka proved to be a successful test and further development of the robot is still being done to enable the robot to operate autonomously with minimal human operators involved.

Currently the detection of flaws and defects in the boiler header is done post inspection of the boiler header separately and requires the human expertise to determine the possibilities of defects. Future development may include online processing of the defects while the robot is still in inspection mode inside the boiler header. Development may involve heavy image processing and algorithms to identify and determine the defects in real time during inspection.

Acknowledgments The authors would like to thank the development team of the Boiler Header Inspection Robot MK-2.0 at Centre of Advanced Mechatronics and Robotics (CAMaRo), UniversitiTenaga National (UNITEN) for their undivided attention and effort in accomplishing the project. The author would like to also thank Tenaga Nasional Berhad (Research) for funding the research project.

References

1. Blais F, Rioux M, Domey J (1991) Optical range image acquisition for the navigation of a mobile robot. In: International conference on robotic and automation, pp 2574–2580
2. Liu Z, Kleiner Y (2013) State of the art review of inspection technologies for condition assessment of water pipes. Measurement 46(1):1–15
3. Newman TS, Jain AK (1995) A survey of automated visual inspection. Comput Vis Image Underst 61(2):231–262
4. Sahari KSM, Anuar A, Mohideen SSK, Baharuddin MZ, Ismail IN, Basri NMH, Roslin NS, Azwan B, Ahmad M (2012) Development of robotic boiler header inspection device. In: SCIS-ISIS, pp 769–773
5. Lim H, Choi JY, Yi BJ (2007) Development of semi-automatic inspection system for indoor pipeline. In: Proceedings of the 2007 IEEE international conference on mechatronics and automation, pp 3640–3645
6. Suzumari K, Miyagawa T, Kimura M, Hasegawa Y (1999) Micro inspection robot for 1-in pipes. IEEE/ASME Trans Mechatron 4(3):286–292
7. Ohya A, Yuta S, Yoshida T, Koyanagi E, Imai T, Kitamura S, Takeuchi A, Minamikawa T (2009) Development of inspection robot for under floor of house. In: IEEE international conference on robotics and automation, May 2009, pp 1429–1434

Chapter 60
Some New Findings on Gauss–Seidel Technique for Load Flow Analysis

Lea Tien Tay, Tze Hoe Foong and Janardan Nanda

Abstract In this paper, research is conducted using conventional Gauss–Seidel (GS), GS with acceleration factor (AF), first iteration of Newton–Raphson (NR) and followed by Gauss–Seidel (with and without acceleration factor) for load flow (LF) solution on three IEEE test systems, i.e. IEEE-30 bus system, IEEE-57 bus system and IEEE-118 bus system. Besides, GS with and without acceleration factor were also tested on different loadings of these IEEE bus systems to ascertain the robustness of the acceleration factors. The results are compared in terms of number of iterations and computation time taken for convergence.

Keywords Power flow · Gauss–Seidel method · Newton–Raphson method · Acceleration factor

60.1 Introduction

In today's world, many aspects of the electrical power industry have to be taken into consideration. This is due to the ever increasing power generation cost, scarcity of energy resources available, environmental issues and also the growing demand for electrical energy. This led to the necessity to produce and use electricity economically and wisely [1].

Janardan Nanda—Fellow IEEE

L. T. Tay (✉) · T. H. Foong
School of Electrical and Electronic Engineering, Engineering Campus, Universiti Sains Malaysia, 14300 Nibong Tebal, Penang, Malaysia
e-mail: tay@eng.usm.my

J. Nanda
Department of Electrical Engineering, Indian Institute of Technology, Delhi, India

Load flow studies have played an important role in power system analysis and design throughout these years. They are highly essential for the optimized planning, operation and economic scheduling of power between the utilities [2].

In power flow analysis, the key is to determine the state variables of a system i.e. the magnitude and phase angle of voltage at each bus. Once these variables are determined, it is then possible to determine the real and reactive power flowing in each transmission line [3]. A closed form solution for the load flow problem is very difficult since it involves the solution of many non-linear algebraic equations with constraints especially in a large system. Therefore, numerical methods that make use of iterative techniques are used to approach these problems [1]. These problems can be solved with the help of powerful computers available now.

There are many methods to solve a particular load flow problem [2]. The best method that should be chosen to solve a particular problem depends on the size of problems to be solved, availability of computing facility, and also the precision of the desired solution [4–10].

60.2 Power Flow Overview

Power flow analysis is very crucial especially in the planning, expanding or design stage of power system because it helps determine the addition of generator sites, transmission line sites and also increment in load demand. It develops an operating condition where the voltages of some particular buses are kept within a specified tolerance level. It also determines the nodal voltage magnitudes and phase angles that enable the computation of power injection at all buses and also power flows through the transmission lines. The line flows can be computed and hence can be made sure not to exceed its stability or thermal limits. This is important in order to prevent damage to the system.

The aim of this paper is to solve load flow problems in four ways, i.e. using conventional GS, conventional GS with acceleration factor, hybrid of first iteration of NR at the beginning and followed by GS with and without acceleration factor for convergence. The analysis is implemented using MATLAB program and applied on three standard IEEE test systems, i.e. IEEE-30 bus system, IEEE-57 bus system and IEEE-118 bus system. Besides, GS with and without acceleration factors were tested on different loading of these IEEE test systems to explore the robustness of the acceleration factor used for nominal loading for other credible loadings of the systems. Such studies to the best of the author's knowledge have not been yet conducted. The results obtained are compared in terms of number of iterations and computation time taken for convergence.

60.3 Power Flow Analysis

The real and reactive power at bus i is given by

$$I_i = \frac{P_i - jQ_i}{V_i^*}$$

Consider a large bus system, this equation can be rewritten to be

$$\frac{P_i - jQ_i}{V_i^*} = V_i \sum_{j=0}^{n} y_{ij} - \sum_{j=1}^{n} y_{ij} V_j \quad j \neq i$$

where n is the total number of buses.

This is also known as the power flow equation and is the fundamental equation in load flow analysis.

The MATLAB program was developed to simulate the results of conventional GS, conventional GS with optimum acceleration factor, hybrid of first iteration of NR and followed by GS and hybrid of first iteration of NR with GS including optimum acceleration factor in solving load flow problems on three standard test systems, i.e. IEEE-30 bus system, IEEE-57 bus system and IEEE-118 bus system for their nominal loadings.

The voltage at each bus can be formulated by rearranging the power flow equations shown earlier.

$$V_i^{(k+1)} = \frac{\frac{P_i^{sch} - jQ_i^{sch}}{V_i^{*k}} - \sum_{j \neq i} Y_{ij} V_j^{(k)}}{Y_{ii}}$$

where P_i^{sch} and Q_i^{sch} are the net real and reactive powers in per unit. The real and reactive power equations are expressed as follows:

$$P_i^{(k+1)} = \text{Real } \{V_i^{*(k)}[V_i^{(k)} Y_{ii} + \sum_{\substack{j=1 \\ j \neq i}}^{n} Y_{ij} V_j^{(k)}]\} \quad j \neq i$$

$$Q_i^{(k+1)} = -\text{Imag } \{V_i^{*(k)}[V_i^{(k)} Y_{ii} + \sum_{\substack{j=1 \\ j \neq i}}^{n} Y_{ij} V_j^{(k)}]\} \quad j \neq i$$

For voltage-controlled buses, $|V_i|$ is already specified, therefore only the imaginary part of $V_i^{(k+1)}$ is retained and the real part is selected based on the equation below:

$$(e_i^{(k+1)})^2 + (f_i^{(k+1)})^2 = |V_i|^2$$

where $(e_i^{(k+1)})$ and $(f_i^{(k+1)})$ are the real and imaginary components of the voltage $V_i^{(k+1)}$ in the iterative sequence.

The reactive power of the buses must be kept within the reactive limits of the generator capability. The iteration process is carried on until the specified tolerance level is met.

In order to improve the convergence rate of the method, acceleration factor (α) can be added to the solution as shown in the following equation.

$$x_i^{(k+1)} = x_i^{(k)} + \alpha \left(x_{ical}^{(k+1)} - x_i^{(k)} \right)$$

Acceleration factors tried in the analysis are in the range of 1–1.8 with the increment step of 0.001 to search for the best acceleration factor. Further, conventional GS with and without acceleration factor were also applied on 30, 50, 75 and 125 % of nominal loading of the three standard test systems—IEEE-30 bus system, IEEE-57 bus system and IEEE-118 bus test system. Different loadings are used to test the robustness of the optimum acceleration factor evaluated for the nominal loading condition. The results are compared in terms of number of iterations and computation time taken for convergence. Computational time for ten simulations was recorded and the average computational time was taken over these ten values.

60.4 Results and Discussions

Table 60.1 presents the results on the computational time obtained for the three test systems. From the results obtained, it is observed that the best method for all three test systems is the GS method with optimum acceleration factor as it shows the fastest computation time. As compared to the conventional GS method, the introduction of the optimum acceleration factor drastically decreases the amount of time taken for convergence. As the system size increases, this improvement becomes more pronounced.

On the other hand, the hybrid method of performing an iteration of NR before proceeding with GS method tends to increase the computation time. However, this is not true for the 30 bus system as this method shows faster computation time as compared to conventional GS. For all three cases, the introduction of acceleration factor to this hybrid method shows improvement. Although not as good as the Gauss–Seidel method with acceleration factor, this hybrid method is also another good alternative approach as compared to the conventional Gauss–Seidel method.

Apart from that, it can also be observed in Fig. 60.1 that the computation time for all the methods is proportional to the size of the system involved. As system size increases, the computation time needed for convergence also increases. The computation time needed for Newton–Raphson method increases with a bigger extend compared to Gauss–Seidel method as the system bus size increases.

60 Some New Findings on Gauss–Seidel Technique

Table 60.1 Results of three test systems

Test systems		GS	NR	GS	Time (s)
IEEE-30 Bus	GS	158	–	–	0.08443
	GS + AF 1.75	26	–	–	0.05188
	NR + GS	–	1	38	0.07487
	NR + GS + AF 1.7	–	1	16	0.06881
IEEE-57 Bus	GS	218	–	–	0.17068
	GS + AF 1.715	43	–	–	0.07582
	NR + GS	–	1	127	0.20273
	NR + GS + AF 1.725	–	1	29	0.14773
IEEE-118 Bus	GS	416	–	–	0.85361
	GS + AF 1.84	69	–	–	0.21642
	NR + GS	–	1	315	0.94189
	NR + GS + AF 1.85	–	1	59	0.49185

Fig. 60.1 Comparison of methods in computational time

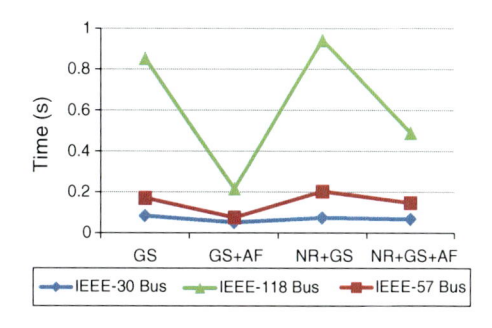

The test system parameter is changed to a loading of 30, 50, 75 and 125 %. This is to test the robustness of the acceleration factor (AF) selected using the conventional GS method. Table 60.2 shows the results obtained on IEEE-30, IEEE-57 and IEEE-118 bus systems. For IEEE-30 bus system, the previously selected AF for conventional GS method is 1.75 with 26 iterations. For all the different loadings, the AF of 1.75 also yielded 26 iterations. For 30 and 50 % loading, the least number of iterations possible is 23 with AF of 1.73 and 1.735, respectively. For 75 % loading, the least number of iterations possible is 25 with AF of 1.73. For 125 %, the AF is the same as 100 % loading. Therefore, it can be concluded that the AF selected for conventional GS method is applicable for different loading and will produce a result that is almost as good as possible.

Based on the previous analysis on IEEE-57 bus system, the selected AF for conventional GS method is 1.715 with 43 iterations. For 30 % loading, AF of 1.715 required 40 iterations compared to 39 iterations with best AF of 1.72. For 50 and 75 % loading, GS with AF of 1.715 which is also the best AF, take 42 and 43 iterations. For 125 % loading, AF of 1.715 required 44 iterations compared 43 iterations with best AF of 1.72. As the number of iterations using AF of conventional GS method does not differ much from their own best AF, the original AF

Table 60.2 Summary of three IEEE bus systems with different loading

Loading %	IEEE-30 bus system			IEEE-57 bus system			IEEE-118 bus system		
	AF	N	Time	AF	N	Time	AF	N	Time
100	W/out AF	158	0.084	W/out AF	218	0.171	W/out AF	416	0.854
	AF 1.75	26	0.052	AF 1.715	43	0.076	AF 1.84	69	0.216
30	W/out AF	116	0.071	W/out AF	184	0.157	W/out AF	397	0.832
	AF 1.75	26	0.049	AF 1.715	40	0.076	AF 1.84	67	0.210
	AF 1.73	23	0.046	AF 1.72	39	0.075	AF 1.875	54	0.206
50	W/out AF	135	0.078	W/out AF	201	0.163	W/out AF	407	0.855
	AF 1.75	26	0.049	AF 1.715	42	0.075	AF 1.84	69	0.214
	AF 1.735	23	0.047	AF 1.715	42	0.075	AF 1.874	56	0.202
75	W/out AF	148	0.080	W/out AF	212	0.169	W/out AF	415	0.867
	AF 1.75	26	0.049	AF 1.715	43	0.077	AF 1.84	69	0.213
	AF 1.73	25	0.047	AF 1.715	43	0.077	AF 1.858	62	0.212
125	W/out AF	164	0.087	W/out AF	225	0.179	W/out AF	415	0.869
	AF 1.75	26	0.049	AF 1.715	44	0.079	AF 1.84	86	0.253
	AF 1.75	26	0.049	AF 1.72	43	0.079	AF 1.815	76	0.252

N Number of iterations
W/out without

is deemed to be robust for all cases of loading. Based on the previous results on IEEE-118 bus system, the selected AF for conventional GS method is 1.84 which required 69 iterations. For 30 % loading, AF of 1.84 requires 67 iterations compared to 54 iterations with the best AF of 1.875. For 50 % loading, AF of 1.84 requires 69 iterations compared to 56 iterations with the best AF of 1.874. For 75 % loading, AF of 1.84 required 69 iterations compared to 62 iterations with the best AF of 1.858. For 125 % loading, AF of 1.84 requires 86 iterations compared to 76 iterations with the best AF of 1.815. As can be observed in the results in Table 60.2, the best AF and original AF for each loading cases differ more compared to the previous two test systems. However, the difference is still acceptable and the original AF chosen is still applicable for different loading cases as the computational time taken are very close.

60.5 Conclusion

The number of iterations for load flow solutions using Gauss–Seidel method increases with the system size. Hybrid methods of combining the two conventional methods of GS and NR do not show significant improvement over the GS method and is therefore not appealing to the utilities.

The best possible method that can be concluded from this research is the Gauss–Seidel method with optimum acceleration factor. It can also be concluded that the optimum acceleration factor selected for the nominal loading condition using conventional GS method holds good for different loadings.

References

1. Freris LL, Sasson AM (1968) Investigation of the load-flow problem. In: Proceedings of the institution of electrical engineers, vol 115. pp 1459–1470
2. Laughton MA, Humphrey Davies MW (1964) Numerical techniques in solution of power-system load-flow problems. In: Proceedings of the institution of electrical engineers, vol 111. pp 1575–1588
3. Stott B (1974) Review of load-flow calculation methods. In: Proceedings of the IEEE, vol 62. pp 916–929
4. Alqadi R, Khammash M (2007) An efficient parallel Gauss–Seidel algorithm for the solution of load flow problems. Int Arab J Inf Technol 4
5. Gilbert G, Bouchard D, Chikhani A (1998) A comparison of load flow analysis using DistFlow, Gauss–Seidel, and optimal load flow algorithms. In: IEEE Canadian conference on in electrical and computer engineering, pp 850–853
6. Nguyen TT (1995) Neural network load-flow. In: IEE Proceedings-generation, transmission and distribution, vol 142. pp 51–58
7. Nguyen TT, Vu CT (2006) Complex-variable Newton–Raphson load-flow analysis with FACTS devices. In: Transmission and distribution conference and exhibition, 2005/2006 IEEE PES, pp 183–190
8. Soeprijanto A (2012) Regular paper combination of generator capability curve constraint and statistic-fuzzy load clustering algorithm to improve NN-OPF performance. J Electr Syst 8:198–208
9. Treece JA (1969) Bootstrap Gauss–Seidel load flow. In: Proceedings of the institution of electrical engineers, vol 116. pp 866–870
10. Vlachogiannis JG (2001) Fuzzy logic application in load flow studies. In: IEE Proceedings of generation, transmission and distribution, vol 148. pp 34–40

Author Index

A

Abas, Fazly Salleh, 239
Abd Manaf, Asrulnizam, 393, 409
Abdul Ghani, Muhammad Fahmi, 521
Abdullah, Jiwa, 265
Abdullah, Mohd, 457, 463
Abdullah, Mohd Zaid, 457
Abdul Rahman, Nurulhamimi, 303
Abdul Razak, Siti Fatimah, 321
Abu Bakar, Juliana Aida, 23
Abu Bakar, Mohd Nazrin Afzan, 119
Abu Bakar, Norazhar, 57
Adnan, Nazrul Hamizi, 23
Ahmad, Noor, 73
Ahmad, Omar, 343
Ahmad, R Badlishah, 145
Ahmad, Zainal Arifin, 457, 463
Ain, Mohd Fadzil, 457, 463, 495
Ali, Fariz, 15, 57
Alsewari, AbdulRahman, 255
Ameri Eshghabadi, Hamidreza, 409
Anjum, Muhammad Latif, 343
Anuar, Adzly, 31, 81, 119, 521
Astaraki, Mehdi, 285
Aziz, Mohd Azwan, 31, 521
Azmadi Hussin, Fawnizu, 137

B

Bahar, Mohd Bazli, 57
Besar, Rosli, 239
Bin Hj Shukor, Ahmad Zaki, 15, 57
Bin Md Saat, Mohd Shakir, 15
Bohari, Azuwien Aida, 501
Bona, Basilio, 343
Burian, Frantisek, 47

C

Chan, Eric, 197
Cheong, Chee Han, 427
Ch'ng, Siew Sin, 365
Chua, Kah Keong, 181
Chuan, Seah Yee, 23
Connie, Tee, 189
Cui, Xuenan, 89

D

Dahari, Zuraini, 393
Durairajah, Vickneswari, 507

E

Eliahim Jeevaraj, P. S., 229

F

Farokhi, Sajad, 129
Flusser, Jan, 129
Foong, Tze Hoe, 529

G

Ghani, Farid, 145
Ghazali, Rozaimi, 3
Goh, Patrick, 401

H

Hafidz, Muhammad, 327
Hajjaj, Sami Salama Hussen, 107
Hamedi, Mahyar, 285
Hamzah, Raseeda, 311

Haron, Zainal Alam, 501
Harris, Arief, 285
Hashemi, Mohammadali, 495
Huh, Uk-Youl, 99
Hutagalung, Sabar, 463

I
Ibrahim, Muhammad Nasir, 205
Idroas, Mariani, 205
Ismail, Widad, 449

J
Jaafar, Haryati, 153
Jafar, Fairul Azni, 327
Jamil, Nursuriati, 303, 311
Javed, Shazia, 73
Jiang, Guannan, 169
Jilek, Tomas, 47
Jit Singh, Mandeep, 449
Joanne Neoh, Wan Shi, 393

K
Kader, Md. Abdul, 145
Karimi, Mohammad Taghi, 39
Keyvanara, Mahboubeh, 39
Khan, Abdullah, 295, 335
Khoo, Bee Ee, 161
Khosa, Ikramullah, 343
Kim, Hakil, 89
Kim, Hyoungrae, 89
Kobayashi, Yoshinori, 65
Kocmanova, Petra, 47
Kuno, Yoshinori, 65
Kwok, Ngaiming, 169

L
Le, Duc, 137
Lee, Jaehong, 89
Lee, Seungjun, 89
Lim, Chee Peng, 355
Lim, Way Soong, 321
Lim, Zong Zheng, 365
Lin, Stephen Ching-Feng, 169

M
Mahyuddin, Nor Muzlifah, 473
Marzuki, Arjuna, 457, 463
Maslan, Mohd Nazmin, 327
Masri, Syafrudin, 385

Mat Ariff, Roslina, 501
Mat Yazid, Mohd Shahrul Amran, 373
Mazlan Huzairi, Mohd Helmi, 373
Miskon, Muhammad Fahmi, 15, 57
Mohamad Hipiny, Irwandi Hipni, 245
Mohamad, Norizah, 385, 449
Mohamed Hariri, Muhammad Hafeez, 385
Mohamed Nor, Mohd Khairi, 15
Mohamed Sahari, Khairul Salleh, 31, 81, 107, 119, 521
Mohamed, Julie Juliewatty, 457
Mohammed, Julie, 463
Mohammed Falah Mohammed, 355
Mohd Ali, Mohd Hazwan Hafiz, 23
Mohd Ali, Nursabillilah, 277
Mohd-Mokhtar, Rosmiwati, 435, 449
Mohd Najib, Suhaila, 205
Mohd. Nawi, Nazri, 295, 335
Mohd Noh, Norlaili, 409, 427
Mohd Sobran, Nur Maisarah, 277
Mojaddarasil, Marzieh, 39
Molloy, Derek, 483
Muhammad Zin, Nooradzianie, 501
Mulyo Utomo, Wahyu, 501
Munalih, Ahmad Syarif, 189
Mustaffa, Mohd Tafir, 365, 409

N
Nanda, Janardan, 529
Neoh Tze How, Dickson, 81, 521
Ngah, Umi Kalthum, 355
Ng, Lee Teng, 221
Ngoc Tran, Duc, 137
Nik Anwar, Nik Syahrim, 419
No, Jin Hong, 99
Noor, Alias Mohd, 285

O
Olokede, Seyi, 457, 463
Ong Yee, Lau, 327
Ong, Thian Song, 189
Osman, Anwar, 373
Othman, Mohamadariff, 457

P
Park, Jong Hun, 99

R
Rafiee, Majid, 495
Ramiah, Harakrishnan, 427

Ramli, Dzati Athiar, 153
Rehman, Muhammad Zubair, 295, 335
Roslin, Nur Shahida, 119

S

Sadigh, Mohammed, 39
Salleh, Sheikh hussain, 285
Salleh, Siti Salwa, 303
Seman, Noraini, 311
Semire, Folasade, 449
Shamsuddin, Siti Mariyam, 129
Shanmugavadivu, P., 213, 229
Sheikh, Usman Ullah, 129
Sidek, Othman, 409
Sim, Kok Swee, 239
Sim, Sy Yi, 501
Spann, Michael, 245
Sreekantan, Srimala, 457, 463
Suandi, Shahrel Azmin, 221
Suresh, Gobee, 507
Syazwani, Nur Farah, 473

T

Tan, Shing Chiang, 321
Tay, Lea-Tien, 529
Tay, Yong Haur, 181
Teoh, Soo Siang, 221
Thabit, Rasha, 161
Ting, Huong Yong, 239
Tomari, Mohd Razali MD, 65
Tung, Desmond, 435

U

Ujir, Hamimah, 245
Ullah, Ubaid, 457, 463

V

Vengusamy, Sivakumar, 213

W

Wan Ali, Wan Fahmin Faiz, 457
Wan, Khairunizam, 23
Wong, Adrian, 507
Wong, Chin Hong, 393
Wong, Chin Yeow, 169

Y

Yahaya, Nor Zaihar, 373
Yahaya, Saifudin Hafiz, 327
Yu, Sha, 483
Yusoff, Mohd Zuki, 137

Z

Za'ba, Nurmaisara, 303
Zalud, Ludek, 47
Zamli, Kamal Z., 255
Zulkifli, Nadiah Amalina, 31

Printed by Publishers' Graphics LLC
DBT140228.15.18.340